GRADUATE TEXTS IN PHYSICS

Graduate Texts in Physics

Graduate Texts in Physics publishes core learning/teaching material for graduate- and advanced-level undergraduate courses on topics of current and emerging fields within physics, both pure and applied. These textbooks serve students at the MS- or PhD-level and their instructors as comprehensive sources of principles, definitions, derivations, experiments and applications (as relevant) for their mastery and teaching, respectively. International in scope and relevance, the textbooks correspond to course syllabi sufficiently to serve as required reading. Their didactic style, comprehensiveness and coverage of fundamental material also make them suitable as introductions or references for scientists entering, or requiring timely knowledge of, a research field

Series Editors

Professor Richard Needs
Cavendish Laboratory
JJ Thomson Avenue
Cambridge CB3 0HE, UK
E-mail: rn11@cam.ac.uk

Professor William T. Rhodes
Florida Atlantic University
Imaging Technology Center
Department of Electrical Engineering
777 Glades Road SE, Room 456
Boca Raton, FL 33431, USA
E-mail: wrhodes@fau.edu

Professor H. Eugene Stanley
Boston University
Center for Polymer Studies
Department of Physics
590 Commonwealth Avenue, Room 204B
Boston, MA 02215, USA
E-mail: hes@bu.edu

For further volumes:
http://www.springer.com/series/8431

Florian Scheck

Classical Field Theory

On Electrodynamics, Non-Abelian Gauge Theories and Gravitation

Springer

Florian Scheck
Johannes Gutenberg-University Mainz
Germany

ISSN 1868-4513
ISBN 978-3-642-43128-9
DOI 10.1007/978-3-642-27985-0
Springer Heidelberg Dordrecht London New York

ISSN 1868-4521 (electronic)
ISBN 978-3-642-27985-0 (eBook)

Softcover reprint of the hardcover 1st edition 2012

Printed on acid-free paper

Springer is part of Springer Science+Business Media (www.springer.com)

To Dörte who showed infinite patience and gave much encouragement during the genesis of this book.

Preface

Traditionally one begins a course or a textbook on electrodynamics with an extensive discussion of electrostatics, of magnetostatics, and of stationary currents, before turning to the full time-dependent Maxwell theory in local form. In this book I choose a somewhat different approach: Starting from Maxwell's equations in integral form, that is to say, from the phenomenological and experimentally verified basis of electrodynamics, the local equations are formulated and discussed with their general time and space dependence right from the start. Static or stationary situations appear as special cases for which Maxwell's equations split into two more or less independent groups and thus are decoupled to a certain extent.

Great importance is attached to the symmetries of the Maxwell equations and, in particular, their covariance with respect to Lorentz transformations. Another central issue is the treatment of electrodynamics in the framework of classical field theory by means of a Lagrange density and Hamilton's principle. General principles that were developed for mechanics, appear in a deeper and more general application that can serve as a model and prototype for any classical field theory. The fact that the fields of Maxwell theory, in general, depend on space *and* time makes it necessary to enlarge the framework of traditional tensor analysis in $\mathbb{R}^3$ to exterior calculus on $\mathbb{R}^4$. The venerable vector and tensor analysis that was designed for three-dimensional Euclidean spaces, does not suffice and must be generalized to higher dimensions and to Minkowski signature. While the exterior product is the generalization of the vector product in $\mathbb{R}^3$, Cartan's exterior derivative is the natural generalization of the curl in $\mathbb{R}^3$ and, by the same token, encompasses the familiar operations of gradient and divergence.

Among the many applications of Maxwell theory I chose some characteristic and, I felt, nowadays particularly relevant examples such as an extensive discussion of polarization of electromagnetic waves, the description of Gaussian beams (these are analytic solutions of the Helmholtz equation in paraxial approximation), and optics of metamaterials with negative index of refraction. Regarding other, more traditional applications I refer to the well-known, excellent textbooks by J. D. Jackson, by L. D. Landau and E. M. Lifshits, and others.

As a novel feature I take up in the fifth chapter a further direction of great importance for present-day physics: The construction of non-Abelian gauge theories. These Yang–Mills theories as they are called[1], are essential and indispensable for our present understanding of the fundamental interactions of nature. Although these theories which are at the basis of the so-called standard model of elementary particle physics, lead us far into *quantized* field theory, their construction and their essential features are of a *classical* nature, at least as long as one considers only the radiation fields, i.e. the analogues of the Maxwell fields, and classical scalar fields, but leaves out fermionic matter particles. Non-Abelian gauge theories are constructed following the example of Maxwell theory. They bear some similarities to the latter but exhibit also significant differences from it. Even the phenomenon of spontaneous symmetry breaking that preserves us from the appearance of too many massless fields, in essence, is a classical mechanism. In view of the great impact of gauge theories on our understanding of the fundamental interactions it would be a loss not to do this step which builds on Maxwell theory in a most natural manner.

Chapter 6 gives an extensive phenomenological and geometric introduction to general relativity and, hence, rounds off the description of all fundamental interactions in the framework of classical field theory. Here too, I use consistently a modern geometric language which – after some investment in differential geometry – allows for a transparent formulation of Einstein's equations which is better focused to its essentials than the older tensor analysis formulated in components only.

Much of the material included in this book was tried out in numerous lectures that I gave at Johannes Gutenberg-University over the years. I am grateful to the students who have followed these courses, for their questions and comments, and to the teaching assistants who took good care of exercise classes, for their stimulating questions and critical comments.

I owe special thanks to Immanuel Bloch for the discussions we had on Gaussian beams and the fascinating topic of metamaterials with negative index of refraction, and for his encouragement to include these modern applications. Special thanks also to Mario Paschke who more than once brought up original ideas and pointed out some almost forgotten but relevant references.

The cooperation with Springer-Verlag in Heidelberg and with the production team of LE-TeX in Leipzig was excellent and very efficient. The great encouragement and editorial care offered by Dr. Thorsten Schneider from Springer-Verlag is gratefully acknowledged.

Mainz, November 2011 *Florian Scheck*

[1] First ideas were published by Oskar Klein, Z. Physik **37** (1926) 895. It is reported that Wolfgang Pauli developed them independently but did not publish them.

Contents

Maxwell's Equations 1

1.1 Introduction

The empirical basis of electrodynamics is defined by Faraday's law of induction, by Gauss' law, by the law of Biot and Savart and by the Lorentz force and the principle of universal conservation of electric charge. These laws can be tested – confirmed or falsified – in realistic experiments. The integral form of the laws deals with physical objects that are one-dimensional, two-dimensional, or three-dimensional, that is to say, objects such as linear wires, conducting loops, spatial charge distributions, etc. Thus, the integral form depends, to some extent, on the concrete experimental set-up. To unravel the relationships between seemingly different phenomena, one must switch from the integral form of the empirically tested laws to a set of *local* equations which are compatible with the former. This reduction to local phenomena frees the laws from any specific laboratory arrangement and yields what we call Maxwell's equations proper. These local equations describe an extremely wide range of electromagnetic phenomena.

The mathematical tools needed for this transition from *integral* to *local* equations are taken, initially, "only" from vector analysis over Euclidian space $\mathbb{R}^3$ and from the well-known differential calculus on this space. However, since electromagnetic fields in general also depend on time and, hence, are defined on spacetime $\mathbb{R}^4$, this calculus must be generalized to more than three dimensions. The necessary generalization becomes particularly transparent and simple if one makes use of *exterior calculus*.

This chapter develops the phenomenology of Maxwell's equations, first by means of the full, space-dependent and time-dependent equations, then, in a second step, by reduction to *stationary* or *static* situations. The formulation of Maxwell's equations requires some knowledge of elementary vector analysis as well as some theorems on integrals over paths, surfaces and volumes. Therefore, we start by recalling for the reader these matters in the case of $\mathbb{R}^3$ before embarking on more general situations, and we illustrate matters with a few examples which will be useful for the sequel.

F. Scheck, *Classical Field Theory*, Graduate Texts in Physics,
DOI 10.1007/978-3-642-27985-0_1,

1.2 Gradient, Curl and Divergence

Electrodynamics and a great deal of general classical field theories are defined on flat spaces $\mathbb{R}^n$ of dimension n. In cases of static or stationary processes, an adequate framework is provided by the ordinary space $\mathbb{R}^3$ and in all other cases by four-dimensional spacetime with one time and three space components, $\mathbb{R}^4$, or, more precisely, $\mathbb{R}^{(1,3)}$. These spaces are special cases of smooth manifolds. They are endowed with various geometric objects and with a natural differential calculus which allows one to set up relations between the former and thus to formulate physical equations of motion. For example, if $\Phi(x) = \Phi(x^1, x^2, \dots, x^n)$ is a smooth function on $\mathbb{R}^n$, then one defines a *gradient field* by

$$\mathbf{grad}\,\Phi(x) = \left(\frac{\partial \Phi(x)}{\partial x^1}, \frac{\partial \Phi(x)}{\partial x^2}, \dots \frac{\partial \Phi(x)}{\partial x^n}\right)^T \tag{1.1}$$

(with superscript T standing for "transposed"). In the case of $\mathbb{R}^3$, **grad** is the familiar differential operator

$$\boldsymbol{\nabla} = \left(\frac{\partial}{\partial x^1}, \frac{\partial}{\partial x^2}, \frac{\partial}{\partial x^3}\right)^T ,$$

often called the *nabla operator.*

Example 1.1

A small probe of mass m is placed in the gravitational field of two equal, pointlike masses M whose positions are $\boldsymbol{x}^{(i)}$, $i = a, b$. The potential created by these masses at the position of the probe is

$$\Phi(\boldsymbol{x}) \equiv U(\boldsymbol{x}) = -G_{\mathrm{N}} m M \left\{\frac{1}{|\boldsymbol{x} - \boldsymbol{x}^{(a)}|} + \frac{1}{|\boldsymbol{x} - \boldsymbol{x}^{(b)}|}\right\} .$$

Without loss of generality, one can choose a system of reference such that $\boldsymbol{x}^{(b)} = -\boldsymbol{x}^{(a)}$. The force field that follows from this potential is

$$\begin{aligned} \boldsymbol{F}(\boldsymbol{x}) = -\boldsymbol{\nabla}_x \Phi(\boldsymbol{x}) &= -G_{\mathrm{N}} m M \left\{\frac{\boldsymbol{x} - \boldsymbol{x}^{(a)}}{|\boldsymbol{x} - \boldsymbol{x}^{(a)}|^3} + \frac{\boldsymbol{x} - \boldsymbol{x}^{(b)}}{|\boldsymbol{x} - \boldsymbol{x}^{(b)}|^3}\right\} \\ &= -G_{\mathrm{N}} m M \left\{\frac{\boldsymbol{x} - \boldsymbol{x}^{(a)}}{|\boldsymbol{x} - \boldsymbol{x}^{(a)}|^3} + \frac{\boldsymbol{x} + \boldsymbol{x}^{(a)}}{|\boldsymbol{x} + \boldsymbol{x}^{(a)}|^3}\right\} . \end{aligned}$$

This is a conservative force field. It is instructive to sketch this field for the choice $\boldsymbol{x}^{(a)} = (d, 0, 0)^T$, $\boldsymbol{x}^{(b)} = (-d, 0, 0)^T$.

If $\hat{e}_i$, $i = 1, \ldots, n$, is a basis, and $V = \sum_{i=1}^{n} V^i(x)\hat{e}_i$ a vector field, then its divergence is defined by

$$\mathbf{div}\, V = \sum_{i=1}^{n} \frac{\partial}{\partial x^i} V^i(x)\,. \tag{1.2}$$

This is a well-known construction in $\mathbb{R}^3$. In particular, if $\boldsymbol{V}$ is a gradient field, i.e. if $\boldsymbol{V} = \nabla\Phi(\boldsymbol{x})$, then its divergence is

$$\mathbf{div}\,\mathbf{grad}\,\Phi = \sum_{i=1}^{3} \frac{\partial^2 \Phi(\boldsymbol{x})}{\partial (x^i)^2} = \boldsymbol{\Delta}\Phi(\boldsymbol{x})\,,$$

where $\boldsymbol{\Delta}$ is the Laplace operator.

In dimension 3, the curl of a vector field under proper rotations behaves again like a vector field. Note, however, that this is a pecularity of dimension 3. Indeed, in $\mathbb{R}^3$,

$$\mathbf{curl}\, V = \nabla \times V \tag{1.3a}$$

has three components. When expressed in Cartesian coordinates, these are

$$\left(\nabla \times V\right)_1 = \frac{\partial V^3}{\partial x^2} - \frac{\partial V^2}{\partial x^3} \qquad \text{(with cyclic completion)}\,, \tag{1.3b}$$

or, making use of the ε-tensor in dimension 3,

$$\left(\nabla \times V\right)^i = \frac{1}{2}\sum_{j,k=1}^{3} \varepsilon^{ijk}\left(\frac{\partial V_k}{\partial x^j} - \frac{\partial V_j}{\partial x^k}\right) = \sum_{j,k=1}^{3} \varepsilon^{ijk}\frac{\partial V_k}{\partial x^j}. \tag{1.3c}$$

We will return to this property in Sect. 2.2.2 below. One should note that $\nabla \times \boldsymbol{V}$ cannot be a "genuine" vector field because the vector field $\boldsymbol{V}$ and its curl have the opposite behaviour under space reflection: While $\boldsymbol{V}$ changes sign under P, its curl $\nabla \times \boldsymbol{V}$ stays invariant.

Remarks

1. On $\mathbb{R}^3$, which admits the metric $g_{ik} = \delta_{ik}$, there is no difference between the contravariant components V^i of $\boldsymbol{V}$ and the covariant components V_i. Therefore, instead of (1.3b), one may as well write

$$\left(\nabla \times V\right)^1 = \frac{\partial V_3}{\partial x^2} - \frac{\partial V_2}{\partial x^3} \qquad \text{(plus cyclic completion)}\,.$$

This was anticipated in (1.3c). The following remark shows that this slightly modified definition of the curl is, in fact, the correct one.

2. On $\mathbb{R}^n$ or, more generally, on the smooth manifold M^n with dimension n and metric tensor $\mathrm{g} = \{g_{ik}\}$, one may associate to the *contra*variant components V^i of the vector field V the *co*variant components $V_i = \sum_k g_{ik} V^k$. By generalization of the curl on $\mathbb{R}^3$ one defines a skew-symmetric tensor field of degree 2 by

$$\mathbf{curl}\, V \equiv \mathrm{C}\,, \quad \text{with } C_{ik} = \frac{\partial}{\partial x^i} V_k - \frac{\partial}{\partial x^k} V_i\,.$$

As $C_{ki} = -C_{ik}$, this tensor has $\frac{1}{2}n(n-1)$ components, i.e. in dimension $n = 2$ it has one component, in dimension $n = 3$ it has three components, in dimension $n = 4$ it has six components, etc. Obviously, only on $\mathbb{R}^3$ does the curl (1.3c) have the right number of components to be akin to a vector field.

3. The preceding remark shows that it is meaningful only in dimension 3 to take the divergence of a curl. In this case, one has

$$\mathbf{div}\,\mathbf{curl}\, \boldsymbol{A} \equiv \nabla \cdot (\nabla \times \boldsymbol{A}) = \sum_{i,j,k} \varepsilon^{ijk} \frac{\partial}{\partial x^i} \frac{\partial}{\partial x^j} A_k = 0\,. \qquad (1.4)$$

This expression vanishes because the ε tensor, which is antisymmetric in i and j, is multiplied by the symmetric product of the two derivatives. More generally, the contraction of a tensor which is *symmetric* in two of its indices with another tensor which is *antisymmetric* in the same indices vanishes. This is confirmed by direct calculation.

4. The curl of a smooth gradient field is zero. In $\mathbb{R}^3$, this is the well-known formula $\nabla \times (\nabla \Phi(\boldsymbol{x})) = 0$. It generalizes to

$$\mathbf{curl}\,\mathbf{grad}\, \Phi(x) = 0 \qquad (1.5)$$

and is a consequence of the equality of the mixed second derivatives of $\Phi(x)$.

5. In any other dimension not equal to 3, the combined application of divergence and gradient yields (the obvious generalization of) the Laplace operator, too. For example, in $\mathbb{R}^n$, it reads

$$\mathbf{div}\,\mathbf{grad}\, \Phi = \sum_{i=1}^{n} \frac{\partial^2 \Phi(x)}{\partial (x^i)^2} = \mathbf{\Delta} \Phi(x)\,.$$

On a smooth manifold M^n which is not flat, or when non-Cartesian coordinates are used on $\mathbb{R}^n$, there holds a somewhat more general formula containing the metric tensor and second derivatives. We return to this in a later section.

Example 1.2

Potential of a spherically symmetric charge distribution:
Let $\varrho(r)$ be a piecewise continuous, local charge distribution with total charge Q. Locality means that there is a sphere S^2_R with radius R about the origin (the

centre of symmetry of the charge distribution) outside of which $\varrho(r)$ vanishes. With $\mathrm{d}^3x = r^2\,\mathrm{d}r\,\mathrm{d}(\cos\theta)\,\mathrm{d}\phi$, the normalization condition reads

$$\iiint \mathrm{d}^3x\ \varrho(r) = 4\pi \int_0^\infty r^2\,\mathrm{d}r\ \varrho(r) = 4\pi \int_0^R r^2\,\mathrm{d}r\ \varrho(r) = Q\ .$$

Construct, then, the following differentiable function from $\varrho(r)$:

$$U(r) = 4\pi \left\{ \frac{1}{r} \int_0^r r'^2\,\mathrm{d}r'\ \varrho(r') + \int_r^\infty r'\,\mathrm{d}r'\ \varrho(r') \right\}\ .$$

For $r \geqslant R$, taking account of the normalization condition, $U(r)$ reduces to $U(r) = Q/r$, that is, to the Coulomb potential of a pointlike charge Q. For values of the radial variable smaller than R the potential $U(r)$ differs from this simple form. For example, in the case of a homogeneous charge distribution,

$$\varrho(r) = \frac{3Q}{4\pi R^3}\Theta(R-r)\,, \quad \text{with}$$
$$\Theta(x) = 1 \quad \text{for } x \geqslant 0\,, \quad \Theta(x) = 0 \quad \text{for } x < 0$$

denoting the Heaviside function, the above integrals give

$$U_{\text{inner}}(r) = \frac{Q}{R^3}\left(\frac{3}{2}R^2 - \frac{1}{2}r^2\right) \qquad \text{for } r \leqslant R\,,$$
$$U_{\text{outer}}(r) = \frac{Q}{r} \qquad \text{for } r > R\ .$$

In the interior region, the potential $U(r)$ has the shape of a parabola; in the exterior region it falls off like $1/r$. At the value $r = R$, both the potential $U(r)$ and its first derivative are continuous, but this is not true for its second derivative. Calculating the (negative) gradient field of $U(r)$ and noting that the operator ∇ in spherical polar coordinates is given by

$$\nabla \equiv (\nabla_r, \nabla_\phi, \nabla_\theta) = \left(\frac{\partial}{\partial r}, \frac{1}{r\sin\theta}\frac{\partial}{\partial\phi}, \frac{1}{r}\frac{\partial}{\partial\theta}\right)\,,$$

one obtains for $\boldsymbol{E} = -\nabla U(r)$

$$\boldsymbol{E}_{\text{inner}}(\boldsymbol{x}) = \frac{Q}{R^3} r\,\hat{\boldsymbol{e}}_r\,, \qquad \boldsymbol{E}_{\text{outer}}(\boldsymbol{x}) = \frac{Q}{r^2}\,\hat{\boldsymbol{e}}_r\ .$$

Field $\boldsymbol{E}$ is oriented radially outwards, and its absolute value is shown in Fig. 1.1. In the outer region, this is the electric field around a pointlike charge Q which

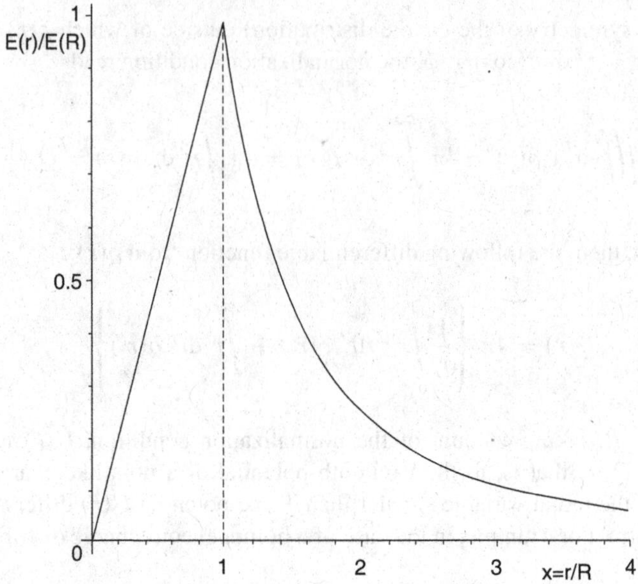

Fig. 1.1 The modulus $E(r)$ of the electric field $\boldsymbol{E}(\boldsymbol{x}) = E(r)\,\hat{\boldsymbol{e}}_r$ for the case of a homogeneous charge distribution with radius R

decreases like the inverse square of the radial variable. In the inner region, depending on the sign of Q, the field increases or decreases linearly from zero at the origin to the value Q/R^2 at R.

Calculating the divergence of $\boldsymbol{E}$ for this example and using

$$\boldsymbol{\Delta} U(r) = \frac{1}{r^2}\frac{\mathrm{d}}{\mathrm{d}r}\left(r^2 \frac{\mathrm{d}U(r)}{\mathrm{d}r}\right),$$

one obtains

$$\mathbf{div}\,\boldsymbol{E} = \boldsymbol{\nabla}\cdot\boldsymbol{E} = -\boldsymbol{\Delta} U(r) = 4\pi\varrho(r)\,,$$

both in the inner and in the outer domain. Of course, this is nothing but the Poisson equation to which we will turn in Sect. 1.8 in more detail. Note that it is written in terms of Gauss' system of units.

Example 1.3

Vector potential of a magnetic dipole:
Suppose we are given the static vector field

$$\boldsymbol{A}(\boldsymbol{x}) = \frac{\boldsymbol{m}\times\boldsymbol{x}}{r^3}\,, \qquad (r = |\boldsymbol{x}|)\,,$$

in which $\boldsymbol{m}$ is a given constant vector. Calculate the quantity $\boldsymbol{B} = \textbf{curl}\,\boldsymbol{A}$. As an example we give here the 1-component,

$$\begin{aligned}
B^1(\boldsymbol{x}) &= \frac{\partial}{\partial x^2} A_3(\boldsymbol{x}) - \frac{\partial}{\partial x^3} A_2(\boldsymbol{x}) \\
&= \frac{\partial}{\partial x^2}\left(\frac{m^1x^2 - m^2x^1}{r^3}\right) - \frac{\partial}{\partial x^3}\left(\frac{m^3x^1 - m^1x^3}{r^3}\right) \\
&= 2\frac{m^1}{r^3} + \frac{3}{r^5}\left[-m^1(x^2)^2 + m^2x^1x^2 + m^3x^1x^3 - m^1(x^3)^2\right] \\
&= -\frac{m^1}{r^3} + \frac{3x^1}{r^5}\left[m^1x^1 + m^2x^2 + m^3x^3\right] \\
&= \frac{1}{r^3}\left(-m^1 + 3(\hat{\boldsymbol{x}} \cdot \boldsymbol{m})\hat{x}^1\right),
\end{aligned}$$

where in the second to last step the term $3m^1(x^1)^2/r^5$ was added and subtracted and $\hat{\boldsymbol{x}} = \boldsymbol{x}/r$ was used. The result then becomes

$$\boldsymbol{B}(\boldsymbol{x}) = \frac{1}{r^3}(3(\hat{\boldsymbol{x}} \cdot \boldsymbol{m})\hat{\boldsymbol{x}} - \boldsymbol{m}) .$$

If $\boldsymbol{m}$ is a static magnetic dipole, then $\boldsymbol{B}(\boldsymbol{x})$ describes the induction field in the outer space that is created by this dipole.

1.3 Integral Theorems for the Case of $\mathbb{R}^3$

Using a simplified notation, which is perfectly acceptable for the case of the space $\mathbb{R}^3$, the most important integral theorems of use in electrodynamics read as follows:

Gauss' theorem
Let F be a smooth, orientable, closed surface embedded in $\mathbb{R}^3$. The surface F is taken to be *localized* (this means that one can find a sphere with a finite radius which encloses F completely). Let $V(F)$ be the volume enclosed by this surface, and let $\boldsymbol{V}$ be a smooth vector field. Then

$$\iiint\limits_{V(F)} \mathrm{d}^3x\, \boldsymbol{\nabla} \cdot \boldsymbol{V} = \iint\limits_{F} \mathrm{d}\sigma\, \boldsymbol{V} \cdot \hat{\boldsymbol{n}} . \tag{1.6}$$

In the surface integral on the right-hand side, $\hat{\boldsymbol{n}}$ denotes the outward directed normal at the surface element $\mathrm{d}\sigma$.

Equation (1.6) relates a volume integral of the divergence of a vector field with the integral of its outward normal component over the surface which encloses the volume. It makes no difference whether the vector field $\boldsymbol{V}$ is static, i.e. depends only on the $\boldsymbol{x}$ coordinate, or is not, i.e. is a function $\boldsymbol{V}(t, \boldsymbol{x})$ of time and space. The right-hand side of equation (1.6) may be understood as the net balance of the flow across surface F, which is defined by the normal component of $\boldsymbol{V}$. The divergence on the left-hand side characterizes the strength of the *source* which feeds this flow. What follows represents an example for the application of Gauss' theorem.

Example 1.4

Electric field of a homogeneous charge distribution. The charge distribution is assumed to be localized, spherically symmetric and homogeneous. If it is normalized to the total charge Q, then its functional form is given by $\varrho(\boldsymbol{x}) = 3Q/(4\pi R^3)\,\Theta(R-r)$. The divergence of the electric field is proportional to the charge distribution, $\boldsymbol{\nabla}\cdot\boldsymbol{E} = 4\pi\varrho$. As there is no preferred direction in space, the field must point in a radial direction, that is to say, it must have the form $\boldsymbol{E} = E(r)\hat{\boldsymbol{e}}_r$. Inserting this ansatz for $\boldsymbol{V}$ in Gauss' theorem and choosing F to be the sphere S_r^2 with radius r about the origin, the right-hand side of (1.6) yields the scalar function $E(r)$ times the surface of S_r^2, i.e. $E(r)4\pi r^2$. The left-hand side, in turn, requires the distinction

$$r \leqslant R: \quad 4\pi \iiint \mathrm{d}^3x\,\varrho(r) = (4\pi)^2 \int_0^r r'^2\,\mathrm{d}r' \left(\frac{3Q}{4\pi R^3}\right)\Theta(R-r')$$

$$= 4\pi \frac{Q}{R^3} r^3 ,$$

$$r > R: \quad 4\pi \iiint \mathrm{d}^3x\,\varrho(r) = 4\pi Q .$$

A comparison with the right-hand side and using example 1.2 yields the expected result

$$E_{\text{inner}}(r) = \frac{Q}{R^3} r , \qquad E_{\text{outer}}(r) = \frac{Q}{r^2} , \qquad (\boldsymbol{E} = E(r)\hat{\boldsymbol{e}}_r) .$$

Figure 1.2 illustrates the geometry of this simple example.

Stokes' theorem

Let C be a smooth closed path and $F(C)$ a smooth orientable surface bounded by C. For any smooth vector field $\boldsymbol{V}$ which is defined on F including its

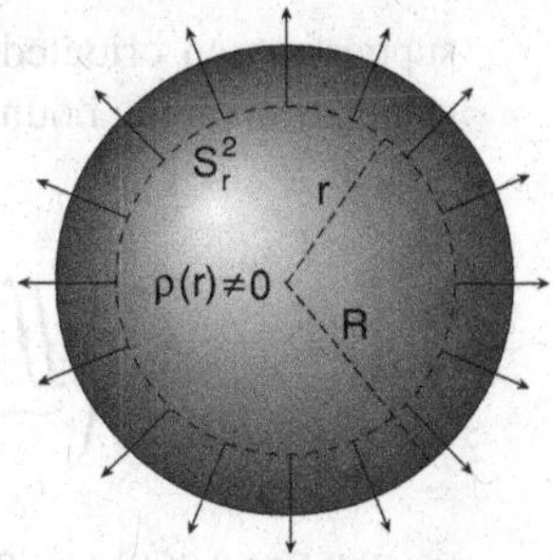

Fig. 1.2 The centre of symmetry of a spherically symmetric charge distribution $\varrho(r)$ is the centre of the auxiliary sphere S_r^2. One integrates $\varrho(r)$ over the volume enclosed by that sphere. The electric field on S_r^2 has a radial direction; its magnitude follows from Gauss' theorem

boundary, one has

$$\iint_{F(C)} \mathrm{d}\sigma \, (\nabla \times \boldsymbol{V}) \cdot \hat{\boldsymbol{n}} = \oint_C \mathrm{d}\boldsymbol{s} \cdot \boldsymbol{V} . \tag{1.7}$$

Here $\hat{\boldsymbol{n}}$ denotes the oriented normal on the surface $F(C)$, and $\mathrm{d}\boldsymbol{s}$ signifies the oriented line element on C. The two orientations are correlated such that the closed path C and $\hat{\boldsymbol{n}}$ form a right helix.

Remarks

1. For Gauss' theorem to be applicable, it suffices that surface F, which is the boundary of the volume $V(F)$, be piecewise smooth. For example, it may look like the surface of a soccer ball: This surface is everywhere continuous and consists of a finite number of smooth patches. Similarly, one may assume the closed path C in Stokes' theorem to be only piecewise smooth.
2. A common feature of the two integral theorems expressed by (1.6) and (1.7) is that they relate an integral over a compact manifold M with a boundary to an integral over its boundary, denoted by ∂M in differential geometry. In Gauss' theorem, M is a compact domain $V(F)$ in $\mathbb{R}^3$, whereas ∂M is its surface F. In Stokes' theorem, M is a two-dimensional surface with a boundary embedded in $\mathbb{R}^3$, whereas ∂M is its boundary curve. Furthermore, the integral over ∂M contains a function of the vector field $\boldsymbol{V}$ itself, whereas the integral over M contains a function of the first derivatives of $\boldsymbol{V}$: In the case of Gauss' theorem, this is the divergence; in the case of Stokes' theorem, it is the curl of $\boldsymbol{V}$. In fact, a further analysis shows that both (1.6) and (1.7) are special cases of the same theorem, though applying to different dimensions. Indeed, in Sect. 2.2.2, one will learn that this important theorem can be formulated for any dimension n. Anticipating the results of Sect. 2.2.2, the general integral theorem says the following: If ω is a smooth $(n-1)$-form with compact

support on an oriented manifold M with boundary ∂M, then, adopting the orientation on the boundary induced by the orientation of M, one has

$$\Bigg(\underbrace{\iint\cdots\int}_{(n)}\Bigg)_M \mathrm{d}\omega = \Bigg(\underbrace{\iint\cdots\int}_{(n-1)}\Bigg)_{\partial M} \omega\,, \tag{1.8a}$$

or, in a more compact notation,

$$\int_M \mathrm{d}\omega = \int_{\partial M} \omega\,. \tag{1.8b}$$

In theorem (1.6), the exterior form ω is a two-form and $\mathrm{d}\omega$ is a three-form, whereas in theorem (1.7), the form ω is a one-form, and its exterior derivative $\mathrm{d}\omega$ is a two-form.

3. There is a particularly simple example for Stokes' theorem that illustrates well the general equation (1.8b). Let the manifold be a smooth curve, $M = \gamma$, which runs from a point a to a point b, and let ω be a function or, in the language of exterior forms, a zero-form, $\omega = f$. The boundary ∂M of M consists of points a and b, $\partial M = \{a, b\}$, and $\mathrm{d}\omega = \mathrm{d}f$ is the total derivative of f. Applying the general equation (1.8b) and translating back to a more familiar notation, Stokes' theorem yields for this case

$$\int_{M=\gamma} \mathrm{d}f = \int_a^b \mathrm{d}t\,\frac{\mathrm{d}f}{\mathrm{d}t} = f(b) - f(a) = \int_{\partial M} f\,,$$

whose inner part will look familiar to the reader.

Green's theorems

Gauss' theorem, (1.6), has two variants, called Green's theorems, both of which are useful in the analysis of boundary value problems in $\mathbb{R}^3$. They are as follows.

Green's first theorem

Let the functions $\Phi(t, \boldsymbol{x})$ and $\Psi(t, \boldsymbol{x})$ be C^2 functions in the argument $\boldsymbol{x}$. Let $V(F)$ be a finite volume, with surface $F \equiv \partial V$ (see Gauss' theorem above). Then

$$\iiint_{V(F)} \mathrm{d}^3x\,\left(\Phi\,\Delta\Psi + \nabla\Phi\cdot\nabla\Psi\right) = \iint_F \mathrm{d}\sigma\,\Phi\frac{\partial\Psi}{\partial\hat{n}}\,. \tag{1.9}$$

This theorem is a direct application of Gauss' theorem, (1.6), provided one chooses the vector field to be

$$\boldsymbol{V}(t,\boldsymbol{x}) = \Phi(t,\boldsymbol{x})\ (\nabla\Psi(t,\boldsymbol{x}))$$

and makes use of the product rule for differentiation,

$$\nabla\cdot(\Phi\nabla\Psi) = \nabla\Phi\cdot\nabla\Psi + \Phi\Delta\Psi\,.$$

Furthermore, rewriting theorem (1.9) with the functions Φ and Ψ interchanged and subtracting the two equations thus obtained from one another yields Green's second theorem:

Green's second theorem
Using the same assumptions as in Gauss' theorem one has

$$\iiint\limits_{V(F)} \mathrm{d}^3x\ (\Phi\Delta\Psi - \Psi\Delta\Phi) = \iint\limits_{F} \mathrm{d}\sigma\ \left(\Phi\frac{\partial\Psi}{\partial\hat{\boldsymbol{n}}} - \Psi\frac{\partial\Phi}{\partial\hat{\boldsymbol{n}}}\right)\,. \tag{1.10}$$

In both cases, $\partial\Psi/\partial\hat{\boldsymbol{n}}$ and $\partial\Phi/\partial\hat{\boldsymbol{n}}$, respectively, denote the normal derivatives of these functions, that is, the directional derivatives along the normal $\hat{\boldsymbol{n}}$ at a given point of the surface F. These derivatives may also be written as $\hat{\boldsymbol{n}}\cdot\nabla\Psi$ and $\hat{\boldsymbol{n}}\cdot\nabla\Phi$, respectively.

1.4 Maxwell's Equations in Integral Form

This section summarizes Maxwell's equations in integral form, i.e. in the form in which they are tested, directly or indirectly, in a great variety of macroscopic experiments. It is assumed that the reader has seen most of the important classical experiments of electrodynamics and knows the essentials of their analysis.

1.4.1 The Law of Induction

Let C be a smooth curve of finite length, $\mathrm{d}\boldsymbol{s}$ the line element along the curve, and $\boldsymbol{E}(t,\boldsymbol{x})$ an electric field. The path integral $\int_C \mathrm{d}\boldsymbol{s}\cdot\boldsymbol{E}(t,\boldsymbol{x})$ is said to be the *electromotive force*.

Consider a magnetic induction field $\boldsymbol{B}(t,\boldsymbol{x})$ which may depend on space as well as on time, and let C be a smooth, *closed* curve in $\mathbb{R}^3$ which is the boundary of a surface F. Note that both the surface F and its boundary C may be variable in time, with the restriction that all variations must be at least continuous, if not smooth. The surface is assumed to be oriented, the local normal to the surface being denoted by $\hat{\boldsymbol{n}}(t, \boldsymbol{x})$. Then the *magnetic flux $\Phi(t)$ across the surface F* is defined to be the surface integral

$$\Phi(t) := \iint_F \mathrm{d}\sigma\, \boldsymbol{B}(t, \boldsymbol{x}) \cdot \hat{\boldsymbol{n}}(t, \boldsymbol{x}) . \tag{1.11}$$

Faraday's law of induction relates the temporal change of the magnetic flux with the electromotive force induced along the boundary curve.

Faraday's law of induction (1831)

$$\oint_C \mathrm{d}\boldsymbol{s} \cdot \boldsymbol{E}'(t, \boldsymbol{x}') = -f_{\mathrm{F}} \frac{\mathrm{d}}{\mathrm{d}t} \iint_F \mathrm{d}\sigma\, \boldsymbol{B}(t, \boldsymbol{x}) \cdot \hat{\boldsymbol{n}}(t, \boldsymbol{x}) , \tag{1.12}$$

$$x' \in C , \; x \in F .$$

The factor f_{F} is a real and positive constant. Its value depends on the system of physical units one has chosen: In the rational MKSA system, called the SI system *(système international des unités)*, it is $f_{\mathrm{F}} = 1$. In the Gauss system of units, it is $f_{\mathrm{F}} = \frac{1}{c}$, with c the velocity of light.

Remarks

1. The integrand of the left-hand side is the component of the electric field *tangent* to the boundary curve. The integrand on the right-hand side is the *normal* component of the induction field at point $\boldsymbol{x}$ on the surface. The negative sign on the right-hand side characterizes a physical principle: The electric current which is induced in curve C is directed such that the magnetic flux created by this current is *opposite* to the time variation of the flux on the right-hand side of (1.12). This is what is called *Lenz's rule*.
2. Law (1.12) summarizes many different experimental situations. For example, a surface and its boundary may be fixed relative to the inertial system of an observer, while the induction field varies with time. A simple example is provided by a circular annulus across which one moves a permanent magnet such that the magnetic flux increases or decreases. Alternatively, given a fixed induction field $\boldsymbol{B}(\boldsymbol{x})$ (which may even be homogeneous), a closed loop is moved through this field such that the flux $\Phi(t)$ changes with time (principle of generators and of electric motors).
3. The previous remark raises a problem which needs to be discussed in more detail. Indeed, there may be experimental situations where on the left-hand side, the electric field at the spacetime point $(t, \boldsymbol{x}')$ refers to a different frame

of reference than the induction field $\boldsymbol{B}(t, \boldsymbol{x})$ of the right-hand side. This is why we are careful in writing $\boldsymbol{E}'$ in lieu of $\boldsymbol{E}$ on the left-hand side. To this question we give a first answer here but refer the reader to a deeper and more detailed analysis below.

Suppose the shape of a conducting loop and of the surface of which it is a boundary are given and fixed. This fixed set-up is assumed to move relative to the inertial frame with respect to which the induction field $\boldsymbol{B}$ is defined. As seen from a reference system which is comoving with C (this is the instantaneous rest system of the set-up), one has

$$\frac{\mathrm{d}}{\mathrm{d}t} = \frac{\partial}{\partial t} + \boldsymbol{v} \cdot \nabla$$

and, when applied to the right-hand side of (1.12),

$$\frac{\mathrm{d}\boldsymbol{B}}{\mathrm{d}t} = \frac{\partial \boldsymbol{B}}{\partial t} + (\boldsymbol{v} \cdot \nabla)\boldsymbol{B} = \frac{\partial \boldsymbol{B}}{\partial t} + \nabla \times (\boldsymbol{B} \times \boldsymbol{v}) + (\nabla \cdot \boldsymbol{B})\boldsymbol{v} .$$

(The terms $(\boldsymbol{B} \cdot \nabla)\boldsymbol{v} - \boldsymbol{B}(\nabla \cdot \boldsymbol{v})$ vanish if $\boldsymbol{v}$ is held fixed.)

Here we anticipate that the induction field always has a vanishing divergence, $\nabla \cdot \boldsymbol{B} = 0$. Inserting this expansion into (1.12) and using Stokes' theorem (1.7), the curl may be converted to a path integral over boundary C. This yields

$$\oint_C \mathrm{d}s \cdot \left[\boldsymbol{E}' - f_{\mathrm{F}}(\boldsymbol{v} \times \boldsymbol{B})\right](t, \boldsymbol{x}') = -f_{\mathrm{F}} \iint_F \mathrm{d}\sigma \; \frac{\partial \boldsymbol{B}(\partial t, \boldsymbol{x})}{t} \cdot \hat{\boldsymbol{n}}(t, \boldsymbol{x}) \; . \quad (1.13\mathrm{a})$$

Obviously, the integrands on both sides now refer to the same system of reference. It is then suggestive to interpret

$$\left[\boldsymbol{E}' - f_{\mathrm{F}}(\boldsymbol{v} \times \boldsymbol{B})\right] =: \boldsymbol{E} \quad (1.13\mathrm{b})$$

as an electric field which should be compared with the induction field $\boldsymbol{B}$. The differential operators now act on the integrands only, not on the integral of the right-hand side as a whole.

1.4.2 Gauss' Law

Besides the electric field the *dielectric displacement* $\boldsymbol{D}(t, \boldsymbol{x})$ is an important element of electrodynamics. In a vacuum, this vector field is proportional to the electric field, $\boldsymbol{D}(t, \boldsymbol{x}) \propto \boldsymbol{E}(t, \boldsymbol{x})$, and hence contains the same information as the latter. In polarizable media, the two fields are related by the formula $\boldsymbol{D} = \varepsilon \boldsymbol{E}$, where $\varepsilon(\boldsymbol{x})$ is a tensor of rank 2 which describes the electric polarizability of the medium.

Gauss' law relates the flux of the dielectric displacement across a *closed* surface to the total charge contained in the volume enclosed by the surface.

Gauss' law

Let F be a closed, smooth or piecewise smooth surface, and let $V(F)$ be the spatial volume which is defined and enclosed by F. Then, with $\varrho(t,\boldsymbol{x})$ being a given electric charge distribution, one has

$$\iint_F \mathrm{d}\sigma \left(\boldsymbol{D}(t,\boldsymbol{x}') \cdot \hat{\boldsymbol{n}}\right) = f_{\mathrm{G}} \iiint_{V(F)} \mathrm{d}^3x \, \varrho(t,\boldsymbol{x}) = f_{\mathrm{G}} Q_V \, . \tag{1.14}$$

The constant f_{G}, which is real and positive, is universal but depends on the system of physical units one has chosen. Unit vector $\hat{\boldsymbol{n}}$ is the outer normal of the surface, and Q_V is the total charge enclosed in the volume $V(F)$.

Remarks

1. The left-hand side shows the net balance of the flux of vector field $\boldsymbol{D}$ across the surface. This integral quantity can be positive, negative, or zero. Figure 1.3 shows the example of two identical spheres that carry equal and opposite charges $q_1 = q$ and $q_2 = -q$, respectively. As the total charge vanishes, $Q = q_1 + q_2 = 0$, the net balance of the flux across any surface that encloses the two spheres completely is equal to zero.
2. The constant on f_{G} in (1.14) has the value

$$f_{\mathrm{G}} = 1 \quad \text{in SI units,}$$
$$f_{\mathrm{G}} = 4\pi \quad \text{in Gauss units.}$$

3. If $\boldsymbol{D}$ is proportional to the electric field $\boldsymbol{E}$, i.e. if $\boldsymbol{D} = \varepsilon \boldsymbol{E}$, with ε a constant factor, and if $\boldsymbol{D}$ does not depend on time, (1.14) leads to the Poisson equation. This may be seen as follows. Making use of Gauss' theorem, (1.6), replace the left-hand side by the volume integral over $V(F)$ of the divergence. As the choice of the surface F and, hence, of the enclosed volume is arbitrary, the integrands on the two sides must be equal:

$$\boldsymbol{\nabla} \cdot \boldsymbol{E}(\boldsymbol{x}) = f_{\mathrm{G}} \frac{1}{\varepsilon} \varrho(\boldsymbol{x}) \, . \tag{1.15a}$$

By writing the (static) electric field as a gradient field[1], $\boldsymbol{E} = -\boldsymbol{\nabla}\Phi(\boldsymbol{x})$, one obtains the Poisson equation

$$\boldsymbol{\Delta}\Phi(\boldsymbol{x}) = -f_{\mathrm{G}} \frac{1}{\varepsilon} \varrho(\boldsymbol{x}) \, . \tag{1.15b}$$

[1] In anticipation of subsequent results, we make use of the fact that in time-independent situations, the electric field has a vanishing curl. This is not true in the general case!

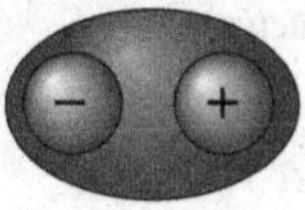

Fig. 1.3 Draw an ellipsoid such that it encloses two geometrically identical spheres with equal and opposite charges. Although locally the flux of the vector field $\boldsymbol{D}$ does not vanish, its net balance over the entire surface is zero because the effects of the two enclosed charge distributions cancel

If the fields are not stationary and depend on both $\boldsymbol{x}$ and t, then one concludes from (1.14) only

$$\nabla \cdot \boldsymbol{D}(t, \boldsymbol{x}) = f_{\mathrm{G}}\varrho(t, \boldsymbol{x}) , \tag{1.15c}$$

and the relation between $\boldsymbol{D}$ and $\boldsymbol{E}$ remains undetermined. In fact, here we have obtained one of Maxwell's equations in local form.

4. Another important result is obtained by applying Gauss' law, (1.14), to *magnetic* charges and to the induction field created by them. Experiment tells us that there are no free magnetic charges. Every permanent magnet has a *north* pole and a *south* pole, and they can in no way be separated and isolated. Whenever one cuts such magnets into smaller pieces, every fraction exhibits both a north pole and a south pole. Hence, the integral on the right-hand side of (1.14) vanishes for every volume $V(F)$ that encloses the total magnetic charge density. Therefore, one expects the general property

$$\iint_F \mathrm{d}\sigma \, \left(\boldsymbol{B}(t, \boldsymbol{x}') \cdot \hat{\boldsymbol{n}}\right) = 0 \tag{1.16}$$

to hold for every smooth or piecewise smooth surface. As in the previous remark, one may apply Gauss' theorem, (1.6), to (1.16), thereby converting the surface integral to a volume integral. As the surface F and, hence, the enclosed volume $V(F)$ are arbitrary, one obtains the local equation

$$\nabla \cdot \boldsymbol{B}(t, \boldsymbol{x}) = 0 . \tag{1.17}$$

Equation (1.16) says that the magnetic induction has no sources anywhere in space.

1.4.3 The Law of Biot and Savart

As is well known, a conductor which carries an electric current creates a magnetic field in the surrounding space. A simple model for a conductor is provided by a thin wire, that is to say, a curve in $\mathbb{R}^3$ carrying the current J. To simplify matters and to obtain a more generally valid concept, it is useful to introduce the current *density* $\boldsymbol{j}(t, \boldsymbol{x}')$, defined as the amount of charge crossing a unit surface per second

Fig. 1.4 Model of a thin cylindrical conductor

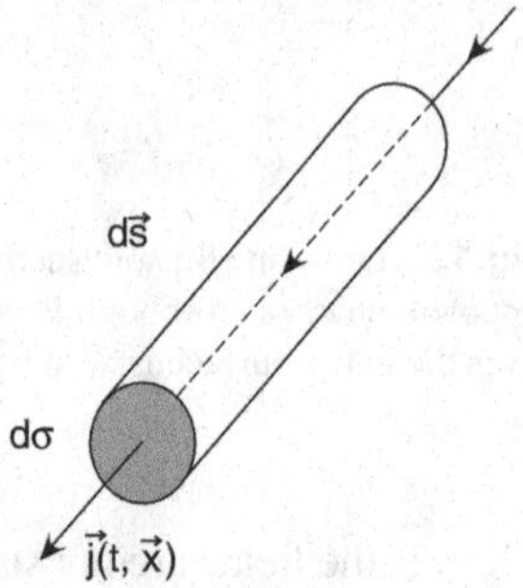

in the direction $\hat{\boldsymbol{j}}$. Loosely speaking, the current J, then, is the integral of $\boldsymbol{j}$ over a section of the conductor. For the sake of illustration, consider a straight cylindrical conductor with cross section F oriented along the 3-direction. The current density is proportional to the direction $\hat{\boldsymbol{e}}_3$; it is different from zero inside the cylinder only. Denoting a segment of the cylinder by $\mathrm{d}\boldsymbol{s}$ and the surface element perpendicular to the 3-direction by $\mathrm{d}\sigma$, one has

$$J\,\mathrm{d}s = \left(\int_F \mathrm{d}\sigma\ |\boldsymbol{j}(t,\boldsymbol{x})|\right)\mathrm{d}s\ .$$

This is illustrated by the wire in Fig. 1.4.

Expressed in differential form, the law of Biot and Savart states that the piece $\mathrm{d}\boldsymbol{s}$ of the conductor yields the contribution

$$\mathrm{d}\boldsymbol{H} = \frac{f_{\mathrm{BS}}}{4\pi} J\,\mathrm{d}\boldsymbol{s} \times \frac{\boldsymbol{x}}{|\boldsymbol{x}|^3}$$

to the magnetic field $\boldsymbol{H}$. By combining these two formulae and assuming the current density to be arbitrary but always localized, the following integral form of the Biot–Savart law becomes plausible.

Law of Biot and Savart (1822)
The current density $\boldsymbol{j}(t,\boldsymbol{x}')$ is assumed to be a localized smooth vector field. The magnetic field created by this distribution is given by

$$\boldsymbol{H}(t,\boldsymbol{x}) = \frac{f_{\mathrm{BS}}}{4\pi}\iiint \mathrm{d}^3x'\ \boldsymbol{j}(t,\boldsymbol{x}') \times \frac{\boldsymbol{x}-\boldsymbol{x}'}{|\boldsymbol{x}-\boldsymbol{x}'|^3}\ . \qquad (1.18)$$

This expression holds outside as well as inside the source density $\boldsymbol{j}(\boldsymbol{x}')$. The value of the positive constant f_{BS} depends on the choice of the system of units.

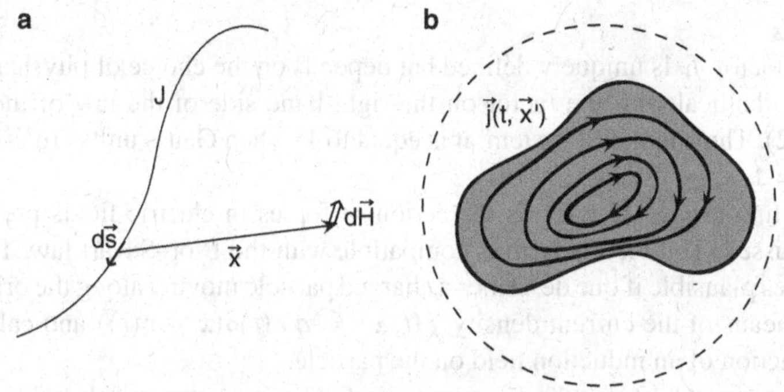

Fig. 1.5 **a** A current J flowing through a thin wire creates a magnetic field in the space around it. The current element $J\,\mathrm{d}s$ contributes the amount $\mathrm{d}\boldsymbol{H}$. **b** A charge density contained in a finite volume. The law of Biot and Savart yields the magnetic field created by this charge density inside and outside the region which contains the current density

The position vector $\boldsymbol{x}$ denotes the point in space at which the field is measured, and $\boldsymbol{x}'$ is the argument by means of which the distribution $\boldsymbol{j}(t, \boldsymbol{x}')$ is probed and integrated over. Figure 1.5a illustrates the differential contribution of the current element $J\,\mathrm{d}s$ of a conductor to the magnetic field, whereas Fig. 1.5b sketches an example of a current density which is localized in the sense that it may be thought of as being enclosable by a sphere with radius R.

1.4.4 The Lorentz Force

In a given electric field $\boldsymbol{E}(t, \boldsymbol{x})$ and a given induction field $\boldsymbol{B}(t, \boldsymbol{x})$, defined with respect to a frame of reference K, a pointlike particle of charge q moving at velocity $\boldsymbol{v}$ with respect to that frame experiences a characteristic force field. This is called the *Lorentz force* and is another important empirical input which is tested experimentally.

Lorentz force acting on charged point particle

A particle of charge q moving at instantaneous velocity $\boldsymbol{v}$ in a pair of fields $\boldsymbol{E}(t, \boldsymbol{x})$ and $\boldsymbol{B}(t, \boldsymbol{x})$, defined with respect to the frame of reference K, is subject to the force field

$$\boldsymbol{F}(t, \boldsymbol{x}) = q\big(\boldsymbol{E}(t, \boldsymbol{x}) + f_{\mathrm{F}}\boldsymbol{v} \times \boldsymbol{B}(t, \boldsymbol{x})\big) . \tag{1.19}$$

In this formula the modulus of the velocity is smaller than or equal to the velocity of light, $|\boldsymbol{v}| \leqslant c$.

▶ **Remarks**

1. The factor f_F is uniquely defined but depends on the choice of physical units. It is identical with the factor on the right-hand side of the law of induction (1.12). Thus, in the SI system, it is equal to 1; when Gauss units are used, it is $f_F = 1/c$.
2. The first term, $q\boldsymbol{E}(t, \boldsymbol{x})$, is the action of forces in electric fields previously discussed. The second term is compatible with the Biot–Savart law. This becomes plausible if one describes a charged particle moving along the orbit $\boldsymbol{r}(t)$ by means of the current density $\boldsymbol{j}(t, \boldsymbol{x}) = q\dot{\boldsymbol{r}}(t)\delta(\boldsymbol{x} - \boldsymbol{r}(t))$ and calculates the action of an induction field on the particle.
3. Expression (1.19) for the Lorentz force field is an exact one. It holds for every velocity that is compatible with Special Relativity. This is an important finding, confirmed by experiment, and we return to it subsequently in more detail.

1.4.5 The Continuity Equation

A further fundamental fact is the conservation of charge:
The electric charge is conserved by all fundamental interactions.
The reader should note that this holds not only globally, i.e. after integration over volumes in $\mathbb{R}^3$, but also locally in any domain of space.

If we try to formulate this law in a general, integral form, the following model seems physically meaningful. Suppose we are given a time-dependent charge distribution $\varrho(t, \boldsymbol{x})$, localized in space, and a current density $\boldsymbol{j}(t, \boldsymbol{x})$ which is due to the motion of the charges within ϱ. For every (piecewise) smooth, closed surface F and the corresponding volume $V(F)$ enclosed by it, the following equation applies:

$$-\frac{\mathrm{d}}{\mathrm{d}t} \iiint\limits_{V(F)} \mathrm{d}^3x\ \varrho(t, \boldsymbol{x}) = \iint\limits_{F} \mathrm{d}\sigma\ \left(\boldsymbol{j}(t, \boldsymbol{x}') \cdot \hat{\boldsymbol{n}}\right) . \qquad (1.20)$$

Expressed in words, this equation says that the negative change in time of the total charge $Q(V)$ which is contained in volume V is equal to the flux of the electric current density, integrated over the surface of that volume. If the total charge $Q(V)$ has diminished, then in the net balance over the surface, more current must have flown outward than inward. If it has increased, then in the integral, more current has flown inward than outward.

By means of Gauss' theorem, (1.6), the right-hand side of (1.20) is replaced by a volume integral of the divergence of $\boldsymbol{j}$. As this expression holds for an arbitrary choice of the volume of integration, the integrands must be equal. This yields the differential form of the continuity equation:

Continuity equation

$$\frac{\partial \varrho(t,\boldsymbol{x})}{\partial t} + \nabla \cdot \boldsymbol{j}(t,\boldsymbol{x}) = 0\,. \tag{1.21}$$

Remarks

1. Note that this equation contains no relative factors which depend on the system of units. This is due to the fact that the physical dimension of the electric current density is already fixed by its definition in terms of the electric charge and the unit one has chosen for the latter. An example is provided by the current density created by a charged particle with charge e moving at velocity $\boldsymbol{v}$:
$$\boldsymbol{j}(t,\boldsymbol{x}) = e\,\boldsymbol{v}(t)\delta(\boldsymbol{x} - \boldsymbol{r}(t))\,.$$
If one measures charge e in cgs-units, then the dimensions of both charge and current density are *derived* units and may be expressed in terms of mass M, length L and time T:
$$[e] = \mathrm{M}^{\frac{1}{2}}\mathrm{L}^{\frac{3}{2}}\mathrm{T}^{-1}\,, \qquad [\boldsymbol{j}] = \mathrm{M}^{\frac{1}{2}}\,\mathrm{L}^{-\frac{1}{2}}\,\mathrm{T}^{-2}\,.$$
More specifically, the unit of charge is then $1\ \mathrm{g}^{1/2}\,\mathrm{cm}^{3/2}\,\mathrm{s}^{-1}$, whereas the unit of current density is $1\ \mathrm{g}^{1/2}\,\mathrm{cm}^{-1/2}\,\mathrm{s}^{-2}$. With the charge density ϱ defined as the charge per unit of volume, it is easily verified that the physical dimensions in (1.21) match.
In the SI system, electric charge is assigned a unit of its own, $[e] = 1\,\mathrm{C}$ (Coulomb), and the current is measured in Ampères, 1 A, so that the dimension of current density becomes $[\boldsymbol{j}] = 1\,\mathrm{C\,m^{-2}\,s^{-1}} = 1\,\mathrm{A\,m^{-2}}$. Here we have used $1\,\mathrm{C} = 1\,\mathrm{A\,s}$, i.e. 1 Coulomb equals 1 (Ampère per second).
2. The operation of divergence $\mathrm{div}(\boldsymbol{a}) = \nabla \cdot \boldsymbol{a}$, which is familiar from the case of $\mathbb{R}^3$, can be generalized to the four-dimensional spacetime of Special Relativity. The analogous operator reads, with $x^0 = ct$, $\{x^i\} \equiv \boldsymbol{x}$,
$$\left\{\frac{\partial}{\partial x^\mu}\right\} = \left(\frac{\partial}{\partial x^0}, \frac{\partial}{\partial x^1}, \frac{\partial}{\partial x^2}, \frac{\partial}{\partial x^3}\right) \equiv \left(\frac{1}{c}\frac{\partial}{\partial t}, \nabla\right)\,. \tag{1.22a}$$
It is not difficult to recall the behaviour of this differential operator under proper, orthochronous Lorentz transformations $\Lambda \in L_+^\uparrow$. Its application to a Lorentz scalar product of a constant four-vector a and of a coordinate x, $a \cdot x = a^0 x^0 - \boldsymbol{a} \cdot \boldsymbol{x} = a_\mu x^\mu$, gives
$$\frac{\partial}{\partial x^\mu}(a \cdot x) = a_\mu\,.$$

The derivative of an invariant such as $(a \cdot x)$ by the (contravariant) argument x^μ gives a_μ, i.e. a covariant quantity. With this behaviour in mind, the operator (1.22a) is best written in the suggestive notation

$$\{\partial_\mu\} = (\partial_0, \partial_1, \partial_2, \partial_3) \,. \tag{1.22b}$$

If one applies to the points x of spacetime a Lorentz transformation such as e.g. the special (boost) transformation

$$L(v\hat{e}_3) = \begin{pmatrix} \gamma & 0 & 0 & \beta\gamma \\ 0 & 1 & 0 & 0 \\ 0 & 0 & 1 & 0 \\ \beta\gamma & 0 & 0 & \gamma \end{pmatrix}, \quad \text{with } \beta = \frac{v}{c}\,, \gamma = \frac{1}{\sqrt{1-\beta^2}}\,,$$

the partial derivatives with respect to t and to x^1 in (1.21) are mixed in a manner that may seem confusing. However, if it were possible to combine the charge and current densities in a four-component current density j,

$$j = \big(c\varrho(x), \boldsymbol{j}(x)\big)^T\,, \qquad \text{with } x = (x^0, \boldsymbol{x})^T\,, \quad x^0 = ct\,, \tag{1.23}$$

and if $j^\mu(x)$ transformed covariantly under $\Lambda \in L_+^\uparrow$, then the invariant

$$\partial_\mu j^\mu(x) = \partial_0 j^0(x) + \nabla\cdot\boldsymbol{j}(x) = \frac{1}{c}\frac{\partial}{\partial t}\big(c\varrho(t,\boldsymbol{x})\big) + \nabla\cdot\boldsymbol{j}(t,\boldsymbol{x}) = 0 \tag{1.24a}$$

would be identical with the continuity Eq. (1.21) and could be written in the compact form

$$\partial_\mu j^\mu(x) = 0\,. \tag{1.24b}$$

If this were true, then what we call a *charge* density and what we call a *current* density would depend on the frame of reference chosen for the observer.

3. One should note that the basic equations of Maxwell theory contain two different sets of physical quantities: on the one hand the electromagnetic fields $\boldsymbol{E}(t,\boldsymbol{x})$, $\boldsymbol{H}(t,\boldsymbol{x})$, $\boldsymbol{D}(t,\boldsymbol{x})$ and $\boldsymbol{B}(t,\boldsymbol{x})$, on the other hand the source terms $\varrho(t,\boldsymbol{x})$ and $\boldsymbol{j}(t,\boldsymbol{x})$. While the fields in the first group may perfectly well "live on their own", i.e. obey testable equations of motion, the quantities of the second group concern the charge carriers which may exist in a great variety of realizations. As an abbreviation that catches the essential difference one may call the first set "radiation", the second set "matter". Matter which is described by equations of motion which differ from those of radiation has its own dynamics, e.g. classical mechanics or quantum mechanics. The question raised in the previous remark, in essence, is whether matter is described by a theory which is invariant under Special Relativity and which respects absolute conservation of electric charge. Only if these conditions are met is there a chance to be able to collect the source terms in Maxwell's equations in a four-vector current density $j(x)$ which has the correct transformation behaviour.

4. A simple example will help to illustrate the previous remark. A charged point particle moving along the world line $x(\tau)$, τ being the proper time, within the framework of Special Relativity has the velocity

$$u(\tau) = \frac{\mathrm{d}}{\mathrm{d}\tau} x(\tau) = (\gamma c, \gamma \boldsymbol{v})^T .$$

At every point of spacetime the world line $x(\tau)$ follows a timelike direction (see, e.g. [ME], Chap. 4). This expresses the fact that the magnitude of the particle's velocity never exceeds the velocity of light. The four-velocity is normalized such that its invariant, squared norm equals c^2,

$$u^2 = (u^0)^2 - \boldsymbol{u}^2 = c^2\gamma^2(1-\beta^2) = c^2 .$$

While $x(\tau)$ and $u(\tau)$ are coordinate-free definitions (remember that proper time τ is a Lorentz-scalar!) the decomposition $u = (\gamma c, \gamma \boldsymbol{v})^T$ presupposes the choice of a frame of reference K.
The particle, which is assumed to carry the charge e, creates the current density

$$j(y) = ec \int \mathrm{d}\tau \, u(y) \, \delta^{(4)}\left(y - x(\tau)\right) . \tag{1.25}$$

This is a Lorentz vector. Indeed, the velocity u is such a vector, the proper time variable and the product of the four δ-distributions being Lorentz scalars. Therefore, j is a Lorentz vector. Furthermore, in any frame of reference K, one recovers the expected expressions for the charge and the current densities by integrating over τ by means of the relation $\mathrm{d}\tau = \mathrm{d}t'/\gamma$ between proper time and coordinate time and using the formula $\delta(y^0 - x^0(\tau)) = \delta(ct - ct') = \delta(t-t')/c$,

$$j^0(t, \boldsymbol{y}) = ce\,\delta^{(3)}\left(\boldsymbol{y} - \boldsymbol{x}(t)\right) \equiv c\varrho(t, \boldsymbol{y}) ,$$
$$j^i(t, \boldsymbol{y}) = e\,v^i(t)\delta^{(3)}\left(\boldsymbol{y} - \boldsymbol{x}(t)\right) , \quad i = 1, 2, 3 .$$

Equation (1.25) describes correctly the densities ϱ and $\boldsymbol{j}$ created by the particle in motion in a form which renders obvious their character as components of a Lorentz vector. We leave as an exercise the verification that $j(y)$ fulfills the continuity equation $\partial_\mu j^\mu(y) = 0$.

1.5 Maxwell's Equations in Local Form

The basic Eqs. (1.12), (1.14), (1.16) and (1.18), in their integral form, have the advantage that they contain genuine observables and, hence, are directly testable in experiments. The drawback is that they refer to specific set-ups such as conducting loops, local volumes and closed surfaces and that they relate phenomena which cannot be simply interpreted as "events", i.e. as physical phenomena which happen at a well-defined position $\boldsymbol{x}$ in space at a well-defined time t. To get rid of concrete and specific experimental set-ups, one makes use of the integral theorems of Sect. 1.3 to transform them into *local* equations. By local equations we mean differential equations which are formulated at the *same* point $(t, \boldsymbol{x})$ of spacetime. This step brings about two advantages. On the one hand, these local equations contain all (historical) experiments from which Maxwell's equations in integral form were obtained by abstraction. On the other hand, the local equations allow one to design new experiments and, hence, perform new tests of the theory.

A famous example is provided by electromagnetic waves in a vacuum: Maxwell's equations in local form allow one to derive a wave equation whose solutions can be calculated for various boundary conditions. This prediction of the theory was brilliantly confirmed in the experiments that Heinrich Hertz carried out in 1887. The same experiments confirmed that Maxwell's displacement current, originally postulated on the ground of theoretical considerations, was real.

1.5.1 Induction Law and Gauss' Law

The law of induction (1.12) is compatible with the relation

$$\nabla \times \boldsymbol{E}(t, \boldsymbol{x}) = -f_{\mathrm{F}} \frac{\partial}{\partial t} \boldsymbol{B}(t, \boldsymbol{x}) \tag{1.26}$$

which is local in the sense described above. To show this apply Stokes' theorem in its form (1.7) to the left-hand side of (1.13a),

$$\oint\limits_{C=\partial F} \mathrm{d}\boldsymbol{s} \cdot \boldsymbol{E}(t, \boldsymbol{x}') = \iint\limits_{F} \mathrm{d}\sigma (\nabla \times \boldsymbol{E}) \cdot \hat{\boldsymbol{n}} \,.$$

The argument then goes as follows: The path C can be contracted continuously such that the surface enclosed by it shrinks to a point. In this limit, the two integrands must be equal, and one recovers the local Eq. (1.26). One must keep in mind, however, that even though the law of induction in its integral form follows from (1.26), the local form does not, a priori, follow uniquely from the former. It does so only if a further hypothesis is made. This assumption may be recognized in Eq. (1.13b), which says that a field which is a pure electric field in a given frame of reference appears as a linear combination of an electric field and a magnetic field with respect to a frame which moves relative to the given frame of reference.

This property may seem surprising, but it is physically plausible. Consider once more the example of a charged pointlike particle in uniform motion along a straight line. In its rest system, it creates only the well-known electric Coulomb field of a point charge. In any other system in which it has the velocity $\boldsymbol{v}$, the particle appears both as a charge density and as a current density which, via the law expressed by Eq. (1.18), creates a magnetic field.

In a similar way, Gauss' law, Eq. (1.14), is converted into a local relation by converting the left-hand side into a volume integral by means of the integral theorem, Eq. (1.7):

$$\iint_F \mathrm{d}\sigma\, (\boldsymbol{D} \cdot \hat{\boldsymbol{n}}) = \iiint_{V(F)} \mathrm{d}^3x\, \boldsymbol{\nabla} \cdot \boldsymbol{D} = f_{\mathrm{G}} \iiint_{V(F)} \mathrm{d}^3x\, \varrho(t, \boldsymbol{x})\,.$$

As the volume is arbitrary and as the surface F is continuously contractible, the integrands must be equal. One obtains the local Eq. (1.15c) derived in Sect. 1.4.2.

1.5.2 Local Form of the Law of Biot and Savart

The aim here is to distill a local equation from the integral law (1.18). One starts by noting the auxiliary formulae

$$\frac{\boldsymbol{x} - \boldsymbol{x}'}{|\boldsymbol{x} - \boldsymbol{x}'|^3} = -\boldsymbol{\nabla}_x \left(\frac{1}{|\boldsymbol{x} - \boldsymbol{x}'|} \right) = +\boldsymbol{\nabla}_{x'} \left(\frac{1}{|\boldsymbol{x} - \boldsymbol{x}'|} \right)\,. \tag{1.27}$$

Inserting the first of these on the right-hand side of (1.18) and noting that the derivatives with respect to $\boldsymbol{x}$ can be taken out of the integral, one has

$$\begin{aligned} \boldsymbol{H}(t, \boldsymbol{x}) &= -\frac{f_{\mathrm{BS}}}{4\pi} \iiint \mathrm{d}^3x'\, \boldsymbol{j}(t, \boldsymbol{x}') \times \boldsymbol{\nabla}_x \left(\frac{1}{|\boldsymbol{x} - \boldsymbol{x}'|} \right) \\ &= +\frac{f_{\mathrm{BS}}}{4\pi} \boldsymbol{\nabla}_x \times \iiint \mathrm{d}^3x'\, \frac{\boldsymbol{j}(t, \boldsymbol{x}')}{|\boldsymbol{x} - \boldsymbol{x}'|}\,. \end{aligned}$$

The change of sign is due to the change of the order of the factors in the cross product. One then calculates the curl of $\boldsymbol{H}$ and makes use of the known identity (see also (1.47c) below)

$$\boldsymbol{\nabla} \times (\boldsymbol{\nabla} \times \boldsymbol{A}) = \boldsymbol{\nabla}(\boldsymbol{\nabla} \cdot \boldsymbol{A}) - \boldsymbol{\Delta} \boldsymbol{A}\,. \tag{1.28}$$

There follows

$$\begin{aligned} \boldsymbol{\nabla} \times \boldsymbol{H}(t, \boldsymbol{x}) &= \frac{f_{\mathrm{BS}}}{4\pi} \boldsymbol{\nabla}_x \times \left(\boldsymbol{\nabla}_x \times \iiint \mathrm{d}^3x'\, \frac{\boldsymbol{j}(t, \boldsymbol{x}')}{|\boldsymbol{x} - \boldsymbol{x}'|} \right) \\ &= \frac{f_{\mathrm{BS}}}{4\pi} \boldsymbol{\nabla}_x \iiint \mathrm{d}^3x' \left(\boldsymbol{j}(t, \boldsymbol{x}') \cdot \boldsymbol{\nabla}_x \left(\frac{1}{|\boldsymbol{x} - \boldsymbol{x}'|} \right) \right) && \text{(1.29a)} \\ &\quad - \frac{f_{\mathrm{BS}}}{4\pi} \iiint \mathrm{d}^3x'\, \boldsymbol{j}(t, \boldsymbol{x}')\, \boldsymbol{\Delta}_x \left(\frac{1}{|\boldsymbol{x} - \boldsymbol{x}'|} \right). && \text{(1.29b)} \end{aligned}$$

By means of the auxiliary formula (1.27) one replaces the derivative with respect to the variable $\boldsymbol{x}$ in the first term (1.29a) by the gradient with respect to $\boldsymbol{x}'$. In a second step one performs a partial integration in this variable.

Regarding the second term (1.29b), one uses the relation

$$\boldsymbol{\Delta}_x \left(\frac{1}{|\boldsymbol{x} - \boldsymbol{x}'|} \right) = -4\pi\delta(\boldsymbol{x} - \boldsymbol{x}') \,, \tag{1.30}$$

(see e.g. [QM], Appendix A.1, example A.3, for a proof of this formula) to find

$$\begin{aligned} \nabla \times \boldsymbol{H}(t, \boldsymbol{x}) = {} & \frac{f_{\mathrm{BS}}}{4\pi} \nabla_x \left(\iiint \mathrm{d}^3 x' \, (\nabla_{x'} \cdot \boldsymbol{j}(t, \boldsymbol{x}')) \frac{1}{|\boldsymbol{x} - \boldsymbol{x}'|} \right) \\ & + f_{\mathrm{BS}} \, \boldsymbol{j}(t, \boldsymbol{x}) \,. \end{aligned}$$

The divergence of $\boldsymbol{j}$ in the integrand, by the continuity equation (1.21), is equal to the negative time derivative of the charge density $\varrho(t, \boldsymbol{x})$. Thus one obtains

$$\nabla \times \boldsymbol{H}(t, \boldsymbol{x}) = -\frac{f_{\mathrm{BS}}}{4\pi} \frac{\partial}{\partial t} \nabla_x \iiint \mathrm{d}^3 x' \, \frac{\varrho(t, \boldsymbol{x}')}{|\boldsymbol{x} - \boldsymbol{x}'|} + f_{\mathrm{BS}} \, \boldsymbol{j}(t, \boldsymbol{x}) \,.$$

In the first term on the right-hand side, the gradient of the integral is proportional to $\boldsymbol{D}(t, \boldsymbol{x})$,

$$\nabla_x \iiint \mathrm{d}^3 x' \, \frac{\varrho(t, \boldsymbol{x}')}{|\boldsymbol{x} - \boldsymbol{x}'|} = -\frac{4\pi}{f_{\mathrm{G}}} \boldsymbol{D}(t, \boldsymbol{x}) \,.$$

This follows from the Maxwell equation (1.15c), by taking its divergence and using relation (1.30) above. Inserting this result, one obtains the equation

$$\nabla \times \boldsymbol{H}(t, \boldsymbol{x}) = \frac{f_{\mathrm{BS}}}{f_{\mathrm{G}}} \frac{\partial}{\partial t} \boldsymbol{D}(t, \boldsymbol{x}) + f_{\mathrm{BS}} \, \boldsymbol{j}(t, \boldsymbol{x}) \,. \tag{1.31}$$

This equation is entirely local in all its parts.

1.5.3 Local Equations in All Systems of Units

To start with, we collect the local equations (1.17), (1.26), (1.15c) and (1.31) without specializing in any particular system of units of use in physics or applied sciences:

$$\nabla \cdot \boldsymbol{B}(t, \boldsymbol{x}) = 0 \,; \tag{1.32a}$$

$$\nabla \times \boldsymbol{E}(t, \boldsymbol{x}) + f_{\mathrm{F}} \frac{\partial}{\partial t} \boldsymbol{B}(t, \boldsymbol{x}) = 0 \,; \tag{1.32b}$$

$$\nabla \cdot \boldsymbol{D}(t, \boldsymbol{x}) = f_{\mathrm{G}} \varrho(t, \boldsymbol{x}) \,; \tag{1.32c}$$

$$\nabla \times \boldsymbol{H}(t, \boldsymbol{x}) - \frac{f_{\mathrm{BS}}}{f_{\mathrm{G}}} \frac{\partial}{\partial t} \boldsymbol{D}(t, \boldsymbol{x}) = f_{\mathrm{BS}} \boldsymbol{j}(t, \boldsymbol{x}) \,. \tag{1.32d}$$

These equations are supplemented by the expression for the Lorentz force (1.19) and by the relation between $\boldsymbol{D}$ and $\boldsymbol{E}$, and between $\boldsymbol{B}$ and $\boldsymbol{H}$, respectively, which apply in a vacuum and which also depend on the system of units that is chosen:

$$\boldsymbol{D}(t,\boldsymbol{x}) = \varepsilon_0 \boldsymbol{E}(t,\boldsymbol{x}), \qquad \boldsymbol{B}(t,\boldsymbol{x}) = \mu_0 \boldsymbol{H}(t,\boldsymbol{x}) . \tag{1.33}$$

The positive constants f_{F}, f_{G} and f_{BS} are labelled such that one is reminded of the fundamental law in which they occur: "F" for Faraday, "G" for Gauss and "BS" for Biot and Savart. The constants ε_0 and μ_0, which are also positive, are called *dielectric constant* and *magnetic permeability,* respectively.

In a first step, one verifies that the continuity equation (1.21) is respected, that is, that it follows from the inhomogeneous equations (1.32c) and (1.32d). From (1.32c) and (1.32d) there follows

$$\frac{\partial \varrho}{\partial t} + \nabla \cdot \boldsymbol{j} = \frac{1}{f_{\mathrm{G}}} \frac{\partial}{\partial t} (\nabla \cdot \boldsymbol{D}) + \frac{1}{f_{\mathrm{BS}}} \nabla \cdot (\nabla \times \boldsymbol{H}) - \frac{1}{f_{\mathrm{G}}} \nabla \cdot \frac{\partial}{\partial t} \boldsymbol{D} = 0 .$$

Indeed, this is equal to zero because the partial derivatives with respect to time and space variables commute and because the divergence of a curl vanishes.

1.5.4 The Question of Physical Units

The Maxwell equations (1.32a)–(1.32d) and the expression (1.19) for the Lorentz force are supplemented by the relations (1.33) relating the dielectric displacement field $\boldsymbol{D}$ with the electric field $\boldsymbol{E}$ and the magnetic induction $\boldsymbol{B}$ with the magnetic field $\boldsymbol{H}$, respectively. Choosing the constants f_{F} of (1.32b) and $f_{\mathrm{BS}}/f_{\mathrm{G}}$ of (1.32d) in such a way that

$$f_{\mathrm{F}} = \frac{f_{\mathrm{BS}}}{f_{\mathrm{G}}} \tag{1.34}$$

holds, the product of $\boldsymbol{E}$ and $\boldsymbol{D}$ obtains the same dimension as the product of $\boldsymbol{H}$ and $\boldsymbol{B}$. Equivalently, by relations (1.33) one has

$$\frac{[\mu_0]}{[\varepsilon_0]} = \frac{[\boldsymbol{E}^2]}{[\boldsymbol{H}^2]} . \tag{1.35}$$

While (1.34) is a convention which fixes relative dimensions such that

$$[\boldsymbol{E}] : [\boldsymbol{B}] = [\boldsymbol{H}] : [\boldsymbol{D}]$$

holds, one learns more about the remaining freedom by deriving from (1.32a)–(1.32d) other known laws of electromagnetic interactions.

Coulomb Force Between Point Charges

The Coulomb force field follows from the third equation (1.32c) with a factor depending on the system of units:

$$\boldsymbol{F}_{\mathrm{C}} = \kappa_{\mathrm{C}} \frac{e_1 e_2}{r^2} \hat{\boldsymbol{r}} , \qquad \text{with} \quad \kappa_{\mathrm{C}} = \frac{f_{\mathrm{G}}}{4\pi\varepsilon_0} . \tag{1.36}$$

This is seen as follows. In any static situation, i.e. one where all fields are independent of time, the group of fields $(\boldsymbol{E}, \boldsymbol{D})$ decouples completely from the group of fields $(\boldsymbol{H}, \boldsymbol{B})$. Regarding the first group, equations (1.32b) and (1.32c), together with relation (1.33), reduce to

$$\nabla \times \boldsymbol{E}(\boldsymbol{x}) = 0 , \qquad \nabla \cdot \boldsymbol{E}(\boldsymbol{x}) = \frac{f_{\mathrm{G}}}{\varepsilon_0} \varrho(\boldsymbol{x}) .$$

The static electric field has a vanishing curl. Therefore, it can be represented as a gradient field $\boldsymbol{E} = -\nabla\Phi(\boldsymbol{x})$, the minus sign being a matter of convention. The second equation then yields the Poisson equation (1.15b) for $\Phi(\boldsymbol{x})$:

$$\boldsymbol{\Delta}\Phi(\boldsymbol{x}) = -f_{\mathrm{G}} \frac{1}{\varepsilon_0} \varrho(\boldsymbol{x}) .$$

Putting the point charge e_1 in the position $\boldsymbol{x}_0$, for example, one has

$$\varrho(\boldsymbol{x}) = e_1 \, \delta(\boldsymbol{x} - \boldsymbol{x}_0) .$$

Relation (1.30) furnishes the corresponding solution of the Poisson equation, viz.

$$\Phi(\boldsymbol{x}) = \frac{f_{\mathrm{G}}}{4\pi\varepsilon_0} \frac{e_1}{|\boldsymbol{x} - \boldsymbol{x}_0|} . \tag{1.37}$$

Then define $\boldsymbol{x} - \boldsymbol{x}_0 = \boldsymbol{r}$ and multiply the negative gradient of Φ by the charge e_2 of the second mass point whose position is $\boldsymbol{x}$. This yields the formula for the Coulomb force given above.

Wave Equation and Velocity of Light

Returning to the full Maxwell equations but omitting any external source terms, equations (1.32a)–(1.32d) imply that every component of the electric and the magnetic fields satisfies the wave equation. We show this for the example of the electric field:

Take the curl of equation (1.32b) and make use of formula (1.28) to obtain

$$-\boldsymbol{\Delta}\boldsymbol{E}(t, \boldsymbol{x}) + \mu_0 f_{\mathrm{F}} \frac{\partial}{\partial t} \nabla \times \boldsymbol{H}(t, \boldsymbol{x}) = 0 .$$

By equation (1.32d) with $\boldsymbol{j}(t,\boldsymbol{x}) \equiv 0$ and relation (1.33), the curl of the H-field can be expressed in terms of the time derivative of $\boldsymbol{E}(t,\boldsymbol{x})$:

$$\nabla \times \boldsymbol{H}(t,\boldsymbol{x}) = \varepsilon_0 \frac{f_{\text{BS}}}{f_{\text{G}}} \frac{\partial}{\partial t} \boldsymbol{E}(t,\boldsymbol{x}) .$$

Inserting this and using convention (1.34), one obtains

$$\left(f_{\text{F}}^2 \varepsilon_0 \mu_0 \frac{\partial^2}{\partial t^2} - \boldsymbol{\Delta} \right) \boldsymbol{E}(t,\boldsymbol{x}) = 0 ,$$

i.e. a partial differential equation that holds for every component of the electric field in a vacuum. The factor multiplying the first term must have the physical dimension of the inverse of a squared velocity, i.e. $[f_{\text{F}}^2 \mu_0 \varepsilon_0] = \text{T}^2\text{L}^{-2}$. For example, assuming the time and space dependence of a solution to be a plane wave, that is

$$\boldsymbol{E}(t,\boldsymbol{x}) = \boldsymbol{\mathcal{E}}\, \text{e}^{-\text{i}\omega t}\, \text{e}^{\text{i}\boldsymbol{k}\cdot\boldsymbol{x}} ;$$

one obtains the relation $(f_{\text{F}}^2 \mu_0 \varepsilon_0)\, \omega^2 = \boldsymbol{k}^2$ between the circular frequency and the wave number. Together with $\omega = 2\pi\nu$ and $|\boldsymbol{k}| = 2\pi/\lambda$ this yields the well-known relation $\nu\lambda = c$ for propagation of light in the vacuum provided

$$f_{\text{F}}^2 \mu_0 \varepsilon_0 = \frac{1}{c^2} \tag{1.38}$$

holds. This yields a further condition which the constants of the system of unities must obey.

Force Between Parallel Currents (Ampère)

We note that one reaches the same conclusion by calculating Ampère's force per line element $\text{d}l$ acting between two parallel straight conductors which carry the stationary currents J_1 and J_2, respectively, and which are situated at a distance a from each other. From formula (1.19) for the Lorentz force and from (1.32d) the modulus of this force is found to be

$$\frac{\text{d}}{\text{d}l} |\boldsymbol{F}_{\text{A}}| = 2\kappa_{\text{A}} \frac{J_1 J_2}{a} , \qquad \text{with} \quad \kappa_{\text{A}} = \frac{f_{\text{F}}^2 f_{\text{G}} \mu_0}{4\pi} . \tag{1.39}$$

A simple comparison of physical dimensions shows that the ratio $\kappa_{\text{C}}/\kappa_{\text{A}}$ has the dimension of a squared velocity, that is L^2T^{-2}. Experiments by Weber and Kohlrausch in 1856 gave an intriguing result: The velocity that appears in this force was found to have the numerical value of c, the velocity of light – even though one was dealing here with a *stationary* situation where only static forces were measured! Thus, it was found that

$$\frac{\kappa_{\text{C}}}{\kappa_{\text{A}}} = c^2 , \qquad \text{i.e. again} \quad f_{\text{F}}^2 \mu_0 \varepsilon_0 = \frac{1}{c^2} .$$

In summary, these are the essential conditions which must be fulfilled in fixing a system of physical units.

1.5.5 Equations of Electromagnetism in SI System

The SI system, also called a *rational* MKSA *system,* is characterized by the choice

$$f_\mathrm{F} = f_\mathrm{G} = 1 \tag{1.40}$$

and by introducing, in addition to the units {m, kg, s}, a unit for the electric current. This is called the Ampère. It is defined by means of formula (1.39) as follows:

> Given two equal currents $J_1 = J_2 \equiv J$ flowing in two parallel, infinitely long wires whose distance is $a = 1\,\mathrm{m}$, the current J has an intensity of 1 A if the Ampère force per unit of length, i.e. per metre, equals $2 \cdot 10^{-7}\,\mathrm{N} = 2 \cdot 10^{-7}\,\mathrm{kg\,m\,s^{-2}}$. By the definition of the Ampère, the unit of charge, called the Coulomb, is fixed, too. The relation is 1 C = 1 A s.

Convention (1.40) fixes the value of μ_0:

$$\mu_0 = 4\pi \cdot 10^{-7}\,\mathrm{NA^{-2}}\,. \tag{1.41a}$$

Relation (1.38), together with $f_\mathrm{F} = 1$, then gives

$$\varepsilon_0 = \frac{1}{4\pi c^2} \cdot 10^7\,. \tag{1.41b}$$

In the SI system, the physical units of these two quantities differ from each other. Their relation to the units of mass M, length L, current I and time T reads

$$[\varepsilon_0] = \mathrm{M^{-1}L^{-3}\,I^2\,T^4}\,, \qquad [\mu_0] = \mathrm{M\,L\,I^{-2}T^{-2}}\,.$$

The constants appearing in Coulomb's law and in the Ampère force are fixed, respectively, as follows:

$$\kappa_\mathrm{C} = \frac{1}{4\pi\varepsilon_0} = c^2\,10^{-7}\,, \qquad \kappa_\mathrm{A} = \frac{\mu_0}{4\pi} = 10^{-7}\,, \tag{1.41c}$$

their dimensions following from those of ε_0 and μ_0. In the second step, the expressions for ε_0 and μ_0, found above, were inserted.

As a result, the local Maxwell equations in SI units read as follows:

$$\nabla \cdot \boldsymbol{B}(t, \boldsymbol{x}) = 0\,; \tag{1.42a}$$

$$\nabla \times \boldsymbol{E}(t, \boldsymbol{x}) + \frac{\partial}{\partial t}\boldsymbol{B}(t, \boldsymbol{x}) = 0\,; \tag{1.42b}$$

$$\nabla \cdot \boldsymbol{D}(t, \boldsymbol{x}) = \varrho(t, \boldsymbol{x})\,; \tag{1.42c}$$

$$\nabla \times \boldsymbol{H}(t, \boldsymbol{x}) - \frac{\partial}{\partial t}\boldsymbol{D}(t, \boldsymbol{x}) = \boldsymbol{j}(t, \boldsymbol{x})\,. \tag{1.42d}$$

The Lorentz force takes the form

$$F(t,x) = q\big(E(t,x) + v \times B(t,x)\big)\,, \tag{1.42e}$$

and relations (1.33) remain unchanged. Electromagnetic fields in the vacuum, i.e. outside of the sources, satisfy the wave equation

$$\left(\frac{1}{c^2}\frac{\partial^2}{\partial t^2} - \Delta\right) g(t,x) = 0\,, \tag{1.42f}$$

where $g(t,x)$ stands for any component of the field under consideration.

One verifies easily that electric fields and magnetic induction fields have the following units in the SI system:

$$[E] = 1\,\text{kg m A}^{-1}\text{s}^{-3}\,, \qquad [B] = 1\,\text{kg A}^{-1}\text{s}^{-2}\,.$$

The unit of electric tension, or voltage, called the *volt*, is

$$[V] = 1\,\text{kg m}^2\,\text{A}^{-1}\,\text{s}^{-3}\,,$$

so that one recovers the well-known convention that electric fields are measured in volt per metre:

$$[E] = 1\,\text{V m}^{-1}\,.$$

For magnetic induction fields, the *Tesla* was introduced as a unit, i.e.

$$[B] = 1\,\text{Tesla} = 1\,\text{Vsm}^{-2}\,,$$

whereas magnetic fields are measured in Ampère turns per metre: $[H] = 1\,\text{A t m}^{-1}$.

1.5.6 The Gaussian System of Units

In the Gaussian system of units, no new unit of charge or current is introduced. Rather, these units are derived units, following from the mechanical units fixed earlier. The factor multiplying the Coulomb force is taken to be 1, $\kappa_C = 1$. Furthermore, the fields E and H, as well as the fields B and D, will all have the same dimension. This means that f_F and $f_{BS}/f_G = f_F$ have the dimension T L^{-1}. Inspection of (1.35) and of (1.38) shows that ε_0 and μ_0 not only have the same dimension but are, in fact, dimensionless. This suggests setting them both equal to 1:

$$\varepsilon_0 = 1\,, \qquad \mu_0 = 1\,, \tag{1.43a}$$

so that in the vacuum one has $D = E$ and $B = H$. This also gives, from (1.38),

$$f_F = \frac{1}{c}\,. \tag{1.43b}$$

One chooses the factor multiplying the charge density on the right-hand side of (1.32c) to be

$$f_G = 4\pi \,, \tag{1.43c}$$

so that the factor κ_C in the expression for the Coulomb force (1.36) is indeed equal to 1. Note that this is in accord with formula (1.30).

With these choices one obtains the following values:

$$f_{BS} = \frac{4\pi}{c} \,, \quad \kappa_C = 1 \,, \quad \kappa_A = \frac{1}{c^2} \,. \tag{1.43d}$$

Up to exceptions that will be mentioned explicitly, we will be using Gaussian units in what follows; thus, it is useful to repeat here the fundamental equations in these units:

Maxwell's equations in Gaussian units

$$\nabla \cdot \boldsymbol{B}(t, \boldsymbol{x}) = 0 \,; \tag{1.44a}$$

$$\nabla \times \boldsymbol{E}(t, \boldsymbol{x}) + \frac{1}{c}\frac{\partial}{\partial t}\boldsymbol{B}(t, \boldsymbol{x}) = 0 \,; \tag{1.44b}$$

$$\nabla \cdot \boldsymbol{D}(t, \boldsymbol{x}) = 4\pi\varrho(t, \boldsymbol{x}) \,; \tag{1.44c}$$

$$\nabla \times \boldsymbol{H}(t, \boldsymbol{x}) - \frac{1}{c}\frac{\partial}{\partial t}\boldsymbol{D}(t, \boldsymbol{x}) = \frac{4\pi}{c}\boldsymbol{j}(t, \boldsymbol{x}) \,. \tag{1.44d}$$

The expression for the Lorentz force reads

$$\boldsymbol{F}(t, \boldsymbol{x}) = q\left[\boldsymbol{E}(t, \boldsymbol{x}) + \frac{1}{c}\boldsymbol{v} \times \boldsymbol{B}(t, \boldsymbol{x})\right] \,. \tag{1.44e}$$

In the vacuum the electric field and the electric displacement field are equal; likewise, the magnetic induction and the magnetic field are identified, i.e.

$$\boldsymbol{D}(t, \boldsymbol{x}) = \boldsymbol{E}(t, \boldsymbol{x}) \,, \qquad \boldsymbol{B}(t, \boldsymbol{x}) = \boldsymbol{H}(t, \boldsymbol{x}) \quad \text{(in the vacuum)} \,. \tag{1.44f}$$

Finally, we note the wave equation in Gauss units:

$$\left(\frac{1}{c^2}\frac{\partial^2}{\partial t^2} - \boldsymbol{\Delta}\right) g(t, \boldsymbol{x}) = 0 \,, \tag{1.45}$$

where $g(t, \boldsymbol{x})$ stands for an arbitrary component of the electric or magnetic field in the vacuum. (Of course, it has the same form in SI units.)

Table 1.1 Two important systems of units and their comparison: the Gaussian system [or centimetre–gram–second (cgs) system] and the SI (or MKSA) system

	Gaussian system	SI system	Comparison
Length	1 cm	1 m	$1\,\mathrm{m} = 1 \cdot 10^2\,\mathrm{cm}$
Mass	1 g	1 kg	$1\,\mathrm{kg} = 1 \cdot 10^3\,\mathrm{g}$
Time	1 s	1 s	
Force	1 dyn	1 N	$1\,\mathrm{N} = 1 \cdot 10^5\,\mathrm{dyn}$
Energy	1 erg	1 J	$1\,\mathrm{J} = 1 \cdot 10^7\,\mathrm{erg}$
Power	$1\,\mathrm{erg\,s^{-1}}$	1 W	$1\,\mathrm{W} = 1 \cdot 10^7\,\mathrm{erg\,s^{-1}}$
Charge	1 esu	1 C	$1\,\mathrm{C} = 3 \cdot 10^9\,\mathrm{esu}$
Current	1 esc	1 A	$1\,\mathrm{A} = 3 \cdot 10^9\,\mathrm{esc}$
Potential	1 esv	1 V	$1\,\mathrm{V} = 1/300\,\mathrm{esv}$
Electric field	$1\,\mathrm{esv\,cm^{-1}}$	$1\,\mathrm{V\,m^{-1}}$	
Magnetic field	1 Oersted (Oe)	$1\,\mathrm{At\,m^{-1}}$	$1\,\mathrm{At\,m^{-1}} = 4\pi \cdot 10^{-3}\,\mathrm{Oe}$
Magnetic induction	1 Gauss (G)	1 Tesla	$1\,\mathrm{Tesla} = 10^4\,\mathrm{Gauss}$

The following table summarizes the comparison of the SI system and the Gaussian system. The abbreviations "esu", "esc" etc. stand for "electrostatic charge unit", "electrostatic current unit", etc. For example, the electrostatic unit of charge is $1\,\mathrm{esu} = 1\,\mathrm{g^{1/2}\,cm^{3/2}\,s^{-1}}$.

The Gaussian system is a useful system of units for theoretical considerations of principle but is unsuitable for practical purposes in the laboratory. As we are studying the foundations here, we will use the Gaussian system almost exclusively in what follows. If at all needed, the conversion should be easy when making use of the arguments of Sect. 1.5.4 and of Table 1.1. Here are some examples.

The elementary charge, i.e. the absolute value of the charge of an electron, when expressed in cgs and in SI units, respectively, is given by

$$\begin{aligned} e &= 4.803\,204\,20(19) \cdot 10^{-10}\,\mathrm{esu} \\ &= 1.602\,176\,462(63) \cdot 10^{-19}\,\mathrm{C}\,; \end{aligned} \tag{1.46a}$$

the numbers in parentheses give the present experimental error bar in the last two digits. Because in physics, energies are often given in electron volts or powers of ten thereof, it is important to know their conversion to SI units. From the numbers given above one has for the electron volt

$$1\,\mathrm{eV} = 1.602\,176\,462(63) \cdot 10^{-19}\,\mathrm{J}\,. \tag{1.46b}$$

Multiples of an electron volt which are useful in practice are given symbols of their own as follows. For example,

$$1\,\mathrm{meV} = 1 \cdot 10^{-3}\,\mathrm{eV}\,, \quad 1\,\mathrm{keV} = 1 \cdot 10^{3}\,\mathrm{eV}\,, \quad 1\,\mathrm{MeV} = 1 \cdot 10^{6}\,\mathrm{eV}\,,$$
$$1\,\mathrm{GeV} = 1 \cdot 10^{9}\,\mathrm{eV}\,, \quad 1\,\mathrm{TeV} = 1 \cdot 10^{12}\,\mathrm{eV}\,, \quad 1\,\mathrm{PeV} = 1 \cdot 10^{15}\,\mathrm{eV}\,,$$

where "m" stand for "milli", "k" for "kilo", "M" for "mega", "G" for "giga", "T" for "tera" and "P" for "peta".

Also, masses m of atomic or subatomic particles, as a rule, are given in the equivalent form of their rest energy, mc^2, expressed in electron volts or multiples thereof. Converting to SI units one has

$$1\,\mathrm{eV/c^2} = 1.782\,661\,731(70)10^{-36}\,\mathrm{kg}\,. \tag{1.46c}$$

To get a feeling for orders of magnitude, it is instructive to express the mass of a very heavy nucleus in units used in daily trading, or to express electric fields which are typical for atoms, in units which are familiar to an electrician, cf. Exercises 1.3 and 1.4.

Remarks

1. In the theory of relativity and in elementary particle physics, it is customary to use so-called *natural units*. These are chosen such that the velocity of light c and Planck's constant (divided by 2π) take the value 1:

$$c = 1\,, \qquad \hbar \equiv \frac{h}{2\pi} = 1\,.$$

The reader who is not familiar with this choice may wish to consult [QM], Sect. 7.2.1.

One may simplify matters even more by absorbing the factors 4π on the right-hand sides of (1.44c) and (1.44d) in the fields and the sources through the following choice. Let

$$\boldsymbol{E}\big|_{\mathrm{nat}} := \frac{1}{\sqrt{4\pi}}\,\boldsymbol{E}\big|_{\mathrm{Gauss}}\,, \qquad \varrho\big|_{\mathrm{nat}} := \sqrt{4\pi}\,\varrho\big|_{\mathrm{Gauss}}\,.$$

The analogous factors $1/\sqrt{4\pi}$ and $\sqrt{4\pi}$, respectively, are absorbed in the fields $\boldsymbol{D}$, $\boldsymbol{H}$ and $\boldsymbol{B}$, as well as in the current density, such that all factors in the Maxwell's equations take the value 1. Although this book does not make use of it, this convention is very convenient in actual calculations. The conversion back to conventional units is easy and follows simple rules. For instance, even though this is a dimensionless number, the relation of Sommerfeld's fine structure constant to the elementary charge depends on the choice of system of units. One has

$$\alpha = \frac{e^2\big|_{\mathrm{Gauss}}}{\hbar c} = \frac{e^2\big|_{\mathrm{nat}}}{4\pi} = \frac{1}{137.036}\,.$$

This implies that in the result of a calculation one must replace $e^2\big|_{\mathrm{nat}}$ by $4\pi\alpha$ with $\alpha = (137.036)^{-1}$.

2. As remarked earlier, Maxwell's equations (1.44a)–(1.44d), on their left-hand sides, contain exclusively *electromagnetic field* quantities, whereas the *sources* appear on the right-hand side of the two inhomogeneous equations (1.44c)–(1.44d). The quantities of the first group concern and describe the *radiation field*, those of the second group concern *matter* whose building blocks are electrons, ions and atomic nuclei. This distinction is physically meaningful: Matter, a priori, is described by a kind of dynamics other than the Maxwell fields. It is remarkable that while matter can hardly be "seen", that is its dynamics can hardly be probed, without its coupling to the electromagnetic fields, Maxwell's equations describe interesting physical phenomena even without external sources, in the absence of charge and current densities, $\varrho \equiv 0$ and $\boldsymbol{j} \equiv 0$.
3. It is instructive to ponder the signs in Maxwell's equations and to analyze which of these are a matter of convention and which of them are fixed by physical principles. We organize this discussion as follows.
   Electric field and positive charge
   It is common to define the electric field created by a positive charge at rest in such a way that it points outwards, *away* from the centre of the charge. But what is a positive charge? This term dates from the early history of electrostatics when experiments were performed with static charges in various materials such as glass, collophonium, combs, and the like. "Glass electricity" was named positive, "resin electricity" was named negative. With this convention the electron turned out to carry a *negative* elementary charge; its antiparticle, the positron e^+, carries a *positive* elementary charge. Likewise, a proton carries a positive elementary charge. The continuity equation (1.21) says that current density is defined by a flux of *positive* charge. Therefore, the direction of a current density is opposite to the direction of the flow of freely moving electrons.
   Permanent magnets and magnetic field
   The two poles of permanent magnets we played with when we were children are called the north pole N and the south pole S respectively. Given a stationary current J flowing in an infinitely long, straight wire, small probing magnets will be oriented as sketched in Fig. 1.6a. Traditionally, the magnetic field lines outside a permanent magnet are chosen to run from N to S. Thus, the field lines created by the current in the wire are directed so as to form a positive helicity (obeying the right-hand rule), as sketched in Fig. 1.6b. With these conventions the law of Biot and Savart yields equation (1.44d), with the signs as indicated there. Similarly, Faraday's law of induction, (1.44b), is obtained with the signs as in (1.44b).

Of course, one could have chosen the sign conventions differently, thereby obtaining different signs in some places of Maxwell's equations. The interesting question, then, is which of the *relative* signs will *not* change by such redefinitions and which physical property is connected with relative signs found to be essen-

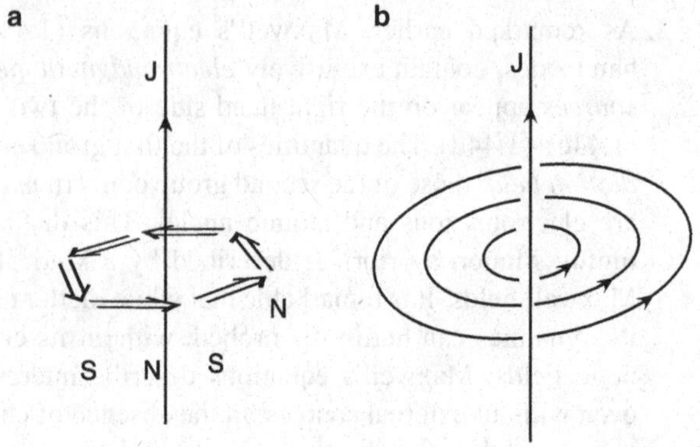

Fig. 1.6 **a** The current J flowing through a thin wire creates a magnetic field in which small probing magnets align as sketched. **b** Magnetic field in space around conductor

tial. If one had defined the direction of the electric field such that for an electron (which carries the charge $-|e|$) this field points outward, or, alternatively, if one had adopted the previous convention for the electric field but had taken the outside magnetic field to run from S to N, instead of from N to S, then on the left-hand sides of equations (1.44b) and (1.44d) the relative signs would be changed. However, all physically relevant information would remain unchanged, such as in the following examples:

- The physical content of (1.44b), which is an expression for the law of induction in the form of Lenz's rule. The induction currents caused by motions of magnets, or of current loops, are directed in such a way that their own magnetic field acts against the motion;
- The orientation of a permanent magnet in the field created by currents in a conductor: the magnet aligns itself such that its magnetic field and the field of the conductor compensate as much as possible. The field energy, being proportional to the space integral of $\boldsymbol{H}^2$, is as small as possible;
- The relative signs of $\boldsymbol{B}$ and $\boldsymbol{H}$, as well as of $\boldsymbol{E}$ and $\boldsymbol{D}$ in (1.44b) and in (1.44d), respectively: as was shown in Sect. 1.5.4, this relative minus sign leads to the wave equation (1.45) for every component of the fields and, thus, guarantees the propagation of electromagnetic waves in the vacuum. Conversely, if this relative sign were not invariant and independent of the conventions, then the differential operator $\left[(1/c^2)\partial^2/\partial t^2 - \boldsymbol{\Delta}\right]$ in (1.45), which is responsible for the propagation of plane waves with velocity c, would compete with the operator

$$\left(\frac{1}{c^2}\frac{\partial^2}{\partial t^2} + \boldsymbol{\Delta}\right),$$

which is relevant for a radically different type of physics.

1.6 Scalar Potentials and Vector Potentials

In static situations, an electric field $\boldsymbol{E}(\boldsymbol{x})$ can be represented as the negative gradient of a scalar function $\Phi(\boldsymbol{x})$ (cf. Sect. 1.4.2) $\boldsymbol{E}(\boldsymbol{x}) = -\nabla\Phi(\boldsymbol{x})$. The electric field, being an observable, is obviously defined in a unique way. The auxiliary function Φ is determined only up to an additive constant function. In this section, we show that also in time-dependent situations both the electric and the magnetic (or induction) fields can be represented in terms of such auxiliary functions and certain auxiliary vector fields. Although these are not direct observables and, hence, are not uniquely defined, they are very useful tools for many reasons. We summarize here their definition and their most important properties. Their deeper significance, as well as their advantages and disadvantages, will become clear in subsequent sections.

1.6.1 A Few Formulae from Vector Analysis

Equations (1.4) and (1.5) of Sect. 1.2 show for dimension 3 that the curl of a gradient field and the divergence of a curl vanish. We derive here these formulae, as well as some other formulae which are useful for the sequel. For the sake of illustration and as a matter of exercise, we do this in different, though equivalent, ways. The most important formulae for the subsequent discussion are as follows:

$$\nabla\cdot\big(\nabla\times\boldsymbol{A}(t,\boldsymbol{x})\big) = 0\,; \tag{1.47a}$$

$$\nabla\times\big(\nabla f(t,\boldsymbol{x})\big) = 0\,; \tag{1.47b}$$

$$\nabla\times\big(\nabla\times\boldsymbol{A}(t,\boldsymbol{x})\big) = \nabla\big(\nabla\cdot\boldsymbol{A}(t,\boldsymbol{x})\big) - \boldsymbol{\Delta}\boldsymbol{A}(t,\boldsymbol{x})\,. \tag{1.47c}$$

The third of these is to be understood as an equation that holds for every component; that is, when written out in Cartesian coordinates, it reads

$$\big(\nabla\times(\nabla\times\boldsymbol{A}(t,\boldsymbol{x}))\big)_i = \frac{\partial}{\partial x^i}\big(\nabla\cdot\boldsymbol{A}(t,\boldsymbol{x})\big) - \boldsymbol{\Delta}A_i(t,\boldsymbol{x})\,, \quad i = 1,2,3\,.$$

Proofs by use of the ε tensor

The ε tensor, or Levi-Civita symbol in dimension 3, is defined by

$$\varepsilon_{ijk} = +1\,, \text{ if } \{i,j,k\} \text{ is an } \textit{even} \text{ permutation of } \{1,2,3\};$$
$$\varepsilon_{ijk} = -1\,, \text{ if } \{i,j,k\} \text{ is an } \textit{odd} \text{ permutation of } \{1,2,3\};$$
$$\varepsilon_{ijk} = 0\,, \quad \text{whenever two or three indices are equal.}$$

In dimension 3, all cyclic permutations of $\{i,j,k\}$ are even, and all anticyclic permutations are odd, i.e. $\varepsilon_{123} = \varepsilon_{231} = \varepsilon_{312} = 1, \varepsilon_{132} = \varepsilon_{321} = \varepsilon_{213} = -1$. The contraction over two indices of a symmetric tensor with the antisymmetric Levi-

Civita symbol gives zero:

$$\sum_{i,j=1}^{3} \varepsilon_{ijk} T^{ij} = 0 \quad \text{if } T^{ij} = T^{ji} .$$

Important formulae for sums of products of two ε tensors are as follows:

$$\sum_{k=1}^{3} \varepsilon_{ijk}\varepsilon_{klm} = \delta_{il}\delta_{jm} - \delta_{im}\delta_{jl} , \tag{1.48a}$$

$$\sum_{j,k=1}^{3} \varepsilon_{ijk}\varepsilon_{jkm} = 2\delta_{im} . \tag{1.48b}$$

The proof of the first of these is the subject of Exercise 1.5, and the second formula follows from the first, viz.

$$\begin{aligned}\sum_{j,k=1}^{3} \varepsilon_{ijk}\varepsilon_{jkm} &= -\sum_{j,k=1}^{3} \varepsilon_{ijk}\varepsilon_{kjm} = -\sum_{j=1}^{3}\left(\delta_{ij}\delta_{jm} - \delta_{im}\delta_{jj}\right) \\ &= -(1-3)\delta_{im} .\end{aligned}$$

It is customary to write partial derivatives in abbreviated notation:

$$\frac{\partial}{\partial x^j} \equiv \partial_j .$$

In this notation, the curl of a vector field reads

$$\left(\nabla \times A\right)_k = \sum_{l,m=1}^{3} \varepsilon_{klm}\partial_l A_m .$$

The divergence of this curl, being a contraction of a symmetric tensor with the ε symbol, vanishes:

$$\sum_k \partial_k \left(\nabla \times A\right)_k = \sum_{k,l,m} \varepsilon_{klm}\partial_k\partial_l A_m = 0 .$$

This proves (1.47a).

Similarly, in (1.47b), the ε tensor is contracted with the symmetric operator $\partial_k\partial_l$ and, hence, gives a vanishing result.

To prove (1.47c), one calculates one Cartesian component on the left-hand side as follows:

$$(\nabla \times (\nabla \times A))_i = \sum_{j,k} \varepsilon_{ijk} \partial_j (\nabla \times A)_k = \sum_{j,k} \sum_{l,m} \varepsilon_{ijk} \varepsilon_{klm} \partial_j \partial_l A_m$$

$$= \sum_{j,l,m} (\delta_{il}\delta_{jm} - \delta_{im}\delta_{jl}) \partial_j \partial_l A_m$$

$$= \partial_i \sum_m \partial_m A_m - \sum_j \partial_j^2 A_i = \partial_i (\nabla \cdot A) - \Delta A_i \, .$$

Proof of (1.47a) and of (1.47b) using integral theorems

Using a somewhat intuitive approach one may choose a very small volume in Gauss' theorem, (1.6), and consider the limit $V \to 0$. This yields a local form of Gauss' theorem:

$$\nabla \cdot V = \lim_{V \to 0} \frac{1}{V} \iint_{F = \partial V} \mathrm{d}\sigma \, V \cdot \hat{n} \, .$$

Stokes' theorem, (1.7), relates a surface integral of the normal component of a curl with the path integral over the boundary curve of the surface:

$$\iint_{F(C)} \mathrm{d}\sigma \, (\nabla \times A) \cdot \hat{n} = \oint_{\partial F} \mathrm{d}s \cdot A \, .$$

If one chooses a *closed* surface, the boundary curve shrinks to such a point where the right-hand side of the equation is zero. Comparison with the local form of Gauss' theorem given above, and inserting there $V = \nabla \times A$, formula (1.47a) follows. If one inserts a gradient field $V = -\nabla f$ into Stokes' theorem (1.7), the path integral becomes, in the general case,

$$\int_a^b \mathrm{d}s \cdot \nabla f = f(b) - f(a) \, .$$

In the case of a closed path integral in Stokes' theorem, the starting point a and the end point b coincide. The right-hand side of (1.7) is zero. As this holds for any choice of surface F, formula (1.47b) is proven.

Proofs using exterior forms

This section makes use of the calculus of exterior forms, though at this point exclusively the formalism on $\mathbb{R}^3$ that is treated e.g. in [ME], Sect. 5.4.5. The reader who is not familiar with this method or is reluctant to repeat it here may skip this part for now. We will return to it later.

The method described above which made use of the ε tensor suggests that (1.47a) and (1.47b) must be closely related and that both might be derivable from one common principle. Indeed, both identities are consequences of the fact that the exterior derivative, when applied twice in succession, gives zero, $d \circ d = 0$. This is shown as follows. To a vector field A on $\mathbb{R}^3$ associate two exterior forms, a one-form and a two-form, by means of the following definitions:

$$\overset{1}{\omega}_A = \sum_{i=1}^{3} A_i \, dx^i , \quad \overset{2}{\omega}_A = A_1 \, dx^2 \wedge dx^3 + \text{cyclic permutations}.$$

Here, dx^1, dx^2 and dx^3 are the base one-forms which correspond to the coordinates of a Cartesian system of reference. The exterior derivative of the first form leads to the curl of the vector field:

$$d \, \overset{1}{\omega}_A = \left(\nabla \times A \right)_3 dx^1 \wedge dx^2 + \text{cyclic permutations} ;$$

the exterior derivative of the second yields the divergence of A:

$$d \, \overset{2}{\omega}_A = \left(\nabla \cdot A \right) dx^1 \wedge dx^2 \wedge dx^3 .$$

To every exterior form of degree k there is an associated, or dual, $(n - k)$-form (in the present case $n = 3$) called the Hogde dual form, which is defined if one knows the duals of the base forms. They are

$$*dx^1 = dx^2 \wedge dx^3 \text{ (cyclic)},$$
$$*\left(dx^1 \wedge dx^2 \right) = dx^3 \text{ (cyclic)},$$
$$*\left(dx^1 \wedge dx^2 \wedge dx^3 \right) = 1 .$$

One concludes that the one-form defined above is the Hodge dual of the two-form, $* \, \overset{2}{\omega}_A = \overset{1}{\omega}_A$. One calculates

$$d \left(d \, \overset{1}{\omega}_A \right) = \nabla \cdot \left(\nabla \times A \right) dx^1 \wedge dx^2 \wedge dx^3 , \text{ or}$$
$$*d \left(d \, \overset{1}{\omega}_A \right) = \nabla \cdot \left(\nabla \times A \right) .$$

As $d \circ d = 0$, relation (1.47a) follows.

If, alternatively, one constructs the one-form which corresponds to the gradient field ∇f,

$$\overset{1}{\omega}_{\text{grad} f} = \sum_i \left(\partial_i f \right) dx^i = df ,$$

its exterior derivative is

$$d \, \overset{1}{\omega}_{\text{grad} f} = \left(\nabla \times \nabla f \right)_3 dx^1 \wedge dx^2 + \text{cyclic permutations}$$
$$= d \circ df = 0 .$$

This is relation (1.47b). Indeed, both relations (1.47a) and (1.47b) are special cases of the general property $\mathrm{d} \circ \mathrm{d} = 0$ of the exterior derivative in $\mathbb{R}^3$.

The third relation, (1.47c), is the most interesting of the three. One starts by taking the exterior derivative of $\overset{1}{\omega}_A$ and then takes the Hodge dual of this:

$$* \mathrm{d} \overset{1}{\omega}_A = \sum_i (\nabla \times A)_i \, \mathrm{d}x^i \, .$$

Next, one takes the exterior derivative of this,

$$\mathrm{d} * \mathrm{d} \overset{1}{\omega}_A = \left(\nabla \times \left(\nabla \times A(t, x)\right)\right)_3 \mathrm{d}x^1 \wedge \mathrm{d}x^2 + \text{cycl. perm.} \, ,$$

and then its Hodge dual,

$$* \mathrm{d} * \mathrm{d} \overset{1}{\omega}_A = \sum_i \left(\nabla \times \left(\nabla \times A(t, x)\right)\right)_i \mathrm{d}x^i \, .$$

In other terms, the left-hand side of (1.47c) is obtained from the one-form $\overset{1}{\omega}_A$ by applying the operator $(* \mathrm{d} *) \mathrm{d}$ to it.

It is then not difficult to show that the operator $\mathrm{d}(* \mathrm{d} *)$, which is different from $(* \mathrm{d} *) \mathrm{d}$, upon application to the same one-form, yields the first term on the right-hand side of (1.47c):

$$\mathrm{d}(* \mathrm{d} *) \overset{1}{\omega}_A = \sum_i \partial_i (\nabla \cdot A) \, \mathrm{d}x^i \, .$$

The combined operation $* \mathrm{d} *$ of exterior derivative and double dualizing appears in the definition of what is called the codifferential. In dimension n, and applied to a k form, it reads as follows.

▶ **Definition 1.1 Codifferential and Laplace–de Rham operator** If d denotes the exterior derivative, and if $*$ denotes the Hodge dual on $\mathbb{R}^n$, then the codifferential applied to an arbitrary smooth k-form is defined by

$$\delta := (-)^{n(k+1)+1} * \mathrm{d} * \, . \tag{1.49}$$

The sum of the combined operations $\mathrm{d} \circ \delta$ and $\delta \circ \mathrm{d}$

$$\boldsymbol{\Delta}_{\mathrm{LdR}} := \mathrm{d} \circ \delta + \delta \circ \mathrm{d} \tag{1.50}$$

is called the *Laplace–de Rham operator.*[2]

Before we return to relation (1.47c) we wish to comment briefly on these definitions. While the exterior derivative d raises by one the degree of the form on which it acts,

$$\mathrm{d} : \Lambda^k \to \Lambda^{k+1} : \overset{k}{\omega} \mapsto \overset{(k+1)}{\eta} = \mathrm{d} \overset{k}{\omega} \, ,$$

[2] These definitions are valid also on more general smooth manifolds if the latter are orientable. This assumption is necessary for the existence of the Hodge dual.

the codifferential δ lowers the degree by one. Indeed, the operator $*$ turns degree k into $(n-k)$, operator d changes it to $(n-k)+1$, and the second Hodge dualizing leads to the degree $n-[(n-k)+1]=k-1$. Combining these steps, one has

$$\delta \Lambda^k \to \Lambda^{k-1} : \overset{k}{\omega} \mapsto \overset{(k-1)}{\lambda} = (-)^{n(k+1)+1} * \mathrm{d} * \overset{k}{\omega} \ .$$

As a consequence, the operator $\boldsymbol{\Delta}_{\mathrm{LdR}}$, which is composed of both, does not change the degree of the form to which it is applied:

$$\boldsymbol{\Delta}_{\mathrm{LdR}} : \Lambda^k \to \Lambda^k \ .$$

If one applies the operator (1.50) to a one-form, then δ in the first term must be taken with $k = 1$ but in the second term with $k = 2$ because the previous action of d turns the one-form into a two-form, so that

$$\boldsymbol{\Delta}_{\mathrm{LdR}} \overset{1}{\omega} = \big(-\mathrm{d}(*\mathrm{d}*) + (*\mathrm{d}*)\mathrm{d}\big) \overset{1}{\omega} \ .$$

Returning now to (1.47c) and applying $\boldsymbol{\Delta}_{\mathrm{LdR}}$ to $\overset{1}{\omega}_{\boldsymbol{A}}$:

$$\begin{aligned}\boldsymbol{\Delta}_{\mathrm{LdR}} \overset{1}{\omega}_{\boldsymbol{A}} &= \sum_i \left[-\partial_i(\nabla \cdot \boldsymbol{A}) + \big(\nabla \times (\nabla \times \boldsymbol{A}(t,\boldsymbol{x}))\big)_i\right] \mathrm{d}x^i \\ &= -\sum_i (\boldsymbol{\Delta} A_i)\, \mathrm{d}x^i \ .\end{aligned}$$

This result is in accord with the statement that the Laplace–de Rham operator, when acting on a function, equals minus the ordinary Laplace operator:

$$\boldsymbol{\Delta}_{\mathrm{LdR}} f = (\mathrm{d} \circ \delta + \delta \circ \mathrm{d}) f = -(*\mathrm{d}*)(\mathrm{d}f) = -\sum_i \partial_i^2 f \ .$$

This latter operator is also called the Laplace–Beltrami operator.

1.6.2 Construction of a Vector Field from Its Source and Its Curl

Suppose the following information on a vector field $\boldsymbol{A}(t,\boldsymbol{x})$ is given. The vector field is smooth, and its divergence is given by the smooth function

$$f(t,\boldsymbol{x}) = \nabla \cdot \boldsymbol{A}(t,\boldsymbol{x}) \ .$$

Its curl is given by the smooth vector field

$$\boldsymbol{g}(t,\boldsymbol{x}) = \nabla \times \boldsymbol{A}(t,\boldsymbol{x})$$

such that both $f(t, \boldsymbol{x})$ and $\boldsymbol{g}(t, \boldsymbol{x})$ are, at all times, localized, i.e. are confined to a finite domain in space. Is it possible to reconstruct the full vector field $\boldsymbol{A}(t, \boldsymbol{x})$ from the data $\big(f(t, \boldsymbol{x}), \boldsymbol{g}(t, \boldsymbol{x})\big)$ and, if so, is the constructed representation unique?

To answer these questions, start from the ansatz

$$\boldsymbol{A}(t, \boldsymbol{x}) = \boldsymbol{A}_1(t, \boldsymbol{x}) + \boldsymbol{A}_2(t, \boldsymbol{x}) \quad \text{such that} \tag{1.51a}$$

$$\begin{aligned} \nabla \cdot \boldsymbol{A}_1(t, \boldsymbol{x}) &= f(t, \boldsymbol{x})\,; & \nabla \times \boldsymbol{A}_1(t, \boldsymbol{x}) &= 0\,; \\ \nabla \cdot \boldsymbol{A}_2(t, \boldsymbol{x}) &= 0\,; & \nabla \times \boldsymbol{A}_2(t, \boldsymbol{x}) &= \boldsymbol{g}(t, \boldsymbol{x})\,. \end{aligned} \tag{1.51b}$$

The first term carries the source but is irrotational (i.e. it has vanishing curl), whereas the second term has a vanishing source but yields the given curl of the vector field. One proceeds in two steps.

The term whose curl is zero may be written as a gradient field $\boldsymbol{A}_1 = -\nabla\Phi$, the minus sign being a matter of convention. Thus, one obtains the Poisson equation

$$\Delta\Phi(t, \boldsymbol{x}) = -f(t, \boldsymbol{x})\,.$$

In the absence of any further condition, a solution is

$$\Phi(t, \boldsymbol{x}) = \frac{1}{4\pi} \iiint \mathrm{d}^3x' \frac{f(t, \boldsymbol{x}')}{|\boldsymbol{x}' - \boldsymbol{x}|}\,. \tag{1.52a}$$

The reader who is unfamiliar with this formula may wish to derive it by means of Exercise 1.6.

The term which has no source is represented in the form of a curl, i.e. $\boldsymbol{A}_2 = \nabla \times \boldsymbol{C}$, where the auxiliary field $\boldsymbol{C}$ can be chosen such that it has a vanishing source, too, that is to say, such that $\nabla \cdot \boldsymbol{C} = 0$. If such a vector field $\boldsymbol{C}$ is already known which has a nonvanishing source, it may be replaced by $\boldsymbol{C}' = \boldsymbol{C} + \boldsymbol{B}$ with $\nabla \times \boldsymbol{B} = 0$, choosing $\boldsymbol{B}$ such that $\nabla \cdot \boldsymbol{C}' = 0$. This is always possible because $\boldsymbol{B}$ can be represented as a gradient field, $\boldsymbol{B} = -\nabla h$, and because the Poisson equation $\Delta h = \nabla \cdot \boldsymbol{C}$ is soluble. By assumption, one then has

$$\nabla \times \big(\nabla \times \boldsymbol{C}(t, \boldsymbol{x})\big) = \boldsymbol{g}(t, \boldsymbol{x})\,.$$

As the divergence of the auxiliary field $\boldsymbol{C}$ vanishes, the left-hand side of this equation equals $-\Delta\boldsymbol{C}$ by (1.47c). Thus, one has $\Delta\boldsymbol{C} = -\boldsymbol{g}(t, \boldsymbol{x})$. This Poisson equation is solved for each component as before, giving

$$\boldsymbol{C}(t, \boldsymbol{x}) = \frac{1}{4\pi} \iiint \mathrm{d}^3x' \frac{\boldsymbol{g}(t, \boldsymbol{x}')}{|\boldsymbol{x}' - \boldsymbol{x}|}\,. \tag{1.52b}$$

Based on these results, one obtains a reconstruction of the vector field $\boldsymbol{A}$ in terms of a gradient field and a rotational field

$$\boldsymbol{A}(t,\boldsymbol{x}) = -\nabla_x \left(\frac{1}{4\pi} \iiint \mathrm{d}^3 x' \, \frac{f(t,\boldsymbol{x}')}{|\boldsymbol{x}'-\boldsymbol{x}|} \right) + \nabla_x \times \left(\frac{1}{4\pi} \iiint \mathrm{d}^3 x' \, \frac{\boldsymbol{g}(t,\boldsymbol{x}')}{|\boldsymbol{x}'-\boldsymbol{x}|} \right) . \tag{1.53}$$

This decomposition of the unknown vector field in terms of its source and its curl is not unique: To the reconstructed field $\boldsymbol{A}$ one can always add a gradient field $\nabla\chi$, without modifying its source or its curl, provided the smooth function $\chi(t,\boldsymbol{x})$ fulfills the Laplace equation $\boldsymbol{\Delta}\chi(t,\boldsymbol{x}) = 0$. All vector fields belonging to the class

$$\{\boldsymbol{A}(t,\boldsymbol{x}) + \nabla\chi(t,\boldsymbol{x}) \mid \chi(t,\boldsymbol{x}) \text{ smooth solution of } \boldsymbol{\Delta}\chi(t,\boldsymbol{x}) = 0\} \tag{1.54}$$

have the same source and the same curl.

1.6.3 Scalar Potentials and Vector Potentials

The results obtained in the previous section can readily be applied to the $\boldsymbol{B}$ and $\boldsymbol{E}$ fields in Maxwell's equations. Equation (1.44a) tells us that the magnetic induction can be represented as the curl of an auxiliary field $\boldsymbol{A}(t,\boldsymbol{x})$, $\boldsymbol{B} = \nabla \times \boldsymbol{A}$. Inserting this ansatz into (1.44b), one concludes

$$\nabla \times \left(\boldsymbol{E} + \frac{1}{c}\frac{\partial}{\partial t}\boldsymbol{A} \right) = 0 .$$

This, in turn, means that the expression in brackets can be written as a gradient field $-\nabla\Phi$ of another auxiliary function $\Phi(t,\boldsymbol{x})$. One thus obtains a representation of the induction field and of the electric field in terms of $\boldsymbol{A}$ and Φ:

$$\boldsymbol{B}(t,\boldsymbol{x}) = \nabla \times \boldsymbol{A}(t,\boldsymbol{x}) , \tag{1.55a}$$

$$\boldsymbol{E}(t,\boldsymbol{x}) = -\frac{1}{c}\frac{\partial}{\partial t}\boldsymbol{A}(t,\boldsymbol{x}) - \nabla\Phi(t,\boldsymbol{x}) . \tag{1.55b}$$

While fields $\boldsymbol{B}$ and $\boldsymbol{E}$ are the genuine observables, neither the function Φ, called the *scalar potential*, nor the vector field $\boldsymbol{A}$, called the *vector potential*, is directly measurable. This is so because these auxiliary quantities can be modified in a way as to be described in a moment, without modifying the Maxwell fields proper. One reason that justifies the introduction of these potentials is that with (1.55a) and (1.55b) the two homogeneous equations (1.44a) and (1.44b) are fulfilled automatically.

In the vacuum and choosing Gaussian units, the fields $\boldsymbol{D}$ and $\boldsymbol{E}$ are equal, as are the fields $\boldsymbol{B}$ and $\boldsymbol{H}$. In this case, one can insert (1.55a) and (1.55b) into the inhomogeneous Maxwell equations (1.44c) and (1.44d), thereby obtaining

$$\Delta \Phi(t,\boldsymbol{x}) + \frac{1}{c}\frac{\partial}{\partial t}\left(\nabla \cdot \boldsymbol{A}(t,\boldsymbol{x})\right) = -4\pi\varrho(t,\boldsymbol{x}) ; \tag{1.56a}$$

$$\Delta \boldsymbol{A}(t,\boldsymbol{x}) - \frac{1}{c^2}\frac{\partial^2}{\partial t^2}\boldsymbol{A}(t,\boldsymbol{x}) - \nabla\left(\frac{1}{c}\frac{\partial \Phi(t,\boldsymbol{x})}{\partial t} + \nabla \cdot \boldsymbol{A}(t,\boldsymbol{x})\right) = -\frac{4\pi}{c}\boldsymbol{j}(t,\boldsymbol{x}) . \tag{1.56b}$$

As noted above, the decomposition of (1.55a) and (1.55b) is not unique. The remaining freedom in the choice of the potentials can be described in more detail as follows. Choosing $\boldsymbol{A}' = \boldsymbol{A} + \nabla\chi$, with $\chi(t,\boldsymbol{x})$ an arbitrary smooth function, the induction field $\boldsymbol{B}$ is unchanged. However, because of (1.55b), the electric field changes unless one replaces Φ *simultaneously* with

$$\Phi'(t,\boldsymbol{x}) = \Phi(t,\boldsymbol{x}) - \frac{1}{c}\frac{\partial}{\partial t}\chi(t,\boldsymbol{x})$$

to compensate the unwanted additional term. This leads to an important new notion:

Definition 1.2 Gauge transformations Let $\chi(t,\boldsymbol{x})$ be an arbitrary function which is at least C^3 in its arguments. If one replaces the scalar potential and the vector potential as follows:

$$\Phi(t,\boldsymbol{x}) \longmapsto \Phi' = \Phi(t,\boldsymbol{x}) - \frac{1}{c}\frac{\partial}{\partial t}\chi(t,\boldsymbol{x}) ; \tag{1.57a}$$

$$\boldsymbol{A}(t,\boldsymbol{x}) \longmapsto \boldsymbol{A}'(t,\boldsymbol{x}) = \boldsymbol{A}(t,\boldsymbol{x}) + \nabla\chi(t,\boldsymbol{x}) , \tag{1.57b}$$

then the electric field and the induction field remain unchanged:

$$\boldsymbol{E}'(t,\boldsymbol{x}) = \boldsymbol{E}(t,\boldsymbol{x}) ; \quad \boldsymbol{B}'(t,\boldsymbol{x}) = \boldsymbol{B}(t,\boldsymbol{x}) . \tag{1.57c}$$

A transformation of this type applied simultaneously to Φ and $\boldsymbol{A}$ is called a *gauge transformation of the Maxwell fields.*

The function $\chi(t,\boldsymbol{x})$, called a *gauge function*, is completely arbitrary. However, it may be subject to restrictions in order to satisfy certain conditions for the potentials. For instance, one may require χ to satisfy the inhomogeneous differential equation

$$\left(\frac{1}{c^2}\frac{\partial^2}{\partial t^2} - \Delta\right)\chi(t,\boldsymbol{x}) = \nabla \cdot \boldsymbol{A}(t,\boldsymbol{x}) + \frac{1}{c}\frac{\partial \Phi(t,\boldsymbol{x})}{\partial t} .$$

With this choice the transformed potentials obey the condition

$$\frac{1}{c}\frac{\partial \Phi'(t,\boldsymbol{x})}{\partial t} + \nabla \cdot \boldsymbol{A}'(t,\boldsymbol{x}) = 0 \,. \tag{1.58}$$

Every choice of the gauge transformation for which this condition holds is called a *Lorenz gauge.*[3] Note, however, that (1.58) fixes no more than a *class* of gauges. Indeed, any subsequent gauge transformation by a gauge function $\psi(t,\boldsymbol{x})$ which is a solution of the differential equation

$$\left(\frac{1}{c^2}\frac{\partial^2}{\partial t^2} - \boldsymbol{\Delta}\right)\psi(t,\boldsymbol{x}) = 0 \,,$$

when applied after the gauge transformation generated by $\chi(t,\boldsymbol{x})$, leaves the Lorenz condition (1.58) unchanged.

Assume that the potentials are chosen such that they satisfy condition (1.58). Then writing Φ instead of Φ', $\boldsymbol{A}$ instead of $\boldsymbol{A}'$, the differential equations (1.56a) and (1.56b) simplify to wave equations of the form of (1.45) with external sources:

$$\left(\frac{1}{c^2}\frac{\partial^2}{\partial t^2} - \boldsymbol{\Delta}\right)\Phi(t,\boldsymbol{x}) = 4\pi\varrho(t,\boldsymbol{x}) \,; \tag{1.59a}$$

$$\left(\frac{1}{c^2}\frac{\partial^2}{\partial t^2} - \boldsymbol{\Delta}\right)\boldsymbol{A}(t,\boldsymbol{x}) = 4\pi\boldsymbol{j}(t,\boldsymbol{x}) \,. \tag{1.59b}$$

Here again, the left-hand sides contain quantities related to "radiation", and the right-hand sides contain "matter" as sources of the equations of motion.

There is a second reason in favour of using potentials: There are situations where it seems simpler to solve the wave equation (1.45) for the auxiliary entities Φ and $\boldsymbol{A}$, with or without source terms, and to derive from them the observable fields rather than to solve the wave equation for the observable fields including the relationships between them that are contained in Maxwell's equations.

At this point it is worthwhile to return to remark 2 in Sect. 1.4.5, where we had assumed that the charge density $\varrho(t,\boldsymbol{x})$ and the current density $\boldsymbol{j}(t,\boldsymbol{x})$ could perhaps be combined into a vector field over $\mathbb{R}^4$, viz. $j(x) = \left(c\varrho, \boldsymbol{j}\right)^T$, with $x = (x^0, \boldsymbol{x})^T$ and $x^0 = ct$, such that $j^\mu(x)$ transforms under $\Lambda \in L_+^\uparrow$ as a contravariant vector field. In this spirit, one may tentatively combine the scalar potential and the vector potential by the following definition.

▶ **Definition 1.3 Four-potential** With $\Phi(t,\boldsymbol{x})$ a scalar potential and $\boldsymbol{A}(t,\boldsymbol{x})$ a vector potential satisfying the differential equations (1.56a) and (1.56b) respec-

[3] After L.V. Lorenz who should not be confused with H.A. Lorentz, see Historical Remarks.

tively, let

$$A(x) := \big(\Phi(x), \boldsymbol{A}(x)\big)^T \;, \text{ i.e.}$$
$$A^0(x) = \Phi(t,\boldsymbol{x}) \;, \; A^i = \big(\boldsymbol{A}(t,\boldsymbol{x})\big)^i \;. \tag{1.60}$$

The importance of this definition will become clearer in a more general framework below. Nevertheless, there are a number of interesting observations to be made already at this point. The time and space derivatives can be combined as in (1.22a) and (1.22b):

$$\{\partial_\mu\} = \big(\partial_0, \nabla\big) \,, \quad \text{with } \partial_0 = \frac{1}{c}\frac{\partial}{\partial t} \,.$$

Taking account of definition (1.60), condition (1.58) takes a simple and invariant form:

$$\partial_\mu A^\mu(x) = \partial_0 A^0(x) + \sum_{i=1}^{3} \partial_i A^i(x) = \partial_0 A^0(t,\boldsymbol{x}) + \nabla \cdot \boldsymbol{A}(t,\boldsymbol{x}) = 0 \,. \tag{1.61}$$

Note that if $A(x)$ transforms like a Lorentz vector, the condition (1.61) will have the same form in every inertial system.

Likewise, in this formulation the general gauge transformation, (1.57a) and (1.57b), takes a simpler and more transparent form. It now reads

$$A^\mu(x) \longmapsto A'^\mu(x) = A^\mu(x) - \partial^\mu \chi(x) \,. \tag{1.62}$$

It contains the contravariant version of the four-gradient, which is obtained from its covariant form in (1.22a) and (1.22b) as follows:

$$\{\partial^\mu\} = \{g^{\mu\nu}\partial_\nu\} = \text{diag}(1,-1,-1,-1)\big(\partial_0, \nabla\big)^T = \big(\partial_0, -\nabla\big)^T \,.$$

The minus sign in front of the second term of (1.62) has no deeper significance because the (arbitrary) gauge function χ can always be replaced by $-\chi$. Here again, the form of (1.62) will be the same in any inertial system. The derivative $\partial^\mu \chi$ transforms like A^μ if $\chi(x)$ is a Lorentz scalar function.

▶ Remark

Another class of gauges is defined by the condition

$$\nabla \cdot \boldsymbol{A}(t,\boldsymbol{x}) = 0 \,. \tag{1.63}$$

A gauge that satisfies this condition is called a *transversal gauge* or *Coulomb gauge*. This class can be useful, for physical reasons, because it emphasizes the transverse nature of the physical radiation field. To make this evident already at this point, consider equations (1.56a) and (1.56b) without external sources, and

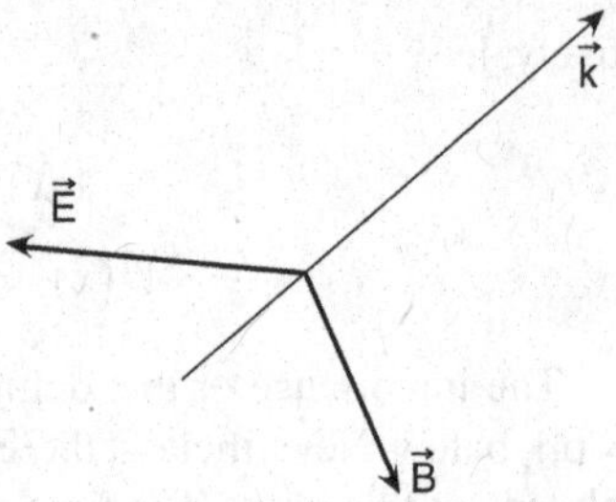

Fig. 1.7 In an electromagnetic plane wave, the electric field and the magnetic induction are perpendicular to each other, and both are perpendicular to the direction of propagation. Vectors $\boldsymbol{E}(t,\boldsymbol{x})$, $\boldsymbol{B}(t,\boldsymbol{x})$ and $\boldsymbol{k}$ form a right-handed system

take the scalar potential Φ to be zero. Then solve the wave equation (1.56b) by means of the ansatz

$$\boldsymbol{A}(t,\boldsymbol{x}) = \boldsymbol{\varepsilon}(\boldsymbol{k})\mathrm{e}^{-\mathrm{i}\omega t}\,\mathrm{e}^{\mathrm{i}\boldsymbol{k}\cdot\boldsymbol{x}} ,$$

where $\boldsymbol{k}$ is the wave number vector and $\omega = c|\boldsymbol{k}|$ the circular frequency. The unit vector $\hat{\boldsymbol{k}}$ defines the direction of propagation, $\boldsymbol{\varepsilon}(\boldsymbol{k})$ is a polarization vector which, in general, depends on $\boldsymbol{k}$. Condition (1.63) gives at once the relation

$$\boldsymbol{\varepsilon}(\boldsymbol{k})\cdot\boldsymbol{k} = 0 ,$$

which tells us that the direction of $\boldsymbol{A}$ is perpendicular to the direction of propagation. The same conclusion also holds for the observable electric and magnetic fields:

It follows from (1.55a) that in a plane wave solution, $\boldsymbol{B}$ is proportional to $\boldsymbol{k}\times\boldsymbol{\varepsilon}$, i.e. is directed perpendicular to the direction of propagation. The electric field contains a transverse part in the first term on the right-hand side of (1.55b) and the contribution $-\nabla\Phi$. The latter vanishes because the potential was assumed to be identically zero. Therefore, the field $\boldsymbol{E}(t,\boldsymbol{x})$ is perpendicular to the direction of propagation, and $\boldsymbol{E} \propto \boldsymbol{\varepsilon}$. This is an important result. In an electromagnetic plane wave, the two fields are perpendicular to each other, and both are perpendicular to the direction of propagation; $\boldsymbol{E}$, $\boldsymbol{B}$ and $\boldsymbol{k}$ span a right-handed system as sketched in Fig. 1.7.

1.7 Phenomenology of the Maxwell Equations

The previous sections clarified the essential features of Maxwell's equations in integral and differential form, whereas this section serves the purpose of exploring some of their phenomenology, with the aim of becoming more familiar with the physical properties coded in these equations. Therefore, we interrupt the formal analysis for a while in favour of some remarks and comments on Maxwell's equations. In doing so we make frequent use of the results and techniques of Sect. 1.6.

1.7.1 The Fundamental Equations and Their Interpretation

Even though some of the following remarks repeat previous ones, it is useful to recall the essential messages of Maxwell's equations and to group them together in one block. We do this, step by step, by following the basic equations (1.44a) to (1.44f).

i The first homogeneous equation (1.44a), $\nabla \cdot \boldsymbol{B}(t, \boldsymbol{x}) = 0$, is a consequence of the statement that the magnetic induction has no isolated source and no source that could be isolated, even in principle. In contrast to the purely electric case, there does not exist a magnetic "charge distribution" which could produce static or nonstatic fields. Magnetism – to invoke a simple picture – always involves the two kinds of poles, the north pole and the south pole.

ii The second homogeneous equation (1.44b)

$$\nabla \times \boldsymbol{E}(t, \boldsymbol{x}) = -\frac{1}{c}\frac{\partial}{\partial t}\boldsymbol{B}(t, \boldsymbol{x})$$

tells us that the vortices of the electric field are due to the time variations of the magnetic induction. In *stationary* cases, i.e. in cases where the induction field does not depend on time, $\boldsymbol{B} = \boldsymbol{B}(\boldsymbol{x})$, the electric field is irrotational. As can be seen from (1.53) with $\boldsymbol{g}(t, \boldsymbol{x}) \equiv 0$, it can be represented as a gradient field. This is true only in stationary situations!

iii The first of the inhomogeneous Maxwell equations, (1.44c), $\nabla \cdot \boldsymbol{D}(t, \boldsymbol{x}) = 4\pi \varrho(t, \boldsymbol{x})$, is an expression of the fact that the given electric charges are the sources of the (electric) displacement field $\boldsymbol{D}$. There is no information about its vortices! This remark, as well as the previous one, shows that, beyond the basic equations, one needs further relations linking $\boldsymbol{D}$ to $\boldsymbol{E}$, and linking $\boldsymbol{B}$ to $\boldsymbol{H}$. In the vacuum these relations are given by (1.44f), with or without constant factors different from 1, depending on the system of units one is using. We return to this in Sect. 1.7.2 and in 1.7.3.

iv The second inhomogeneous Maxwell equation (1.44d),

$$\nabla \times \boldsymbol{H}(t, \boldsymbol{x}) = \frac{4\pi}{c}\left(\boldsymbol{j}(t, \boldsymbol{x}) + \frac{1}{4\pi}\frac{\partial}{\partial t}\boldsymbol{D}(t, \boldsymbol{x})\right),$$

here slightly rewritten in view of the next remark, in any *stationary* case reduces to

$$\nabla \times \boldsymbol{H}(\boldsymbol{x}) = \frac{4\pi}{c}\boldsymbol{j}(\boldsymbol{x}) . \tag{1.64}$$

The vortices of stationary magnetic fields are due solely to the given time-independent current density $\boldsymbol{j}(\boldsymbol{x})$. As the divergence of a curl vanishes, the continuity equation (1.21) holds in the simpler, time-independent version

$\nabla \cdot \boldsymbol{j}(\boldsymbol{x}) = 0$. Suppose that, from the beginning, we had not characterized the continuity equation in its general form, (1.21), as an important basic equation. One could then argue as follows: The stationary equation (1.64) is compatible with the continuity equation. Stationary currents necessarily are always closed currents and, by (1.64), cause magnetic fields which are stationary, too. If the continuity equation holds true also in the *nonstationary*, i.e. time-dependent, case, in its general form, (1.21), the current density must be replaced by

$$\boldsymbol{j}(\boldsymbol{x}) \longmapsto \left(\boldsymbol{j}(t, \boldsymbol{x}) + \frac{1}{4\pi} \frac{\partial}{\partial t} \boldsymbol{D}(t, \boldsymbol{x}) \right) , \tag{1.65a}$$

so that stationary equation (1.64) becomes the Maxwell equation (1.44d). Thus, to the time- and space-dependent current density one must formally add the new "current density":

$$\boldsymbol{j}_{\text{Maxwell}}(t, \boldsymbol{x}) = \frac{1}{4\pi} \frac{\partial}{\partial t} \boldsymbol{D}(t, \boldsymbol{x}) . \tag{1.65b}$$

Maxwell named this additional term the *displacement current.* This new term must be present for Maxwell's equations in order not to contradict the continuity equation. Note that it is only partially possible to visualize this displacement term. Exercise 1.13 gives a simple example in which a dielectric medium is inserted between the plates of a charged capacitor and charges are indeed found to be locally shifted in the medium during discharge of the capacitor. However, in empty space it is not obvious what is being shifted. Whatever justification one chooses, this term is essential for basic properties of physics. It guarantees not only the validity of the continuity equation if the fields become time dependent; it provides the essential basis for the existence of electromagnetic waves. As was shown in Sect. 1.5.4, the wave equation follows from Maxwell's equations only if this term is present. Therefore, the experimental verification of the existence of electromagnetic waves was the touchstone in the proof of the displacement current postulated by Maxwell.

v The Lorentz force, as shown by (1.44e), with its typical dependence on velocity, finally gives an important hint about the spacetime symmetries of Maxwell's equations which we work out in Sect. 2.3 below.

vi In static, i.e. time-independent, situations the Maxwell equations decompose into two independent groups of equations:

$$\nabla \times \boldsymbol{E}(\boldsymbol{x}) = 0 , \tag{1.66a}$$

$$\nabla \cdot \boldsymbol{D}(\boldsymbol{x}) = 4\pi \varrho(\boldsymbol{x}) , \tag{1.66b}$$

$$\boldsymbol{D}(\boldsymbol{x}) = \varepsilon(\boldsymbol{x}) \boldsymbol{E}(\boldsymbol{x}) ; \tag{1.66c}$$

$$\nabla \cdot \boldsymbol{B}(x) = 0\,, \tag{1.67a}$$

$$\nabla \times \boldsymbol{H}(x) = \frac{4\pi}{c}\boldsymbol{j}(x)\,, \tag{1.67b}$$

$$\boldsymbol{B}(x) = \boldsymbol{F}\big(\boldsymbol{H}(x)\big)\,. \tag{1.67c}$$

Only under certain conditions does the last of these, (1.67c), yield a *linear* relation between $\boldsymbol{B}$ and $\boldsymbol{H}$, analogous to (1.66c):

$$\boldsymbol{B}(x) = \mu(x)\boldsymbol{H}(x)\,, \tag{1.67d}$$

a question to which we subsequently return.
The first three are the basic equations of *electrostatics*, whereas the second group defines the basic equations of *magnetostatics*. Note, however, that the decoupling of magnetic and electric phenomena is only an apparent one because currents are due to charges which are in motion. As soon as the electric and magnetic quantities become time dependent, all phenomena intermix. This is why one talks about *electromagnetic* processes, with "electromagnetic" in one term.

vii The wave equation (1.45) plays a fundamental role: It guarantees that electromagnetic oscillations in the vacuum always satisfy the relation $\omega^2 = k^2c^2$ between the circular frequency $\omega = 2\pi\nu = 2\pi/T$ and the wave number $k = 2\pi/\lambda$ of a monochromatic wave. Adding to this the relations $E = \hbar\omega$ and $\boldsymbol{p}^2 = \hbar^2 k^2$ by which quantum theory relates energy and circular frequency, and momentum and wave number, respectively, one finds the relation between energy and momentum

$$E^2 = \boldsymbol{p}^2 c^2\,, \tag{1.68}$$

which is typical for *massless* particles. This is a hint that Maxwell theory, when subject to the postulates of quantum theory, describes massless particles, *photons*.

viii The wave equation (1.45) provides a *necessary*, but certainly not a *sufficient*, condition for field quantities of Maxwell theory. Although it ensures that there are electromagnetic waves at all and fixes the relation between circular frequency and wave number and, thus, after quantization, between the energy and momentum of photons, Maxwell's equations contain more information than that. These partial differential equations determine the correlations between the directions of electric and magnetic fields. The monochromatic plane wave may serve as an example. As was noted earlier, its electric and magnetic fields are transversal and oscillate perpendicularly to each other. When translated into quantum theory, this means that Maxwell's equations contain information about the spin and polarization of photons.

1.7.2 Relation Between Displacement Field and Electric Field

The relation between the electric displacement field $\boldsymbol{D}(t,\boldsymbol{x})$ and the electric field $\boldsymbol{E}(t,\boldsymbol{x})$ is determined by the properties of physical media, which, strictly speaking, can be obtained only from a theory of matter proper. In macroscopic electrostatics, magnetostatics and electrodynamics, where one investigates phenomena at macroscopic scales, it is useful to parametrize the properties of matter in a bulk manner by means of quantities which, though in principle are calculable from a microscopic description, reflect averaged properties of matter. For example, for the purposes of electrostatics, it makes sense to distinguish generally between *electric conductors* and *polarizable media.* In an idealized conductor, there are freely moving charges which, when subject to an electric field, will move until an electrostatic equilibrium is reached. In polarizable media, there are no freely moving charges. Yet, an external electric field can polarize the medium locally, i.e. over microscopic distances.

If the medium is homogeneous and isotropic in its electric properties, then $\boldsymbol{D}(\boldsymbol{x}) = \varepsilon\boldsymbol{E}(\boldsymbol{x})$, where ε is a constant and commutes with all differential operators. If the medium is isotropic but no longer homogeneous, then $\varepsilon(\boldsymbol{x})$ is a function of the position at which one investigates the relation between the two types of fields. In both cases, it is advisable to write $\varepsilon(\boldsymbol{x})\mathbb{1}_3$, i.e.

$$\boldsymbol{D}(\boldsymbol{x}) = \varepsilon(\boldsymbol{x})\mathbb{1}_3\,\boldsymbol{E}(\boldsymbol{x})\,,$$

with $\mathbb{1}_3$ the 3×3 unit matrix. This notation emphasizes the assumed isotropy of the medium. Indeed, if the medium is not isotropic and the response of the medium to an applied external field depends on the direction of that field, then the function $\varepsilon(\boldsymbol{x})$ is replaced by a 3×3 matrix $\boldsymbol{\varepsilon}(\boldsymbol{x})$ whose entries are functions of $\boldsymbol{x}$.

The electric field $\boldsymbol{E}$ is the elementary, microscopic field. The electric displacement field $\boldsymbol{D}$ can only deviate from $\boldsymbol{E}$ inside a medium (except, of course, for possible numerical factors which are due to the system of units!). This will only happen if the electric field induces some polarization in the medium, i.e. if there are locally mobile charges in the medium. To illustrate this, we consider a simple schematic model.

A piece of material which is electrically neutral is divided into cells such that within each cell there exist positive and negative charges which can move inside the cell but cannot leave it. Without an external field these charges are assumed to be homogeneously distributed such that not only the whole piece of matter but also each cell is electrically neutral. If one now applies an external electric field (taken to be homogeneous), equal amounts of positive and negative charges are displaced inside each cell, the positive ones in the direction of the field, the negative ones in the opposite direction (Fig. 1.8b). As a model, every such polarized cell i can be described by an electric dipole $\boldsymbol{d}_i$. Its macroscopic effect will be contained in a *polarizability*:

$$\boldsymbol{P}(\boldsymbol{x}) = \sum_i N_i\,\langle \boldsymbol{d}_i\rangle(\boldsymbol{x})\,, \tag{1.69}$$

where N_i is the average number of dipoles per volume element and $\langle \boldsymbol{d}_i \rangle(\boldsymbol{x})$ is the average dipole acting at point $\boldsymbol{x}$.

A single dipole $\boldsymbol{d}$, located at point $\boldsymbol{x}'$, creates a static potential at some test point $\boldsymbol{x}$:

$$\Phi_{\text{dipole}}(\boldsymbol{x}) = \frac{\boldsymbol{d} \cdot (\boldsymbol{x} - \boldsymbol{x}')}{|\boldsymbol{x} - \boldsymbol{x}'|^3} = \boldsymbol{d} \cdot \nabla_{x'} \frac{1}{|\boldsymbol{x} - \boldsymbol{x}'|} , \tag{1.70a}$$

where the second formula (1.27) was inserted. Denoting by $\varrho(\boldsymbol{x}')$ the distribution of the true charges, the total potential, including the induced polarization in the material, reads

$$\begin{aligned} \Phi(\boldsymbol{x}) &= \iiint \mathrm{d}^3x' \left\{ \frac{\varrho(\boldsymbol{x}')}{|\boldsymbol{x} - \boldsymbol{x}'|} + \boldsymbol{P}(\boldsymbol{x}') \cdot \nabla_{x'} \frac{1}{|\boldsymbol{x} - \boldsymbol{x}'|} \right\} \\ &= \iiint \mathrm{d}^3x' \, \frac{\varrho(\boldsymbol{x}') - \nabla_{x'} \cdot \boldsymbol{P}(\boldsymbol{x}')}{|\boldsymbol{x} - \boldsymbol{x}'|} . \end{aligned} \tag{1.70b}$$

The electric field is given by the negative gradient field, $\boldsymbol{E} = -\nabla\Phi$, its divergence is calculated by means of (1.30) and the first of Eqs. (1.27), giving $\nabla \cdot \boldsymbol{E}(\boldsymbol{x}) = 4\pi[\varrho(\boldsymbol{x}) - \nabla_x \cdot \boldsymbol{P}(\boldsymbol{x})]$. Comparing this to the first inhomogeneous Maxwell equation (1.44c) yields the relation

$$\boldsymbol{D} = \boldsymbol{E} + 4\pi \boldsymbol{P} . \tag{1.71}$$

In the simplest case, the response of the medium to the applied field is *linear* in $\boldsymbol{E}$ and the same in all directions (isotropy). Expressed as a formula, it is

$$\boldsymbol{P}(\boldsymbol{x}) = \chi_{\mathrm{e}}(\boldsymbol{x})\boldsymbol{E}(\boldsymbol{x}) , \tag{1.72}$$

where $\chi_{\mathrm{e}}(\boldsymbol{x})$ is the *electric susceptibility* of the medium. In this case, one obtains the relation

$$\boldsymbol{D}(\boldsymbol{x}) = \varepsilon(\boldsymbol{x}) \, \boldsymbol{E}(\boldsymbol{x}), \quad \text{with} \tag{1.73a}$$

$$\varepsilon(\boldsymbol{x}) = 1 + 4\pi\chi_{\mathrm{e}}(\boldsymbol{x}) . \tag{1.73b}$$

If, in addition, the medium is homogeneous, then ε is a constant which is the same everywhere in the medium. It is called a *dielectric constant.* By use of (1.44c) one obtains the following inhomogeneous differential equation:

$$\nabla \cdot \boldsymbol{E}(\boldsymbol{x}) = \frac{4\pi}{\varepsilon} \varrho(\boldsymbol{x}) . \tag{1.74}$$

An electric dipole points from the negative to the positive charge. Therefore, the polarization $\boldsymbol{P}$ has the same direction as the field $\boldsymbol{E}$, and one finds $\chi_{\mathrm{e}} > 0$ and $\varepsilon > 1$. When this is inserted into (1.73a), one finds that the applied field is *weakened* by

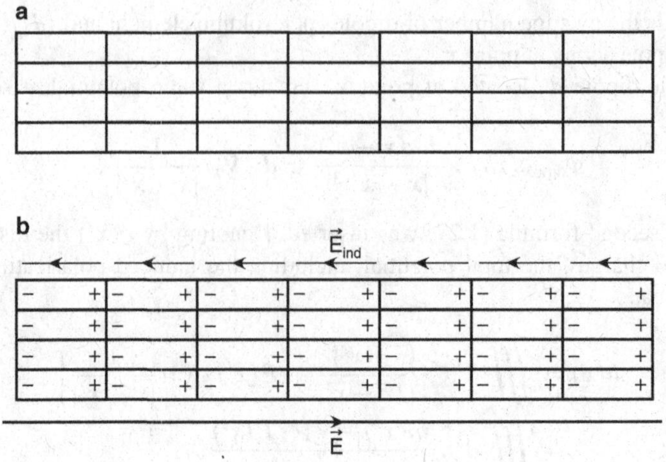

Fig. 1.8 Schematic model of an electrically polarizable medium. **a** The medium consists of elementary cells which are electrically neutral. **b** Under the action of an external electric field $\boldsymbol{E}$, the positive and negative charges separate in the elementary cells, creating an induced electric field which acts against the external field

the dipole fields that it induces in the material, in agreement with the simple model sketched in Fig. 1.8.

In electric conductors, as long as they are assumed to have ideal conductivity, there is no polarization. All charges contained in the metal move freely and, hence, will flow under the action of the applied field until equilibrium is reached. The induced charges will all sit on the surfaces of the conductor being studied. Except for these surfaces, $\boldsymbol{D}$ and $\boldsymbol{E}$ are equal.

1.7.3 Relation Between Induction and Magnetic Fields

Use is made here of a few formulae of magnetostatics taken from Sect. 1.9.3 below which will be derived there but are made plausible at this stage. Their purpose is to derive relations between the magnetic induction $\boldsymbol{B}$ and the magnetic field $\boldsymbol{H}$, which are analogous to (1.73a).

A magnetic pointlike dipole $\boldsymbol{m}$, placed at the origin of the system of reference, gives rise to the vector potential (up to gauge transformations):

$$\boldsymbol{A}_{\text{dipole}}(\boldsymbol{x}) = \frac{\boldsymbol{m} \times \boldsymbol{x}}{|\boldsymbol{x}|^3} . \tag{1.75a}$$

If it is placed at some other point $\boldsymbol{x}'$, then the potential at the test point $\boldsymbol{x}$ reads

$$\boldsymbol{A}_{\text{dipole}}(\boldsymbol{x}) = \frac{\boldsymbol{m} \times (\boldsymbol{x} - \boldsymbol{x}')}{|\boldsymbol{x} - \boldsymbol{x}'|^3} . \tag{1.75b}$$

The formal resemblance to the scalar potential of an electric dipole (1.70a) is obvious.

A stationary current density $\boldsymbol{j}(\boldsymbol{x})$, by equation (1.67b), has vanishing divergence. It creates a magnetic field whose curl is given by (1.67b). Using the general decomposition (1.53), the field $\boldsymbol{H}$ can be represented as follows:

$$\boldsymbol{H}(\boldsymbol{x}) = \nabla_x \times \left(\frac{1}{4\pi} \iiint \mathrm{d}^3 x' \, \frac{(4\pi/c)\,\boldsymbol{j}(\boldsymbol{x}')}{|\boldsymbol{x}' - \boldsymbol{x}|} \right) .$$

In the vacuum, one has $\boldsymbol{H}(\boldsymbol{x}) = \boldsymbol{B}(\boldsymbol{x})$, so that, upon using representation (1.55a) of $\boldsymbol{B}$ in terms of a vector potential, this potential is given by the integral on the right-hand side of the preceding formula:

$$\boldsymbol{A}_{\text{current}}(\boldsymbol{x}) = \frac{1}{c} \iiint \mathrm{d}^3 x' \, \frac{\boldsymbol{j}(\boldsymbol{x}')}{|\boldsymbol{x}' - \boldsymbol{x}|} . \tag{1.75c}$$

In this formula, the test point $\boldsymbol{x}$ lies outside the domain where the current density is different from zero.

Consider a localized piece of matter whose magnetic polarizability can be characterized macroscopically by a *magnetization density*:

$$\boldsymbol{M}(\boldsymbol{x}) = \sum_i N_i \, \langle \boldsymbol{m}_i \rangle (\boldsymbol{x}) , \tag{1.76}$$

where $\langle \boldsymbol{m}_i \rangle (\boldsymbol{x})$ is the average magnetic dipole moment of an elementary cell (e.g. a molecule) at point $(\boldsymbol{x})$ and N_i is the average number of such cells (molecules of a given kind). If, furthermore, there is a nonvanishing free current density $\boldsymbol{j}(\boldsymbol{x})$, it is possible to write down a vector potential. By equations (1.75c) and (1.75a), it has the form

$$\begin{aligned} \boldsymbol{A}(\boldsymbol{x}) &= \boldsymbol{A}_{\text{current}}(\boldsymbol{x}) + \boldsymbol{A}_{\text{dipole}}(\boldsymbol{x}) \\ &= \frac{1}{c} \iiint \mathrm{d}^3 x' \left\{ \frac{\boldsymbol{j}(\boldsymbol{x}')}{|\boldsymbol{x}' - \boldsymbol{x}|} + c \, \frac{\boldsymbol{M}(\boldsymbol{x}') \times (\boldsymbol{x} - \boldsymbol{x}')}{|\boldsymbol{x} - \boldsymbol{x}'|^3} \right\} . \end{aligned}$$

In the second term of this expression, one of the formulae contained in (1.27) is used, and then a partial integration with respect to variable $\boldsymbol{x}'$ is performed to find

$$\begin{aligned} \iiint \mathrm{d}^3 x' \, \frac{\boldsymbol{M}(\boldsymbol{x}') \times (\boldsymbol{x} - \boldsymbol{x}')}{|\boldsymbol{x} - \boldsymbol{x}'|^3} &= \iiint \mathrm{d}^3 x' \, \boldsymbol{M}(\boldsymbol{x}') \times \nabla_{x'} \frac{1}{|\boldsymbol{x} - \boldsymbol{x}'|} \\ &= \iiint \mathrm{d}^3 x' \, \left(\nabla_{x'} \times \boldsymbol{M}(\boldsymbol{x}') \right) \frac{1}{|\boldsymbol{x} - \boldsymbol{x}'|} . \end{aligned}$$

There is no minus sign following the partial integration because the order of the nabla operator and the magnetization density was changed in the cross-product. As by assumption $\boldsymbol{M}(\boldsymbol{x})$ is localized, there are no surface terms. With this intermediate result the vector potential takes a form which exhibits an obvious similarity to (1.70b):

$$\boldsymbol{A}(\boldsymbol{x}) = \frac{1}{c} \iiint \mathrm{d}^3 x' \, \frac{\boldsymbol{j}(\boldsymbol{x}') + c\,\nabla_{x'} \times \boldsymbol{M}(\boldsymbol{x}')}{|\boldsymbol{x} - \boldsymbol{x}'|} \,. \tag{1.77}$$

Taking the curl of $\boldsymbol{B}$ and using formulae (1.47c) and (1.30), one has

$$\nabla \times \boldsymbol{B}(\boldsymbol{x}) = \nabla \times \left(\nabla \times \boldsymbol{A}(\boldsymbol{x})\right) = -\boldsymbol{\Delta}\boldsymbol{A}(\boldsymbol{x}) = \frac{4\pi}{c}\boldsymbol{j}(\boldsymbol{x}) + 4\pi \nabla \times \boldsymbol{M}(\boldsymbol{x}) \,.$$

(A term in (1.47c) containing the divergence of $\boldsymbol{A}$ vanishes because the current density $\boldsymbol{j}$ has vanishing divergence.) Rewriting this in the form of the second inhomogeneous Maxwell equation (1.67b), one sees that one should write

$$\boldsymbol{H}(\boldsymbol{x}) = \boldsymbol{B}(\boldsymbol{x}) - 4\pi \boldsymbol{M}(\boldsymbol{x}) \tag{1.78a}$$

to obtain the familiar form of the basic equations (1.67a) and (1.67b).

The relationship between the magnetic induction $\boldsymbol{B}$ and the magnetic field $\boldsymbol{H}$ is an issue concerning the magnetic properties of matter. For isotropic, *diamagnetic* and *paramagnetic* media the relation is a linear one:

$$\boldsymbol{B}(\boldsymbol{x}) = \mu(\boldsymbol{x})\boldsymbol{H}(\boldsymbol{x}) \,. \tag{1.78b}$$

The function $\mu(\boldsymbol{x})$ is called the *magnetic permeability*. As in the case of dielectric media, the response of the medium to the applied magnetic field $\boldsymbol{H}$ is linear:

$$\boldsymbol{M}(\boldsymbol{x}) = \chi_{\mathrm{m}}(\boldsymbol{x})\boldsymbol{H}(\boldsymbol{x}) \,, \tag{1.78c}$$

with χ_{m} the *magnetic susceptibility*, so that the permeability is given by

$$\mu(\boldsymbol{x}) = 1 + 4\pi\chi_{\mathrm{m}}(\boldsymbol{x}) \,, \tag{1.78d}$$

a formula which is analogous to (1.73b).

Diamagnetic substances can be modelled by an assembly of atoms whose total angular momentum is equal to zero and, hence, which have no static magnetic moment. The applied magnetic field induces magnetic moments which are directed opposite to the applied field; the induced magnetic moments weaken the external field by their own magnetic field. Regarding the macroscopic parameters, this means that $\chi_{\mathrm{m}} < 0$ and, hence, $\mu < 1$.

Paramagnetic substances consist of atoms which have a nonvanishing total angular momentum and a nonvanishing primary magnetic moment. This magnetic moment, which stems from the unpaired electron of the atomic shells, aligns parallel to the applied field, the magnetic susceptibility is positive, $\chi_{\mathrm{m}} > 0$ and therefore the permeability is $\mu > 1$.

In both cases, for diamagnetic and paramagnetic substances, the susceptibility χ_{m} is found to be very small, and the permeability μ is very close to 1.

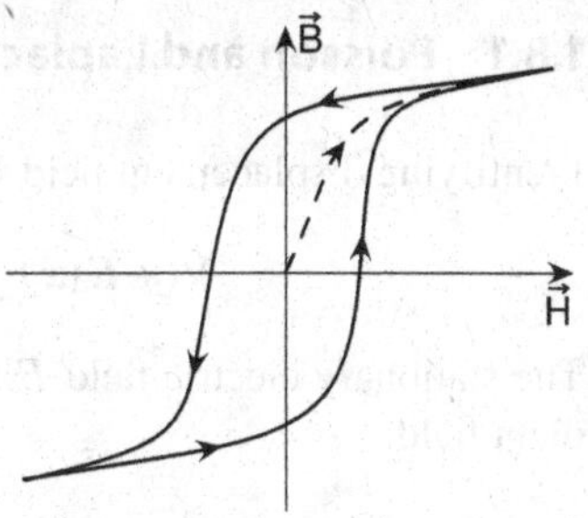

Fig. 1.9 Qualitative relation between the applied magnetic field and the resulting induced field for a ferromagnetic substance (hysteresis in steel)

In *ferromagnetic* substances, finally, the response of the medium is no longer linear. Furthermore, the function $\boldsymbol{F}(\boldsymbol{H}(\boldsymbol{x}))$ is *multivalued,* i.e. the value of the induction $\boldsymbol{B}$, for a given value of $\boldsymbol{H}$, depends on how field $\boldsymbol{H}$ was built up. This is the phenomenon of hysteresis, which is illustrated qualitatively in Fig. 1.9.

Remark

So far we have introduced the polarizability and the magnetization density as phenomenological, semimacroscopic, averaged properties of matter to obtain an impression of the nature of these quantities. This was certainly not meant to indicate that they exist only in this sense and that the fields $\boldsymbol{D}$ and $\boldsymbol{H}$ are necessarily macroscopic fields. Indeed, electric polarizability and magnetization are also well defined microscopically, i.e. for a single atom or even an elementary particle. The electric displacement $\boldsymbol{D}(t, \boldsymbol{x})$ and the magnetic field $\boldsymbol{H}(t, \boldsymbol{x})$ are entities which are as fundamental as the electric field $\boldsymbol{E}(t, \boldsymbol{x})$ and the induction field $\boldsymbol{B}(t, \boldsymbol{x})$. (For a more detailed discussion, see Hehl–Obukhov 2003 and the literature quoted therein.)

1.8 Static Electric States

The basic equations of electrostatics are given in (1.66a)–(1.66c), where in the vacuum, and upon using Gaussian units, the function ε is the constant function 1. In view of the general remarks in Sect. 1.7.2, it is meaningful to leave aside for the moment polarizable media and to study electrostatic phenomena only in conducting media and in the vacuum. Except for the surfaces of ideal conductors, fields $\boldsymbol{D}$ and $\boldsymbol{E}$ can be identified. The essential problems of electrostatics are then well defined: Establish the relationship between given charges or charge distributions and the electric fields created by them, define surface charges on the surfaces of ideally conducting bodies, and study the discontinuities of fields across surfaces.

1.8.1 Poisson and Laplace Equations

Identifying displacement field $\boldsymbol{D}$ and electric field $\boldsymbol{E}$, one has

$$\nabla \times \boldsymbol{E}(\boldsymbol{x}) = 0, \qquad \nabla \cdot \boldsymbol{E}(\boldsymbol{x}) = 4\pi\varrho(\boldsymbol{x}).$$

The stationary electric field $\boldsymbol{E}$ is irrotational and can therefore be written as a gradient field:

$$\boldsymbol{E}(\boldsymbol{x}) = -\nabla\Phi(\boldsymbol{x}), \tag{1.79}$$

where $\Phi(\boldsymbol{x})$ is a real, piecewise continuous and differentiable function. The equation $\Phi(\boldsymbol{x}) = c$, with c a constant, defines a locally smooth surface in $\mathbb{R}^3$, as sketched in Fig. 1.10. Let $\hat{\boldsymbol{v}}_n$ denote the normal to the surface at point P, the direction of $\hat{\boldsymbol{v}}_n$ being chosen such that the function grows in that direction. Let $\hat{\boldsymbol{v}}$ be an arbitrary unit vector in P. As shown in Fig. 1.10, it is decomposed in terms of $\hat{\boldsymbol{v}}_n$ and of a unit vector $\hat{\boldsymbol{v}}_t$ in the tangent space at P:

$$\hat{\boldsymbol{v}} = a_n\hat{\boldsymbol{v}}_n + a_t\hat{\boldsymbol{v}}_t, \quad \text{with} \quad a_n, a_t \in \mathbb{R} \quad \text{and} \quad a_n^2 + a_t^2 = 1.$$

Calculating the directional derivative of $\Phi(\boldsymbol{x})$ in the direction of $\hat{\boldsymbol{v}}$, written symbolically as $\partial\Phi/\partial v$,

$$\frac{\partial\Phi}{\partial v} \equiv \nabla\Phi(\boldsymbol{x}) \cdot \hat{\boldsymbol{v}},$$

and inserting the decomposition of $\hat{\boldsymbol{v}}$, only the normal component is seen to contribute. The value of $\Phi(\boldsymbol{x})$ is constant along a tangent vector. Therefore, one has

$$\frac{\partial\Phi}{\partial v} = a_n\left(\nabla\Phi(\boldsymbol{x}) \cdot \hat{\boldsymbol{v}}_n\right) \quad \text{with} \quad -1 \leqslant a_n \leqslant +1.$$

This result shows that the growth of the function is maximal in the direction of the positive normal, i.e. for the choice $\hat{\boldsymbol{v}} = \hat{\boldsymbol{v}}_n$. Thus, the gradient field defines the directions along which the potential grows or decreases the fastest. With the choice of sign in definition (1.79), one draws the following conclusion:
In every point of the surface $\Phi(\boldsymbol{x}) = c$, the electric field is perpendicular to the surface and points in the direction along which the potential falls off the most rapidly.

A surface of the kind $\Phi(\boldsymbol{x}) = c$ is called an *equipotential surface*. Every curve whose tangent vector field coincides with the electric field is orthogonal to this surface. In other terms, a charged particle which follows the electric field moves on an orthogonal trajectory to the surfaces $\Phi(\boldsymbol{x}) = c$.

With $\boldsymbol{D}(\boldsymbol{x}) = \boldsymbol{E}(\boldsymbol{x})$ and by ansatz (1.79), the potential $\Phi(\boldsymbol{x})$ obeys the following ordinary differential equation:

Poisson equation

$$\Delta\Phi(\boldsymbol{x}) = -4\pi\varrho(\boldsymbol{x}). \tag{1.80a}$$

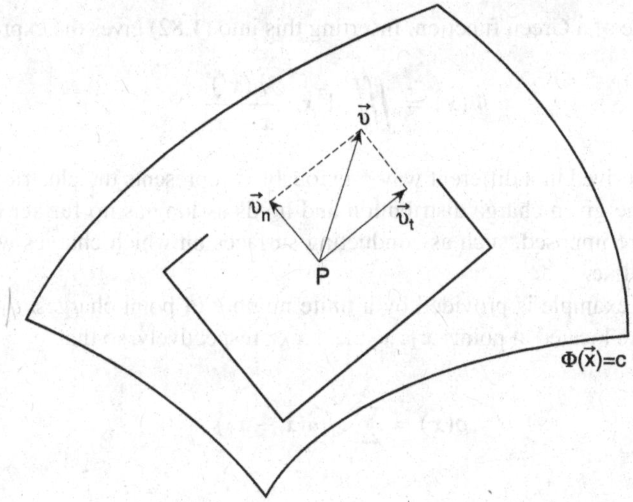

Fig. 1.10 Locally the equation $\Phi(\boldsymbol{x}) = c$ defines a surface in $\mathbb{R}^3$ whose tangent plane in the point P is also shown

The Poisson equation, supplemented by boundary conditions which are usually defined by experimental set-ups, is another basic equation of electrostatics. Outside of given point charges or charge distributions, that is everywhere where $\varrho(\boldsymbol{x})$ vanishes locally, the Poisson equation is replaced by the following equation:

Laplace equation

$$\Delta \Phi(\boldsymbol{x}) = 0 . \tag{1.80b}$$

This differential equation is called the *Laplace equation*.

Generally speaking the Poisson equation is solved formally by means of Green functions. A Green function, which in fact is not a function but a distribution, depends on two arguments, say $\boldsymbol{x}$ and $\boldsymbol{x}'$, and obeys the differential equation

$$\Delta_x G(\boldsymbol{x}, \boldsymbol{x}') = \delta(\boldsymbol{x} - \boldsymbol{x}') , \tag{1.81}$$

with $\delta(z)$ denoting the Dirac δ-distribution. Indeed, if the source $\varrho(\boldsymbol{x})$ in (1.80a) is given, then

$$\Phi(\boldsymbol{x}) = -4\pi \iiint \mathrm{d}^3x' \; G(\boldsymbol{x}, \boldsymbol{x}') \, \varrho(\boldsymbol{x}') \tag{1.82}$$

solves the Poisson equation. Relation (1.30), whose proof may be found, for example, in [QP], Appendix A.1, shows that

$$G(\boldsymbol{x}, \boldsymbol{x}') = -\frac{1}{4\pi} \frac{1}{|\boldsymbol{x} - \boldsymbol{x}'|} \tag{1.83}$$

is an example of a Green function. Inserting this into (1.82) gives the expression

$$\Phi(\boldsymbol{x}) = \iiint \mathrm{d}^3x' \, \frac{\varrho(\boldsymbol{x}')}{|\boldsymbol{x}-\boldsymbol{x}'|} \,, \tag{1.84a}$$

which was derived in a different way previously. It represents the electric potential created by the given charge distribution and holds as long as no further boundary conditions are imposed, such as conducting surfaces on which charges will be induced or the like.

A simple example is provided by a finite number of point charges, $q_1, q_2, \ldots, q_N$, which are located at points $\boldsymbol{x}_1, \boldsymbol{x}_2, \ldots, \boldsymbol{x}_N$, respectively, so that

$$\varrho(\boldsymbol{x}) = \sum_{i=1}^{N} q_i \delta(\boldsymbol{x}-\boldsymbol{x}_i) \,,$$

the corresponding potential being

$$\Phi(\boldsymbol{x}) = \sum_{i=1}^{N} \frac{q_i}{|\boldsymbol{x}-\boldsymbol{x}_i|} \,. \tag{1.84b}$$

In electrostatics the force $\boldsymbol{K}(\boldsymbol{x}) = q_0 \boldsymbol{E}(\boldsymbol{x})$ acting on a test charge q_0 is a potential force. Therefore, the work done or gained by moving this charge from A ("alpha") to Ω ("omega") is independent of the path linking A and Ω:

$$W = -\int \mathrm{d}\boldsymbol{s} \cdot \boldsymbol{K}(\boldsymbol{x}) = -q_0 \int \mathrm{d}\boldsymbol{s} \cdot \boldsymbol{E}(\boldsymbol{x}) = q_0[\Phi(\Omega) - \Phi(A)] \,.$$

It is given by the difference of the potential energies of the test charge at these points.

Example 1.5 Spherically symmetric charge distribution

Assume the charge density to be spherically symmetric, i.e. $\varrho(\boldsymbol{x}) = \varrho(r)$, and to be localized, which means that one may draw a sphere with radius R about the origin outside of which there are no further charges. The radial part of the Laplace operator is taken from (1.96) below. When this is inserted into the Poisson equation, this differential equation reduces to a differential equation in the radial variable alone:

$$\frac{1}{r^2}\frac{\mathrm{d}}{\mathrm{d}r}\left(r^2 \frac{\mathrm{d}\Phi}{\mathrm{d}r}\right) = -4\pi\,\varrho(r) \,.$$

A first integral of this equation yields

$$\frac{\mathrm{d}\Phi}{\mathrm{d}r} = -\frac{4\pi}{r^2}\int_0^r \mathrm{d}r' \, r'^2 \varrho(r') + \frac{c_1}{r^2} \,.$$

The integration constant c_1 must be chosen to be zero. If it were not zero, the electric field strength would become infinite at $r = 0$. The first term has a physical interpretation: $4\pi \int_0^r \mathrm{d}r' \, r'^2 \varrho(r')$ is the charge $Q(r)$ contained in the sphere with radius r. Depending on whether $r \leqslant R$ or $r > R$, this charge is part or all of the total charge contained in the charge distribution, respectively. By integrating once more, and using partial integration, one finds

$$\begin{aligned}\Phi(r) &= \int_0^r \mathrm{d}r' \left(-\frac{1}{r'^2}\right) Q(r') + c_2 \\ &= \left.\frac{Q(r')}{r'}\right|_0^r - 4\pi \int_0^r \mathrm{d}r' \, r' \varrho(r') + c_2 \\ &= \frac{Q(r)}{r} + 4\pi \int_r^\infty \mathrm{d}r' \, r' \varrho(r') + \left[-4\pi \int_0^\infty \mathrm{d}r' \, r' \varrho(r') + c_2 \right] .\end{aligned}$$

The third term (in square brackets) is a constant which can be taken to be zero by a suitable choice of the integration constant c_2, without loss of generality. There results an important formula which was derived previously in Example 1.2:

$$\Phi(r) = 4\pi \left\{ \frac{1}{r} \int_0^r \mathrm{d}r' \, r'^2 \varrho(r') + \int_r^\infty \mathrm{d}r' \, r' \varrho(r') \right\} . \qquad (1.85)$$

By means of this formula one can derive the potential for any localized, spherically symmetric charge distribution. Examples are as follows:
(i) the homogeneous density of Example 1.2 in Sect. 1.2;
(ii) a sphere of radius R made of an ideal conductor material: All charges sit on the surface of the sphere and are equally distributed, i.e. we have $Q(r) = 0$ for all $r < R$. With Q denoting the total charge carried by the sphere, one has

$$\boldsymbol{E}_{\text{inner}} = 0 , \qquad \boldsymbol{E}_{\text{outer}} = \frac{Q}{r^2} \hat{\boldsymbol{r}} .$$

The component of the electric field parallel to directions tangent to the sphere vanishes inside and outside the sphere. The radially oriented normal component has a discontinuity at $r = R$;
(iii) a distribution which is often used as a model for the charge distribution of atomic nuclei:

$$\varrho_{\text{Fermi}}(r) = \frac{N}{1 + \exp[(r - c)/z]} \qquad \text{with}$$

$$N = \frac{3}{4\pi c^3} \left[1 + \left(\frac{\pi z}{c} \right)^2 - 6 \left(\frac{z}{c} \right)^3 \sum_{n=1}^{\infty} \frac{(-)^n}{n^3} \mathrm{e}^{-nc/z} \right]^{-1} ;$$

it is called a *Fermi distribution*. Parameter c is the distance from the origin at which ϱ has decreased to about half its value at $r = 0$, and parameter z characterizes the width of the fall-off of ϱ near the surface. Although this is somewhat more tedious, the potential can also be calculated analytically for this example. In applications to electromagnetic processes with nuclei (electron scattering, muonic atoms, etc.), one often constructs $\Phi(r)$ by numerical integration of (1.85).

Example 1.6 A function-theoretic method

Suppose an electrostatic set-up is homogeneous in one space direction, so that its nontrivial physical features are confined to the planes perpendicular to that direction. In such a situation, the analysis can be restricted to a plane, i.e. the given problem is reduced to two spatial dimensions. Denoting Cartesian coordinates in $\mathbb{R}^3$ by x, y, ζ and choosing them such that the ζ-axis is taken to be the direction of homogeneity, neither the charge distribution ϱ nor the potential Φ depends on ζ. The Poisson equation reduces to

$$\Delta\Phi(x,y) = \left(\frac{\partial^2}{\partial x^2} + \frac{\partial^2}{\partial y^2}\right)\Phi(x,y) = -4\pi\varrho(x,y) .$$

The corresponding homogeneous differential equation ($\varrho(x,y) \equiv 0$) is well known from function theory. With z denoting a complex variable and $w(z)$ an analytic function, and with the decompositions into real and imaginary parts,

$$z = x + \mathrm{i}y \longrightarrow w(z) = u(x,y) + \mathrm{i}v(x,y) ,$$

respectively, the functions u and v satisfy the Cauchy–Riemann differential equations:

$$\frac{\partial u}{\partial x} = \frac{\partial v}{\partial y} , \tag{1.86a}$$

$$\frac{\partial u}{\partial y} = -\frac{\partial v}{\partial x} . \tag{1.86b}$$

These equations have two consequences. (i) Both $u(x,y)$ and $v(x,y)$ satisfy the Laplace equation in two dimensions:

$$\left(\frac{\partial^2}{\partial x^2} + \frac{\partial^2}{\partial y^2}\right)u(x,y) = 0 , \qquad \left(\frac{\partial^2}{\partial x^2} + \frac{\partial^2}{\partial y^2}\right)v(x,y) = 0 .$$

This is shown for the function $u(x,y)$ by taking the partial derivative of (1.86a) with respect to x and the partial derivative of (1.86b) with respect to y and adding the results. Similarly for $v(x,y)$, one takes the partial derivative of (1.86a) by y

and the partial derivative of (1.86b) by x and subtracts the resulting equations. (ii) Equations (1.86a) and (1.86b) tell us that the curves $u(x,y) = \text{const}$ and $v(x,y) = \text{const}$ are orthogonal to each other in the (x,y)-plane. This is seen by taking the scalar product of the tangent vectors at the point (x,y):

$$(\partial u/\partial x \quad \partial u/\partial y) \begin{pmatrix} \partial v/\partial x \\ \partial v/\partial y \end{pmatrix} = 0 .$$

These curves are said to be orthogonal trajectories of one another.

In the two-dimensional sections of our electrostatic problem, we encounter a similar situation: The intersection curves of the equipotential surfaces with the (x,y)-plane and the electric field lines are orthogonal to each other. One can make use of this observation to generate further solutions from a given solution of the Laplace equation. Every analytic function $f(z)$ generates a *conformal mapping* at any point where its derivative is different from zero, i.e. a mapping from the complex z-plane to the plane $w = f(z)$ under which the angles at intersections of curves are preserved both in modulus and orientation. For example, consider the two sets of curves

$$x^2 - y^2 = a \quad \text{and} \quad 2xy = b\,, \quad a \in \mathbb{R}_+,\ b \in \mathbb{R}\,,$$

as a model for equipotential lines (Fig. 1.11a) and for field lines (Fig. 1.11b), respectively. Obviously, these two sets are the real and imaginary parts, respectively, of the function $w = f(z) = z^2$. The curves $x^2 - y^2 = a$ on the z-plane are mapped to straight lines $w = a + iv$ parallel to the v-axis, and the curves $2xy = b$ are mapped to straight lines $w = u + ib$ parallel to the u-axis. The image of the given system is found to be a homogeneous electric field parallel to the u-axis, together with its equipotential lines, which are parallel to the v-axis, as illustrated in (Fig. 1.11c).

Remarks

1. If the function is chosen to be of the type

$$z \longmapsto w = \frac{az+b}{cz+d}\,, \quad \text{with} \quad ad - bc \neq 0\,, \quad c \neq 0\,,$$

then the mapping is not only conformal but also bijective. Therefore, it may be used in either direction.

2. The class of problems defined previously still live in $\mathbb{R}^3$, even though one dimension is irrelevant. In a "genuine" $\mathbb{R}^2$, the Laplace operator

$$\begin{aligned}\mathbf{\Delta}^{(\text{Dim } 2)}\, \Phi(\boldsymbol{x}) &= \left(\frac{\partial^2}{\partial x^2} + \frac{\partial^2}{\partial y^2}\right) \Phi(\boldsymbol{x}) \\ &= \frac{1}{r}\frac{\partial}{\partial r}\left(r\frac{\partial \Phi(r,\phi)}{\partial r}\right) + \frac{1}{r^2}\frac{\partial^2 \Phi(r,\phi)}{\partial \phi^2} = 0\,,\end{aligned}$$

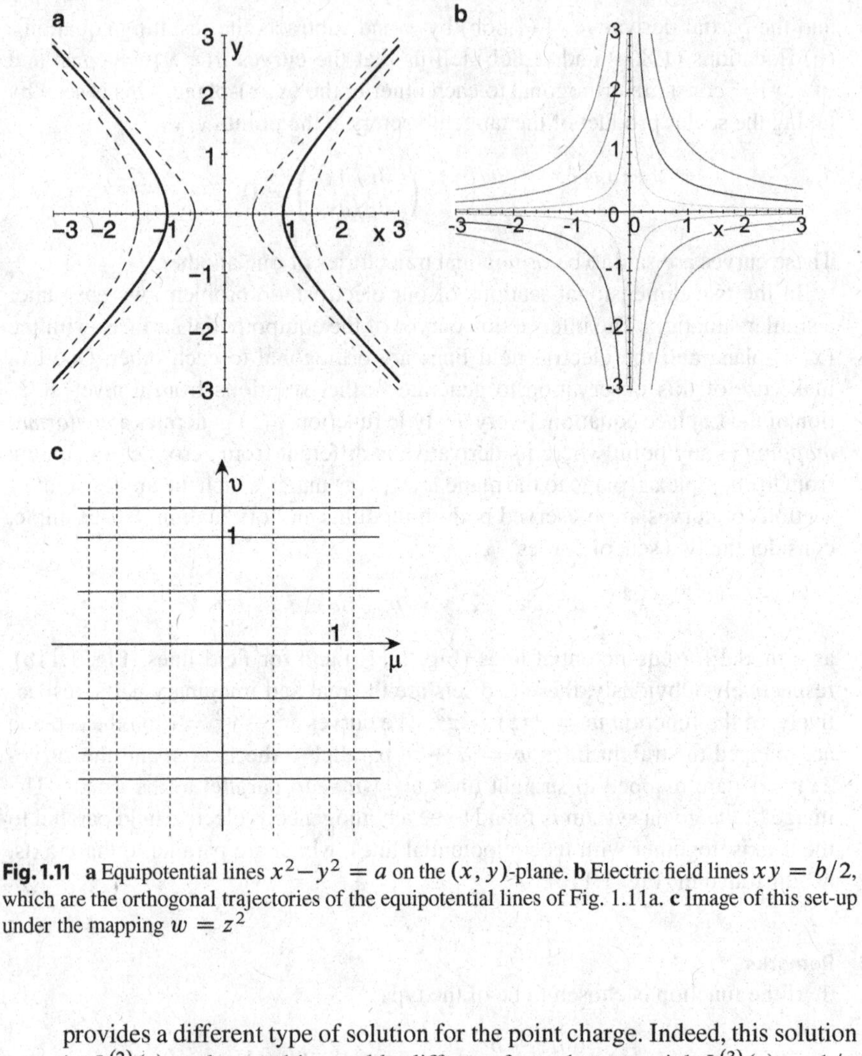

Fig. 1.11 **a** Equipotential lines $x^2 - y^2 = a$ on the (x, y)-plane. **b** Electric field lines $xy = b/2$, which are the orthogonal trajectories of the equipotential lines of Fig. 1.11a. **c** Image of this set-up under the mapping $w = z^2$

provides a different type of solution for the point charge. Indeed, this solution is $\Phi^{(2)}(r) = \ln r$ and is notably different from the potential $\Phi^{(3)}(r) = 1/r$ of a point charge in $\mathbb{R}^3$!

1.8.2 Surface Charges, Dipoles and Dipole Layers

Example 1.5 (ii) may serve as an inspiration to consider the mathematical limit of a more general surface charge. On a localized ideal conductor whose boundary is a smooth surface in $\mathbb{R}^3$, the charge will be distributed exclusively on its surface. With this idealization the *spatial* charge density ϱ (whose physical dimension is charge/volume) is replaced by a *surface* charge density η whose dimension is charge/surface. The potential which is created by a surface charge is given by

$$\Phi(\boldsymbol{x}) = \iint \mathrm{d}\sigma' \, \frac{\eta(\boldsymbol{x}')}{|\boldsymbol{x} - \boldsymbol{x}'|} \, .$$

Through application of Gauss' theorem, (1.6), to a suitably chosen volume that encloses a piece of the surface, and with $\hat{\boldsymbol{n}}$ denoting the normal directed *outwards*, one finds the following relation between the difference of electric field strengths outside and inside the surface carrying the charge density η (cf. Exercise 1.8):

$$(\boldsymbol{E}_{\mathrm{o}} - \boldsymbol{E}_{\mathrm{i}}) \cdot \hat{\boldsymbol{n}} = 4\pi\eta \, . \tag{1.87a}$$

The normal component of the electric field is discontinuous; the discontinuity is 4π times the charge density per unit area. Applying now Stokes' theorem, (1.7), to a closed loop which joins inside and outside (for details see Exercise 1.9), one finds that the component *tangent* to the surface is continuous, i.e.

$$(\boldsymbol{E}_{\mathrm{a}} - \boldsymbol{E}_{\mathrm{i}}) \cdot \hat{\boldsymbol{t}} = 0 \, . \tag{1.87b}$$

These considerations hold for any piecewise smooth surface, independently of whether or not this is a conductor. The tangential component of $\boldsymbol{E}$ is continuous, and the normal component is discontinuous by the amount $4\pi\eta$. In the case of an electric conductor, however, the tangential component must vanish since otherwise charges would flow along the surface until equilibrium is reached. Inside the conductor the potential is constant, and the electric field strength is equal to zero, $\boldsymbol{E}_{\mathrm{i}} = 0$. In the exterior region, there is only a normal component whose modulus is determined by the surface charge density:

$$\boldsymbol{E}_{\mathrm{a}} = 4\pi\eta\,\hat{\boldsymbol{n}} \, .$$

A static electric dipole is obtained by the mathematical limit of a system consisting of two equal pointlike charges of opposite sign whose distance is taken to zero. More precisely, let the charges $+q$ and $-q$ be given in the positions $(0, 0, (a/2)\hat{\boldsymbol{e}}_3)$ and $(0, 0, -(a/2)\hat{\boldsymbol{e}}_3)$, respectively, as sketched in Fig. 1.12. They give rise to the electrostatic potential

$$\Phi(\boldsymbol{x}) = q \left[\frac{1}{|\boldsymbol{x} - (a/2)\hat{\boldsymbol{e}}_3|} - \frac{1}{|\boldsymbol{x} + (a/2)\hat{\boldsymbol{e}}_3|} \right] .$$

As a is assumed to be small compared to $|\boldsymbol{x}|$, the two terms can be expanded up to terms which are linear in a. With $r = |\boldsymbol{x}|$ and $x^3 = r\cos\theta$ one has

$$\frac{1}{|\boldsymbol{x} \mp (a/2)\hat{\boldsymbol{e}}_3|} \simeq \frac{1}{r} \left[1 \pm \frac{1}{2}\frac{a}{r^2} x^3 \right] = \frac{1}{r} \pm \frac{1}{2}\frac{a}{r^3} x^3 = \frac{1}{r} \pm \frac{1}{2}\frac{a}{r^2} \cos\theta \, .$$

Upon insertion of these formulae, the first term is seen to cancel out while the second term is proportional to the product qa. Taking, then, the simultaneous limit ($q \to \infty$,

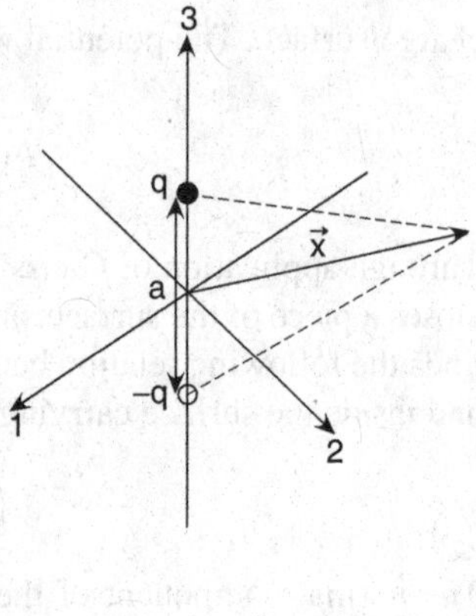

Fig. 1.12 Shrinking the distance a of two opposite point charges $\pm q$ while simultaneously increasing the value of q such that the product qa stays finite yields an idealized dipole

$a \to 0$) performed such that the value of the product is kept fixed,

$$q \to \infty\,, \quad a \to 0\,, \quad \text{with} \quad qa =: d \quad \text{fixed,}$$

one obtains an expression which is exact in this limit, viz.

$$\Phi_{\text{dipole}}(\boldsymbol{x}) = \frac{d}{r^2}\cos\theta\,. \tag{1.88a}$$

It is not difficult to generalize this expression to a situation where the dipole is still located in the origin but does not point along the 3-axis. Let $\boldsymbol{a}$ be the vector which points from the negative to the positive charge (in the figure this is $\boldsymbol{a} = a\hat{\boldsymbol{e}}_3$) and let $\boldsymbol{d} = \lim(q\boldsymbol{a})$ as defined above for $q \to \infty$ and $|\boldsymbol{a}| \to 0$. Then we have $d\cos\theta = \boldsymbol{d}\cdot\boldsymbol{x}/r$, and (1.88a) takes the form

$$\Phi_{\text{dipole}}(\boldsymbol{x}) = \frac{\boldsymbol{d}\cdot\boldsymbol{x}}{r^3} = -\boldsymbol{d}\cdot\nabla\left(\frac{1}{|\boldsymbol{x}|}\right)\,. \tag{1.88b}$$

When the dipole's position is $\boldsymbol{x}' \neq \boldsymbol{0}$ and is not the origin, the potential (1.88b) reads

$$\Phi_{\text{dipole}}(\boldsymbol{x}) = \frac{\boldsymbol{d}\cdot(\boldsymbol{x}-\boldsymbol{x}')}{|\boldsymbol{x}-\boldsymbol{x}'|^3} = -\boldsymbol{d}\cdot\nabla_x\left(\frac{1}{|\boldsymbol{x}-\boldsymbol{x}'|}\right)\,. \tag{1.88c}$$

An electric *dipole layer* is another idealization which consists in superposing two charged surfaces, F_1 and F_2, on top of each other, as sketched in Fig. 1.13. The surface densities are chosen such that $\eta_1\,\mathrm{d}\sigma_1 = -\eta_2\,\mathrm{d}\sigma_2$ and, when the two surfaces approach each other, the surface charge density η is inversely proportional to their distance $a(\boldsymbol{x})$ such that the product $\eta(\boldsymbol{x})a(\boldsymbol{x})$ stays finite when $a \to 0$. Qualitatively speaking, this system consists locally of very many small dipoles. There emerges a position-dependent dipole moment of the double layer which is aligned along the positive normal of the surfaces:

$$\boldsymbol{D}(\boldsymbol{x}) = D(\boldsymbol{x})\cdot\boldsymbol{n} \quad \text{with} \quad D(\boldsymbol{x}) = \lim_{a\to 0,\eta\to\infty}(\eta(\boldsymbol{x})a(\boldsymbol{x}))\,.$$

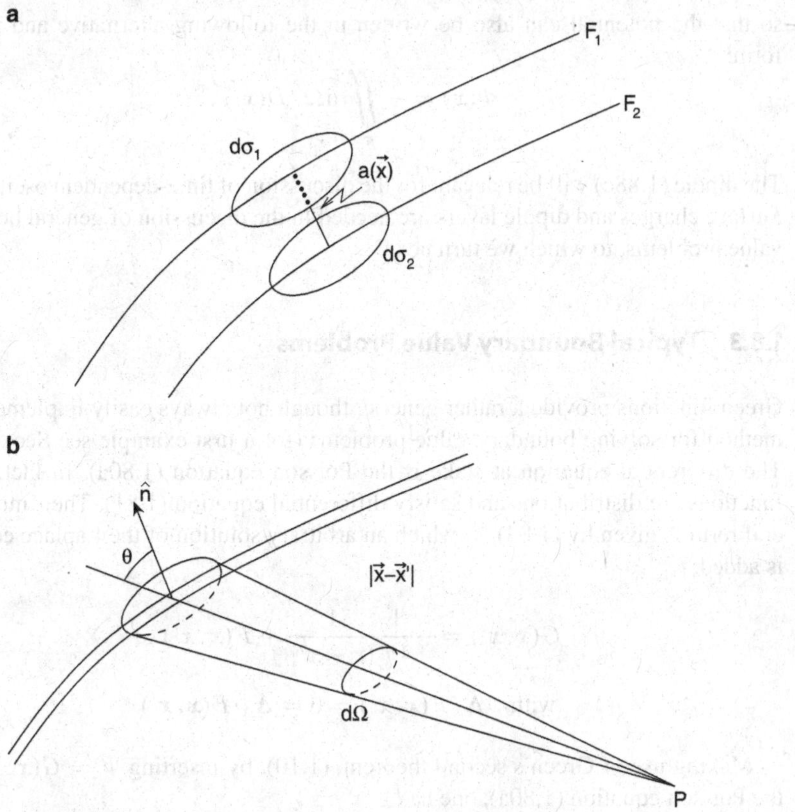

Fig. 1.13 **a** A dipole layer which is composed of equal and opposite surface charges on surface elements facing each other is generated if one lets the distance of the surfaces go to zero while sending the surface charge density to infinity. **b** At a test point P outside the dipole layer, the potential can be expressed in terms of the solid angle from P, cf. (1.89b)

The potential created by an idealized dipole layer can easily be inferred from the results for the point dipole. It reads

$$\Phi(\boldsymbol{x}) = -\iint\limits_F \mathrm{d}\sigma'\, \boldsymbol{D}(\boldsymbol{x}') \cdot \nabla_x \left(\frac{1}{|\boldsymbol{x}-\boldsymbol{x}'|}\right) \tag{1.89a}$$

$$= \iint\limits_F \mathrm{d}\sigma'\, \boldsymbol{D}(\boldsymbol{x}') \cdot \nabla_{x'} \left(\frac{1}{|\boldsymbol{x}-\boldsymbol{x}'|}\right) .$$

At a test point P far from the surface F one observes the surface element $\mathrm{d}\sigma$ under the solid angle $\mathrm{d}\Omega$, and one has

$$\hat{\boldsymbol{n}} \cdot \nabla_x \left(\frac{1}{|\boldsymbol{x}-\boldsymbol{x}'|}\right) = \frac{\cos\theta\, \mathrm{d}\sigma}{|\boldsymbol{x}-\boldsymbol{x}'|^2} = \mathrm{d}\Omega ,$$

so that the potential can also be written in the following alternative and simpler form:

$$\Phi(\boldsymbol{x}) = - \iint\limits_{F} \mathrm{d}\Omega' \, D(\boldsymbol{x}') \, . \tag{1.89b}$$

The dipole (1.88c) will be relevant for the discussion of time-dependent oscillations. Surface charges and dipole layers are needed in the discussion of general boundary value problems, to which we turn next.

1.8.3 Typical Boundary Value Problems

Green functions provide a rather general, though not always easily implementable, method for solving boundary value problems (for a first example see Sect. 1.8.1). The differential equation at stake is the Poisson equation (1.80a). In fact, Green functions are distributions and satisfy differential equation (1.81). Their most general form is given by (1.83), to which an arbitrary solution of the Laplace equation is added:

$$G(\boldsymbol{x}, \boldsymbol{x}') = -\frac{1}{4\pi} \frac{1}{|\boldsymbol{x} - \boldsymbol{x}'|} + F(\boldsymbol{x}, \boldsymbol{x}') \, , \tag{1.90}$$

$$\text{with} \quad \boldsymbol{\Delta}_x F(\boldsymbol{x}, \boldsymbol{x}') = 0 = \boldsymbol{\Delta}_{x'} F(\boldsymbol{x}, \boldsymbol{x}') \, .$$

Making use of Green's second theorem, (1.10), by inserting $\Psi = G(\boldsymbol{x}, \boldsymbol{x}')$ and the Poisson equation (1.80a), one has

$$\begin{aligned} \Phi(\boldsymbol{x}) = & -4\pi \iiint\limits_{V(F)} \mathrm{d}^3 x' \, G(\boldsymbol{x}, \boldsymbol{x}') \varrho(\boldsymbol{x}') \\ & + \iint\limits_{F} \mathrm{d}\sigma' \left(\Phi(\boldsymbol{x}') \frac{\partial}{\partial \hat{\boldsymbol{n}}'} G(\boldsymbol{x}, \boldsymbol{x}') - G(\boldsymbol{x}, \boldsymbol{x}') \frac{\partial}{\partial \hat{\boldsymbol{n}}'} \Phi(\boldsymbol{x}') \right) , \end{aligned} \tag{1.91}$$

whenever $\boldsymbol{x}$ lies inside volume $V(F)$, which is defined by the closed surface F. If, in turn, $\boldsymbol{x}$ lies outside this volume, then the left-hand side is equal to zero.

To obtain a physical interpretation of the additional terms on the right-hand side of (1.91), it is appropriate to choose $F(\boldsymbol{x}, \boldsymbol{x}') = 0$ in a first step. One then finds

$$\begin{aligned} \left. \begin{array}{ll} \boldsymbol{x} \in V(F): & \Phi(\boldsymbol{x}) \\ \boldsymbol{x} \notin V(F): & 0 \end{array} \right\} = & \iiint\limits_{V(F)} \mathrm{d}^3 x' \, \frac{\varrho(\boldsymbol{x})}{|\boldsymbol{x} - \boldsymbol{x}'|} \\ & + \frac{1}{4\pi} \iint\limits_{F} \mathrm{d}\sigma' \left(-\Phi(\boldsymbol{x}') \frac{\partial}{\partial \hat{\boldsymbol{n}}'} \frac{1}{|\boldsymbol{x} - \boldsymbol{x}'|} + \frac{1}{|\boldsymbol{x} - \boldsymbol{x}'|} \frac{\partial}{\partial \hat{\boldsymbol{n}}'} \Phi(\boldsymbol{x}') \right) . \end{aligned} \tag{1.92a}$$

The interpretation of the two terms in the surface integral is as follows. Upon comparison with (1.89a) the first term is seen to be the potential of a dipole layer whose dipole density is given by

$$D(\boldsymbol{x}) = -\frac{1}{4\pi}\,\Phi(\boldsymbol{x})|_F \,. \tag{1.92b}$$

The second term is the potential of a surface charge density which sits on surface F and is described by the function

$$\eta(\boldsymbol{x}) = \frac{1}{4\pi}\,\frac{\partial\Phi}{\partial\hat{\boldsymbol{n}}}\bigg|_F \,. \tag{1.92c}$$

Finally, one may study the limiting case whereby the closed surface is shifted to infinity. If the directional derivatives of the potential, for $\boldsymbol{x}' \to \infty$, decrease faster than the inverse distance function, then the surface integral does not contribute in this limit and one recovers the familiar formula

$$\Phi(\boldsymbol{x}) = \iiint\limits_{V(F)} \mathrm{d}^3x' \,\frac{\varrho(\boldsymbol{x})}{|\boldsymbol{x}-\boldsymbol{x}'|} \,.$$

In discussing boundary value problems, i.e. problems where certain potentials or fields are given, it seems natural to ask which kind of data determine the solution of an electrostatic problem in a unique manner. For example, one will realize that it is not possible to fix both the potential $\Phi(\boldsymbol{x})|_F$ and its normal derivative $\partial\Phi/\partial\hat{\boldsymbol{n}}|_F$ on a given closed surface. Admissible, i.e. well-posed, boundary value problems are as follows:

(a) On some closed surfaces, the values of the potentials are given. An example is a set-up of conductors with known constant potentials. A specification of this kind is called a *Dirichlet boundary condition.*

(b) On certain surfaces supporting known surface charges, the normal component of the electric field is specified. A problem of this kind is called a *Neumann boundary condition.*

These conditions, either (a) *or* (b), lead to a unique solution of the given problem. This can be shown by using the integral theorems as follows. Suppose either set-up (a) *or* set-up (b) admitted two solutions, $\Phi_1(\boldsymbol{x}) \neq \Phi_2(\boldsymbol{x})$. Then define $U(\boldsymbol{x}) := \Phi_2(\boldsymbol{x}) - \Phi_1(\boldsymbol{x})$. Inside the volume defined by closed surface F, the Laplace equation $\boldsymbol{\Delta} U(\boldsymbol{x}) = 0$ applies, whereas on the surface F, which need not be singly connected, we must specify

$$\text{in case (a):}\quad U(\boldsymbol{x})|_F = 0 \,,$$

$$\text{in case (b):}\quad \frac{\partial U(\boldsymbol{x})}{\partial\hat{\boldsymbol{n}}}\bigg|_F = 0 \,.$$

Applying now Green's first theorem, (1.9), with $\Phi = \Psi = U$, one obtains

$$\iiint\limits_{V(F)} \mathrm{d}^3x \left(U\Delta U + (\nabla U)^2\right) = \iint\limits_F \mathrm{d}\sigma\, U\frac{\partial U}{\partial \hat{\boldsymbol{n}}} .$$

(Recall that $\partial U/\partial \hat{\boldsymbol{n}} = \hat{\boldsymbol{n}} \cdot \nabla U$ is the directional derivative along the normal to the surface. Up to a sign this is the normal component of the electric field.) The two terms on the left-hand side, by partial integration, give the same contribution. On the right-hand side, either U or its normal derivative vanishes. One concludes that in either case,

$$\iiint\limits_{V(F)} \mathrm{d}^3x\, (\nabla U)^2 = 0 ,$$

and, hence, that $\nabla U \equiv 0$ or $U(\boldsymbol{x}) = 0$, respectively, for all $\boldsymbol{x} \in V(F)$. In the case of the Dirichlet condition, this means that $U(\boldsymbol{x}) = 0$; in the case of the Neumann condition, it means that U and the solution $\Phi(\boldsymbol{x})$ are determined up to an additive constant. Such a constant is physically irrelevant. This result shows, on the one hand, that in either class of problems the solution, if it exists, is unique (possibly up to an additive constant). On the other hand, it shows that a set-up in which both the potentials *and* their normal derivatives were specified on a closed surface would in general be overdetermined.[4]

Formally, but rarely implementable in practice, one can choose the solution $F(\boldsymbol{x}, \boldsymbol{x}')$ of the Laplace equation in a Dirichlet problem such that the Green function vanishes for all $\boldsymbol{x}'$ on the surface:

$$G_{\mathrm{D}}(\boldsymbol{x}, \boldsymbol{x}') = 0 \quad \text{for all } \boldsymbol{x}' \in F .$$

Example 1.7 Dirichlet condition on the sphere

Let the closed surface be a sphere with radius R about the origin, $F = S^2_R \subset \mathbb{R}^3$. It is a matter of a simple geometric construction (cf. Exercise 1.14) to find an additive term in (1.90) which guarantees the vanishing of $G_{\mathrm{D}}(\boldsymbol{x}, \boldsymbol{x}')$ on the surface of the sphere. The result is found to be

$$G_{\mathrm{D}}(\boldsymbol{x}, \boldsymbol{x}') = -\frac{1}{4\pi}\frac{1}{|\boldsymbol{x} - \boldsymbol{x}'|} + \frac{R|\boldsymbol{x}'|}{4\pi}\frac{1}{\left||\boldsymbol{x}'|^2\boldsymbol{x} - R^2\boldsymbol{x}'\right|} .$$

The required property $G_{\mathrm{D}}(\boldsymbol{x}, \boldsymbol{x}') = 0$ for all $|\boldsymbol{x}'| = R$ is easily verified.

Having found this solution, we obtain the representation

$$\Phi(\boldsymbol{x}) = -4\pi \iiint\limits_{V(F)} \mathrm{d}^3x'\, G_{\mathrm{D}}(\boldsymbol{x}, \boldsymbol{x}')\varrho(\boldsymbol{x}') + \iint\limits_F \mathrm{d}\sigma'\, \Phi(\boldsymbol{x}')\frac{\partial}{\partial \hat{\boldsymbol{n}}'} G_{\mathrm{D}}(\boldsymbol{x}, \boldsymbol{x}') \quad (1.93)$$

of the potential. This solution for the potential is unique.

[4] The closed surface F may be localized. However, with suitable care it may be continued entirely or partially to infinity.

In the case of a Neumann problem, one proceeds as follows. Like every Green function, the distribution which satisfies the Neumann boundary condition obeys the differential equation (1.81):

$$\Delta G_{\mathrm{N}}(\boldsymbol{x}, \boldsymbol{x}') = \delta(\boldsymbol{x} - \boldsymbol{x}') \, .$$

Let the piecewise smooth surface F be given and let $V(F)$ be the volume which it defines. Inserting the gradient field $\boldsymbol{V} = \boldsymbol{\nabla} G_{\mathrm{N}}$ into Gauss' theorem, (1.6), and recalling the definition

$$\frac{\partial G_{\mathrm{N}}}{\partial \hat{\boldsymbol{n}}} = (\boldsymbol{\nabla} G_{\mathrm{N}}) \cdot \hat{\boldsymbol{n}} \, ,$$

Gauss' formula (1.6) implies

$$1 = \iint_F \mathrm{d}\sigma \frac{\partial G_{\mathrm{N}}}{\partial \hat{\boldsymbol{n}}} \, .$$

Therefore, one can choose the normal derivative of G_{N} such that it is constant on the surface F and equal to $1/F$:

$$\frac{\partial G_{\mathrm{N}}}{\partial \hat{\boldsymbol{n}}} = \frac{1}{F} \, .$$

Then, from the general formula (1.91) one obtains the representation

$$\Phi(\boldsymbol{x}) = -4\pi \iiint_{V(F)} \mathrm{d}^3 x' \, G_{\mathrm{N}}(\boldsymbol{x}, \boldsymbol{x}') \varrho(\boldsymbol{x}') + \langle \Phi \rangle_F - \iint_F \mathrm{d}\sigma' \, G_{\mathrm{N}}(\boldsymbol{x}, \boldsymbol{x}') \frac{\partial}{\partial \hat{\boldsymbol{n}}'} \Phi(\boldsymbol{x}') \, , \qquad (1.94)$$

which contains the average of the potential over the surface F,

$$\langle \Phi \rangle_F = \frac{1}{F} \iint_F \mathrm{d}\sigma \, \Phi(\boldsymbol{x}) \, .$$

This average is zero if one adds a closed surface placed at spatial infinity. Of course, the implicit assumption is that all physical quantities are localized and, hence, vanish at infinity.

1.8.4 Multipole Expansion of Potentials

There is an important, practicable and very useful technique which makes use of expansions in terms of *spherical harmonics*. Before developing the relevant formulae and techniques, let us summarize the essential ideas of this method.

The aim of the method is to find a set of fundamental solutions of the Laplace equation (1.80b) which are *complete*, in a sense yet to be specified. Every physically relevant solution which cannot be formulated directly and analytically should nevertheless be written as a series in these fundamental solutions. This program consists of the following steps:

1. One solves the Laplace equation (1.80b) for a specific ansatz in spherical polar coordinates (r, θ, ϕ) such that the dependences on r, on θ and on ϕ are factorized:

$$\Phi_{\text{ansatz}}(\boldsymbol{x}) \equiv \frac{1}{r} R(r) P(\theta) f(\phi) \,. \tag{1.95}$$

   Although this is not the most general solution, it has the advantage that the three types of functions $R(r)$, $Y(\theta)$ and $f(\phi)$ are known explicitly.
2. One then discovers that the set of all functions $Y(\theta) f(\phi)$ which are regular in the interval $(0 \leqslant \theta \leqslant \pi, 0 \leqslant \phi \leqslant 2\pi)$, are *orthogonal*, in a generalized sense, and *complete*. Therefore, these product solutions span a basis for functions which are regular on S^2, the unit sphere in $\mathbb{R}^3$ which is described by coordinates θ and ϕ.
3. More general, and in particular nonfactorizing, solutions of the Laplace equation are expanded in terms of this base system. The completeness of this basis is essential for this procedure. The method of expansion in terms of spherical harmonics is also very useful in finding solutions of the Poisson equation if it is combined with the technique of Green functions.

When expressed in terms of spherical polar coordinates the Laplace operator reads

$$\begin{aligned}\boldsymbol{\Delta}\Phi(r,\theta,\phi) = &\frac{1}{r^2}\frac{\partial}{\partial r}\left(r^2\frac{\partial \Phi}{\partial r}\right) + \frac{1}{r^2 \sin\theta}\frac{\partial}{\partial \theta}\left(\sin\theta \frac{\partial \Phi}{\partial \theta}\right) \\ &+ \frac{1}{r^2 \sin^2\theta}\frac{\partial^2 \Phi}{\partial \phi^2} \,.\end{aligned} \tag{1.96}$$

Upon inserting the factorization ansatz (1.95) one obtains separate *ordinary* differential equations for the radial part and for the parts depending on θ and ϕ only. Without developing this method in detail we write down the differential equations for the functions $P(\theta)$ and $f(\phi)$ of the angular variables θ and ϕ, respectively, as well as the conditions on their parameters:

$$\frac{\mathrm{d}^2}{\mathrm{d}\phi^2} f(\phi) + m^2 f(\phi) = 0 \,, \quad m = 0, 1, 2, \dots \,,$$

$$\frac{1}{\sin\theta}\frac{\mathrm{d}}{\mathrm{d}\theta}\left(\sin\theta \frac{\mathrm{d}P(\theta)}{\mathrm{d}\theta}\right) + \left(\ell(\ell+1) - \frac{m^2}{\sin^2\theta}\right) P(\theta) = 0 \,,$$

$$\ell = 0, 1, 2, \dots \,.$$

The solutions of these differential equations, which are regular on the unit sphere

$$Y(\theta, \phi) = P(\theta) f(\phi) ,$$

are called *spherical harmonics*. The $f(\phi)$ solutions are labelled by the real number m, the $P(\theta)$ solutions by the reals ℓ and m:

$$Y_{\ell m}(\theta, \phi) = N_{\ell m} P_\ell^m(\theta) f_m(\phi) ,$$

The requirement that $f_m(\phi)$ should be single valued, which is to say that a complete rotation of the frame of reference acts as the identity, $f_m(\phi + 2\pi) = f_m(\phi)$, implies that m must be a (positive or negative) *integer*. The differential equation for $P_\ell^m(\theta)$ has solutions which are regular in the whole interval $\theta \in [0, \pi]$ and $\cos\theta \in [-1, 1]$ only if $\ell \in \mathbb{N}_0$ and if the modulus of m does not exceed ℓ.

Spherical harmonics belong to the realm of *special functions*, which one should learn to make use of, in physics and in many other disciplines, like trigonometric functions, logarithms, the exponential function and many other elementary functions of analysis and function theory. Therefore, I restrict this presentation to their definition and their essential properties, as well as to a few examples.[5]

Spherical harmonics on the unit sphere S^2

Spherical harmonics $Y_{\ell m}(\theta, \phi)$ are products of exponential functions in the azimuthal variable ϕ and of Legendre functions of the first kind $P_\ell^m(\theta)$:

$$Y_{\ell m}(\theta, \phi) = \sqrt{\frac{2\ell + 1}{4\pi}} \sqrt{\frac{(\ell - m)!}{(\ell + m)!}} \, P_\ell^m(\cos\theta) \, e^{im\phi} . \tag{1.97a}$$

The indices ℓ and m take their values in the sets

$$\ell \in \mathbb{N}_0 , \quad m \in [-\ell, +\ell] , \quad \text{i.e. more explicitly,}$$
$$\ell = 0, 1, 2, \ldots , \quad \text{and} \quad m = -\ell, -\ell + 1, \ldots, +\ell . \tag{1.97b}$$

The Legendre functions of the first kind depend on $\cos\theta =: z$ only and are generated by differentiation of Legendre polynomials:

$$P_\ell^m(z) = (-)^m (1 - z^2)^{\frac{m}{2}} \frac{d^m}{dz^m} P_\ell(z) , \quad (z \equiv \cos\theta) , \tag{1.97c}$$

where, in turn, the Legendre polynomials may be defined, for example, by the formula of Rodrigues:

$$P_\ell(z) = \frac{1}{2^\ell \ell!} \frac{d^\ell}{dz^\ell} (z^2 - 1)^\ell . \tag{1.97d}$$

[5] The differential equations for spherical harmonics derive from (1.96). For details, consult, for example, [QP], Sect. 1.9.1, in the context of orbital angular momentum in quantum mechanics. A proof of formula (1.97a) can also be found there.

The first five Legendre polynomials read explicitly as follows:

$$P_0(z) = 1\,, \qquad P_1(z) = z\,,$$
$$P_2(z) = \frac{1}{2}\left(3z^2 - 1\right)\,, \qquad P_3(z) = \frac{1}{2}\left(5z^3 - 3z\right)\,,$$
$$P_4(z) = \frac{1}{8}\left(35z^4 - 30z^2 + 3\right)\,. \tag{1.97e}$$

The spherical harmonics with $\ell = 0, 1, 2, 3$ are given by

$$Y_{0,0} = \frac{1}{\sqrt{4\pi}}\,,$$
$$Y_{1,0} = \sqrt{\frac{3}{4\pi}}\cos\theta\,, \qquad Y_{1,\pm1} = \mp\sqrt{\frac{3}{8\pi}}\sin\theta\,\mathrm{e}^{\pm i\phi}\,,$$
$$Y_{2,0} = \sqrt{\frac{5}{16\pi}}\left(3\cos^2\theta - 1\right)\,,$$
$$Y_{2,\pm1} = \mp\sqrt{\frac{15}{8\pi}}\cos\theta\sin\theta\,\mathrm{e}^{\pm i\phi}\,, \qquad Y_{2,\pm2} = \sqrt{\frac{15}{32\pi}}\sin^2\theta\,\mathrm{e}^{\pm 2i\phi}\,, \tag{1.97f}$$

$$Y_{3,0} = \sqrt{\frac{7}{16\pi}}\left(5\cos^3\theta - 3\cos\theta\right)\,,$$
$$Y_{3,\pm1} = \mp\sqrt{\frac{21}{64\pi}}\,,\left(4\cos^2\sin\theta - \sin^3\theta\right)\mathrm{e}^{\pm i\phi}\,,$$
$$Y_{3,\pm2} = \sqrt{\frac{105}{32\pi}}\cos\theta\sin^2\theta\,\mathrm{e}^{\pm 2i\phi}\,, \qquad Y_{3,\pm3} = \sqrt{\frac{35}{64\pi}}\sin^3\theta\,\mathrm{e}^{\pm 3i\phi}\,.$$

Some of their important properties concern the relations between $Y_{\ell m}(\theta,\phi)$ and its complex conjugate $Y^*_{\ell m}(\theta,\phi)$, as well as between $Y_{\ell m}(\theta,\phi)$ at the point (θ,ϕ), the S^2 and the same function at the antipode $(\pi-\theta,\phi+\pi)$ of this point. [The reader may wish to check these relations by means of the general formula (1.97a) and the examples of (1.97f)]:

$$Y^*_{\ell m}(\theta,\phi) = (-)^m Y_{\ell,-m}(\theta,\phi)\,, \tag{1.97g}$$
$$Y_{\ell m}(\pi-\theta,\pi+\phi) = (-)^\ell Y_{\ell m}(\theta,\phi)\,. \tag{1.97h}$$

As the points (θ,ϕ) and $(\pi-\theta,\phi+\pi)$ are related by reflection about the origin, relation (1.97h) says that spherical harmonics with *even* ℓ are *even* under space reflection, whereas spherical harmonics with *odd* ℓ change sign under space reflection.

The spherical harmonics are orthogonal in a generalized sense. They satisfy the following *orthogonality relation*:

$$\int_{S^2} \mathrm{d}\Omega \; Y^*_{\ell' m'}(\theta,\phi) Y_{\ell m}(\theta,\phi) = \delta_{\ell' \ell} \delta_{m' m} \,. \tag{1.98a}$$

The volume element is $\mathrm{d}\Omega = \mathrm{d}\phi \, \sin\theta \, \mathrm{d}\theta$, and the integration extends over the entire sphere. Furthermore, the spherical harmonics form a complete set of base functions on S^2. This means that every function $f(\theta,\phi)$ which is continuous on S^2 can be expanded in terms of this basis:

$$f(\theta,\phi) = \sum_{\ell=0}^{\infty} \sum_{m=-\ell}^{+\ell} Y_{\ell m}(\theta,\phi) a_{\ell m}, \quad \text{with} \tag{1.98b}$$

$$a_{\ell m} = \int_{S^2} \mathrm{d}\Omega \; Y^*_{\ell m}(\theta,\phi) f(\theta,\phi) \,. \tag{1.98c}$$

The second of these may also be written as a *completeness relation* which in the present case reads as follows:

$$\sum_{\ell=0}^{\infty} \sum_{m=-\ell}^{+\ell} Y_{\ell m}(\theta,\phi) Y^*_{\ell m}(\theta',\phi') = \delta(\phi-\phi')\delta(\cos\theta - \cos\theta') \,. \tag{1.98d}$$

Having clarified the angular dependence of the factorized ansatz (1.95), one needs to identify the differential equation for the radial function $R(r)$. For this purpose, it is useful to extract a factor $1/r$ from the radial function. Indeed, the nested differentiation by r in (1.96) then turns into the second derivative by r multiplied by $1/r$:

$$\frac{1}{r^2}\frac{\mathrm{d}}{\mathrm{d}r}\left(r^2 \frac{\mathrm{d}}{\mathrm{d}r}\left(\frac{1}{r}R(r)\right)\right) = \frac{1}{r}\frac{\mathrm{d}^2 R(r)}{\mathrm{d}r^2} \,.$$

Making use of the differential equations for $f_m(\phi)$ and for $P_\ell^m(\theta)$, equation (1.96) reduces to the ordinary differential equation

$$\frac{\mathrm{d}^2 R(r)}{\mathrm{d}r^2} - \frac{\ell(\ell+1)}{r^2} R(r) = 0 \tag{1.99a}$$

whose general solution is easily seen to be

$$R(r) = c^{(1)} \, r^{\ell+1} + c^{(2)} \, \frac{1}{r^\ell} \,. \tag{1.99b}$$

The first term is the one to be chosen if $R(r)$ must be regular at $r = 0$. The second term is relevant if $R(r)$ is required to be regular at infinity.

Electrostatic Problems with Axial Symmetry

In a problem which exhibits axial symmetry about some direction in space, it is useful to choose the 3-axis along that direction. As the potential does not depend on ϕ, the solutions of the Laplace equation can only depend on r and θ. When expanding a solution of this kind, $\Phi(r,\theta)$, in terms of spherical harmonics, only base functions with $m = 0$ can contribute. In this case, Legendre functions (1.97c) reduce to Legendre polynomials (1.97d) so that the potential may be expanded directly in terms of the latter, viz.

$$\Phi(r,\theta) = \sum_{\ell=0}^{\infty} \left(c_\ell^{(1)} r^\ell + c_\ell^{(2)} \frac{1}{r^{\ell+1}} \right) P_\ell(z) , \quad (z = \cos\theta) . \tag{1.100}$$

The expansion coefficients $c_\ell^{(1)}$ and $c_\ell^{(2)}$ are determined by taking account of the boundary conditions that are given. Note, however, that although the Legendre polynomials are orthogonal, they are not normalized to 1. Instead, by definitions (1.97a) and (1.97c), and by the orthogonality relation (1.98a), one has

$$\int_0^\pi \sin\theta \, d\theta \; P_\ell^2(\cos\theta) = \int_{-1}^{+1} dz \; P_\ell^2(z) = \frac{2}{2\ell+1} .$$

The normalization of the Legendre polynomials is such that at $\theta = 0$, i.e. at $z = 1$, they have the value 1 for all ℓ, $P_\ell(1) = 1$. This is a useful condition when they are being generated from a generating function:

$$\frac{1}{\sqrt{1 - 2zt + t^2}} = \sum_{\ell=0}^{\infty} t^\ell P_\ell(z) , \tag{1.101}$$

a relation which for $z = 1$ goes over into the well-known geometric series.

As an example of expansion (1.100) we study the following application: Solution (1.100) is uniquely determined. Therefore, it is sufficient to calculate the coefficients for a fixed value of the argument z such as e. g. $z = 1$. We set out to determine the inverse distance function $|\boldsymbol{x} - \boldsymbol{x}'|$, where $\boldsymbol{x}$ and $\boldsymbol{x}'$ are two vectors

Fig. 1.14 The difference between test point $\boldsymbol{x}$ and source point $\boldsymbol{x}'$ of a concrete electrostatic problem, expressed in terms of the two vectors and the angle α sustained by them

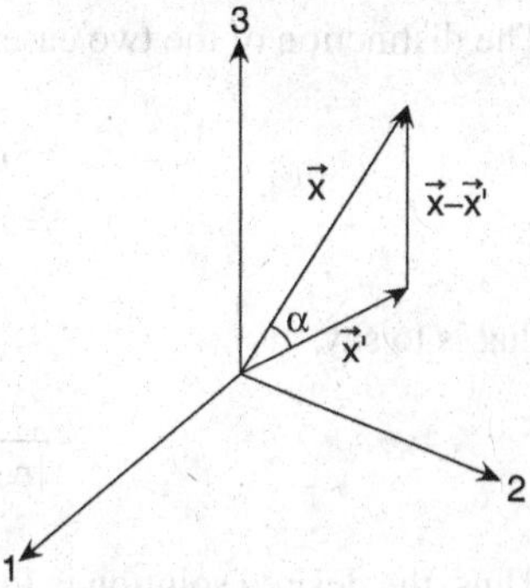

defining the angle α (cf. Fig. 1.14). Choosing the 3-axis along $\boldsymbol{x}'$, one has

$$\frac{1}{|\boldsymbol{x}-\boldsymbol{x}'|}=\frac{1}{\sqrt{r^2+r'^2-2rr'\cos\alpha}} .$$

To determine the expansion

$$\frac{1}{|\boldsymbol{x}-\boldsymbol{x}'|}=\sum_{\ell=0}^{\infty}\left(c_\ell^{(1)}\,r^\ell+c_\ell^{(2)}\frac{1}{r^{\ell+1}}\right)P_\ell(z)\,, \qquad (z=\cos\theta)\,,$$

consider the case $\cos\alpha = 1$, i.e. choose vector $\boldsymbol{x}$ along the 3-direction as well, without loss of generality. The distance function, then, is the absolute value of the difference of the two radial variables, and the expansion coefficients are deduced from the simpler equation

$$\frac{1}{|r-r'|}=\sum_{\ell=0}^{\infty}\left(c_\ell^{(1)}\,r^\ell+c_\ell^{(2)}\frac{1}{r^{\ell+1}}\right) .$$

The well-known series expansion $1/(1-x) = 1 + x + x^2 \cdots$ converges only if $|x| < 1$. Therefore, one has

$$\text{for } r' > r : \frac{1}{|r-r'|}=\frac{1}{r'}\sum_{\ell=0}^{\infty}\frac{r^\ell}{r'^\ell}, \qquad \text{and hence}$$

$$c_\ell^{(1)}=\frac{1}{r'^{\ell+1}}, \quad c_\ell^{(2)}=0\,;$$

$$\text{for } r' < r : \frac{1}{|r-r'|}=\frac{1}{r}\sum_{\ell=0}^{\infty}\frac{r'^\ell}{r^\ell}, \qquad \text{and hence}$$

$$c_\ell^{(1)}=0\,, \quad c_\ell^{(2)}=r'^\ell\,.$$

The distinction of the two cases can be written in a more compact notation:

$$r_< = r\,, \quad r_> = r' \quad \text{for} \quad r < r'\,,$$
$$r_> = r\,, \quad r_< = r' \quad \text{for} \quad r > r'\,,$$

that is to say,

$$\frac{1}{|r - r'|} = \frac{1}{r_>} \sum_{\ell=0}^{\infty} \left(\frac{r_<}{r_>} \right)^{\ell} .$$

Thus, the desired solution is found to be

$$\frac{1}{|\boldsymbol{x} - \boldsymbol{x}'|} = \sum_{\ell=0}^{\infty} \frac{r_<^{\ell}}{r_>^{\ell+1}} P_\ell(\cos\alpha)\,. \qquad (1.102)$$

As a remark, we note that this series may also be obtained from the generating function (1.101). Conversely, the latter may be obtained from the result (1.102) above.

Example 1.8 Potential of a spherically symmetric charge distribution

Choose the 3-direction along the direction of $\boldsymbol{x}$ so that the angle α becomes the polar angle of $\boldsymbol{x}'$. Inserting the expansion (1.102) and integrating over $\mathrm{d}^3x' = r'^2\,\mathrm{d}r'\,\mathrm{d}\Omega'$, the Legendre polynomials which have $\ell \neq 0$ give vanishing contributions:

$$\iint_{S^2} \mathrm{d}\Omega' P_\ell(\cos\theta) = 2\pi \int_{-1}^{+1} \mathrm{d}z\; P_\ell(z) P_0(z) = 0 \quad \text{for all} \quad \ell \neq 0\,.$$

For $\ell = 0$ there remains the integration over the radial variable keeping in mind the distinction between $r' < r$ and $r' > r$:

$$\Phi(\boldsymbol{x}) = \iiint \mathrm{d}^3x' \frac{\varrho(r)}{|\boldsymbol{x} - \boldsymbol{x}'|} = 4\pi \int_0^{\infty} \mathrm{d}r' r'^2 \frac{1}{r_>} \varrho(r')$$
$$= 4\pi \left\{ \frac{1}{r} \int_0^{r} \mathrm{d}r' r'^2\, \varrho(r') + \int_r^{\infty} \mathrm{d}r' r'\, \varrho(r') \right\} .$$

Note that we obtained this formula in (1.85) by direct integration of the Poisson equation.

More General Configurations Without Axial Symmetry

If there is no axial symmetry about some distinct direction in $\mathbb{R}^3$, then formula (1.100) must be replaced by the more general ansatz

$$\Phi(r,\theta,\phi) = \sum_{\ell=0}^{\infty} \sum_{m=-\ell}^{+\ell} \left(c_{\ell m}^{(1)} r^{\ell} + c_{\ell m}^{(2)} \frac{1}{r^{\ell+1}} \right) Y_{\ell m}(\theta,\phi) \,. \tag{1.103}$$

By making use of the important *addition theorem* for spherical harmonics

$$P_\ell(\cos\alpha) = \frac{4\pi}{2\ell+1} \sum_{m=-\ell}^{+\ell} Y_{\ell m}^*(\theta',\phi') Y_{\ell m}(\theta,\phi) \,, \tag{1.104}$$

expansion (1.102) of the inverse distance function is written in a more general form:

$$\frac{1}{|\boldsymbol{x}-\boldsymbol{x}'|} = \sum_{\ell=0}^{\infty} \frac{4\pi}{2\ell+1} \frac{r_<^{\ell}}{r_>^{\ell+1}} \sum_{m=-\ell}^{+\ell} Y_{\ell m}^*(\theta',\phi') Y_{\ell m}(\theta,\phi) \,. \tag{1.105}$$

In this formula, (θ,ϕ) are the spherical polar coordinates of the unit vector $\hat{\boldsymbol{x}}$, (θ',ϕ') are the coordinates of $\hat{\boldsymbol{x}}'$, and α is the angle between these two vectors.

The proof of the addition theorem (1.104) provides a nice illustration of the technique of expansion in terms of spherical harmonics. For this reason, and although this belongs rather to the subject of a monograph on special functions, the proof is included here.

Proof of the addition theorem (1.104)

The Legendre polynomial $P_\ell(\cos\theta)$ is a function which is regular on S^2. Therefore, it may be expanded in terms of spherical harmonics:

$$P_\ell(\cos\alpha) = \sum_{\ell'=0}^{\infty} \sum_{m=-\ell'}^{m=+\ell'} c_{\ell' m} Y_{\ell' m}(\theta,\phi) \,.$$

The angles α, θ and ϕ are defined as shown in Fig. 1.15, which means that the expansion coefficients $c_{\ell' m}$ will depend on θ' and ϕ'. Obviously, one has the freedom to choose the vector $\boldsymbol{x}'$ in Fig. 1.15 on the cone with opening angle α around $\boldsymbol{x}$. As the left-hand and right-hand sides of this ansatz must obey the same differential equation, one concludes that only terms with $\ell' = \ell$ can contribute. The coefficients are calculated by means of the orthogonality relation (1.98a):

$$\begin{aligned} c_{\ell m}(\theta',\phi') &= \iint \mathrm{d}\Omega \; Y_{\ell m}^*(\theta,\phi) P_\ell(\cos\alpha) \\ &= \sqrt{\frac{4\pi}{2\ell+1}} \iint \mathrm{d}\Omega \; Y_{\ell m}^*(\theta,\phi) Y_{\ell m=0}(\alpha,\beta) \,. \end{aligned}$$

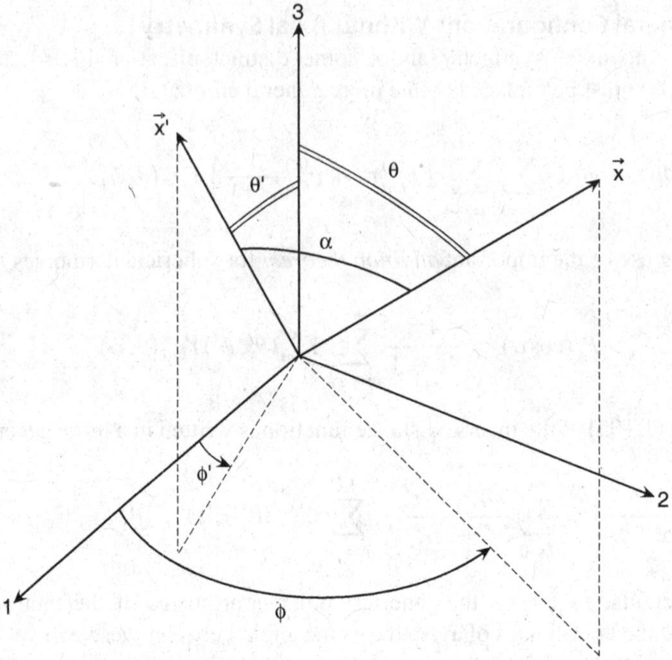

Fig. 1.15 The unit vectors $\hat{\boldsymbol{x}}$ and $\hat{\boldsymbol{x}}'$ define the angle α, whereas (θ, ϕ) are the polar angles of the first, (θ', ϕ') those of the second vector

The second spherical harmonic formally contains a second azimuth β, which, however, is irrelevant because the function $Y_{\ell m=0}$ does not depend on this argument. If we then knew how to expand the function $\sqrt{4\pi/(2\ell+1)}Y^*_{\ell m}(\theta, \phi)$ in terms of the basis $Y_{\ell m=0}(\alpha, \beta)$, we would have found the sought-after formula. We now show that this is indeed possible without any further calculation. Because of the uniqueness of such an expansion

$$Y^*_{\ell m}(\theta(\alpha, \beta), \phi(\alpha, \beta)) = \sum_m b_{\ell m} Y_{\ell m}(\alpha, \beta) ,$$

it is sufficient to consider special cases for which the spherical harmonics on the right-hand side are obvious. Furthermore, one must keep in mind that the expansion assumes the direction $\hat{\boldsymbol{x}}'$ of Fig. 1.15 to be the 3-axis. At $\alpha = 0$ only the term $b_{\ell 0} Y_{\ell 0}(0, \beta) = \sqrt{(2\ell+1)/4\pi} b_{\ell 0}$ survives on the right-hand side. This follows directly from (1.97a) and the property $P_\ell(\cos\alpha = 1) = 1$. This yields

$$b_{\ell 0} = \sqrt{\frac{4\pi}{2\ell+1}}\, Y^*_{\ell m}(\theta(\alpha, \beta), \phi(\alpha, \beta))\big|_{\alpha=0} , \quad \text{or}$$

$$c_{\ell m}(\theta',\phi') = \sqrt{\frac{4\pi}{2\ell+1}}\, b_{\ell 0} = \frac{4\pi}{2\ell+1}\, Y^*_{\ell m}(\theta(\alpha,\beta),\phi(\alpha,\beta))\big|_{\alpha=0} \,.$$

On the other hand, if $\alpha = 0$, then inspection of Fig. 1.15 shows that

$$\theta(\alpha,\beta)|_{\alpha=0} = \theta', \quad \phi(\alpha,\beta)|_{\alpha=0} = \phi' \,.$$

This completes the proof of formula (1.104).

Assume a localized charge distribution $\varrho(\boldsymbol{x}')$ to be given. The potential that it creates at a test point $\boldsymbol{x}$ (this point may lie inside or outside the given density) reads

$$\Phi(\boldsymbol{x}) = \iiint \mathrm{d}^3x' \frac{\varrho(\boldsymbol{x}')}{|\boldsymbol{x}-\boldsymbol{x}'|} \tag{1.106a}$$

$$= \sum_{\ell=0}^{\infty} \frac{4\pi}{2\ell+1} \sum_{m=-\ell}^{m=+\ell} Y_{\ell m}(\hat{\boldsymbol{x}}) \int_0^\infty r'^2\,\mathrm{d}r' \,\frac{r_<^\ell}{r_>^{\ell+1}} \iint \mathrm{d}\Omega'\, Y^*_{\ell m}(\hat{\boldsymbol{x}}')\varrho(\boldsymbol{x}') \,.$$

In this formula, the polar coordinates of $\boldsymbol{x}$ and $\boldsymbol{x}'$ are symbolized by the corresponding unit vectors. The radial integral requires the distinction of cases of smaller or larger radial argument, viz.

$$\int_0^\infty r'^2\,\mathrm{d}r' \,\frac{r_<^\ell}{r_>^{\ell+1}} \cdots = \frac{1}{r^{\ell+1}} \int_0^r r'^2\,\mathrm{d}r'\, r'^\ell \cdots + r^\ell \int_r^\infty r'^2\,\mathrm{d}r' \,\frac{1}{r'^{\ell+1}} \cdots \,. \tag{1.106b}$$

Matters simplify considerably if one considers the exterior region only, i.e. if $r \equiv |\boldsymbol{x}| > r' \equiv |\boldsymbol{x}'|$, because then only the first term in (1.106b) contributes. In this case one has

$$\Phi(\boldsymbol{x})|_{\mathrm{outer}} = \sum_{\ell=0}^{\infty} \frac{4\pi}{2\ell+1} \sum_{m=-\ell}^{m=+\ell} Y_{\ell m}(\hat{\boldsymbol{x}}) \frac{q_{\ell m}}{r^{\ell+1}}, \quad \text{with} \tag{1.106c}$$

$$q_{\ell m} = \int_0^r r'^2\,\mathrm{d}r'\, r'^\ell \iint \mathrm{d}\Omega'\, Y^*_{\ell m}(\hat{\boldsymbol{x}}')\varrho(\boldsymbol{x}') \,. \tag{1.106d}$$

By assumption, the charge distribution is localized, that is to say, one can draw a sphere of radius R about the origin such that $\varrho(\boldsymbol{x}') = 0$ for all $r' > R$. Therefore, the upper limit of the radial integral in (1.106d) may be taken to R, or even to $+\infty$. Formula (1.106c) tells us that the properties of the source which are contained in the *multipole moments* (1.106d), and the functional dependence of the potential on the test point $\boldsymbol{x}$ *factorize* – in contrast to the more general case of formula (1.106a), where the two contributions are folded and intermix. This simplification will be useful and sufficient in many applications. Clearly, one will obtain more information on the spatial structure of the charge density ϱ if it can be probed *outside* as well as in the *inside*.

The multipole moments expressed by (1.106d) have the following properties. Although the charge distribution and the potential are real functions, the moments $q_{\ell m}$ are complex numbers. However, these complex quantities satisfy the symmetry relation

$$q^*_{\ell m} = (-)^m \, q_{\ell -m} \, . \tag{1.107}$$

This is a consequence of the property (1.97g) of spherical harmonics and of the fact that ϱ is a real function. For $\ell = 0$, there is only one multipole moment, the monopole moment, which reads

$$q_{00} = \sqrt{\frac{1}{4\pi}} \iiint \mathrm{d}^3x \, \varrho(\boldsymbol{x}) = \sqrt{\frac{1}{4\pi}} Q \, , \tag{1.108}$$

and where Q is the total charge.[6]

There are three multipole moments for $\ell = 1$, only two of which, by the property of (1.107), need to be calculated:

$$\begin{aligned} q_{11} &= \iiint \mathrm{d}^3x \; rY^*_{11}(\hat{\boldsymbol{x}})\varrho(\boldsymbol{x}) \\ &= -\sqrt{\frac{3}{8\pi}} \iiint \mathrm{d}^3x \; (x^1 - \mathrm{i}x^2)\varrho(\boldsymbol{x}) \equiv -\sqrt{\frac{3}{8\pi}}(d^1 - \mathrm{i}d^2) \, , \end{aligned} \tag{1.109a}$$

$$\begin{aligned} q_{10} &= \iiint \mathrm{d}^3x \; rY_{10}(\hat{\boldsymbol{x}})\varrho(\boldsymbol{x}) \\ &= \sqrt{\frac{3}{4\pi}} \iiint \mathrm{d}^3x \; x^3\varrho(\boldsymbol{x}) \equiv \sqrt{\frac{3}{4\pi}} d^3 \, . \end{aligned} \tag{1.109b}$$

We inserted the spherical harmonics of (1.97f), expressed them in terms of Cartesian components of $\boldsymbol{x}$, and finally introduced the Cartesian components of the dipole moment

$$\boldsymbol{d} = \int \mathrm{d}^3\boldsymbol{x} \; \boldsymbol{x}\varrho(\boldsymbol{x}) \, . \tag{1.109c}$$

As an example, insert moment q_{10} into formula (1.106c):

$$\Phi_{\text{dipole}}(\boldsymbol{x}) = \frac{4\pi}{3} \sqrt{\frac{3}{4\pi}} d^3 \frac{1}{r^2} Y_{10}(\hat{\boldsymbol{x}}) = \frac{1}{r^2} d^3 \cos\theta \, .$$

Here the explicit expression for Y_{10} from (1.97f) was used. This result agrees with formula (1.88b), which was derived in Sect. 1.8.2.

[6] As the contribution from the source and the dependence on the test point factorize completely, we have suppressed the prime on the integration variable $\boldsymbol{x}'$.

In the case $\ell = 2$, there are $2\ell + 1 = 5$ multipole moments, four of which are related via relation (1.107). Hence, it is sufficient to know the following formulae:

$$q_{22} = \iiint \mathrm{d}^3x\, r^2 Y^*_{22}(\hat{\boldsymbol{x}})\varrho(\boldsymbol{x})$$

$$= \sqrt{\frac{15}{32\pi}} \iiint \mathrm{d}^3x\, (x^1 - \mathrm{i}x^2)^2 \varrho(\boldsymbol{x})\,, \tag{1.110a}$$

$$q_{21} = \iiint \mathrm{d}^3x\, r^2 Y^*_{21}(\hat{\boldsymbol{x}})\varrho(\boldsymbol{x})$$

$$= -\sqrt{\frac{15}{8\pi}} \iiint \mathrm{d}^3x\, x^3(x^1 - \mathrm{i}x^2)\varrho(\boldsymbol{x})\,, \tag{1.110b}$$

$$q_{20} = \iiint \mathrm{d}^3x\, r^2 Y_{20}(\hat{\boldsymbol{x}})\varrho(\boldsymbol{x})$$

$$= \sqrt{\frac{5}{16\pi}} \iiint \mathrm{d}^3x\, \left(3(x^3)^2 - r^2\right)\varrho(\boldsymbol{x})\,. \tag{1.110c}$$

At this point of the analysis it is instructive to pause and calculate the same monopole, dipole and quadrupole terms in yet another way. Suppose the test point $\boldsymbol{x}$, which is the point at which the potential or its gradient field is measured, lies far outside the charge density $\varrho(\boldsymbol{x}')$. In this case, the multipole expansion is nothing but a Taylor expansion of

$$\Phi(\boldsymbol{x}) = \iiint \mathrm{d}^3x'\, \frac{\varrho(\boldsymbol{x}')}{|\boldsymbol{x} - \boldsymbol{x}'|}$$

around $\boldsymbol{x}' = \boldsymbol{0}$. We set $|\boldsymbol{x}| =: r$ and expand in the variable $\boldsymbol{x}'$. In doing so, we start from the formulae

$$|\boldsymbol{x} - \boldsymbol{x}'| = \sqrt{\sum_{i=1}^{3}(x'^i - x^i)^2}\,,$$

$$\frac{\partial}{\partial x'^i}|\boldsymbol{x} - \boldsymbol{x}'| = \frac{x'^i - x^i}{|\boldsymbol{x} - \boldsymbol{x}'|}\,,$$

$$\frac{\partial}{\partial x'^i}\frac{1}{|\boldsymbol{x} - \boldsymbol{x}'|} = -\frac{x'^i - x^i}{|\boldsymbol{x} - \boldsymbol{x}'|^3} = +\frac{x^i - x'^i}{|\boldsymbol{x} - \boldsymbol{x}'|^3}\,,$$

which allow us to determine the mixed second derivatives:

$$\frac{\partial^2}{\partial x'^k \partial x'^i}\frac{1}{|\boldsymbol{x} - \boldsymbol{x}'|} = -\frac{1}{|\boldsymbol{x} - \boldsymbol{x}'|^3}\delta^{ik} + \frac{3}{|\boldsymbol{x} - \boldsymbol{x}'|^5}(x^i - x'^i)(x^k - x'^k)\,.$$

These derivatives appear in the Taylor series at $\boldsymbol{x}' = \boldsymbol{0}$, which reads

$$\frac{1}{|\boldsymbol{x}-\boldsymbol{x}'|} \simeq \frac{1}{r} + \frac{1}{r^2}\sum_{i=1}^{3}\frac{x^i}{r}x'^i + \frac{1}{2!}\sum_{i,k}\frac{3x^i x^k - r^2\delta^{ik}}{r^5}x'^i x'^k \,.$$

Inserting this expression into the potential $\Phi(\boldsymbol{x})$, one obtains the terms with $\ell = 0$, $\ell = 1$ and $\ell = 2$ of expansion (1.106c), but now in the form

$$\Phi(\boldsymbol{x}) \simeq \frac{Q}{r} + \frac{\boldsymbol{d}\cdot\boldsymbol{x}}{r^3} + \frac{1}{2}\sum_{i,k} Q^{ik}\frac{x^i x^k}{r^5}\,, \quad \text{with} \tag{1.111a}$$

$$\boldsymbol{d} = \iiint \mathrm{d}^3x'\, \boldsymbol{x}'\varrho(\boldsymbol{x}')\,, \tag{1.111b}$$

$$Q^{ik} = \iiint \mathrm{d}^3x'\, \left(3x'^i x'^k - r'^2\delta^{ik}\right)\varrho(\boldsymbol{x}')\,. \tag{1.111c}$$

The dipole term (1.111b) is the familiar one derived earlier. The quadrupole term, expressed here in Cartesian components, is symmetric, i.e. $Q^{ki} = Q^{ik}$, and has vanishing trace $\mathrm{tr}\, \mathrm{Q} \equiv \sum_i Q^{ii} = 0$. Therefore, it has five independent entries only, in accordance with the fact that for $\ell = 2$ there are five values of m. It is not difficult to establish the exact relationship between the moments q_{2m} (spherical basis) and the (Cartesian) Q^{ik}. One finds

$$q_{22} = \frac{\sqrt{5}}{4\sqrt{6\pi}}\left(Q^{11} - 2\mathrm{i}Q^{12} - Q^{22}\right),$$

$$q_{21} = \frac{\sqrt{5}}{2\sqrt{6\pi}}\left(-Q^{13} + \mathrm{i}Q^{23}\right), \quad q_{20} = \frac{\sqrt{5}}{4\sqrt{\pi}}Q^{33}\,.$$

The physical significance of terms (1.111b) and (1.111c) is further clarified by calculating the energy of some charge distribution $\eta(\boldsymbol{x})$ in the electric field due to an external potential $\Phi(\boldsymbol{x})$. In the expression for the energy

$$W = \iiint \mathrm{d}^3x\, \eta(\boldsymbol{x})\Phi(\boldsymbol{x})\,, \tag{1.112a}$$

one inserts the Taylor expansion of the potential around the value $\boldsymbol{0}$ of the argument:

$$\begin{aligned}\Phi(\boldsymbol{x}) &= \Phi(\boldsymbol{0}) + \boldsymbol{x}\cdot\nabla\Phi(\boldsymbol{0}) + \frac{1}{2}\sum_{i,k} x^i x^k \frac{\partial^2\Phi}{\partial x^i\partial x^k}(\boldsymbol{0}) + \ldots \\ &= \Phi(\boldsymbol{0}) - \boldsymbol{x}\cdot\boldsymbol{E}(\boldsymbol{0}) - \frac{1}{2}\sum_{i,k} x^i x^k \frac{\partial E^k}{\partial x^i}(\boldsymbol{0}) + \frac{1}{6}r^2\nabla\cdot\boldsymbol{E}(\boldsymbol{0}) + \ldots\,.\end{aligned}$$

In the last step, we added a term proportional to $\nabla \cdot \boldsymbol{E}$ which is zero. Indeed, as long as the source ϱ of the potential Φ lies outside the charge distribution η, field $\boldsymbol{E}$ has a vanishing divergence everywhere that the integrand in (1.112a) differs from zero. The added term then vanishes, and the energy becomes

$$W = Q\Phi(\mathbf{0}) - \boldsymbol{d} \cdot \boldsymbol{E}(\mathbf{0}) - \frac{1}{6}\sum_{i,k} Q^{ik} \frac{\partial E^k}{\partial x^i}(\mathbf{0}) + \dots . \tag{1.112b}$$

As expected, the first term is simply the product of the charge and the potential at the origin. The second term is the energy of an electric dipole in the external electric field. The third term is new and contains the inner product of the quadrupole tensor Q and the *field gradient* $\{\partial E^k / \partial x^i\}$.

1.9 Stationary Currents and Static Magnetic States

The basic equations (1.67a)–(1.67c) describe all phenomena involving permanent magnets at rest or stationary, i.e. time-independent, electric currents. We repeat the first two of these here:

$$\nabla \cdot \boldsymbol{B}(\boldsymbol{x}) = 0 , \tag{1.113a}$$

$$\nabla \times \boldsymbol{H}(\boldsymbol{x}) = \frac{4\pi}{c}\boldsymbol{j}(\boldsymbol{x}) . \tag{1.113b}$$

In the vacuum, outside of magnetically polarizable media, and using Gaussian units, one has $\boldsymbol{B}(\boldsymbol{x}) = \boldsymbol{H}(\boldsymbol{x})$.

For stationary processes, the continuity equation reduces to

$$\nabla \cdot \boldsymbol{j}(\boldsymbol{x}) = 0 . \tag{1.113c}$$

The first of these, (1.113a), which holds also for nonstatic and nonstationary processes, follows from the fact that there are no isolated magnetic monopoles. The second equation is the time-independent version of the Maxwell equation (1.44d) and, in its integral form, yields *Ampère's law:* With a given finite, (possibly piecewise) smooth surface F in $\mathbb{R}^3$ whose boundary is denoted by C, integrate the left-hand side of (1.113b) over this surface. Using Stokes' theorem (1.7) this yields

$$\iint_{F(C)} \mathrm{d}\sigma\, (\nabla \times \boldsymbol{H} \cdot \hat{\boldsymbol{n}}) = \oint_C \mathrm{d}\boldsymbol{s} \cdot \boldsymbol{H} ,$$

where $\hat{\boldsymbol{n}}$ is the local positive normal to the surface (positive with respect to the orientation of boundary curve C). When integrated over F, the right-hand side of (1.113b) yields

$$\frac{4\pi}{c} \iint_{F(C)} \mathrm{d}\sigma\, \boldsymbol{j} \cdot \hat{\boldsymbol{n}} = \frac{4\pi}{c} J ,$$

with J the total electric current flowing across the surface. Thus, one obtains Ampère's law in the form

$$\oint_C \mathrm{d}\boldsymbol{s} \cdot \boldsymbol{H} = \frac{4\pi}{c} J . \tag{1.114}$$

In analogy to the electromotive force in (1.12), the left-hand side of (1.114) represents a *magnetomotive force.* Furthermore, there is an interesting analogy to Gauss' law (1.14): The integral over a closed surface on the left-hand side of (1.14) is replaced by an integral over a closed curve, and the integral over the volume on the right-hand side of (1.14) becomes a surface integral in (1.114).

1.9.1 Poisson Equation and Vector Potential

As before, the field $\boldsymbol{B}$ may alternatively be described in terms of a vector potential $\boldsymbol{A}$, $\boldsymbol{B} = \nabla \times \boldsymbol{A}$. In cases where the magnetic field and the magnetic induction can be identified, one obtains from (1.47c) and from (1.113b)

$$\nabla \times (\nabla \times \boldsymbol{A}) = -\boldsymbol{\Delta A} + \nabla(\nabla \cdot \boldsymbol{A}) = \frac{4\pi}{c}\boldsymbol{j} .$$

When using a Coulomb gauge, (1.63), this differential equation reduces to a Poisson equation for the components of $\boldsymbol{A}(\boldsymbol{x})$:

$$\boldsymbol{\Delta A}(\boldsymbol{x}) = -\frac{4\pi}{c}\boldsymbol{j}(\boldsymbol{x}) , \tag{1.115}$$

whose general, time-dependent form is given by (1.59b). With the experience gained in Sect. 1.8.1, and if there are no special boundary conditions, one obtains a solution at once. It reads

$$\boldsymbol{A}(\boldsymbol{x}) = \frac{1}{c}\iiint \mathrm{d}^3x' \, \frac{\boldsymbol{j}(\boldsymbol{x}')}{|\boldsymbol{x} - \boldsymbol{x}'|} . \tag{1.116}$$

This result suggests a certain analogy between electrostatics and magnetostatics. However, this analogy should not be taken too literally because it may conceal the more profound differences in their physics.

1.9.2 Magnetic Dipole Density and Magnetic Moment

We assume the spatial domain in which the current density $\boldsymbol{j}(\boldsymbol{x})$ is different from zero to be localized, i.e. to be contained inside a sphere about the origin, $|\boldsymbol{x}'| \leqslant R$. In the space outside this sphere and with $|\boldsymbol{x}| \gg |\boldsymbol{x}'|$, the inverse distance function in (1.116) is expanded as follows:

$$\frac{1}{|\boldsymbol{x} - \boldsymbol{x}'|} \simeq \frac{1}{|\boldsymbol{x}|} + \frac{\boldsymbol{x} \cdot \boldsymbol{x}'}{|\boldsymbol{x}|^3} ,$$

so that the ith component of the vector potential is obtained in the form

$$A^i(\boldsymbol{x}) \simeq \frac{1}{c|\boldsymbol{x}|} \iiint \mathrm{d}^3x' \, j^i(\boldsymbol{x}') + \frac{1}{c|\boldsymbol{x}|^3} \sum_{k=1}^{3} x^k \iiint \mathrm{d}^3x' \, x'^k j^i(\boldsymbol{x}') \,. \quad (1.117)$$

To analyse the second term and to render it more transparent, one makes use of the following formula.

Auxiliary Formula
Let f and g be smooth functions on $\mathbb{R}^3$. For a smooth vector field $\boldsymbol{v}(\boldsymbol{x})$ which is localized and does not possess any sources, one has

$$\iiint \mathrm{d}^3x \, \{f(\boldsymbol{x}) \, \boldsymbol{v}(\boldsymbol{x}) \cdot \boldsymbol{\nabla} g(\boldsymbol{x}) + g(\boldsymbol{x}) \, \boldsymbol{v}(\boldsymbol{x}) \cdot \boldsymbol{\nabla} f(\boldsymbol{x})\} = 0 \,. \quad (1.118)$$

The proof of this formula is easy: In the second term, perform a partial integration:

$$\iiint \mathrm{d}^3x \, \{\cdots\} = \iiint \mathrm{d}^3x \, \{f \, \boldsymbol{v} \cdot \boldsymbol{\nabla} g - \boldsymbol{\nabla} \cdot (g\boldsymbol{v}) \, f\}$$
$$= \iiint \mathrm{d}^3x \, \{f \, \boldsymbol{v} \cdot \boldsymbol{\nabla} g - (\boldsymbol{\nabla} g)\boldsymbol{v} f - g(\boldsymbol{\nabla} \cdot \boldsymbol{v}) f\} \,.$$

As the vector field $\boldsymbol{v}$ is localized, there will be no surface terms from the partial integration. The first two terms cancel, and the third term is proportional to the divergence $\boldsymbol{\nabla} \cdot \boldsymbol{v}$, which vanishes, by assumption. This proves (1.118).

With $\boldsymbol{v}$ replaced by the current density $\boldsymbol{v}$, two applications are relevant here:
(i) Choose f to be the constant function 1, g to be the ith coordinate, $f = 1$, $g = x^i$. Then (1.118) yields

$$\iiint \mathrm{d}^3x \, j^i(\boldsymbol{x}) = 0 \,. \quad (1.119a)$$

Taking account of (1.115), this is equivalent to the integral version of (1.113a): There are no magnetic monopoles.
(ii) Choose $f = x^i$, $g = x^k$, whereby (1.118) yields the relation

$$\iiint \mathrm{d}^3x \, \left\{x^i j^k(\boldsymbol{x}) + x^k j^i(\boldsymbol{x})\right\} = 0 \,.$$

This is used to transform the second term in (1.117). We have

$$\sum_{k=1}^{3} x^k \iiint \mathrm{d}^3x' \, x'^k j^i(\boldsymbol{x}')$$
$$= \frac{1}{2} \sum_k x^k \iiint \mathrm{d}^3x' \, \left\{x'^k j^i(\boldsymbol{x}') - x'^i j^k(\boldsymbol{x}')\right\}$$

$$= -\frac{1}{2}\sum_{k,l} \varepsilon_{ikl} x^k \iiint \mathrm{d}^3x' \, (\boldsymbol{x}' \times \boldsymbol{j}(\boldsymbol{x}'))^l$$

$$= -\frac{1}{2}\left(\boldsymbol{x} \times \left(\iiint \mathrm{d}^3x' \, \boldsymbol{x}' \times \boldsymbol{j}(\boldsymbol{x}')\right)\right)^i .$$

One inserts this into (1.117) and defines

$$\boldsymbol{m}(\boldsymbol{x}) := \frac{1}{2c}\boldsymbol{x} \times \boldsymbol{j}(\boldsymbol{x}) \tag{1.120a}$$

to be the *magnetic dipole density*. The space integral of this density is the *magnetic moment*:

$$\boldsymbol{\mu} := \frac{1}{2c}\iiint \mathrm{d}^3x \, \boldsymbol{x} \times \boldsymbol{j}(\boldsymbol{x}) . \tag{1.120b}$$

With these results and definitions, the second term of (1.117) is seen to be a dipole term and takes the form

$$\boldsymbol{A}_{\text{dipole}}(\boldsymbol{x}) = \frac{1}{|\boldsymbol{x}|^3}\boldsymbol{\mu} \times \boldsymbol{x} . \tag{1.121}$$

Remarks

1. Consider briefly the physical dimensions of the quantities being discussed here: Denoting the dimension of electric charge summarily by $[q]$, the dimension of length by L, and the dimension of time by T, one has

$$[\varrho] = [q]\mathrm{L}^{-3} , \quad [\boldsymbol{j}] = [\varrho]\mathrm{L}\mathrm{T}^{-1} = [q]\mathrm{L}^{-2}\mathrm{T}^{-1} ,$$
$$\text{and, therefore,} \quad [\boldsymbol{\mu}] = [q]\mathrm{L} .$$

Students of atomic physics know the Bohr magneton:

$$\mu_{\mathrm{B}} = \frac{e\hbar}{2mc} , \tag{1.122}$$

where e denotes the elementary charge and m is the mass of the electron. The magnetic moment of the electron is expressed in terms of this unit. One then verifies immediately that the $\boldsymbol{\mu}$ thus defined has the correct physical dimension:

$$\left[\frac{e\hbar}{mc}\right] = [q]\left[\frac{\hbar c}{mc^2}\right] = [q]L .$$

2. When using Gaussian units, the fields involved in Maxwell's equations all have the same dimension:

$$[\boldsymbol{E}] = [\boldsymbol{D}] = [\boldsymbol{H}] = [\boldsymbol{B}] = [q]L^{-2} .$$

The product of an electric dipole moment and an electric field, very much like the product of a magnetic moment and a magnetic field, have the dimension of an energy:

$$[d \cdot E] = [q^2]L^{-1} = [\mu \cdot H] .$$

Note that this confirms the result expressed by (1.112b) above.

Example 1.9 Magnetic moment of a plane current loop

Consider a stationary current density $j(x)$ flowing in the planar, closed and smooth loop of Fig. 1.16. If the wire is ideally thin, then one has

$$\iiint \mathrm{d}^3x\, x \times j(x) = J \oint x \times \mathrm{d}s .$$

The magnetic moment generated by this loop in the space around it is given by

$$\mu = \frac{1}{2c} J \oint x \times \mathrm{d}s .$$

The integral in this formula is seen to be twice that of the surface F enclosed by the loop as can be inferred from Fig. 1.16. The magnetic moment is directed perpendicular to the surface and has the absolute value $|\mu| = JF/c$.

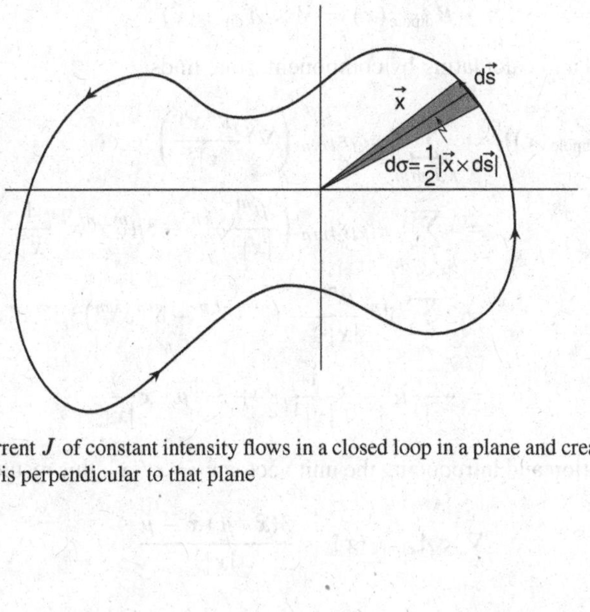

Fig. 1.16 A current J of constant intensity flows in a closed loop in a plane and creates a magnetic moment which is perpendicular to that plane

Example 1.10 Magnetic moment of a flux of particles
Given an ensemble of N pointlike particles, $i = 1, 2, \ldots N$, carrying the charges q_i and moving with velocities $\boldsymbol{v}^{(i)}$, respectively, these particles generate the current density

$$\boldsymbol{j}(\boldsymbol{x}) = \sum_{i=1}^{N} q_i \boldsymbol{v}^{(i)} \delta(\boldsymbol{x} - \boldsymbol{x}^{(i)}) .$$

Inserting this expression into (1.120b), one obtains

$$\begin{aligned} \boldsymbol{\mu}(\boldsymbol{x}) &= \frac{1}{2c} \sum_{i=1}^{N} q_i \boldsymbol{x}^{(i)} \times \boldsymbol{v}^{(i)} \\ &= \sum_{i=1}^{N} \frac{q_i}{2m_i c} \boldsymbol{\ell}^{(i)} = \sum_{i=1}^{N} \frac{q_i \hbar}{2m_i c} \left(\frac{1}{\hbar} \boldsymbol{\ell}^{(i)} \right) . \end{aligned} \tag{1.123}$$

The right-hand side of this result exhibits the analogue of Bohr's magneton (1.122) and contains the dimensionless vector $\boldsymbol{\ell}^{(i)}/\hbar$, which in quantum mechanics becomes the operator of orbital angular momentum.

1.9.3 Fields of Magnetic and Electric Dipoles

In a first step, we calculate the induction field which follows from the vector potential (1.121):

$$\boldsymbol{B}_{\text{dipole}}(\boldsymbol{x}) = \nabla \times \boldsymbol{A}_{\text{dipole}}(\boldsymbol{x}) .$$

In some detail and calculating by components, one finds

$$\begin{aligned} \left(\nabla \times \boldsymbol{A}_{\text{dipole}}(\boldsymbol{x})\right)^i &= \sum_{k,l,m,n} \varepsilon_{ikl} \varepsilon_{lmn} \left(\nabla^k \frac{\mu^m x^n}{|\boldsymbol{x}|^3} \right) \\ &= \sum_{k,l,m,n} \varepsilon_{ikl} \varepsilon_{lmn} \left(\frac{\mu^m}{|\boldsymbol{x}|^3} \delta^{kn} - 3\mu^m x^n x^k \frac{1}{|\boldsymbol{x}|^5} \right) \\ &= 2 \sum_m \delta^{im} \frac{\mu^m}{|\boldsymbol{x}|^3} - (\delta^{im}\delta^{kn} - \delta^{in}\delta^{km}) 3\mu^m x^n x^k \frac{1}{|\boldsymbol{x}|^5} \\ &= \frac{2}{|\boldsymbol{x}|^3} \mu^i - 3 \frac{1}{|\boldsymbol{x}|^3} \mu^i + 3x^i \boldsymbol{\mu} \cdot \boldsymbol{x} \frac{1}{|\boldsymbol{x}|^5} . \end{aligned}$$

In vector notation and introducing the unit vector $\hat{\boldsymbol{x}} = \boldsymbol{x}/|\boldsymbol{x}|$, this result is rewritten as follows:

$$\nabla \times \boldsymbol{A}_{\text{dipole}}(\boldsymbol{x}) = \frac{3(\hat{\boldsymbol{x}} \cdot \boldsymbol{\mu})\, \hat{\boldsymbol{x}} - \boldsymbol{\mu}}{|\boldsymbol{x}|^3} . \tag{1.124a}$$

This is not yet the full answer. As will be shown subsequently, the correct expression for the induction field contains an additional, distribution-valued term in the origin:

$$\boldsymbol{B}_{\text{dipole}}(\boldsymbol{x}) = \frac{3(\hat{\boldsymbol{x}} \cdot \boldsymbol{\mu})\,\hat{\boldsymbol{x}} - \boldsymbol{\mu}}{|\boldsymbol{x}|^3} + \frac{8\pi}{3}\boldsymbol{\mu}\,\delta(\boldsymbol{x})\,. \tag{1.124b}$$

This formula and (1.121) refer to cases where the dipole is located in the origin. If the (pointlike) dipole is located in $\boldsymbol{x}'$ instead, $\boldsymbol{x}$ must be replaced by $\boldsymbol{x} - \boldsymbol{x}'$, and $\hat{\boldsymbol{x}}$ by the unit vector $\hat{\boldsymbol{n}}$, which points from $\boldsymbol{x}'$ to $\boldsymbol{x}$:

$$\boldsymbol{B}_{\text{dipole}}(\boldsymbol{x}) = \frac{3(\hat{\boldsymbol{n}} \cdot \boldsymbol{\mu})\,\hat{\boldsymbol{n}} - \boldsymbol{\mu}}{|\boldsymbol{x} - \boldsymbol{x}'|^3} + \frac{8\pi}{3}\boldsymbol{\mu}\,\delta(\boldsymbol{x} - \boldsymbol{x}')\,, \text{ with } \hat{\boldsymbol{n}} = \frac{\boldsymbol{x} - \boldsymbol{x}'}{|\boldsymbol{x} - \boldsymbol{x}'|}\,. \tag{1.124c}$$

Before the analogous calculation is performed for the electric dipole, a discussion of the origin of the additional term in (1.124b) or (1.124c) will be presented.

Derivation of the Distributional Term in (1.124b)
Let the dipole be located in the origin, for the sake of simplicity, and let the 3-axis be chosen along the direction of $\boldsymbol{\mu}$. The dipole generates a magnetization density:

$$\boldsymbol{m} = \boldsymbol{\mu}\,\delta(\boldsymbol{x}) = \mu\,\delta(\boldsymbol{x})\hat{\boldsymbol{e}}_3 \equiv m(r)\hat{\boldsymbol{e}}_3\,.$$

The notation introduced in the last step is simply an abbreviation. It leaves open the alternative of the dipole having a finite spatial extension, localized around the origin. This will be the case, for example, when the dipole is an atomic nucleus with a nonvanishing magnetic moment in interaction with the dipole moment of the electrons of the atom.

Equation (1.78a) yields $\boldsymbol{B} = \boldsymbol{H} + 4\pi\boldsymbol{m}$ and, thus, as $\nabla \cdot \boldsymbol{B} = 0$, $\nabla \cdot \boldsymbol{H} = -4\pi\nabla \cdot \boldsymbol{m}$. Furthermore, field $\boldsymbol{H}$ is irrotational in magneto*statics*, $\nabla \times \boldsymbol{H} = 0$. Therefore, one can write it as the gradient field of a magnetic potential $\Psi(\boldsymbol{x})$ and obtain a Poisson equation for the latter:

$$\boldsymbol{H} = -\nabla\Psi\,, \qquad \Delta\Psi(\boldsymbol{x}) = 4\pi\big(\nabla \cdot \boldsymbol{m}\big)(\boldsymbol{x})\,.$$

Comparing with the Poisson equation (1.80a) and accounting for the sign on the right-hand side, one immediately obtains a solution:

$$\Psi(\boldsymbol{x}) = -\iiint \mathrm{d}^3 y\,\frac{\nabla \cdot \boldsymbol{m}(\boldsymbol{y})}{|\boldsymbol{x} - \boldsymbol{y}|}\,.$$

One then calculates the divergence in the numerator of the integrand. With r or $s := |\boldsymbol{y}|$ and with the choice of the 3-axis above, using the explicit form of the spherical harmonic Y_{10}, one has

$$\nabla \cdot \boldsymbol{m} = \left(\frac{\mathrm{d}}{\mathrm{d}r}m(r)\right)\frac{\partial r}{\partial z} = m'(r)\cos\theta = m'(r)\frac{x^3}{r} = \sqrt{\frac{4\pi}{3}}m'(r)Y_{10}(\hat{\boldsymbol{y}}).$$

One inserts the multipole expansion (1.105) of the inverse distance function, of which only the term with $\ell = 1$ and $m = 0$ contributes, by the orthogonality relation (1.98a):

$$\Psi = -\sqrt{\frac{4\pi}{3}}\left(\int_0^\infty s^2\,\mathrm{d}s\;\mathrm{d}\Omega_y\; m'(s)\frac{4\pi}{3}\frac{r_<}{r_>^2}Y_{10}^*(\hat{\boldsymbol{y}})Y_{10}(\hat{\boldsymbol{y}})\right)Y_{10}(\hat{\boldsymbol{x}})\,.$$

As long as one stays outside the source, one must take $r_< = s$ and $r_> = r$, so that

$$\Psi = -\left(\frac{4\pi}{3}\right)^{\frac{3}{2}}\frac{1}{r^2}Y_{10}(\hat{\boldsymbol{x}})\int_0^r s^3\,\mathrm{d}s\,\frac{\mathrm{d}}{\mathrm{d}s}m(s)$$

$$= +\left(\frac{4\pi}{3}\right)^{\frac{3}{2}}\frac{3Y_{10}(\hat{\boldsymbol{x}})}{r^2}\int_0^r s^2\,\mathrm{d}s\;m(s) \equiv \frac{4\pi}{3}x^3\,f(r)\,. \tag{1.125a}$$

Here, a partial integration was performed, x^3 was replaced by $r\cos\theta$, and the short-hand notation

$$f(r) := \frac{3}{r^3}\int_0^r s^2\,\mathrm{d}s\;m(s) = \frac{3}{4\pi r^3}\iiint \mathrm{d}^3y\;\boldsymbol{m}(\boldsymbol{y})\cdot\hat{\boldsymbol{e}}_3 \tag{1.125b}$$

was introduced. In calculating the fields $\boldsymbol{H}$ and $\boldsymbol{B}$, one makes use of the expression

$$f'(r) = \frac{3}{r}\big(m(r) - f(r)\big) \tag{1.125c}$$

for the derivative of $f(r)$. With

$$\boldsymbol{H} = -\boldsymbol{\nabla}\Psi = -\frac{4\pi}{3}\left(x^3 f'\frac{x^1}{r}, x^3 f'\frac{x^2}{r}, x^3 f'\frac{x^3}{r} + f(r)\right)^T,$$

one has

$$\begin{aligned}\boldsymbol{B}(\boldsymbol{x}) &= \boldsymbol{H}(\boldsymbol{x}) + 4\pi\boldsymbol{m}(\boldsymbol{x})\\ &= 4\pi\left\{\left[f(r) - m(r)\right]\left[\frac{x^3}{r^2}\boldsymbol{x} - \frac{1}{3}\hat{\boldsymbol{e}}_3\right] + \frac{2m(r)}{3}\hat{\boldsymbol{e}}_3\right\}\\ &= \frac{8\pi}{3}m(r)\hat{\boldsymbol{e}}_3 + \frac{4\pi}{3}\left[f(r) - m(r)\right]\left[3\frac{x^3}{r}\hat{\boldsymbol{x}} - \hat{\boldsymbol{e}}_3\right].\end{aligned}$$

Inserting now (1.125b), and noticing that the ideal dipole is different from zero only in the origin and that the part of $m(r)$ which lies outside the origin does not contribute, one finds the result

$$B(x) = \frac{8\pi}{3}\mu\,\delta(x)\hat{e}_3 + \mu\frac{3x^3\hat{x}/r - \hat{e}_3}{r^3}$$
$$\widehat{=} \frac{8\pi}{3}\mu\,\delta(x) + \frac{3(\mu\cdot\hat{x})\hat{x} - \mu}{r^3}\,. \qquad (1.125d)$$

In the last step, the dipole was taken in an arbitrary direction.

The contact term proportional to the δ-distribution plays an important role in the description of hyperfine structure with atomic s-states (see, e. g. [QP], Sect. 5.2.4). For the case of hydrogen, for example, this concerns the interaction of the spin of the pointlike proton at the origin with the spin of the electron.[7] In this case, the function $m(r)$ is proportional to the probability of finding the electron in the origin, i.e. the position where the proton is at rest, in good approximation, as it is much heavier than the electron:

$$m(r) = \mu_P\,|\psi(r)|^2\,, \quad \text{with} \quad \psi(r) = \frac{1}{\sqrt{4\pi}}R_{1s}(r)\,.$$

$R_{1s}(r)$ is the radial function in the 1s-state, the factor $1/\sqrt{4\pi}$ stems from the angular wave function Y_{00}.

One will have noticed a certain, yet not complete, analogy of these results to the *electric* dipole which is described by the scalar potential of (1.88c):

$$\Phi_{\text{dipole}}(x) = \frac{d\cdot(x - x')}{|x - x'|^3}\,.$$

In calculating the electric field as the negative gradient of this function, one finds a form akin to (1.124a), viz.

$$-\nabla\Phi_{\text{dipole}}(x) = \frac{3(\hat{n}\cdot d)\,\hat{n} - d}{|x - x'|^3}\,,$$

where d is the electric dipole moment, as in (1.88b). However, this answer is incomplete as well: The electric field of the dipole at the position x' is supplemented by a contact term as well so that it reads

$$E_{\text{dipole}}(x) = \frac{3(\hat{n}\cdot d)\,\hat{n} - d}{|x - x'|^3} - \frac{4\pi}{3}d\,\delta(x - x')\,. \qquad (1.126)$$

[7] The derivation given here follows essentially R.A. Sorensen, Am. J. Phys. **35** (1967) 1078. Another, rather natural, approach which does not exhibit the singularity of the contact term derives from the relativistic treatment of hyperfine structure by means of the Dirac equation. It yields the nonrelativistic contact term in the approximation $v/c \ll 1$. Both aspects are worked out in J. Hüfner, F. Scheck, and C.S. Wu, *Muon Physics I*, Chap. 3.

For all $\boldsymbol{x} \neq \boldsymbol{x}'$ this is the same expression as the previous one. The term proportional to the δ-distribution may be derived in a way similar to the case of the magnetic dipole. It guarantees that the integral over a sphere V which encloses the dipole yields

$$\iiint\limits_V \mathrm{d}^3x\, \boldsymbol{E}_{\text{dipole}}(\boldsymbol{x}) = -\frac{4\pi}{3}\boldsymbol{d} \, .$$

(Note that one must integrate the first term of (1.126) over the angular variables first!) That this must be so is the subject of Exercise 1.12.

1.9.4 Energy and Energy Density

There is an aspect of field theory that may be new for some readers: Static or time-dependent electric and magnetic fields possess an energy content. This can be demonstrated by means of a simple example which relates directly to mechanics.

Suppose in vacuum $N-1$ point charges $q_1, q_2, \ldots, q_{N-1}$ are given, and none of them is located at an infinite distance from the others. Their positions are $\boldsymbol{x}_1, \ldots, \boldsymbol{x}_{N-1}$. At an arbitrary test point $\boldsymbol{x}$ the potential created by this set-up is given by

$$\Phi(\boldsymbol{x}) = \sum_{k=1}^{N-1} \frac{q_k}{|\boldsymbol{x} - \boldsymbol{x}^{(k)}|} \, .$$

Imagine now another point charge q_N being shifted from infinity to the position $\boldsymbol{x}_N$. The work involved in doing this is $W = q_N \Phi(\boldsymbol{x}_N)$. This is also the potential energy that is built up by the given $N-1$ charges. This argument is the same as with a pointlike mass that is brought into a gravitational potential, well known from mechanics. Thus, the potential energy contained in the set-up of the N charges in total is

$$W_{\mathrm{E}} = \sum_{i=2}^{N} \sum_{k=1}^{i-1} \frac{q_i q_k}{|\boldsymbol{x}^{(i)} - \boldsymbol{x}^{(k)}|} = \frac{1}{2} \sum_{\substack{i,k=1 \\ i \neq k}}^{N} \frac{q_i q_k}{|\boldsymbol{x}^{(i)} - \boldsymbol{x}^{(k)}|} \, . \tag{1.127a}$$

In the second expression, only $k \neq i$ is required. The original restriction $k < i$ is taken care of by the factor $1/2$.

Only the mutual energy was calculated here, not the energy that would be needed to concentrate the charge q_i in the point $\boldsymbol{x}^{(i)}$. This so-called *self-energy*, which is infinitely large, is ignored here. The problem of self-energy in classical field theory is a difficult one and becomes manageable only in its quantized version.

If the pointlike charges are replaced by continuous localized charge distributions, then the obvious generalization of formula (1.127a) reads

$$W_{\mathrm{E}} = \frac{1}{2} \iiint \mathrm{d}^3x \iiint \mathrm{d}^3x' \, \frac{\varrho(\boldsymbol{x})\varrho(\boldsymbol{x}')}{|\boldsymbol{x} - \boldsymbol{x}'|} \, . \tag{1.127b}$$

This formula can be transformed such that the energy is expressed in terms of the electric field. Introducing the potential expressed by (1.84a) and making use of the Poisson equation (1.80a), one finds

$$W_{\mathrm{E}} = \frac{1}{2} \iiint \mathrm{d}^3x\, \varrho(\boldsymbol{x})\Phi(\boldsymbol{x}) = -\frac{1}{8\pi} \iiint \mathrm{d}^3x\, \Phi(\boldsymbol{x})\boldsymbol{\Delta}\Phi(\boldsymbol{x})$$
$$= \frac{1}{8\pi} \iiint \mathrm{d}^3x\, (\boldsymbol{\nabla}\Phi)^2 .$$

In going from the second to the third formula, a partial integration was performed. As the charge density is localized, implying that the potential tends to zero at infinity, there is no boundary term from the partial integration. At this point the electric field $\boldsymbol{E}(\boldsymbol{x}) = -\boldsymbol{\nabla}\Phi(\boldsymbol{x})$ comes into play. There follows an expression for the energy:

$$W_{\mathrm{E}} = \frac{1}{8\pi} \iiint \mathrm{d}^3x\, \boldsymbol{E}^2(\boldsymbol{x}) , \tag{1.127c}$$

which is remarkable for two reasons. On the one hand, one has succeeded in expressing the total energy contained in the vector field $E(\boldsymbol{x})$ by the electric field and no longer by auxiliary functions such as the potential. On the other hand, the integrand

$$u_{\mathrm{E}}(\boldsymbol{x}) := \frac{1}{8\pi} \boldsymbol{E}^2(\boldsymbol{x}) \tag{1.127d}$$

may be interpreted as the *energy density* of the electric field, that is as a *local* quantity like the field itself. This offers the possibility to define the energy content of a partial domain V of the space $\mathbb{R}^3$ and to determine it by integration of (1.127d) over that volume V.

Remarks

1. There is an important difference between (1.127a) and (1.127b). While in expression (1.127a) the energy W can be positive, negative or zero, in expression (1.127b) one will always have $W \geqslant 0$. The reason for this difference stems from the self-energies which are contained in the second expression, (1.127b), but not in the first.
2. In a medium with a nontrivial dielectric constant ε (i.e. ε is not equal to one, in Gaussian units), expressions (1.127c) and (1.127d) are modified. Let Φ be the potential generated by the given charge density. If the charge density is changed by a small amount $\delta\varrho$, the energy changes by the amount

$$\delta W_{\mathrm{E}} = \iiint \mathrm{d}^3x\, \Phi(\boldsymbol{x})\, \delta\varrho(\boldsymbol{x}) .$$

Through the Maxwell equation (1.44c) the change in charge density $\delta\varrho$ is related to a change in the displacement field $\boldsymbol{D}$,

$$\delta\varrho(\boldsymbol{x}) = \frac{1}{4\pi} \boldsymbol{\nabla} \cdot (\delta \boldsymbol{D}) ,$$

so that the change in energy can also be expressed in terms of $\delta \boldsymbol{D}$. Inserting this into the integrand of δW_{E} and performing a partial integration, one obtains with $\boldsymbol{E} = -\nabla \Phi$

$$\delta W_{\mathrm{E}} = \frac{1}{4\pi} \iiint \mathrm{d}^3x \, \boldsymbol{E} \cdot \delta \boldsymbol{D} \,. \tag{1.128a}$$

If the relation between $\boldsymbol{E}$ and $\boldsymbol{D}$ is *linear*, i.e. if the response of the medium to the applied field is a linear one, then one has $\boldsymbol{E} \cdot \delta \boldsymbol{D} = \delta(\boldsymbol{E} \cdot \boldsymbol{D})/2$. The total energy content of the field configuration may be thought of as the addition of many small contributions of this kind. By formal integration over the space variables one obtains the following expression for the energy of the configuration:

$$W_{\mathrm{E}} = \frac{1}{8\pi} \iiint \mathrm{d}^3x \, \boldsymbol{E}(\boldsymbol{x}) \cdot \boldsymbol{D}(\boldsymbol{x}) \,. \tag{1.128b}$$

As before, the integrand may be interpreted as an energy density:

$$u_{\mathrm{E}}(\boldsymbol{x}) = \frac{1}{8\pi} \boldsymbol{E}(\boldsymbol{x}) \cdot \boldsymbol{D}(\boldsymbol{x}) \,. \tag{1.128c}$$

In the vacuum, where one can select $\boldsymbol{D} = \boldsymbol{E}$ by the appropriate choice of units, the two expressions go over into formulae (1.127c) and (1.127d), respectively.

The energy content of a magnetic field configuration is calculated in close analogy to the electrostatic case. Let $\boldsymbol{j}(\boldsymbol{x})$ be a stationary, localized current density, and let $\boldsymbol{A}(\boldsymbol{x})$ be the vector potential by means of which the induction $\boldsymbol{B}(\boldsymbol{x})$ is derived. With reference to (1.127b) one would guess the relation

$$W_{\mathrm{M}} = \frac{1}{2} \iiint \mathrm{d}^3x \, \frac{1}{c} \, \boldsymbol{j}(\boldsymbol{x}) \cdot \boldsymbol{A}(\boldsymbol{x}) \,. \tag{1.129a}$$

The charge density is replaced by the current density (multiplied by $1/c$), and the scalar potential is replaced by the vector potential. If this is true, then the magnetic energy can be expressed in terms of fields $\boldsymbol{H}$ and $\boldsymbol{B}$, making use of the Maxwell equations (1.67a) and (1.67b),

$$W_{\mathrm{M}} = \frac{1}{8\pi} \iiint \mathrm{d}^3x \, \boldsymbol{H}(\boldsymbol{x}) \cdot \boldsymbol{B}(\boldsymbol{x}) \,. \tag{1.129b}$$

As in the previous case, the total magnetic energy is the integral of the *magnetic energy density*

$$u_{\mathrm{M}}(\boldsymbol{x}) := \frac{1}{8\pi} \boldsymbol{H}(\boldsymbol{x}) \cdot \boldsymbol{B}(\boldsymbol{x}) \,, \tag{1.129c}$$

in close and obvious analogy to (1.128c). Of course, in a more rigorous derivation, one must first calculate the change of the magnetic energy caused by a change of the vector potential:

$$\delta W_{\mathrm{M}} = \frac{1}{c} \iiint \mathrm{d}^3x \, \delta \boldsymbol{A}(\boldsymbol{x}) \cdot \boldsymbol{j}(\boldsymbol{x}) \, .$$

Under the assumption that all quantities contained in the integrand are localized, one uses (1.67b) to obtain

$$\delta W_{\mathrm{M}} = \frac{1}{4\pi} \iiint \mathrm{d}^3x \, \delta \boldsymbol{B}(\boldsymbol{x}) \boldsymbol{H}(\boldsymbol{x}) \, ,$$

i.e. an expression which is the analogue of (1.128a). One now understands the condition under which one obtains the result expressed by (1.129b): The above infinitesimal expression can be integrated to obtain formula (1.129b) only when the relation between $\boldsymbol{B}$ and $\boldsymbol{H}$ is linear, i.e. if one is dealing with a paramagnetic or a diamagnetic substance.

In summary, one obtains the following result for the total electromagnetic energy density and for the total energy, respectively:

$$u(\boldsymbol{x}) = u_{\mathrm{E}}(\boldsymbol{x}) + u_{\mathrm{M}}(\boldsymbol{x}) \, , \tag{1.130}$$

$$W = W_{\mathrm{E}} + W_{\mathrm{M}} = \frac{1}{8\pi} \iiint \mathrm{d}^3x \, \left(\boldsymbol{E}(\boldsymbol{x}) \cdot \boldsymbol{D}(\boldsymbol{x}) + \boldsymbol{H}(\boldsymbol{x}) \cdot \boldsymbol{B}(\boldsymbol{x})\right) \, . \tag{1.131}$$

These formulae hold for linear relationships between $\boldsymbol{D}$ and $\boldsymbol{E}$, as well as between $\boldsymbol{B}$ and $\boldsymbol{H}$. Although here we sketched their derivation only for static or stationary situations, we will see later that these formulae apply also to time-dependent electromagnetic processes.

1.9.5 Currents and Conductivity

The current density $\boldsymbol{j}$ in matter is, as a rule, proportional to the force density $\boldsymbol{f}$,

$$\boldsymbol{j}(\boldsymbol{x}) = \sigma \boldsymbol{f}(\boldsymbol{x}) \, . \tag{1.132}$$

(We assume the current density to be stationary here.) The function σ is called the *conductivity*, and its reciprocal is the specific resistance. The force density may be obtained from the Lorentz force, (1.44e), so that

$$\boldsymbol{j}(\boldsymbol{x}) = \sigma \left[\boldsymbol{E}(\boldsymbol{x}) + \frac{1}{c} \boldsymbol{v} \times \boldsymbol{B}(\boldsymbol{x}) \right] . \tag{1.132a}$$

Typically, in electric circuits, the action of magnetic forces is negligible. In this approximation, one obtains the simple relation

$$\boldsymbol{j}(\boldsymbol{x}) = \sigma \boldsymbol{E}(\boldsymbol{x}) \, , \tag{1.132b}$$

which contains the well-known Ohm's law

$$V = RI \ . \tag{1.133}$$

Here V denotes the tension, I the electric current, and R the Ohm resistance. Obviously, Ohm's law in the form of (1.133) is of great importance for the practitioner. However, the basic relation is contained in equation (1.132). As a simple example, consider a homogeneous cylinder of length L and cross-section F made from a material whose conductivity is σ. As a result, the resistance is given by the formula $R = L/(\sigma F)$.

2 Symmetries and Covariance of the Maxwell Equations

2.1 Introduction

Already within a given, fixed division of four-dimensional spacetime into the space where experiments are performed, and the laboratory time variable, Maxwell's equations show interesting transformation properties under continuous and discrete spacetime transformations. However, only the action of the whole Lorentz group on them reveals their full symmetry structure. A good example that illustrates the covariance of Maxwell's equations is provided by the electromagnetic fields of a point charge uniformly moving along a straight line.

A reformulation of Maxwell theory in the language of exterior forms over $\mathbb{R}^4$, on the one hand, sheds light on some of its properties which are less transparent in the framework of the older vector analysis. On the other hand, it reveals the geometric character of this example of a simple gauge theory and, hence, prepares the ground for the understanding of non-Abelian gauge theories which are essential for the description of the fundamental interactions of nature.

2.2 The Maxwell Equations in a Fixed Frame of Reference

In a fixed inertial system in which $\boldsymbol{x}$ are coordinates in ordinary space $\mathbb{R}^3$, t the coordinate time that an observer at rest reads on his clock, Maxwell's equations (1.44a–1.44d) read

$$\nabla \cdot \boldsymbol{B}(t, \boldsymbol{x}) = 0\,, \tag{2.1a}$$

$$\nabla \times \boldsymbol{E}(t, \boldsymbol{x}) + \frac{1}{c}\frac{\partial}{\partial t}\boldsymbol{B}(t, \boldsymbol{x}) = 0\,, \tag{2.1b}$$

$$\nabla \cdot \boldsymbol{D}(t, \boldsymbol{x}) = 4\pi\varrho(t, \boldsymbol{x})\,, \tag{2.1c}$$

$$\nabla \times \boldsymbol{H}(t, \boldsymbol{x}) - \frac{1}{c}\frac{\partial}{\partial t}\boldsymbol{D}(t, \boldsymbol{x}) = \frac{4\pi}{c}\boldsymbol{j}(t, \boldsymbol{x})\,. \tag{2.1d}$$

F. Scheck, *Classical Field Theory*, Graduate Texts in Physics,
DOI 10.1007/978-3-642-27985-0_2, © Springer-Verlag Berlin Heidelberg 2012

They are supplemented by the relationships

$$\boldsymbol{D} = \varepsilon \boldsymbol{E} , \qquad \boldsymbol{B} = \mu \boldsymbol{H} \tag{2.2}$$

between the displacement field and the electric field, and between the induction field and the magnetic field, respectively, ε being the dielectric constant, and μ the magnetic permeability. (In vacuum and using Gaussian units both constants are equal to 1.) The force that acts on a particle carrying the charge q and moving with the velocity $\boldsymbol{v}$ relative to the observer, is the Lorentz force (1.44e)

$$\boldsymbol{F}(t, \boldsymbol{x}) = q \left[\boldsymbol{E}(t, \boldsymbol{x}) + \frac{1}{c} \boldsymbol{v} \times \boldsymbol{B}(t, \boldsymbol{x}) \right] , \tag{2.3}$$

the second, velocity dependent, term of which is particularly remarkable. Finally, we note the relation between the current density in a given medium and the applied electric field

$$\boldsymbol{j}(t, \boldsymbol{x}) = \sigma \boldsymbol{E}(t, \boldsymbol{x}) , \tag{2.4}$$

where σ describes the conductivity of the medium.

The frame of reference in $\mathbb{R}^3 \times \mathbb{R}_t$ with respect to which these equations are formulated, for the time being, is defined by the observer who interprets his position as the origin of the frame, chooses appropriate coordinates in $\mathbb{R}^3$ and uses his clock for measuring time. An experimenter measures *electric* fields with specific instruments which differ from those he uses for measuring *magnetic* fields. In this sense the specific nature of the two types of vector fields is established empirically. This remark which seems to be a matter of course, will be important when one asks whether an electric field and a magnetic field, when measured by a second observer who moves relative to the first at constant velocity, will continue to be an electric or a magnetic field, respectively.

2.2.1 Rotations and Discrete Spacetime Transformations

Before following up the question raised above let us remain for a while in the inertial frame chosen by the observer and let us analyze the covariance of the equations (2.1a–2.4) with respect to rotations, space reflection, and time reversal, as well as charge conjugation.

Rotations of the Frame of Reference in $\mathbb{R}^3$

Rotations $\mathsf{R} \in \mathrm{SO}(3)$ of the frame of reference are coordinate transformations

$$(t, \boldsymbol{x})^T \longmapsto (t' = t, \boldsymbol{x}' = \mathsf{R}\boldsymbol{x})^T , \quad \text{with} \quad \mathsf{R}^T \mathsf{R} = \mathbb{1} , \; \det \mathsf{R} = +1 .$$

A scalar field φ, by definition, stays invariant,

$$\varphi(t, \boldsymbol{x}) \longmapsto \varphi'(t', \boldsymbol{x}') = \varphi(t, \boldsymbol{x}) , \tag{2.5a}$$

while a vector field transforms according to

$$\boldsymbol{A}(t,\boldsymbol{x}) \longmapsto \boldsymbol{A}'(t',\boldsymbol{x}') = \mathsf{R}\boldsymbol{A}(t,\boldsymbol{x}) . \tag{2.5b}$$

(Here we have made use of the fact that in the orthogonal group SO(3) the inverse of the transposed equals the original matrix, $(\mathsf{R}^T)^{-1} = \mathsf{R}$.) If instead of SO(3) one admits the full group O(3) then also transformations $\tilde{\mathsf{R}} \in \mathrm{O}(3)$ must be studied whose determinant equals -1. These can be written as the product of a proper rotation $\mathsf{R} \in \mathrm{SO}(3)$ and space reflection Π. There are fields $\tilde{\varphi}$ of the first kind (2.5a) which though invariant under rotations, obtain a factor $\det \tilde{\mathsf{R}} = -1$ under space reflection. Likewise, in the second category there are fields $\tilde{\boldsymbol{A}}$ which beyond the transformation behaviour (2.5b) receive the same factor $\det \tilde{\mathsf{R}}$. Thus, with $\mathsf{R} \in \mathrm{SO}(3)$ and $\tilde{\mathsf{R}} = \mathsf{R}\Pi$ they transform according to

$$\tilde{\varphi}(t,\boldsymbol{x}) \mapsto \tilde{\varphi}'(t',\boldsymbol{x}') = \left(\det \tilde{\mathsf{R}}\right) \tilde{\varphi}(t,\boldsymbol{x}) , \tag{2.6a}$$

$$\tilde{\boldsymbol{A}}(t,\boldsymbol{x}) \mapsto \tilde{\boldsymbol{A}}'(t',\boldsymbol{x}') = \left(\det \tilde{\mathsf{R}}\right) \mathsf{R}\tilde{\boldsymbol{A}}(t,\boldsymbol{x}) . \tag{2.6b}$$

Although in geometric terms $\tilde{\varphi}$ is not a scalar field, and $\tilde{\boldsymbol{A}}$ is not a vector field, the customary nomenclature in physics for them is *pseudoscalar field* for $\tilde{\varphi}(t,\boldsymbol{x})$, and *axial vector field* for $\tilde{\boldsymbol{A}}(t,\boldsymbol{x})$. A few examples over the space $\mathbb{R}^3$ will illustrate these definitions:

(i) The velocity $\boldsymbol{v}$, very much like the momentum $\boldsymbol{p}$, is a genuine vector, i.e. it transforms under rotations $\mathsf{R} \in \mathrm{SO}(3)$ as indicated in (2.5b). If these vectors are defined as smooth functions over $\mathbb{R}^3$ they become vector fields. In contrast, the orbital angular momentum $\boldsymbol{\ell} = \boldsymbol{x} \times \boldsymbol{p}$ is an axial vector. Indeed, under a space reflection both $\boldsymbol{x}$ and $\boldsymbol{p}$ change sign, while $\boldsymbol{\ell}$ does not.

(ii) The scalar product $\boldsymbol{x} \cdot \boldsymbol{p}$ is a scalar. Likewise the scalar product of a spin and an orbital angular momentum $\boldsymbol{s} \cdot \boldsymbol{\ell}$ is a genuine scalar. However, the products $\boldsymbol{x} \cdot \boldsymbol{\ell}$ and $\boldsymbol{x} \cdot \boldsymbol{s}$ are pseudoscalars.

The geometric interpretation of the quantities (2.6a) and (2.6b) in the language of exterior forms will be clarified in Sect. 2.5.3 below. For the moment we will stick to the terminology defined above.

Inspection of Maxwell's equations (2.1a–2.1b) shows that they are covariant under rotations from SO(3) provided the fields $\boldsymbol{E}$, $\boldsymbol{D}$, $\boldsymbol{H}$, $\boldsymbol{B}$, and the current density $\boldsymbol{j}$ transform according to (2.5b) and the charge density ϱ transforms like in (2.5a). The first equation (2.1a) contains the divergence of $\boldsymbol{B}$ and is a scalar with respect to $\mathsf{R} \in \mathrm{SO}(3)$. Regarding the second equation (2.1b) we have for the first term

$$\left(\nabla' \times \boldsymbol{E}'\right) = (\mathsf{R}\nabla) \times (\mathsf{R}\boldsymbol{E}) = \mathsf{R}(\nabla \times \boldsymbol{E}) ,$$

and, obviously, for the second term

$$\frac{1}{c}\frac{\partial}{\partial t}\boldsymbol{B}' = \frac{1}{c}\frac{\partial}{\partial t}(\mathsf{R}\boldsymbol{B}) = \mathsf{R}\left(\frac{1}{c}\frac{\partial}{\partial t}\boldsymbol{B}\right),$$

so that covariance of (2.1b) is established. A similar reasoning proves the covariance of the two inhomogeneous Maxwell equations (2.1c) and (2.1d). All terms which are related by Maxwell's equations exhibit the same transformation behaviour.

Space Reflection of the Frame of Reference

The behaviour of Maxwell's equations under a reflection of the spatial coordinates about the origin,

$$(t, \boldsymbol{x})^T \longmapsto (t' = t, \boldsymbol{x}' = -\boldsymbol{x})^T$$

is less obvious. In a first step one asserts that the curl of a genuine vector field (in $\mathbb{R}^3$) is an axial vector field,

$$\boldsymbol{A}'(t', \boldsymbol{x}') = -\boldsymbol{A}(t, \boldsymbol{x}) \Longleftrightarrow \nabla' \times \boldsymbol{A}'(t', \boldsymbol{x}') = +\nabla \times \boldsymbol{A}(t, \boldsymbol{x}),$$

while the curl of an axial vector field is again a vector field. Equipped with this knowledge one sees that the Maxwell equations are invariant under space reflection if

$\boldsymbol{E}$, $\boldsymbol{D}$, and $\boldsymbol{j}$ are *vector* fields,
$\boldsymbol{B}$ and $\boldsymbol{H}$ are *axial vector* fields,
ϱ is a *scalar* field.

This becomes plausible if one recalls some concrete experimental situations involving electric and magnetic fields. For instance, the electric field of a point charge at rest

$$\boldsymbol{E}(\boldsymbol{x}) = \frac{q}{r^2}\hat{\boldsymbol{r}}$$

is proportional to the position vector $\boldsymbol{r}$, up to factors which are invariant under Π, and, hence, is a vector field. A current density $\boldsymbol{j}$ may be thought of as a flux of point-like charged particles which flow through space with the velocity $\boldsymbol{v}$. This is a genuine vector field, too. The magnetic dipole density (1.120a) is proportional to the cross product of $\boldsymbol{x}$ and $\boldsymbol{j}(\boldsymbol{x})$ and, therefore, is an axial vector field. The same statement holds for the induction field (1.124b). Finally, the charge density must be a scalar, be it only for the reason that the continuity equation (1.21) relates the time derivative of ϱ with the divergence of the current density and, as a whole, must be invariant.

Once more we refer the reader to the geometric formulation of Maxwell theory if he or she wishes to work out more clearly the noted difference between the electric quantities $\boldsymbol{E}$ and $\boldsymbol{D}$ on the one hand, and the magnetic quantities $\boldsymbol{B}$ and $\boldsymbol{H}$ on the other. One will then find out that the first two are equivalent to exterior *one*-forms, while the second group are equivalent to exterior *two*-forms over $\mathbb{R}^3$.

Behaviour Under Time Reversal

It is certainly reasonable to expect that the charge density $\varrho(t, \boldsymbol{x})$ does not depend on the direction of time, whether time runs towards the future or towards the past. That is to say to require that it be invariant under time reversal T,

$$\varrho'(t', \boldsymbol{x}') = \varrho(t, \boldsymbol{x}), \qquad t' = -t, \ \boldsymbol{x}' = \boldsymbol{x}.$$

The continuity equation which contains the first derivative of the charge density with respect to time, then implies that the current density must be *odd*, $\boldsymbol{j}'(t', \boldsymbol{x}') = -\boldsymbol{j}(t, \boldsymbol{x})$. Note that this property was to be expected on the basis of the simple model developed above. In order for the two Maxwell equations (2.1c) and (2.1d) to be invariant, one must have

$$\boldsymbol{H}'(t', \boldsymbol{x}') = -\boldsymbol{H}(t, \boldsymbol{x}), \qquad \boldsymbol{D}'(t', \boldsymbol{x}') = +\boldsymbol{D}(t, \boldsymbol{x}).$$

The electric field $\boldsymbol{E}$ transforms like the displacement field $\boldsymbol{D}$, the induction field $\boldsymbol{B}$ transforms like the magnetic field $\boldsymbol{H}$.

Charge Conjugation

A particularly interesting question which is new as compared to mechanics concerns the behaviour of Maxwell's equations when the signs of all charges are reversed. This is the operation of *charge conjugation* C which plays a central role in the quantum dynamics of elementary systems. For example, when applied to a hydrogen atom this means that the proton p is replaced by an antiproton $\overline{p}$, the electron e^- is replaced by a positron e^+.

By their definition both the charge density and the current density reverse their signs. Written symbolically, $\mathsf{C}\varrho(t, \boldsymbol{x}) = -\varrho(t, \boldsymbol{x})$, $\mathsf{C}\boldsymbol{j}(t, \boldsymbol{x}) = -\boldsymbol{j}(t, \boldsymbol{x})$. From (2.1c) and from the first of these relations one concludes that the displacement field $\boldsymbol{D}$ changes sign, too. This behaviour then also applies to the electric field. The second relation, together with (2.1d), requires that $\boldsymbol{H}$ and thus also $\boldsymbol{B}$ be odd as well. In summary,

$$\mathsf{C}\boldsymbol{D}(t, \boldsymbol{x}) = -\boldsymbol{D}(t, \boldsymbol{x}), \qquad \mathsf{C}\boldsymbol{E}(t, \boldsymbol{x}) = -\boldsymbol{E}(t, \boldsymbol{x}),$$
$$\mathsf{C}\boldsymbol{H}(t, \boldsymbol{x}) = -\boldsymbol{H}(t, \boldsymbol{x}), \qquad \mathsf{C}\boldsymbol{B}(t, \boldsymbol{x}) = -\boldsymbol{B}(t, \boldsymbol{x}).$$

As before these transformation rules are plausible: If the charges which are the sources of the electric field change sign (without modifying their absolute value) the electric field changes everywhere from $\boldsymbol{E}(t, \boldsymbol{x})$ to $-\boldsymbol{E}(t, \boldsymbol{x})$. As all current densities change sign, too, this applies also to the magnetic fields they give rise to.

In summary, we note that Maxwell's equations are covariant under rotations in the given frame of reference, as well as under the discrete transformations Π, T, and C. However, whether or not the discrete transformations are symmetries in the sense of quantum mechanics is a question about the interactions other than electrodynamics which are acting between the building blocks of matter. The electromagnetic interaction, taken in isolation, is indeed invariant under space reflection and time reversal, as well as under charge conjugation. In a world where all protons are replaced by antiprotons, all electrons are replaced by positrons, the atoms have the same bound states and the spectral lines of atomic physics are the same as in our familiar environment.

2.2.2 Maxwell's Equations and Exterior Forms

This section deals for the first, but not the last, time with the geometric nature of the physical quantities that are involved in Maxwell's equations. In particular, we elucidate what in the intuitive language of physics is called pseudoscalar and axial vector. We do this by means of a short summary of exterior differential calculus on Euclidean spaces $\mathbb{R}^n$, but refer to [ME], Chap. 5, for an extensive and more general presentation.

Exterior Forms on $\mathbb{R}^n$

Exterior one-forms $\overset{1}{\omega}$ in the point $x \in M = \mathbb{R}^n$ are linear maps of tangent vectors on M in x, i.e. of elements of the tangent space $T_x M$, into the reals,

$$\overset{1}{\omega} : T_x M \longrightarrow \mathbb{R} : v \longmapsto \overset{1}{\omega}(v) . \tag{2.7a}$$

An important example which illustrates well this notion is the total differential $\mathrm{d}f$ of a smooth function on $\mathbb{R}^n$, in which case

$$\mathrm{d}f(v)|_x = v(f)(x) = \sum_{i=1}^{n} v^i \frac{\partial f}{\partial x^i}\bigg|_x \equiv \sum_{i=1}^{n} v^i \partial_i f\big|_x \tag{2.7b}$$

represents the directional derivative of the function f at the point x along the direction of v. The action of $\mathrm{d}f$ on the tangent vector v is equal to the action $v(f)$ of this vector on the function and coincides with the derivative of f in the direction defined by v. Indeed, the directional derivative is a real number. In the formulation of (2.7b) we introduced the compact notation

$$\partial_i f := \frac{\partial f}{\partial x^i} \tag{2.7c}$$

for the derivative by the contravariant component x^i which, in turn, is *co*variant.

The set of linear maps from $T_x M$ to $\mathbb{R}$ (by definition) spans the dual vector space $T_x^* M$, called *cotangent space* which is attached to the point x, like $T_x M$ is attached to x.

▶ **Remark**
If M is a smooth n-manifold which is not a Euclidean space $\mathbb{R}^n$, one must construct a complete atlas composed of local charts (φ, U) (also called coordinate

systems) where U is an open neighbourhood of the point $p \in M$ and

$$\varphi : M \to \mathbb{R}^n : U \mapsto \varphi(U)$$

is a homeomorphism from U on M to the image $\varphi(U) \subset \mathbb{R}^n$. Denoting local coordinates in this chart by $\{x^i\}, i = 1, \ldots, n$, the partial derivative of a function f is given by

$$\partial_i^{(\varphi)}\Big|_p (f) = \frac{\partial(f \circ \varphi^{-1})}{\partial x^i}(\varphi(p)) . \tag{2.8}$$

Only the composition of $\varphi^{-1} : \mathbb{R}^n \to M$ and of $f : M \to \mathbb{R}$ is a real function on $\mathbb{R}^n$ which can be differentiated according to the rules of analysis. In case the manifold is an $\mathbb{R}^n$ matters simplify: For $M = \mathbb{R}^n$ only one single chart $U = M$ is needed, up to diffeomorphisms, and the corresponding map can be chosen to be $\varphi = \mathrm{id}$, the identical mapping. In this case the formerly local expression (2.8) holds on the whole of M and reduces to the usual partial derivative (2.7c) well known from real analysis.

With $v(f)$ the directional derivative of the function f in the point x, and $\partial_i f$ the partial derivative with respect to the coordinate x^i the expression

$$v = \sum_{i=1}^{n} v^i \partial_i$$

is the decomposition of the vector v in terms of the base fields $\{\partial_i\}, i = 1, \ldots, n$. These base fields span the tangent space $T_x M$. In the case of an $\mathbb{R}^n$, however, one can identify all tangent spaces with one another and with the base manifold. This means that every smooth vector field V on $M = \mathbb{R}^n$ can be decomposed

$$V = \sum_{i=1}^{n} v^i(x)\partial_i , \tag{2.9}$$

with coefficients $v^i(x)$ which are smooth functions.

Of course, the coordinates x^i are smooth functions on M themselves: x^i associates to the point $x \in M$ its i-th coordinate. The differentials $\mathrm{d}x^i$ of these functions are one-forms and are called *base one-forms.* The set of all $\{\mathrm{d}x^i\}, i = 1, \ldots, n$ is dual to the basis $\{\partial_i\}$, since

$$\mathrm{d}x^i(\partial_k) = \partial_k(x^i) \equiv \frac{\partial}{\partial x^k} x^i = \delta^i_k .$$

Therefore, every one-form $\overset{1}{\omega} \in T_x M$ can be expanded in this basis, $\overset{1}{\omega} = \sum \omega_i \, \mathrm{d}x^i$.

A one-form $\overset{1}{\omega}$ is said to be smooth if it is defined on all of M and if $\overset{1}{\omega}(V)$ is a smooth function for all smooth vector fields $V \in \mathcal{V}(M)$. When applied to $M = \mathbb{R}^n$ this means that every one-form $\overset{1}{\omega}$ can be written as an expansion

$$\overset{1}{\omega} = \sum_{i=1}^{n} \omega_i(x)\,\mathrm{d}x^i \tag{2.10}$$

where the coefficients $\omega_i(x)$ are smooth functions. The coefficient functions are calculated from the action of the one-form on the base vector fields, viz.

$$\omega_i(x) = \overset{1}{\omega}(\partial_i)\,.$$

Thus, the action on an arbitrary smooth vector field is given by

$$\overset{1}{\omega}(V) = \sum_{i=1}^{n} V^i(x)\omega_i(x)\,,$$

where

$$V = \sum_j V^j(x)\partial_j \quad \text{and} \quad \overset{1}{\omega} = \sum_k \omega_k(x)\,\mathrm{d}x^k\,.$$

There exists a skew-symmetric, associative product of exterior forms, called *exterior product*, or *wedge product*, whose definition is most simply given for base one-forms and their action on vectors as follows

$$\left(\mathrm{d}x^i \wedge \mathrm{d}x^j\right)(v, w) = v^i w^j - v^j w^i = \det\begin{pmatrix} v^i & w^i \\ v^j & w^j \end{pmatrix}, \tag{2.11a}$$

where use was made of its antisymmetry,

$$\mathrm{d}x^i \wedge \mathrm{d}x^j = -\mathrm{d}x^j \wedge \mathrm{d}x^i\,. \tag{2.11b}$$

The following example shows that this definition is the direct generalization of the well-known cross product in $\mathbb{R}^3$:
In the space $\mathbb{R}^3$ there are three base one-forms, $\mathrm{d}x^1$, $\mathrm{d}x^2$ and $\mathrm{d}x^3$. If one applies the wedge product of the second and the third of these to two vectors $\boldsymbol{a}$ and $\boldsymbol{b}$,

$$\left(\mathrm{d}x^2 \wedge \mathrm{d}x^3\right)(\boldsymbol{a}, \boldsymbol{b}) = a^2 b^3 - a^3 b^2 = (\boldsymbol{a} \times \boldsymbol{b})_1 \quad (\text{on } \mathbb{R}^3)\,,$$

the result is seen to be the first component of the cross product. Adding to this formula the two formulae obtained by cyclic permutation of the indices one obtains the full cross product $\boldsymbol{a} \times \boldsymbol{b}$.

The exterior product is easily extended to three or more factors. For example, with three base one-forms and three tangent vectors one has

$$\left(\mathrm{d}x^i \wedge \mathrm{d}x^j \wedge \mathrm{d}x^k\right)(u, v, w) = \det \begin{pmatrix} u^i & v^i & w^i \\ u^j & v^j & w^j \\ u^k & v^k & w^k \end{pmatrix} \tag{2.11c}$$

This formula illustrates the associativity of the exterior product. No parentheses need be written in a product of more than two factors. For example, the product $(\mathrm{d}x^i \wedge \mathrm{d}x^j) \wedge \mathrm{d}x^k$ is the same as $\mathrm{d}x^i \wedge (\mathrm{d}x^j \wedge \mathrm{d}x^k)$. (Note that in the second example the position of parentheses corresponds to the expansion of the determinant along the first row.)

The products $\mathrm{d}x^i \wedge \mathrm{d}x^j$ with $i < j$, of which there are $n(n-1)/2 = \binom{n}{2}$, are elements of $T_x^* \times T_x^*$, which, in addition, are antisymmetric. The whole set for all i and j provides a basis for arbitrary smooth *two-forms* so that

$$\overset{2}{\omega} = \sum_{i<j} \omega_{ij}(x)\, \mathrm{d}x^i \wedge \mathrm{d}x^j \,. \tag{2.12}$$

The coefficients $\omega_{ij}(x)$ are smooth functions on $M = \mathbb{R}^n$. In the language of classical tensor analysis such an object $\overset{2}{\omega}$ is a tensor field of type $(0, 2)$

$$\overset{2}{\omega} \in \mathfrak{T}^0_2(M) \,,$$

which, in addition, is antisymmetric. The set of all coefficients ω_{ij} gives its representation in coordinates and in the form of an antisymmetric tensor of degree 2.

The chain of base forms can be continued, in a finite number of steps, up to the wedge product of n base one-forms. This procedure yields the base k-forms $\mathrm{d}x^{i_1} \wedge \mathrm{d}x^{i_2} \wedge \cdots \wedge \mathrm{d}x^{i_k}$, $k = 3, \ldots, n$, of which there are $\binom{n}{k}$ for every k. With these tools at hand one can construct smooth k-forms

$$\overset{k}{\omega} = \sum_{i_1 < \cdots < i_k} \omega_{i_1 \ldots i_k}(x)\, \mathrm{d}x^{i_1} \wedge \ldots \wedge \mathrm{d}x^{i_k} \,, \tag{2.13}$$

with coefficients $\omega_{i_1 \ldots i_k}(x)$ which are again smooth functions on $\mathbb{R}^n$. The exterior form $\overset{k}{\omega}$ is an element of $\mathfrak{T}^0_k(M)$, the space of covariant tensor fields of degree k, and is antisymmetric in the k vector fields to which it is applied. It is customary to denote the space of antisymmetric, covariant tensor fields of degree k by

$$\overset{k}{\omega} \in \Lambda^k(M) \,. \tag{2.14}$$

It is not difficult to determine the dimension of theses spaces. By counting the base elements $\mathrm{d}x^{i_1} \wedge \mathrm{d}x^{i_2} \wedge \ldots \wedge \mathrm{d}x^{i_k}$ one finds that the dimension of $\Lambda^k(M)$ is (s. Exercise 2.1)

$$\dim \Lambda^k(M) = \binom{n}{k} = \frac{n!}{k!(n-k)!} \,.$$

Thus, Λ^1 has dimension n, very much like Λ^{n-1}. Λ^n has dimension 1, while there is no space Λ^m whose dimension is greater than n.

The Exterior Derivative

The exterior derivative is the generalization of the total differential for functions, of the gradient of a function and of the curl and the divergence for vector fields in $\mathbb{R}^3$. It has the following properties:
It maps k-forms to $(k+1)$-forms (which may be zero),

$$\mathrm{d} : \Lambda^k(M) \to \Lambda^{k+1}(M) : \overset{k}{\omega} \mapsto \mathrm{d}\,\overset{k}{\omega} \,. \tag{2.15a}$$

When applied to a smooth function it yields the total differential

$$\mathrm{d} : f \mapsto \mathrm{d}f = \sum_i \frac{\partial f}{\partial x^i}\mathrm{d}x^i \,. \tag{2.15b}$$

The exterior derivative fulfills a graded Leibniz rule with specific signs as follows: When applied to the exterior product of an r-form and an s-form ($r, s = 0, 1, \ldots, n$) the result is

$$\mathrm{d}(\overset{r}{\omega} \wedge \overset{s}{\omega}) = (\mathrm{d}\,\overset{r}{\omega}) \wedge \overset{s}{\omega} + (-)^r\, \overset{r}{\omega} \wedge (\mathrm{d}\,\overset{s}{\omega}) \,. \tag{2.15c}$$

This resembles the familiar product rule of differential calculus except for the fact that the second term keeps its plus sign only if the first factor is an *even* form, but receives a minus sign if the degree r of the first factor is *odd*. As a rule of thumb one may remember that "shifting" the operator d past an r-form produces a sign $(-)^r$. Obviously, the exterior product $\overset{r}{\omega} \wedge \overset{s}{\omega}$ is an element of Λ^{r+s}, its exterior derivative is in Λ^{r+s+1}.

If d is applied twice the result is always zero

$$\mathrm{d} \circ \mathrm{d} = 0 \,. \tag{2.15d}$$

The following formula for the exterior derivative of a k-form in the representation (2.13) is useful in practice

$$\begin{aligned}\mathrm{d}\,\overset{k}{\omega} &= \sum_{i_1<\ldots<i_k} \mathrm{d}\omega_{i_1\ldots i_k}(x) \wedge \mathrm{d}x^{i_1} \wedge \mathrm{d}x^{i_2} \wedge \cdots \wedge \mathrm{d}x^{i_k} \\ &= \sum_{j=1}^{n} \sum_{i_1<\ldots<i_k} \frac{\partial \omega_{i_1\ldots i_k}(x)}{\partial x^j} \mathrm{d}x^j \wedge \mathrm{d}x^{i_1} \wedge \mathrm{d}x^{i_2} \wedge \cdots \wedge \mathrm{d}x^{i_k} \,.\end{aligned} \tag{2.15e}$$

It contains in the first step the total differential of the functions $\omega_{i_1,\ldots,i_k}(x^1, \ldots, x^n)$ which is to be calculated following the rule (2.15b) for functions. At the end of a calculation of this type one must reorder the base one-forms in order to arrange them in increasing order and keep track of the signs that this may produce.

Remarks

1. For functions the property (2.15d) is nothing but the fact that the mixed second derivatives of a smooth function are equal. Indeed, taking the exterior derivative of $\mathrm{d}f$, one has

$$\begin{aligned} \mathrm{d}(\mathrm{d}f) &= \sum_i \left(\mathrm{d}\frac{\partial f}{\partial x^i}\right) \wedge \mathrm{d}x^i && \text{(according to rule (2.15e))} \\ &= \sum_{k \neq i} \frac{\partial^2 f}{\partial x^k \partial x^i}\, \mathrm{d}x^k \wedge \mathrm{d}x^i && \text{(with formula (2.15b) for total differentials)} \\ &= \sum_{k<i} \left\{ \frac{\partial^2 f}{\partial x^k \partial x^i} - \frac{\partial^2 f}{\partial x^i \partial x^k} \right\} \mathrm{d}x^k \wedge \mathrm{d}x^i = 0 \\ & && \text{(antisymmetry of base-forms)} . \end{aligned}$$

For forms of higher degree the rule (2.15d) follows from the Leibniz rule (2.15c).

2. Contemplating the series of spaces $\Lambda^1(M), \ldots, \Lambda^k(M), \ldots, \Lambda^n(M)$, one notices that their dimensions follow the binomial series $\binom{n}{1} = n, \ldots, \binom{n}{n} = 1$ but that the series of numbers in Pascal's triangle is incomplete. The number $\binom{n}{0} = 1$, i.e. the dimension of $\Lambda^0(M)$ is missing. Conversely, the exterior derivative $\mathrm{d}f$ of a function f is a one-form and, according to (2.15a), the operator d leads from $\Lambda^k(M)$ to $\Lambda^{k+1}(M)$. This suggests interpretation of the smooth functions in the framework of exterior forms as *zero-forms,*

$$f \in \mathfrak{F}(M) \quad \text{(smooth functions on } M\text{)}, \qquad f \in \Lambda^0(M) .$$

3. If the application of d to an exterior k-form is zero this form is said to be *closed,*

$$\mathrm{d}\omega = 0 , \qquad \omega \in \Lambda^k(M) .$$

For instance, the total differential of a function is a closed form because $\mathrm{d} \circ \mathrm{d}f = \mathrm{d}(\mathrm{d}f) = 0$.

In turn, it may happen that a $(k+1)$-form η can be written as the exterior derivative of a k-form ω, i.e.

$$\eta = \mathrm{d}\omega , \quad \eta \in \Lambda^{k+1}(M) , \quad \omega \in \Lambda^k(M) .$$

A form of this kind is said to be an *exact* form. Clearly, every exact form is also a closed form. Regarding the converse one has: On $M = \mathbb{R}^n$ every closed k-form can be written as the derivative of a $(k-1)$-form. Note that on more general manifolds this holds only locally. This is the content of Poincaré's lemma[1].

Hodge Dual Forms

The space $\mathbb{R}^n$ not only is a smooth manifold but is also orientable. In other terms, an ordered basis $(\hat{e}_1, \hat{e}_2, \ldots, \hat{e}_n)$ spans a generalized parallelepiped which can be assigned a sign. The spaces $\Lambda^k(M)$ and $\Lambda^{(n-k)}(M)$ have the same dimension because of the equality of binomial coefficients

$$\binom{n}{k} = \binom{n}{n-k} = \frac{n!}{k!(n-k)!}$$

and they are isomorphic. One defines a bijective mapping, the so-called $\star$-operation, which associates to every k-form a $(n-k)$-form. The image of a k-form ω under Hodge dualism is denoted by $\star\omega$. Defining it by means of the action of forms onto unit vectors, the k-form ω is mapped to the $(n-k)$-form $\star\omega$ by

$$(\star\omega)\left(\hat{e}_{i_{k+1}}, \ldots, \hat{e}_{i_n}\right) = \varepsilon_{i_1 \ldots i_k i_{k+1} \ldots i_n}\, \omega\left(\hat{e}_{i_1}, \ldots, \hat{e}_{i_k}\right) . \tag{2.16}$$

For example, in $\mathbb{R}^3$ one has

$$\star\, \mathrm{d}x^i = \frac{1}{2} \sum_{j,k} \varepsilon_{ijk}\, \mathrm{d}x^j \wedge \mathrm{d}x^k , \tag{2.17a}$$

$$\star \left(\mathrm{d}x^i \wedge \mathrm{d}x^j\right) = \varepsilon_{ijk}\, \mathrm{d}x^k , \tag{2.17b}$$

$$\star \left(\mathrm{d}x^1 \wedge \mathrm{d}x^2 \wedge \mathrm{d}x^3\right) = 1 . \tag{2.17c}$$

In this example the spaces of one-forms and of two-forms are isomorphic because of the equality $\binom{3}{1} = \binom{3}{2} = 3$ and because $\mathbb{R}^3$ is orientable. Likewise, the three-forms and the functions are related one-to-one. (Note that in this case there is only one base three-form.)

In the general case of an $\mathbb{R}^n$ bijectivity is established by the relation

$$\star\left((\star\omega)\right) = (-)^{k(n-k)}\omega , \quad (\omega \in \Lambda^k) . \tag{2.18}$$

Applying the $\star$-operation to a k-form twice takes it back to the original form, up to a sign which depends on its degree.

The star operation and the exterior derivative can be combined to a new and very interesting operator. As anticipated in equations (1.49) and (1.50) define

$$\delta = (-)^{n(k+1)+1} \star \mathrm{d} \star , \tag{2.19a}$$

$$\mathbf{\Delta}_{\mathrm{LdR}} = \mathrm{d} \circ \delta + \delta \circ \mathrm{d} . \tag{2.19b}$$

[1] The Poincaré lemma applies to any open neighbourhood $U \subset M$ of the point $p \in M$ which can be contracted to p without leaving the manifold M.

The first of these operators, in some sense, is the counterpart of the exterior derivative. Indeed, one easily verifies that δ *lowers* the degree of the exterior form by one,

$$\delta : \Lambda^k(M) \longrightarrow \Lambda^{(k-1)}(M) .$$

The first mapping $\star$ leads from degree k to degree $(n-k)$, the application of d converts it to an $(n-k+1)$-form, and a second $\star$-mapping yields a $(n-(n-k+1)) = (k-1)$-form. By the same token one sees that the Laplace-de-Rham operator $\mathbf{\Delta}_{\mathrm{LdR}}$ does not change the degree of the form on which it acts.

Example 2.1

On the space $\mathbb{R}^3$ there exist the spaces Λ^0 and Λ^3 both of which have the dimension $\binom{3}{0} = 1 = \binom{3}{3}$, as well as the spaces Λ^1 and Λ^2 which both have dimension $\binom{3}{1} = 3 = \binom{3}{2}$. Regarding the base forms one has

$$\star\, \mathrm{d}x^i = \frac{1}{2}\sum_{j,k} \varepsilon_{ijk}\, \mathrm{d}x^j \wedge \mathrm{d}x^k , \tag{2.20a}$$

$$\star \left(\mathrm{d}x^i \wedge \mathrm{d}x^j\right) = \sum_{k} \varepsilon_{ijk}\, \mathrm{d}x^k , \tag{2.20b}$$

$$\star \left(\mathrm{d}x^1 \wedge \mathrm{d}x^2 \wedge \mathrm{d}x^3\right) = 1 . \tag{2.20c}$$

Example 2.2

This example is particularly important for electrodynamics, though it repeats an example whose details are worked out, for example in [ME], Sect. 5.4.5. Note that we made use of it in Sect. 1.6.1 above. Let $\boldsymbol{a}$ be a vector field on $M = \mathbb{R}^3$. Define the covariant components $a_i = a^i$ (with $\boldsymbol{a} \equiv \boldsymbol{a}(\boldsymbol{x})$), a one-form and a two-form, respectively, by

$$\overset{1}{\omega}_{\boldsymbol{a}} = \sum_{i=1}^{3} a_i(\boldsymbol{x})\, \mathrm{d}x^i , \tag{2.21a}$$

$$\overset{2}{\omega}_{\boldsymbol{a}} = \frac{1}{2}\sum_{i,j,k} \varepsilon_{ijk} a_i(\boldsymbol{x})\, \mathrm{d}x^j \wedge \mathrm{d}x^k . \tag{2.21b}$$

(The numerical factor in (2.21b) accounts for the antisymmetric permutations of (i, j, k).) On account of the relation (2.20b) and the formula (1.48b) one sees that

$$\star\, \overset{2}{\omega}_{\boldsymbol{a}} = \frac{1}{2}\sum_{i,j,k} \varepsilon_{ijk} a_i(\boldsymbol{x}) \varepsilon_{jkl}\, \mathrm{d}x^l = \overset{1}{\omega}_{\boldsymbol{a}} ,$$

i.e. the Hodge dual of the two-form (2.21b) coincides with the original one-form (2.21a).

The exterior derivative of the first form (2.21a) yields the two-form corresponding to the curl of $\boldsymbol{a}$, and then, after application of $\star$, the one-form constructed with the curl, viz.

$$\mathrm{d}\overset{1}{\omega}_{\boldsymbol{a}} = \frac{1}{2}\sum_{i,j,k} \varepsilon_{ijk}\left(\boldsymbol{\nabla}\times\boldsymbol{a}\right)_i \mathrm{d}x^j \wedge \mathrm{d}x^k , \quad \text{or} \quad \star\,\mathrm{d}\overset{1}{\omega}_{\boldsymbol{a}} = \overset{1}{\omega}_{\boldsymbol{\nabla}\times\boldsymbol{a}} . \tag{2.22}$$

The exterior derivative of the two-form (2.21b) yields the divergence of $\boldsymbol{a}$,

$$\mathrm{d}\overset{2}{\omega}_{\boldsymbol{a}} = \left(\boldsymbol{\nabla}\cdot\boldsymbol{a}\right)\mathrm{d}x^1 \wedge \mathrm{d}x^2 \wedge \mathrm{d}x^3 , \quad \text{or} \quad \star\,\mathrm{d}\overset{2}{\omega}_{\boldsymbol{a}} = \boldsymbol{\nabla}\cdot\boldsymbol{a} . \tag{2.23}$$

The action of the Laplace–de-Rham operator on a function or on a one-form of the type (2.21a) gives the results, respectively,

$$\boldsymbol{\Delta}_{\mathrm{LdR}} f = -\boldsymbol{\Delta} f(\boldsymbol{x}) , \tag{2.24a}$$

$$\boldsymbol{\Delta}_{\mathrm{LdR}}\,\overset{1}{\omega}_{\boldsymbol{a}} = -\sum_{i=1}^{3}\left(\boldsymbol{\Delta} a_i(\boldsymbol{x})\right)\mathrm{d}x^i , \tag{2.24b}$$

where $\boldsymbol{\Delta}$ denotes the customary Laplace(–Beltrami) operator, i.e. $\boldsymbol{\Delta} = \partial_1^2 + \partial_2^2 + \partial_3^2$, acting on smooth functions in either case.

Fields and Sources in Maxwell's Equations

We still remain in a fixed reference frame where spacetime has the structure $\mathbb{R}_t \times \mathbb{R}^3$ with a fixed division of spacetime into the factor to be called time and the remainder describing the well-known laboratory space of an experimenter. All quantities which are related by Maxwell's equations are defined as geometric objects over $\mathbb{R}^3$ but depend parametrically on time. Surely, this is a restricted perspective because it rests on a subjective perception of time and space. Nevertheless, the fundamental laws of electrodynamics in integral form give direct hints at the geometric role of fields and densities.

Faraday's law (1.12) contains, on the one side, the path integral of the tangential component of the electric field, and, on the other side, the integral of the magnetic flux over a surface whose boundary is that path. The path $\mathcal{C}$, by itself, is a smooth, closed manifold with dimension $\dim \mathcal{C} = 1$. The surface whose boundary is $\mathcal{C}$, is also a smooth manifold with dimension $\dim F(\mathcal{C}) = 2$. Quite generally, on an orientable manifold M with metric g and $\dim M = n$, the exterior n-form

$$\Omega^{(n)} = \sqrt{|g|}\,\mathrm{d}x^1 \wedge \mathrm{d}x^2 \wedge \cdots \wedge \mathrm{d}x^n \tag{2.25}$$

defines the so-called volume form. Here $|g|$ is the determinant, or more precisely, the absolute value of the determinant of the metric, $|g| = |\det\{g_{ik}\}|$. The exterior form which is proportional to the only base element that exists in $\Lambda^n(M)$, carries the orientation of the basis through the order of factors in the product $\mathrm{d}x^1 \wedge \cdots \wedge \mathrm{d}x^n$. It is independent of the choice of the coordinate system. This is shown as follows: Let Φ be a diffeomorphism which relates the coordinates $(x^1, \ldots, x^n)$ to new coordinates $(y^1, \ldots, y^n)$. The metric tensor which in the original coordinates has the form $\{g_{ij}(x)\}$, is replaced by $\bar{g}_{kl}(y)$ in the new coordinates, and we have

$$(x^1, \ldots, x^n) \longleftrightarrow (y^1, \ldots, x^n)\,, \quad \bar{g}_{kl}(y) = \sum_{i,j=1}^{n} \frac{\partial x^i}{\partial y^k} \frac{\partial x^j}{\partial y^l} g_{ij}(x)\,.$$

Hence, the determinants of the metric tensors g and $\bar{g}$ are related by

$$|\bar{g}|\,(y) = \left(\det\left(\frac{\partial x^i}{\partial y^k}\right)\right)^2 |g|\,(x)\,.$$

If the map Φ preserves the orientation, i.e. if the two coordinate systems have the same orientation, one takes the square root and obtains

$$\sqrt{|\bar{g}|} = \det\left(\frac{\partial x^i}{\partial y^k}\right)\sqrt{|g|}\,.$$

Therefore, the expansions of an arbitrary smooth n-form in the first and in the second coordinate system, respectively, are related by

$$\overset{n}{\omega} = a(x)\,\mathrm{d}x^1 \wedge \cdots \wedge \mathrm{d}x^n = \bar{a}(y)\,\mathrm{d}y^1 \wedge \cdots \wedge \mathrm{d}y^n \quad \text{with}$$

$$\bar{a}(y) = a(x)\det\left(\frac{\partial x^i}{\partial y^k}\right).$$

This shows that, indeed, the volume form (2.25), $\Omega^{(n)}$, is invariant.

Example 2.3

This is an example in dimension 2 where calculations are particularly simple:

$$\begin{aligned}
\Omega^{(2)}(x) &= \sqrt{|g|}\,\mathrm{d}x^1 \wedge \mathrm{d}x^2 \\
&= \sqrt{|g|}\left(\frac{\partial x^1}{\partial y^1}\mathrm{d}y^1 + \frac{\partial x^1}{\partial y^2}\mathrm{d}y^2\right) \wedge \left(\frac{\partial x^2}{\partial y^1}\mathrm{d}y^1 + \frac{\partial x^2}{\partial y^2}\mathrm{d}y^2\right) \\
&= \sqrt{|g|}\left\{\frac{\partial x^1}{\partial y^1}\frac{\partial x^2}{\partial y^2} - \frac{\partial x^1}{\partial y^2}\frac{\partial x^2}{\partial y^1}\right\}\mathrm{d}y^1 \wedge \mathrm{d}y^2 \\
&= \sqrt{|g|}\det\left(\frac{\partial x^i}{\partial y^k}\right)\mathrm{d}y^1 \wedge \mathrm{d}y^2 \\
&= \sqrt{|\bar{g}|}\,\mathrm{d}y^1 \wedge \mathrm{d}y^2 \equiv \Omega^{(2)}(y)\,.
\end{aligned}$$

One sees that not only the volume element but also the orientation of the coordinate system is conserved. The transition between x and y contains the Jacobi determinant of the transformation which carries a well-defined sign. Liouville's theorem on the conservation of a domain of initial conditions, of its volume and orientation, provides a good illustration.

These arguments and the example show that integration over an n-dimensional manifold M must have the form

$$\int_M (\text{integrand})\, \Omega^{(n)} \,, \quad \text{with } \Omega^{(n)} \text{ the volume form on } M \,.$$

Expressed differently this means that only integration of n-forms over the whole of M is meaningful.

A detailed discussion of integration on smooth manifolds would go beyond the scope of this section and also of this book. Therefore, I concentrate here on some plausibility arguments which emerge from Maxwell's equations in integral form, and refer to the literature on differential geometry for a more rigourous presentation. (For a concise, though short introduction and, in particular, a proof of Stokes' theorem in the general form of the equation (1.8b) see, e. g. [Arnol'd 1988, Sect. 36].)

Let us return for a moment to the original integral form of Faraday's law (1.12). The closed curve C over which one integrates, geometrically speaking, is a one-dimensional manifold embedded in $\mathbb{R}^3$. It inherits an induced metric from $g_{ik} = \text{diag}(1, 1, 1)$. In the spirit of what was noticed above the path integral over the tangential component $\mathrm{d}\boldsymbol{s} \cdot \boldsymbol{E}(t, \boldsymbol{x})$ of the electric field must be the integral of a *one*-form on C, and, therefore, on $\mathbb{R}^3$, by the embedding of the curve in space. Thus, it seems natural to associate to the electric field a one-form of the kind of (2.21a), viz.

$$\overset{1}{\omega}_{\boldsymbol{E}} := E_1(t, \boldsymbol{x})\, \mathrm{d}x^1 + E_2(t, \boldsymbol{x})\, \mathrm{d}x^2 + E_3(t, \boldsymbol{x})\, \mathrm{d}x^3 \,. \tag{2.26a}$$

On the right-hand side of Faraday's law (1.12) the normal component of the field $\boldsymbol{B}$ is integrated over a surface F which is also embedded in $\mathbb{R}^3$. If one compares this with the definition of the two-form (2.21b) including its characteristic ordering of indices, and takes account of the statement that only two-forms can be integrated consistently over surfaces ($\dim F = 2$), one realizes that, from a geometric point of view, $\boldsymbol{B}$ must be associated to a *two*-form of the type of (2.21b),

$$\begin{aligned} \overset{2}{\omega}_{\boldsymbol{B}} := {} & B_1(t, \boldsymbol{x})\, \mathrm{d}x^2 \wedge \mathrm{d}x^3 + B_2(t, \boldsymbol{x})\, \mathrm{d}x^3 \wedge \mathrm{d}x^1 \\ & + B_3(t, \boldsymbol{x})\, \mathrm{d}x^1 \wedge \mathrm{d}x^2 \,. \end{aligned} \tag{2.26b}$$

Fig. 2.1 Spherical calotte in 3-space. Faraday's law is formulated by exterior forms on this surface

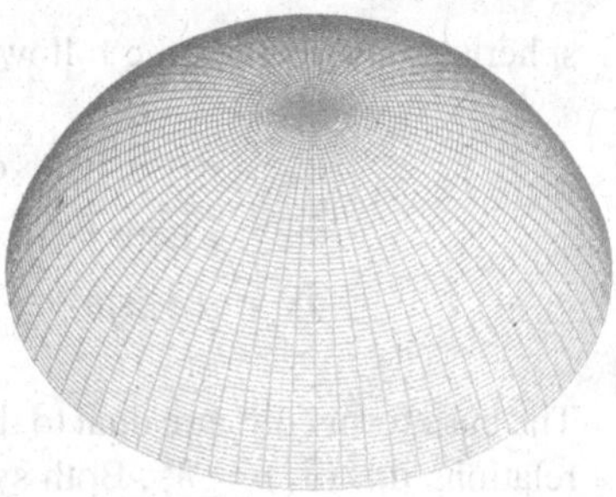

Example 2.4

A simple, though physically unrealistic example may be helpful. Consider the rectangle in the (1, 2)-plane which is defined by the vectors $\boldsymbol{v} = v\hat{\boldsymbol{e}}_1$ and $\boldsymbol{w} = w\hat{\boldsymbol{e}}_2$. The integral of the one-form (2.26a) over the edges of the rectangle is seen to be the integral of the tangential component of $\boldsymbol{E}$ along this curve. Regarding the restriction of the two-form (2.26b) to the surface of the rectangle, conversely, only the third term survives whose coefficient, indeed, is B_3.

Example 2.5

The following example is closer to physics and it should be studied carefully because, on the one hand, it illustrates the assertion that only the integration of an n-form over an n-dimensional manifold is meaningful. On the other hand, it shows that the integral form of Faraday's law written in terms of exterior forms,

$$\int_{\partial F} \overset{1}{\omega}_{\boldsymbol{E}} = -\frac{1}{c}\frac{\mathrm{d}}{\mathrm{d}t}\int_{F} \overset{2}{\omega}_{\boldsymbol{B}} ,$$

is identical with the customary form (1.12) of that law.

Let the surface F be the calotte of a sphere with radius $r = R$ shown in Fig. 2.1 which is enclosed between the latitude defined by the angle θ_0 and the north pole ($\theta = 0$). Its boundary ∂F is the circle of latitude with fixed $\theta = \theta_0$ and azimuth in the interval $\phi \in [0, 2\pi]$. In this example it is useful to use spherical polar coordinates r, θ, ϕ instead of the cartesian coordinates x^1, x^2, x^2,

$$x^1 = r \sin\theta \cos\phi , \quad x^2 = r \sin\theta \sin\phi , \quad x^3 = r \cos\theta .$$

The first problem consists in determining the base one-forms $\mathrm{d}u^k$ in polar coordinates and to expand the two exterior forms in terms of these. Denoting by $\hat{\boldsymbol{e}}_i$ the cartesian unit vectors, by $\hat{\boldsymbol{a}}_1 \equiv \hat{\boldsymbol{e}}_r$ (pointing in radial direction), $\hat{\boldsymbol{a}}_2 \equiv \hat{\boldsymbol{e}}_\theta$ (tangential to the meridian), and $\hat{\boldsymbol{a}}_3 \equiv \hat{\boldsymbol{e}}_\phi$ (tangential to circle of latitude) the spherical unit vectors, the following relation holds

$$\hat{\boldsymbol{a}}_1 = \hat{\boldsymbol{e}}_1 \sin\theta\cos\phi + \hat{\boldsymbol{e}}_2 \sin\theta\sin\phi + \hat{\boldsymbol{e}}_3\cos\theta\ ,$$
$$\hat{\boldsymbol{a}}_2 = \hat{\boldsymbol{e}}_1 \cos\theta\cos\phi + \hat{\boldsymbol{e}}_2 \cos\theta\sin\phi - \hat{\boldsymbol{e}}_3\sin\theta\ ,$$
$$\hat{\boldsymbol{a}}_3 = -\hat{\boldsymbol{e}}_1 \sin\phi + \hat{\boldsymbol{e}}_2\cos\phi\ .$$

The base forms $\mathrm{d}u^k$ are dual to the base vectors $\hat{\boldsymbol{a}}_j$ and, therefore, must fulfill the relations $\mathrm{d}u^k(\hat{\boldsymbol{a}}_j) = \delta^k_j$. Both systems refer to real and orthogonal coordinates. Hence, the same transformation formulae hold for base one-forms and for base vectors,

$$\mathrm{d}u^1 = \mathrm{d}x^1 \sin\theta\cos\phi + \mathrm{d}x^2 \sin\theta\sin\phi + \mathrm{d}x^3\cos\theta\ ,$$
$$\mathrm{d}u^2 = \mathrm{d}x^1 \cos\theta\cos\phi + \mathrm{d}x^2 \cos\theta\sin\phi - \mathrm{d}x^3\sin\theta\ ,$$
$$\mathrm{d}u^3 = -\mathrm{d}x^1 \sin\phi + \mathrm{d}x^2\cos\phi\ .$$

This partial result may be interpreted in two different ways: One calculates the action of $\mathrm{d}u^k$ on the unit vectors $\hat{\boldsymbol{a}}_j$, makes use of the relation $\mathrm{d}x^p(\hat{\boldsymbol{e}}_q) = \delta^p_q$ and deduces the expected relation $\mathrm{d}u^k(\hat{\boldsymbol{a}}_j) = \delta^k_j$. Alternatively, one utilizes the differentials

$$\mathrm{d}x^1 = \sin\theta\cos\phi\ \mathrm{d}r + r\cos\theta\cos\phi\ \mathrm{d}\theta - r\sin\theta\sin\phi\ \mathrm{d}\phi\ ,$$
$$\mathrm{d}x^2 = \sin\theta\sin\phi\ \mathrm{d}r + r\cos\theta\sin\phi\ \mathrm{d}\theta + r\sin\theta\cos\phi\ \mathrm{d}\phi\ ,$$
$$\mathrm{d}x^3 = \cos\theta\ \mathrm{d}r - r\sin\theta\ \mathrm{d}\theta\ ,$$

to determine the line element

$$(\mathrm{d}s)^2 = \sum_{i=1}^{3}(\mathrm{d}x^i)^2 = \sum_{k=1}^{3}(\mathrm{d}u^k)^2$$
$$= (\mathrm{d}r)^2 + (r\,\mathrm{d}\theta)^2 + (r\sin\theta\,\mathrm{d}\phi)^2\ .$$

As a result one obtains the representation of the base one-forms in spherical polar coordinates:

$$\mathrm{d}u^1 = \mathrm{d}r\ ,\quad \mathrm{d}u^2 = r\,\mathrm{d}\theta\ ,\quad \mathrm{d}u^3 = r\sin\theta\,\mathrm{d}\phi\ .$$

One then applies these results to the exterior forms of electrodynamics to obtain on the circle of latitude $\theta = \theta_0$

$$\overset{1}{\omega}_{\boldsymbol{E}} = \sum_{i=1}^{3} E_i\,\mathrm{d}x^i = [-E_1\sin\phi + E_2\cos\phi]\,R\sin\theta\,\mathrm{d}\phi = E_\phi\,\mathrm{d}u^3\ .$$

As expected, the one-form appears as the tangential component oriented along the circle of latitude. Its integral over that circle is

$$\int_{\partial F} \overset{1}{\omega}_{\boldsymbol{E}} = \int_0^{2\pi} R \sin\theta \, \mathrm{d}\phi \, E_\phi \, .$$

Regarding the two-form of the magnetic induction, its restriction to the spherical calotte implies that $\mathrm{d}u^1 = 0$ so that only the base two-form $\mathrm{d}u^2 \wedge \mathrm{d}u^3$ contributes. Inserting the results given above one finds

$$B_1 \, \mathrm{d}x^2 \wedge \mathrm{d}x^3 + B_2 \, \mathrm{d}x^3 \wedge \mathrm{d}x^1 + B_3 \, \mathrm{d}x^1 \wedge \mathrm{d}x^2$$
$$= [B_1 \sin\theta \cos\phi + B_2 \sin\theta \sin\phi + B_3 \cos\theta] \; \mathrm{d}u^2 \wedge \mathrm{d}u^3 \, .$$

The base form $\mathrm{d}u^2 \wedge \mathrm{d}u^3$ fixes the orientation of the normal to the surface and is equal to $R^2 \sin\theta \, \mathrm{d}\theta \, \mathrm{d}\phi$. The expression in square brackets is the normal component $B_n = \boldsymbol{B} \cdot \hat{\boldsymbol{n}}$ of the magnetic induction, the normal $\hat{\boldsymbol{n}}$ being directed *outwards* on the calotte. Thus, the integral of the two-form is

$$\int_F \overset{2}{\omega}_{\boldsymbol{B}} = R^2 \int_0^{\theta_0} \sin\theta \, \mathrm{d}\theta \int_0^{2\pi} \mathrm{d}\phi \; \boldsymbol{B} \cdot \hat{\boldsymbol{n}}$$

and one recovers the integral form of Faraday's law.

Following similar arguments as for $\boldsymbol{B}$ one deduces from Gauss' law (1.14) that one must associate to the field $\boldsymbol{D}$ a *two*-form – in contrast to the electric field $\boldsymbol{E}$,

$$\overset{2}{\omega}_{\boldsymbol{D}} := D_1(t, \boldsymbol{x}) \, \mathrm{d}x^2 \wedge \mathrm{d}x^3 + D_2(t, \boldsymbol{x}) \, \mathrm{d}x^3 \wedge \mathrm{d}x^1 + D_3(t, \boldsymbol{x}) \, \mathrm{d}x^1 \wedge \mathrm{d}x^2 \, . \tag{2.27a}$$

It should be clear that Maxwell's equations when written in terms of exterior forms, can only relate forms of equal degree. The second inhomogeneous Maxwell equation (2.1d), considered in vacuum, i.e. with $\boldsymbol{j} \equiv 0$, without loss of generality, indicates that the curl of the field $\boldsymbol{H}$ must be equivalent to a two-form. Therefore, the field $\boldsymbol{H}$ itself must be associated to a one-form of the type (2.21a),

$$\overset{1}{\omega}_{\boldsymbol{H}} := H_1(t, \boldsymbol{x}) \, \mathrm{d}x^1 + H_2(t, \boldsymbol{x}) \, \mathrm{d}x^2 + H_3(t, \boldsymbol{x}) \, \mathrm{d}x^3 \, . \tag{2.27b}$$

Regarding the source terms in the inhomogeneous equations (2.1c) and (2.1d) one sees that to the charge density one must associate a three-form, to the current density a two-form, respectively, as follows,

$$\overset{3}{\omega}_{\varrho} := \varrho(t,\boldsymbol{x})\,\mathrm{d}x^1 \wedge \mathrm{d}x^2 \wedge \mathrm{d}x^3\,, \tag{2.28a}$$

$$\overset{2}{\omega}_{\boldsymbol{j}} := j_1(t,\boldsymbol{x})\,\mathrm{d}x^2 \wedge \mathrm{d}x^3 + j_2(t,\boldsymbol{x})\,\mathrm{d}x^3 \wedge \mathrm{d}x^1 + j_3(t,\boldsymbol{x})\,\mathrm{d}x^1 \wedge \mathrm{d}x^2 \tag{2.28b}$$

These assignments were deduced from the inhomogeneous equations but they can also be made plausible from the integral fundamental laws. Indeed, the charge density always appears integrated over three-dimensional volume in order to yield physical charges, while the current density is integrated over cross sections of conductors such as to yield current strengths.

Maxwell's equations can now be formulated in terms of exterior forms such that their local form (2.1a–2.1d) follow by comparison of coefficients for forms of equal degree. Both the two homogeneous Maxwell equations (2.1a) and (2.1b), and the two inhomogeneous equations (2.1c) and (2.1d) take a very simple form. They read

$$\mathrm{d}\,\overset{2}{\omega}_{\boldsymbol{B}} = 0\,, \tag{2.29a}$$

$$\mathrm{d}\,\overset{1}{\omega}_{\boldsymbol{E}} + \frac{1}{c}\frac{\partial}{\partial t}\overset{2}{\omega}_{\boldsymbol{B}} = 0\,, \tag{2.29b}$$

$$\mathrm{d}\,\overset{2}{\omega}_{\boldsymbol{D}} = 4\pi\,\overset{3}{\omega}_{\varrho}\,, \tag{2.29c}$$

$$\mathrm{d}\,\overset{1}{\omega}_{\boldsymbol{H}} - \frac{1}{c}\frac{\partial}{\partial t}\overset{2}{\omega}_{\boldsymbol{D}} = \frac{4\pi}{c}\overset{2}{\omega}_{\boldsymbol{j}}\,. \tag{2.29d}$$

The first of these equations follows from Example 2.2, making use of (2.23), while the second is obtained using (2.22) of the same example. Similar arguments apply to the case of equations (2.29c) and (2.29d). The first and the third equations relate exterior three-forms, the second and the fourth relate exterior two-forms.

The continuity equation expressed in terms of exterior forms reads

$$\frac{\partial}{\partial t}\overset{3}{\omega}_{\varrho} + \mathrm{d}\,\overset{2}{\omega}_{\boldsymbol{j}} = 0\,, \qquad \Longrightarrow \qquad \frac{\partial \varrho(t,\boldsymbol{x})}{\partial t} + \nabla\cdot\boldsymbol{j}(t,\boldsymbol{x}) = 0 \tag{2.30}$$

where again the equation (2.23) was used. The second equation on the right follows by comparison of coefficients.

▶ **Remarks**

1. Contemplating the definitions (2.26a), (2.26b), (2.27a), and (2.27b), an important observation comes to mind: All exterior forms are *invariant* under rotations $\mathsf{R} \in \mathrm{SO}(3)$ of the frame of reference. In contrast to the

components of the original fields they do not depend on the choice of coordinate system. The covariance of the Maxwell equations (2.29a–2.29d) with respect to rotations is obvious and needs not be checked separately.

2. Their behaviour under space reflection Π introduced in Sect. 2.2.1, is also of interest. The one-form (2.26a) of the electric field and the two-form (2.26b) of the induction field are invariant under Π. By the same token the peculiar difference in the behaviour of the original vector fields $\boldsymbol{E}$ and $\boldsymbol{B}$ under Π is understood: The electric field is a genuine vector field and corresponds to a *one*-form, the induction field which has the "wrong" transformation behaviour under space reflection corresponds to a *two*-form.
3. The spacetime on which Maxwell's equations are formulated, is $\mathbb{R}^4$, i.e. an orientable manifold. As long as we study only proper rotations R $\in$ SO(3) the orientation is not changed and all four exterior forms corresponding to the fields remain unchanged. Space reflection reverses the orientation. In contrast to the one-form (2.26a) of the electric field, the one-form (2.27b) of the magnetic field changes its sign. A similar conclusion follows from the comparison of (2.26b) with (2.27a). Exterior forms of this kind are defined on nonorientable manifolds and are called *twisted forms*.
4. Although this cannot be the last word on this topic it is instructive to summarize the behaviour of the exterior forms that are related by (2.29a–2.29d), and by (2.30), under the three discrete transformations Π, T, and C. Table 2.1 is based on the analysis of Sect. 2.2.1.
5. This analysis as a first attempt of a geometric interpretation of the Maxwell fields remains unsatisfactory because the fields depend not only on $\boldsymbol{x} \in \mathbb{R}^3$, but also on the time coordinate $t \in \mathbb{R}_t$, hence, are defined over a *four*-dimensional manifold. In the following section we will quit the fixed frame of reference that was assumed here, and will prove the covariance of Maxwell's equations under Lorentz transformations. This will lead quite naturally to generalizing these definitions in such a way that the fields and the source terms become exterior forms over Minkowski space.
6. If like in Sect. 1.6.3 one wishes to describe the electric field and the magnetic induction by means of potentials – still within a fixed division of spacetime into coordinate space and time – it is useful to define the one-form

$$\overset{1}{\omega}_{A} := \sum_{i=1}^{3} A_i(t,\boldsymbol{x})\,\mathrm{d}x^i \tag{2.31}$$

whose coefficients are the components of the vector potential $\boldsymbol{A}(t,\boldsymbol{x})$. The scalar potential $\Phi(t,\boldsymbol{x})$ is a function and can be interpreted as a zero-form over the space $\mathbb{R}^3$. The representations (1.55a) and (1.55b) of the induction

Table 2.1 Behaviour of the electromagnetic exterior forms under the three discrete transformations. Note, however, that the behaviour under T and, hence, under the product ΠTC will be modified to some extent when these forms are defined over space *and* time

	Π	T	C	ΠTC
$\overset{1}{\omega}_{\boldsymbol{E}}$	+	+	−	−
$\overset{2}{\omega}_{\boldsymbol{B}}$	+	−	−	+
$\overset{1}{\omega}_{\boldsymbol{H}}$	−	−	−	−
$\overset{2}{\omega}_{\boldsymbol{D}}$	−	+	−	+
$\overset{3}{\omega}_{\varrho}$	−	+	−	+
$\overset{2}{\omega}_{\boldsymbol{j}}$	−	−	−	−

field and the electric field, respectively, written by means of exterior forms, become

$$\overset{2}{\omega}_{\boldsymbol{B}} = \mathrm{d}\,\overset{1}{\omega}_{\boldsymbol{A}}\,, \tag{2.32a}$$

$$\overset{1}{\omega}_{\boldsymbol{E}} = -\frac{1}{c}\frac{\partial}{\partial t}\,\overset{1}{\omega}_{\boldsymbol{A}} - \mathrm{d}\Phi\,. \tag{2.32b}$$

Taking the exterior derivative of the first of these, one obtains

$$\mathrm{d}\,\overset{2}{\omega}_{\boldsymbol{B}} = \mathrm{d}\big(\mathrm{d}\,\overset{1}{\omega}_{\boldsymbol{A}}\big) = 0\,, \quad \text{or} \quad \boldsymbol{\nabla}\cdot\boldsymbol{B}(t,\boldsymbol{x}) = 0\,.$$

Here use was made of the property (2.15d) of the exterior derivative. This repeats the well-known fact that an induction field that can be represented by a vector potential has divergence zero automatically. The exterior derivative of the second equation (2.32b), in turn, yields the conclusion that the curl of $\boldsymbol{E}$ is related to the time derivative of $\boldsymbol{A}$,

$$\mathrm{d}\,\overset{1}{\omega}_{\boldsymbol{E}} = -\frac{1}{c}\frac{\partial}{\partial t}\mathrm{d}\,\overset{1}{\omega}_{\boldsymbol{A}}\,, \quad \text{or} \quad \boldsymbol{\nabla}\times\boldsymbol{E}(t,\boldsymbol{x}) = -\frac{1}{c}\frac{\partial}{\partial t}\boldsymbol{\nabla}\times\boldsymbol{A}(t,\boldsymbol{x})\,.$$

If the vector potential is independent of time the electric field is irrotational.

This representation has a further unsatisfactory feature: Although to the vector potential $\boldsymbol{A}$ it associates a *one*-form over $\mathbb{R}^3$, the scalar potential is described by a *zero*-form. Furthermore, no real use is made of the time dependence of these quantities.

Why is there this asymmetry between space and time?

2.3 Lorentz Covariance of Maxwell's Equations

Space and time translations,

$$\boldsymbol{x}' = \boldsymbol{x} + \boldsymbol{a}\,, \quad t' = t + s\,,$$

rotations ($t' = t$, $\boldsymbol{x}' = \mathbf{R}\boldsymbol{x}$) in $\mathbb{R}^3$, space reflection diag $(1, -1, -1, -1)$, and time reversal diag$(-1, 1, 1, 1)$ have the same effect in the Galilei group and in the Poincaré group. Only the Special Galilei transformations

$$\begin{pmatrix} t \\ \boldsymbol{x} \end{pmatrix} \longmapsto \begin{pmatrix} t' = t \\ \boldsymbol{x}' = \boldsymbol{x} + \boldsymbol{v}t \end{pmatrix}, \quad \text{or} \quad \begin{pmatrix} t' \\ |\boldsymbol{x}'\rangle \end{pmatrix} = \begin{pmatrix} 1 & \boldsymbol{0} \\ |\boldsymbol{v}\rangle & \mathbb{1}_3 \end{pmatrix} \begin{pmatrix} t \\ |\boldsymbol{x}\rangle \end{pmatrix} \tag{2.33}$$

differ in an essential way from the Special Lorentz transformations (also called *boosts*)

$$\begin{pmatrix} x^0 \\ |\boldsymbol{x}\rangle \end{pmatrix} \longmapsto \begin{pmatrix} x'^0 \\ |\boldsymbol{x}'\rangle \end{pmatrix} = \begin{pmatrix} \gamma & \frac{1}{c}\gamma\langle\boldsymbol{v}| \\ \frac{1}{c}\gamma|\boldsymbol{v}\rangle & \mathbb{1}_3 + \frac{\gamma^2}{c^2(\gamma+1)}|\boldsymbol{v}\rangle\langle\boldsymbol{v}| \end{pmatrix} \begin{pmatrix} x^0 \\ |\boldsymbol{x}\rangle \end{pmatrix}, \tag{2.34}$$

where the time variable is replaced by the equivalent length $x^0 = ct$. A Galilei transformation of this class, taken along the 1-axis $|\boldsymbol{v}\rangle = v|\hat{\boldsymbol{e}}_1\rangle$, for example, reads

$$t' = t\,, \quad x'^1 = x^1 + vt\,, \quad x'^2 = x^2\,, \quad x'^3 = x^3\,,$$

or, in a somewhat different notation, using $x^0 = ct$ and $\beta = v/c$,

$$x'^0 = x^0\,, \qquad x'^1 = \beta x^0 + x^1\,,$$
$$x'^2 = x^2\,, \qquad x'^3 = x^3\,,$$

while in the case of the Lorentz group one has, instead,

$$x'^0 = \gamma x^0 + \gamma\beta x^1\,, \qquad x'^1 = \gamma\beta x^0 + \gamma x^1\,,$$
$$x'^2 = x^2\,, \qquad x'^3 = x^3\,,$$

with $\beta = |\boldsymbol{v}|/c$ and $\gamma = (1 - \beta^2)^{-1/2}$.

Under a special Lorentz transformation which relates two observers moving with constant velocity $\boldsymbol{v}$ relative to each other, neither the electric field $\boldsymbol{E}$ by itself nor the induction field $\boldsymbol{B}$ by itself, can have a simple transformation behaviour. This can be seen in different ways. A first, intuitive argument is the following:

The argument relates to the Biot–Savart law (1.18) and starts from the model of a single point charge q assumed to be moving with constant velocity $\boldsymbol{v}$ relative to an inertial observer. In his frame of reference K the observer sees the particle moving along a straight line with constant velocity, first approaching and then flying off, so that the strength of the particle's Coulomb field increases and decreases in the course of time. In addition, he perceives the particle which passes by as an electric current density $\boldsymbol{j}(t,\boldsymbol{x}) = q\,\delta\big(\boldsymbol{x} - (\boldsymbol{v}t + \boldsymbol{x}_0)\big)$ which according to (1.18) creates an $\boldsymbol{H}$-field – and, as he is in vacuum, an induction field $\boldsymbol{B} \equiv \boldsymbol{H}$ – which is time and space dependent. Another inertial observer who travels along with the particle sees something radically different: In his frame of reference K' the particle is at rest and creates the spherically symmetric electric field $\boldsymbol{E}' = q\,\hat{\boldsymbol{r}}/r^2$. Since there is no electric current there is also no magnetic or induction field, $\boldsymbol{H}' = \boldsymbol{B}' = 0$. As only *relative* motion is relevant, the two observers are equivalent and the Maxwell equations should have the same physical interpretation in the two frames of reference. Independently of whether the relative velocity is small as compared to, or close to the speed of light, the special Lorentz transformation $\mathsf{L}(\boldsymbol{v}) : (\boldsymbol{E}(t,\boldsymbol{x}), \boldsymbol{B}(t,\boldsymbol{x})) \mapsto (\boldsymbol{E}'(t',\boldsymbol{x}'), \boldsymbol{B}'(t',\boldsymbol{x}'))$ must mix the two types of fields.

A more rigourous analytical argument starts from the Lorentz force (1.44e) and is worked out in more detail in Sect. 2.3.4 below. It leads straightforwardly to the correct transformation behaviour: Maxwell's equations are found to be covariant under the Lorentz group.

We start with a summary of the most important properties of the Poincaré and the Lorentz groups, but refer to [ME] for a more detailed exposition.

2.3.1 Poincaré and Lorentz Groups

A Poincaré transformation is a general affine transformation of the coordinates x of spacetime $M = \mathbb{R}^4$, as well as of tangent vectors[2] $v \in T_x M$,

$$(\Lambda, a) : x \longmapsto x' = \Lambda x + a\,, \quad y \longmapsto y' = \Lambda y + a\,, \tag{2.35a}$$

which leave the generalized (squared) distance

$$(x-y)^2 = \left(x^0 - y^0\right)^2 - (\boldsymbol{x} - \boldsymbol{y})^2 \tag{2.35b}$$

invariant. Here $(x^0 = ct, \boldsymbol{x})$ is the decomposition of x into temporal and spatial parts, respectively, in a given system of reference K which may, but need not be an inertial system. By convention the components are labelled by *Greek* indices whenever one deals with all four of them, for time and for space, and by *Latin* indices if only the spatial components are concerned,

$$x = \{x^\mu \mid \mu = 0,1,2,3\} = \left(x^0, \{x^i \mid i = 1,2,3\}\right)^T = \left(x^0, \boldsymbol{x}\right)^T .$$

When it is equal to zero the invariant (2.35b) describes the causal relationship between emission of a light quantum at x, i.e. at time $t_x = x^0/c$ and position $\boldsymbol{x}$, and

2 As the base manifold M is the flat space $\mathbb{R}^4$ all tangent spaces $T_x\mathbb{R}^4$ can be identified with this space. As a consequence, the points $x \in M$ and the vectors $v \in T_x M$ have the same transformation behaviour.

its detection at the world point y, i.e. at time $t_y = y^0/c$ and position $\boldsymbol{y}$. It expresses the experimentally confirmed *constancy of the speed of light:* In all inertial frames the speed of light has the universal value

$$c = 2.99792458 \cdot 10^8 \text{ m s}^{-1} . \tag{2.36}$$

The squared distance (2.35b) is invariant under Poincaré transformations, independently of whether it is zero (i.e. *lightlike*), positive (i.e. *timelike*), or negative (i.e. *spacelike*). A notation which is equivalent to (2.35b) makes use of the *metric tensor* $\mathsf{g} = \{g_{\mu\nu}\} = \mathrm{diag}(1, -1, -1, -1)$. It reads

$$(x-y)^2 = \sum_{\mu,\nu=0}^{3} (x^\mu - y^\mu) g_{\mu\nu} (x^\nu - y^\nu) \equiv (x^\mu - y^\mu) g_{\mu\nu} (x^\nu - y^\nu) , \tag{2.37}$$

where in the second step Einstein's summation convention was introduced which says that two equal indices, one of which is covariant, while the other is contravariant, are to be summed over from 0 to 3. Conventionally covariant indices are written as *lower* indices, contravariant indices are noted as *upper* ones.

Inserting the transformation (2.35a) into the formula (2.37), and requesting the equality $(x'-y')^2 = (x-y)^2$ for all inertial systems, the translation term a cancels out in the difference of x and y. There remains a condition on the homogeneous part of the transformation (2.35a), viz.

$$\Lambda^T \mathsf{g} \Lambda = \mathsf{g} . \tag{2.38a}$$

This equation is the essential condition for the Lorentz group from which all characteristic properties of Lorentz transformations are deduced. One should note the analogy to the rotation group in $\mathbb{R}^3$: The defining property of the rotation group O(3) in three-dimensional space with the metric $\mathsf{g}|_{\mathbb{R}^3} = \mathrm{diag}(1, 1, 1)$ is

$$\mathsf{R}^T \mathbb{1}_3 \mathsf{R} = \mathbb{1}_3 ,$$

from which one concludes that $(\det \mathsf{R})^2 = 1$ and that $\mathsf{R}^{-1} = \mathsf{R}^T$, i.e. that R is orthogonal.

Equation (2.38a), written out in components, leads to a number of consequences that we describe schematically as follows. With the notation $\Lambda = \{\Lambda^\mu{}_\nu\}$ and using the summation convention the equation (2.38a) reads more explicitly

$$\Lambda^\mu{}_\sigma g_{\mu\nu} \Lambda^\nu{}_\tau = g_{\sigma\tau} . \tag{2.38b}$$

Note that μ and ν are summation indices, while σ and τ assume fixed values on both sides of the equation. The relative position of indices in the left-hand factor of (2.38b) seems in conflict with the rules of matrix multiplication but, in fact, is correct because it is the transpose of Λ which appears here.

Depending on the values of the fixed indices σ and τ, equation (2.38b) yields for

$$\sigma = 0, \tau = 0: \quad (\Lambda^0{}_0)^2 - \sum_{j=1}^{3} (\Lambda^j{}_0)^2 = 1 , \tag{2.38c}$$

$$\sigma = i, \tau = k: \quad \Lambda^0{}_i \Lambda^0{}_k - \sum_{j=1}^{3} \Lambda^j{}_i \Lambda^j{}_k = -\delta_{ik} , \tag{2.38d}$$

$$\sigma = 0, \tau = k: \quad \Lambda^0{}_0 \Lambda^0{}_k - \sum_{j=1}^{3} \Lambda^j{}_0 \Lambda^j{}_k = 0 . \tag{2.38e}$$

One concludes from the first of these the alternatives that

$$\text{either} \quad \text{(a): } \Lambda^0{}_0 \geqslant +1 , \quad \text{or} \quad \text{(b): } \Lambda^0{}_0 \leqslant -1 . \tag{2.39a}$$

Lorentz transformations which have the property (a) map the time coordinate forward, i.e. into the future. They are called *orthochronous*. Calculation of the determinant of the two sides of (2.38a), remembering that Λ is real, yields

$$(\det \Lambda)^2 = 1 , \quad \text{hence, either (c): } \det \Lambda = +1 \quad \text{or (d): } \det \Lambda = -1 . \tag{2.39b}$$

Thus, the four possible combinations of the properties (a) to (d) show that the Lorentz group has four disjoint branches. These are denoted by $\pm$ for the sign of the determinant, and by an arrow which points upward if $\Lambda^0{}_0$ is larger than or equal to $+1$, downward if $\Lambda^0{}_0$ is smaller than or equal to -1. The branch $L_+^\uparrow$, called the *proper, orthochronous Lorentz group*, contains all elements Λ with $\det \Lambda = 1$ and $\Lambda^0{}_0 \geqslant +1$. As one easily verifies, this is a subgroup of the Lorentz group: It contains the identity $\mathbb{1}_4$; the product of two transformations $\Lambda_1, \Lambda_2 \in L_+^\uparrow$ is an element of $L_+^\uparrow$, and so is the inverse Λ^{-1} of every element $\Lambda \in L_+^\uparrow$.

Space reflection $\Lambda = \Pi = \mathrm{diag}(1, -1, -1, -1)$ has $\det \Lambda = -1$, and $\Lambda^0{}_0 = +1$, hence, is an element of the branch $L_-^\uparrow$. Time reversal $\mathsf{T} = \mathrm{diag}(-1, 1, 1, 1)$ belongs to $L_-^\downarrow$, while the product $\Pi\mathsf{T}$ of space reflection and time reversal has determinant $+1$, but $\Lambda^0{}_0 = -1$, hence, belongs to the branch $L_+^\downarrow$. As an essential lesson from this analysis we note that one knows the entire Lorentz group once one understands its subgroup $L_+^\uparrow$, the proper orthochronous Lorentz group. Indeed, every element of $L_-^\uparrow$ can be written as the product of an element of $L_+^\uparrow$ with Π, every element of $L_-^\downarrow$ as the product of an element of $L_+^\uparrow$ with T, and every element of $L_+^\downarrow$ as the product of an element of $L_+^\uparrow$ with $\Pi\mathsf{T}$.

The key to the proper orthochronous Lorentz group is provided by the *decomposition theorem,* which asserts that every element of $L_+^\uparrow$ can be written, in a unique manner, as the product of a rotation

$$\mathcal{R} = \begin{pmatrix} 1 & \mathbf{0} \\ \mathbf{0} & \mathsf{R} \end{pmatrix}, \qquad \text{with} \quad \mathsf{R} \in \mathrm{SO}(3),$$

and a special Lorentz transformation, cf. (2.34). Thus, one has

$$\Lambda = \mathsf{L}(\boldsymbol{v})\,\mathcal{R}, \qquad \Lambda \in L_+^{\uparrow}. \tag{2.40a}$$

The entries of the 4×4-matrix $\mathsf{L}(\boldsymbol{v})$ are determined by the velocity

$$\boldsymbol{v} = \frac{c}{\Lambda^0{}_0}\left(\Lambda^1{}_0, \Lambda^2{}_0, \Lambda^3{}_0\right)^T, \tag{2.40b}$$

hence, by the entries $\Lambda^\mu{}_0$ of the given transformation Λ. The entries of the orthogonal 3×3-matrix R are calculated from the formulae

$$R^{ik} = \Lambda^i{}_k - \frac{\Lambda^i{}_0 \Lambda^0{}_k}{1 + \Lambda^0{}_0}. \tag{2.40c}$$

For a proof of this important theorem see, for example [ME], Sect. 4.5.1.

2.3.2 Relativistic Kinematics and Dynamics

To the best of our knowledge, all *charged* particles of nature are massive particles. Physical trajectories on which such a particle moves at a velocity smaller than c, are described by *world lines* $x(\tau)$ whose tangent vector field is everywhere *time*like. The Lorentz invariant parameter τ denotes proper time, i.e. the time that an observer who travels with the particle reads on his clock. The trajectory is described by the function $x(\tau)$ in a way independent of any choice of specific coordinates. The corresponding velocity is characterized by a four-vector which is given by, in a coordinate-free way,

$$u(\tau) := \frac{\mathrm{d}}{\mathrm{d}\tau} x(\tau). \tag{2.41}$$

Without loss of generality, the invariant square of u can be normalized such that $u^2 = c^2$.

In the rest system the proper time coincides with the coordinate time of K_0 and one has $\mathrm{d}\tau = \mathrm{d}t$. With respect to a moving system K the line element is $(\mathrm{d}s)^2 = c^2(\mathrm{d}\tau)^2 = c^2(\mathrm{d}t)^2 - (\mathrm{d}\boldsymbol{x})^2$ so that

$$(\mathrm{d}\tau)^2 = (\mathrm{d}t)^2 - \tfrac{1}{c^2}(\mathrm{d}\boldsymbol{x})^2 = (1 - \beta^2)(\mathrm{d}t)^2 = (\mathrm{d}t)^2/\gamma^2.$$

In the momentary rest system K_0 of the particle (whose existence is guaranteed when $m \neq 0$) the velocity four-vector is

$$u(\tau)|_{\mathsf{K}_0} = \left(c, \mathbf{0}\right)^T, \tag{2.41a}$$

while in the "laboratory" system K of another observer relative to whom the particle is moving, it is

$$u(\tau)|_{\mathsf{K}} = \left(\gamma c, \gamma \boldsymbol{v}\right)^T . \tag{2.41b}$$

The relativistic variant of the momentum is the four-vector $p := mu$. It comprises the spatial momentum $\boldsymbol{p} = m\gamma\boldsymbol{v}$ and the corresponding energy (divided by c) $p^0 = mc\gamma = E_p/c$. In the frame of reference K of the observer, i.e. in the laboratory system, one has

$$p|_{\mathsf{K}} = \left(\tfrac{1}{c}E, \boldsymbol{p}\right)^T , \text{ with } E = \gamma mc^2 = \sqrt{(\boldsymbol{p}c)^2 + (mc^2)^2} , \quad \boldsymbol{p} = m\gamma\boldsymbol{v} , \tag{2.42a}$$

while in the momentary rest system of the particle one has, of course,

$$p|_{\mathsf{K}_0} = \left(mc, \boldsymbol{0}\right)^T , \quad E|_{\mathsf{K}_0} = mc^2 , \quad \boldsymbol{p}|_{\mathsf{K}_0} = \boldsymbol{0} . \tag{2.42b}$$

One easily verifies that (2.41b) follows from (2.41a) by the action of the special Lorentz transformation $\mathsf{L}(\boldsymbol{v})$, cf. (2.34). In a similar way one verifies that

$$p|_{\mathsf{K}} = \mathsf{L}(\boldsymbol{p})\, p|_{\mathsf{K}_0} ,$$

$$\text{with} \quad \mathsf{L}(\boldsymbol{p}) = \frac{1}{mc^2}\begin{pmatrix} E & c\langle\boldsymbol{p}| \\ c|\boldsymbol{p}\rangle & mc^2\,\mathbb{1}_3 + \frac{c^2}{(E+mc^2)}|\boldsymbol{p}\rangle\langle\boldsymbol{p}| \end{pmatrix}$$

where L was converted from a parametrization in terms of the velocity $\boldsymbol{v}$ to a parametrization in terms of the spatial momentum $\boldsymbol{p}$, using the relations (2.42a). Indeed, one has

$$\begin{pmatrix} E & c\langle\boldsymbol{p}| \\ c|\boldsymbol{p}\rangle & mc^2\,\mathbb{1}_3 + \frac{c^2}{(E+mc^2)}|\boldsymbol{p}\rangle\langle\boldsymbol{p}| \end{pmatrix}\begin{pmatrix} mc \\ |\boldsymbol{0}\rangle \end{pmatrix} = mc^2\begin{pmatrix} (\frac{E}{c}) \\ |\boldsymbol{p}\rangle \end{pmatrix} .$$

The relativistic, Lorentz covariant version of Newton's second law reads

$$m\frac{\mathrm{d}^2}{\mathrm{d}\tau^2}x(\tau) = f(x) . \tag{2.43}$$

It is obtained from the usual nonrelativistic formula $m\ddot{\boldsymbol{x}} = \boldsymbol{F}_{\mathrm{N}}(\boldsymbol{x})$ in which $\boldsymbol{F}_{\mathrm{N}}$ denotes the force field of Newtonian mechanics, by a "boost" from the rest system K_0. Thus, in the rest system

$$m\frac{\mathrm{d}^2}{\mathrm{d}\tau^2}x(\tau)\bigg|_{\mathsf{K}_0} = m(0, \ddot{\boldsymbol{x}})^T \quad \text{and} \quad f|_{\mathsf{K}_0} = (0, \boldsymbol{F}_{\mathrm{N}})^T$$

must hold. The action of the special Lorentz transformation $\mathsf{L}(\boldsymbol{v})$ on these two four-vectors,

$$\begin{pmatrix} f^0 \\ |\boldsymbol{f}\rangle \end{pmatrix} = \begin{pmatrix} \gamma & \frac{1}{c}\gamma\langle\boldsymbol{v}| \\ \frac{1}{c}\gamma|\boldsymbol{v}\rangle & \mathbb{1}_3 + \frac{\gamma^2}{c^2(\gamma+1)}|\boldsymbol{v}\rangle\langle\boldsymbol{v}| \end{pmatrix}\begin{pmatrix} 0 \\ |\boldsymbol{F}_{\mathrm{N}}\rangle \end{pmatrix} , \tag{2.44}$$

yields the individual components of f, using the notation $\langle \boldsymbol{a}|\boldsymbol{c}\rangle \equiv \boldsymbol{a}\cdot\boldsymbol{c}$, as follows

$$f^0 = \frac{1}{c}\gamma\left(\boldsymbol{v}\cdot\boldsymbol{F}_{\mathrm{N}}\right), \tag{2.44a}$$

$$|f\rangle = \boldsymbol{F}_{\mathrm{N}} + \frac{\gamma^2}{c^2(\gamma+1)}\left(\boldsymbol{v}\cdot\boldsymbol{F}_{\mathrm{N}}\right)|\boldsymbol{v}\rangle\ . \tag{2.44b}$$

These expressions can be converted to another form which is instructive: Take the scalar product of (2.44b) with $\boldsymbol{v}$ and use the relation

$$\frac{\boldsymbol{v}^2}{c^2} = \beta^2 = \frac{\gamma^2-1}{\gamma^2} = (\gamma-1)\frac{\gamma+1}{\gamma^2},$$

to obtain

$$(\boldsymbol{v}\cdot\boldsymbol{f}) = \left\{1+\frac{\gamma^2}{\gamma+1}\beta^2\right\}\left(\boldsymbol{v}\cdot\boldsymbol{F}_{\mathrm{N}}\right) = \gamma\left(\boldsymbol{v}\cdot\boldsymbol{F}_{\mathrm{N}}\right).$$

Thus, the zero-component of (2.44a) can be written alternatively $f^0 = (1/c)(\boldsymbol{v}\cdot\boldsymbol{f})$. Regarding the space component (2.44b), make use once more of $\beta^2 = (\gamma^2-1)/\gamma^2$, and insert the identity

$$(\boldsymbol{v}\cdot\boldsymbol{a})\boldsymbol{v} = \boldsymbol{v}^2\boldsymbol{a} + \boldsymbol{v}\times(\boldsymbol{v}\times\boldsymbol{a})$$

to obtain

$$\begin{aligned}\boldsymbol{f} &= \boldsymbol{F}_{\mathrm{N}} + \frac{\gamma^2}{\gamma+1}\left\{\frac{1}{c^2}\boldsymbol{v}\times(\boldsymbol{v}\times\boldsymbol{F}_{\mathrm{N}}) + \beta^2\boldsymbol{F}_{\mathrm{N}}\right\}\\ &= \boldsymbol{F}_{\mathrm{N}}(1+\gamma-1) + \frac{\gamma}{c}\boldsymbol{v}\times\left(\frac{\gamma}{c(\gamma+1)}(\boldsymbol{v}\times\boldsymbol{F}_{\mathrm{N}})\right)\\ &= \gamma\left[\boldsymbol{F}_{\mathrm{N}} + \frac{1}{c}\boldsymbol{v}\times\left(\frac{\gamma}{c(\gamma+1)}(\boldsymbol{v}\times\boldsymbol{F}_{\mathrm{N}})\right)\right].\end{aligned} \tag{2.44c}$$

The spatial part of the left-hand side of the equation of motion (2.43), when expressed in terms of the time derivative of the spatial momentum, is equal to $\gamma\,\mathrm{d}\boldsymbol{p}/\mathrm{d}t$, so that the equation of motion divided by γ reads

$$\frac{\mathrm{d}\boldsymbol{p}}{\mathrm{d}t} = \boldsymbol{F}_{\mathrm{N}} + \frac{1}{c}\boldsymbol{v}\times\left(\frac{\gamma}{c(\gamma+1)}(\boldsymbol{v}\times\boldsymbol{F}_{\mathrm{N}})\right). \tag{2.43a}$$

The time component satisfies the differential equation

$$mc\frac{\mathrm{d}\gamma}{\mathrm{d}t} = \frac{1}{c}\left(\boldsymbol{F}_{\mathrm{N}}\cdot\boldsymbol{v}\right). \tag{2.43b}$$

(It is easily verified that (2.43b) follows from (2.43a).) The first of these equations shows a striking similarity to the equation of motion of a charged particle under the action of the Lorentz force, with $\boldsymbol{F}_{\mathrm{N}}$ taking the role of $q\boldsymbol{E}$, and

$$\text{“}q\boldsymbol{B}\text{”} \equiv \frac{\gamma}{c(\gamma+1)}(\boldsymbol{v}\times\boldsymbol{F}_{\mathrm{N}}) \equiv \frac{q\gamma}{c(\gamma+1)}(\boldsymbol{v}\times\boldsymbol{E})$$

appearing in the magnetic term of the force.

2.3.3 Lorentz Force and Field Strength

The Lorentz force with its characteristic dependence on the velocity

$$\frac{\mathrm{d}\boldsymbol{p}}{\mathrm{d}t} = q\Big(\boldsymbol{E}(t,\boldsymbol{x}) + \frac{1}{c}\boldsymbol{v}\times\boldsymbol{B}(t,\boldsymbol{x})\Big) \tag{2.45}$$

can be cast into the form of the equation of motion (2.43). The space part is obtained by multiplication of (2.45) with a factor $\gamma = 1/\sqrt{1-\boldsymbol{v}^2/c^2}$, the temporal part is obtained from the scalar product of (2.45) with the vector $(\gamma/c)\boldsymbol{v}$:

$$m\gamma\frac{\mathrm{d}}{\mathrm{d}t}(\gamma c) = \gamma\frac{1}{c}q\boldsymbol{E}\cdot\boldsymbol{v}\,, \tag{2.45a}$$

$$m\gamma\frac{\mathrm{d}}{\mathrm{d}t}(\gamma\boldsymbol{v}) = \gamma\Big(q\boldsymbol{E}(t,\boldsymbol{x}) + \frac{q}{c}\boldsymbol{v}\times\boldsymbol{B}(t,\boldsymbol{x})\Big)\,. \tag{2.45b}$$

One sees again the analogy between the differential equations (2.45b) and (2.43a), as well as between the equations (2.45a) and (2.43b). The left-hand sides of (2.45a) and of (2.45b), written covariantly, are $m(\mathrm{d}u^\mu/\mathrm{d}\tau)$; their right-hand sides can be expressed in terms of the four-velocity u as follows. In the frame of reference K define

$$F^{\mu\nu}(x) := \begin{pmatrix} 0 & -E^1(x) & -E^2(x) & -E^3(x) \\ +E^1(x) & 0 & -B^3(x) & +B^2(x) \\ +E^2(x) & +B^3(x) & 0 & -B^1(x) \\ +E^3(x) & -B^2(x) & +B^1(x) & 0 \end{pmatrix}, \quad x = (t,\boldsymbol{x})^T\,, \tag{2.46}$$

and let this field act on $u_\nu = g_{\nu\sigma}u^\sigma = (\gamma c, -\gamma\boldsymbol{v})^T$. Using the summation convention one has

$$F^{\mu\nu}u_\nu = \begin{pmatrix} 0 & -E^1(x) & -E^2(x) & -E^3(x) \\ +E^1(x) & 0 & -B^3(x) & +B^2(x) \\ +E^2(x) & +B^3(x) & 0 & -B^1(x) \\ +E^3(x) & -B^2(x) & +B^1(x) & 0 \end{pmatrix}\begin{pmatrix} \gamma c \\ -\gamma v^1 \\ -\gamma v^2 \\ -\gamma v^3 \end{pmatrix}$$

$$= \begin{pmatrix} \gamma(\boldsymbol{E}(x)\cdot\boldsymbol{v}) \\ \gamma c E^1(x) + \gamma(v^2B^3(x) - v^3B^2(x)) \\ \gamma c E^2(x) + \gamma(v^3B^1(x) - v^1B^3(x)) \\ \gamma c E^3(x) + \gamma(v^1B^2(x) - v^2B^1(x)) \end{pmatrix}.$$

The equation of motion in its general form (2.43) appears here in the specific form

$$m\frac{\mathrm{d}u^\mu}{\mathrm{d}\tau} = \frac{q}{c}\,F^{\mu\nu}u_\nu\,. \tag{2.47}$$

In the frame K it is seen to be identical with the differential equations (2.45a) and (2.45b).

This observation raises an important question:

Does the combination of the electric field and the magnetic induction by the definition (2.46) *have a deeper and more general significance than just to reformulate the Lorentz force* (2.45) *in a compact form with respect to the special frame of reference* K*?*

In other terms the question we are asking is the following: In another frame K′ which differs from K by a Lorentz transformation (which is to say that with K also K′ is an inertial system) the Lorentz force can be expressed in the same compact form, i.e. as $F'^{\mu\nu}u'_\nu$. Are the fields $F'^{\mu\nu}$ and $F^{\mu\nu}$ also related by Lorentz transformations? More precisely, is it true that with

$$u' = \Lambda u \quad \text{also} \quad F' = \Lambda\, F\, \Lambda^T \ \text{holds true?}$$

or, written in components,

$$u'^\sigma = \Lambda^\sigma{}_\mu u^\mu\,, \quad F'^{\sigma\tau}(x') = \Lambda^\sigma{}_\mu \Lambda^\tau{}_\nu F^{\mu\nu}(x)\ ?$$

If this were so then the equation of motion (2.47) would be Lorentz covariant. Its right-hand side $F^{\mu\nu}u_\nu$, summed over ν, is a Lorentz vector and thus transforms with Λ like its left-hand side. The equation of motion has the same form in every inertial system. The question that is raised then narrows down to the question:

Are the Maxwell equations covariant with regard to the transformations $\Lambda \in L_+^\uparrow$ *as suggested by the special form of the Lorentz force?*

The analysis of this question is the subject of the next Section. Before turning to it let us collect the inverse formulae that express the electric and induction fields in terms of $F^{\mu\nu}$. These are

$$E^i = F^{i0} = -F^{0i}\,, \qquad (i = 1,2,3)\,, \tag{2.48a}$$

$$B^i = -\frac{1}{2}\sum_{j,k=1}^{3} \varepsilon_{ijk}F^{jk}\,, \qquad (i = 1,2,3)\,. \tag{2.48b}$$

The object $F^{\mu\nu}(x)$ will turn out to be a tensor field, the *tensor field of electromagnetic field strengths.* It is called, somewhat shorter, *field strength tensor*[3]. Its definition (2.46) shows that this tensor is antisymmetric. In fact, one could have

[3] Its geometric role will be clarified in Chap. 5 below.

deduced this property directly from the equation of motion (2.47): Because of the property $u^2 = c^2 = \text{const.}$ one has

$$\frac{1}{2}\frac{\mathrm{d}u^2}{\mathrm{d}\tau} = u_\mu \frac{\mathrm{d}u^\mu}{\mathrm{d}\tau} = 0\,.$$

Therefore, by contraction of (2.47) with u_μ, one concludes that for all x

$$u_\mu F^{\mu\nu}(x) u_\nu = 0\,.$$

This can only be correct if $F^{\nu\mu}(x) = -F^{\mu\nu}(x)$: The tensor $u_\mu u_\nu$ which is *symmetric* in μ and ν, when contracted with the *antisymmetric* tensor $F^{\mu\nu}$, gives zero. Conversely, if $F^{\mu\nu}$ had a symmetric term this would not give zero upon contraction with $u_\mu u_\nu$.

2.3.4 Covariance of Maxwell's Equations

The homogeneous Maxwell equations (2.1a) and (2.1b) are easily expressed in terms of the tensor field $F^{\mu\nu}(x)$. We show that they read as follows

$$\partial^\lambda F^{\mu\nu} + \partial^\mu F^{\nu\lambda} + \partial^\nu F^{\lambda\mu} = 0\,, \quad \text{with } \lambda \neq \mu \neq \nu \in (0,1,2,3)\,. \tag{2.49}$$

An alternative notation is obtained if one introduces the Levi-Civita symbol in dimension 4 whose properties are:

$$\begin{aligned}
\varepsilon_{0123} &= +1 \\
\varepsilon_{\mu\nu\sigma\tau} &= +1 \quad \text{for } (\mu,\nu,\sigma,\tau) = \text{even permutation of } (0,1,2,3) \\
\varepsilon_{\mu\nu\sigma\tau} &= -1 \quad \text{for } (\mu,\nu,\sigma,\tau) = \text{odd permutation of } (0,1,2,3)
\end{aligned} \tag{2.50}$$

while $\varepsilon_{\mu\nu\sigma\tau} = 0$ in all other cases, i.e. whenever two or more indices are equal. In terms of this totally antisymmetric symbol the equations (2.49) become

$$\varepsilon_{\mu\nu\sigma\tau}\partial^\nu F^{\sigma\tau}(x) = 0\,, \quad (\mu = 0,1,2,3)\,. \tag{2.49a}$$

It is not difficult to verify that (2.49a) summarizes the four homogeneous Maxwell equations. In doing so one must recall that

$$\partial^0 = \partial_0 = \frac{\partial}{\partial x^0}\,, \quad \text{but} \quad \partial^i = -\partial_i = -\frac{\partial}{\partial x^i} = -(\nabla)_i\,.$$

Equation (2.49a) with $\mu = 0$ and with $\varepsilon_{0\nu\sigma\tau} \equiv \varepsilon_{0ijk} = \varepsilon_{ijk}$ (where in the last step one sees the usual ε-symbol in dimension 3) yields

$$\begin{aligned}
0 = \varepsilon_{0\nu\sigma\tau}\partial^\nu F^{\sigma\tau}(x) &= \sum_{i,j,k=1}^{3} \varepsilon_{ijk}\partial^i F^{jk}(t,\boldsymbol{x}) \\
&= 2\left[\varepsilon_{123}\partial^1 F^{23} + \varepsilon_{231}\partial^2 F^{31} + \varepsilon_{312}\partial^3 F^{12}\right] = 2\nabla\cdot\boldsymbol{B}(t,\boldsymbol{x})\,.
\end{aligned}$$

This is the Maxwell equation (2.1a). If the first free index of (2.49a) is taken to be 1, one of the remaining indices ν, σ, and τ, must equal 0, while the other two must be 2 and 3, respectively. In this case (2.49a) yields

$$\begin{aligned} 0 &= \varepsilon_{1023}\partial^0 F^{23} + \varepsilon_{1230}\partial^2 F^{30} + \varepsilon_{1302}\partial^3 F^{02} \\ &= -1\Big(\partial_0(-B^1) + (-\partial_2)E^3 + (-\partial_3)(-E^2)\Big) \\ &= \frac{1}{c}\frac{\partial B^1}{\partial t} + \frac{\partial E^3}{\partial x^2} - \frac{\partial E^2}{\partial x^3}\,. \end{aligned}$$

Obviously, this is the 1-component of the homogeneous Maxwell equation (2.1b). The other two space components are obtained by cyclic permutation of the indices $(1, 2, 3)$.

The *inhomogeneous* equations (2.1c) and (2.1d) are more difficult to translate to a covariant form because they contain source terms which are not genuine elements of Maxwell theory but should follow from a theory of *matter*. We start with a heuristic remark:

The volume element in $\mathbb{R}^4$ is invariant under $\Lambda \in L_+^\uparrow$, $d^4x' = d^4x$, or, with respect to a given frame of reference K, $dx'^0\, d^3x' = dx^0\, d^3x$. If $\varrho(t, \boldsymbol{x})$ is the charge density in that frame then the charge element

$$dq = \varrho(t, \boldsymbol{x})\, d^3x = \varrho'(t', \boldsymbol{x}')\, d^3x'\,,$$

by its very nature is a (physical) invariant. This suggests that under Lorentz transformations the charge density should transform like the time component of a four-vector. This, in turn, is so if the charge density and the current density together build up a Lorentz four-vector, i.e. if

$$j(x) = \big(c\varrho(x), \boldsymbol{j}(x)\big)^T\,, \quad x = \big(x^0, \boldsymbol{x}\big)^T\,, \tag{2.51}$$

is a vector field transforming like a Lorentz vector. As we anticipated in Sect. 1.4.5 the continuity equation then has the compact and Lorentz invariant form (1.24b), $\partial_\mu j^\mu(x) = 0$. Of course, this is a question whose answer must be found outside of Maxwell theory proper. Charged matter which provides the sources of Maxwell's equations, must be described as well by a Lorentz covariant theory and must allow for a four-vector current $j(x)$ which is conserved. This is not obvious!

In what follows let us assume that this is valid and that the charged matter particles, i.e. electrons, atomic nuclei, ions, which compose macroscopic matter, obey a Lorentz covariant theory.

It is suggestive to combine the dielectric displacement field $\boldsymbol{D}$ and the magnetic field $\boldsymbol{H}$ in a tensor field analogous to (2.46),

$$\mathcal{F}^{\mu\nu}(x) := \begin{pmatrix} 0 & -D^1(x) & -D^2(x) & -D^3(x) \\ +D^1(x) & 0 & -H^3(x) & +H^2(x) \\ +D^2(x) & +H^3(x) & 0 & -H^1(x) \\ +D^3(x) & -H^2(x) & +H^1(x) & 0 \end{pmatrix}\,, \tag{2.52a}$$

which is antisymmetric, too. The fields are expressed in terms of $\mathcal{F}^{\mu\nu}$ by formulae analogous to (2.48a) and to (2.48b),

$$D^i = \mathcal{F}^{i0} = -\mathcal{F}^{0i} \,, \qquad (i = 1, 2, 3) \,, \tag{2.52b}$$

$$H^i = -\frac{1}{2} \sum_{j,k=1}^{3} \varepsilon_{ijk} \mathcal{F}^{jk} \,, \qquad (i = 1, 2, 3) \,. \tag{2.52c}$$

The inhomogeneous Maxwell equations now take the compact form

$$\partial_\mu \mathcal{F}^{\mu\nu}(x) = \frac{4\pi}{c} j^\nu(x) \,, \quad (\nu = 0, 1, 2, 3) \,. \tag{2.53}$$

As in the case of the homogeneous equations let us verify this equation in more detail.

For $\nu = 0$ the first index of $\mathcal{F}$ can only take the values 1, 2, and 3, so that (2.53) reduces to

$$\sum_{i=1}^{3} \partial_i \mathcal{F}^{i0}(x) = \nabla \cdot \boldsymbol{D}(t, \boldsymbol{x}) = \frac{4\pi}{c} c\varrho(t, \boldsymbol{x}) = 4\pi\varrho(t, \boldsymbol{x}) \,.$$

Obviously, this is the same as (2.1c).
For a space index, for example $\nu = 1$, the equation (2.53) yields

$$\begin{aligned} \frac{4\pi}{c} j^1(t, \boldsymbol{x}) &= \partial_0 \mathcal{F}^{01}(x) + \partial_2 \mathcal{F}^{21}(x) + \partial_3 \mathcal{F}^{31}(x) \\ &= -\partial_0 D^1(t, \boldsymbol{x}) + \partial_2 H^3(t, \boldsymbol{x}) - \partial_3 H^2(t, \boldsymbol{x}) \\ &= \left(-\frac{1}{c} \frac{\partial \boldsymbol{D}(t, \boldsymbol{x})}{\partial t} + \nabla \times \boldsymbol{H}(t, \boldsymbol{x}) \right)^1 \,. \end{aligned}$$

This is the 1-component of the differential equation (2.1d); the remaining two components follow by cyclic permutation of the space indices $(1, 2, 3)$.

Under the assumption discussed above, i.e. the current density $j(x)$ transforms like a Lorentz vector, the inhomogeneous equations (2.53) are manifestly covariant: Their left-hand sides as well as the right-hand sides transform like vectors under $\Lambda \in L_+^\uparrow$.

▶ **Remarks**

1. The contraction of the Levi-Civita symbol in dimension 4 with the field strength tensor that was used in the homogeneous equations (2.49a) rather naturally leads one to define another covariant tensor field of degree 2, the *dual field strength tensor field,*

$$\star\, F_{\alpha\beta}(x) := \tfrac{1}{2} \varepsilon_{\alpha\beta\mu\nu} F^{\mu\nu}(x) \,. \tag{2.54a}$$

The corresponding contravariant tensor field is

$$\star F^{\sigma\tau}(x) = g^{\sigma\alpha}\left(\star F_{\alpha\beta}(x)\right)g^{\beta\tau} . \tag{2.54b}$$

It is not difficult to calculate $\star F_{\alpha\beta}$ and then $\star F^{\sigma\tau}$. The result (up to completion by virtue of its antisymmetry)

$$\star F^{\sigma\tau} = \begin{pmatrix} 0 & B^1 & B^2 & B^3 \\ & 0 & -E^3 & E^2 \\ & & 0 & -E^1 \\ & & & 0 \end{pmatrix} \tag{2.54c}$$

is interesting because it shows that replacing $F^{\mu\nu}$ by $\star F^{\mu\nu}$ means an exchange of electric field and magnetic induction according to the rule

$$F^{\mu\nu} \longmapsto \star F^{\mu\nu} : (\boldsymbol{E}, \boldsymbol{B}) \longmapsto (-\boldsymbol{B}, \boldsymbol{E}) . \tag{2.55}$$

In the vacuum where $\boldsymbol{D} = \boldsymbol{E}$ and $\boldsymbol{H} = \boldsymbol{B}$, with our choice of units, and in the absence of external sources, the replacement (2.55) is a symmetry of Maxwell's equations (2.1a–2.1d). This symmetry is called *electric-magnetic duality*. It interchanges (2.1a) with (2.1c), as well as (2.1b) with (2.1d). This duality is closely related to the Hodge duality that we studied in Sect. 2.2.2, (2.16). We will return to this in Sect. 2.5.1 below.

2. Of central importance for the covariance of Maxwell's equations was the postulate of the constancy of the speed of light, cf. Sect. 2.3.1. If this were not valid the Maxwell equations would single out a special class of frames of reference whose elements can differ only by translations and rotations but not by special Lorentz transformations. In the early era of electromagnetism this class of frames was called the "ether", its characteristic property being that Maxwell's equations hold in the form given above and that the speed of light has the value (2.36). As is well known, the experiments of A. A. Michelson and E. W. Morley disproved this hypothesis. No effects were found that would show any comotion of light with a frame moving uniformly relative to the hypothetical ether. The speed of light has the same universal value in all inertial systems.
3. The full system of Maxwell equations are Lorentz covariant if and only if the four-current density $j(x)$ is a four-vector field. As was emphasized previously this is a condition concerning the sources in (2.53). The reformulation of Maxwell's equations by means of the tensor fields $F^{\mu\nu}(x)$ and $\mathcal{F}^{\mu\nu}(x)$, and of the current density $j^{\mu}(x)$, in the differential equations (2.49a) and (2.53), renders the covariance explicit.This observation is termed *manifest Lorentz covariance*. In the equivalent differential equations (2.49a) and (2.53) covariance is not evident.

4. The conservation of the four-current density $j(x)$ now is manifest, too. Indeed, calculating the (four-)divergence of the inhomogeneous equations (2.53), one finds

$$\partial_\nu \partial_\mu \mathcal{F}^{\mu\nu}(x) = 0 \,.$$

The differential operator $\partial_\nu \partial_\mu$ being symmetric in μ and ν is contracted with the antisymmetric tensor field $\mathcal{F}^{\mu\nu}(x)$. This is compatible with (2.53) only if

$$\partial_\nu j^\nu(x) = \frac{\partial}{\partial t}\varrho(t, \boldsymbol{x}) + \nabla \cdot \boldsymbol{j}(t, \boldsymbol{x}) = 0 \tag{2.56}$$

holds. Thus, the continuity equation is essential for (2.53) to hold true. It guarantees the universal law of conservation of electric charge.

2.3.5 Gauge Invariance and Potentials

As anticipated in Sect. 1.6.3 the scalar potential $\Phi(t, \boldsymbol{x})$ and the vector potential $\boldsymbol{A}(t, \boldsymbol{x})$ can be combined in the definition

$$A(x) := \left(\Phi(t, \boldsymbol{x}), \boldsymbol{A}(t, \boldsymbol{x})\right)^T \,. \tag{2.57}$$

The representation of the electric field and the induction field by the potentials Φ and $\boldsymbol{A}$ is equivalent to the representation of the field strength tensor in terms of the four-potential (2.57)

$$F^{\mu\nu}(x) = \partial^\mu A^\nu(x) - \partial^\nu A^\mu(x) \,. \tag{2.58}$$

It is not difficult to verify this assertion: For $\nu = 0$ we have

$$\begin{aligned} F^{i0}(x) &= E^i(t, \boldsymbol{x}) = \partial^i A^0(t, \boldsymbol{x}) - \partial^0 A^i(t, \boldsymbol{x}) \\ &= -\partial_i \Phi(t, \boldsymbol{x}) - \frac{1}{c}\partial_t A^i(t, \boldsymbol{x}) \,, \end{aligned}$$

in agreement with (1.55b). Considering a space-space component, say $\mu = 3$ and $\nu = 2$, one has

$$\begin{aligned} F^{32} &= B^1(t, \boldsymbol{x}) = \partial^3 A^2(t, \boldsymbol{x}) - \partial^2 A^3(t, \boldsymbol{x}) \\ &= -\partial_3 A^2(t, \boldsymbol{x}) + \partial_2 A^3(t, \boldsymbol{x}) = \left(\nabla \times \boldsymbol{A}(t, \boldsymbol{x})\right)^1 \,. \end{aligned}$$

The components $B^2(t, \boldsymbol{x})$ and $B^3(t, \boldsymbol{x})$ are obtained in an analogous manner, in agreement with the noncovariant representation (1.55a).

By definition, gauge transformations of potentials are space- and time-dependent transformations which do not change the observable fields. Solving for the scalar and vector potentials they have the somewhat complicated form (1.57a) and (1.57b). In a Lorentz covariant formulation their appearance is simpler, they read

$$A^\mu(x) \longmapsto A'^\mu(x) = A^\mu(x) - \partial^\mu \chi(x) . \tag{2.59}$$

The covariant derivative is related to the time derivative and the gradient in $\mathbb{R}^3$ by

$$\partial^\mu = \left(\frac{1}{c}\frac{\partial}{\partial t}, -\nabla\right) .$$

One sees at once that (2.59) is identical to the equations (1.57a) and (1.57b). As $\chi(x)$ is a smooth function, its mixed second derivatives $\partial_\mu \partial_\nu \chi(x)$ and $\partial_\nu \partial_\mu \chi(x)$ are equal. Thus, they cancel in the difference on the right-hand side of (2.58) so that the tensor field $F^{\mu\nu}(x)$ stays invariant. For the same reason, the homogeneous equations (2.49a) are fulfilled automatically when using the definition (2.58). Thus, one has the choice: Either one works exclusively with *observables*, i.e. with the fields $\boldsymbol{E}$ and $\boldsymbol{B}$ and imposes the homogeneous equations (2.49a), or one expresses the field strength tensor in terms of potentials $A^\mu(x)$. In this case the homogenous equations are redundant.

The inhomogeneous equations in vacuum become

$$\Box A^\nu(x) - \partial^\nu\left(\partial_\mu A^\mu(x)\right) = \frac{4\pi}{c} j^\nu(x) . \tag{2.60}$$

As one would have expected, the first term of the left-hand side contains the differential operator $\Box = \partial_\mu \partial^\mu = (1/c^2)\partial_t^2 - \boldsymbol{\Delta}$ which is characteristic for the wave equation. The second term on the left-hand side depends on the choice of the gauge. The right-hand side, finally, is the source term. For example, if one uses the Lorenz gauge (2.61) below, then (2.60) is precisely the inhomogeneous wave equation.

Remarks

1. As we just noted the covariance of Maxwell's equations is guaranteed only if $F^{\mu\nu}(x)$ is a contravariant tensor field of degree 2 and $j^\mu(x)$ is a contravariant vector field with respect to Lorentz transformations. The four-potential $A^\mu(x)$ *may be* a Lorentz vector field. However, it is always possible to hide the manifest Lorentz covariance without modifying the covariance of the Maxwell equations and of its physical content. For example, instead of the Lorentz invariant condition
$$\partial_\mu A^\mu(x) = 0 \tag{2.61}$$
(this is the Lorenz condition (1.58)), one may choose classes of noncovariant gauges. One may wish to impose, for example, the Coulomb gauge (1.63) and single out a special class of gauges by this choice. Lorentz covariance of

Maxwell's equations is then no longer manifestly visible, even though it is not lost. One has made use of the freedom in the choice of gauge, though hiding the covariance, for the purpose of emphasizing other properties of the theory. In the case of the Coulomb gauges, for instance, this was the transversality of electromagnetic waves. As no physical prediction of the theory is changed by gauge transformations, the theory is the same, no matter which formulation one has chosen and independently of the theory appearing in different disguises.

2. As will become clear in a later section the gauge freedom (2.59), in essence, means invariance of Maxwell theory under the group of local U(1)-transformations,

$$\mathcal{U}(1):$$
$$\{g \in \mathcal{F}(\mathbb{R}^4)\,,\ \text{smooth function} \,|\, g(x) = \mathrm{e}^{\mathrm{i}\alpha(x)}, \alpha(x) \text{ smooth, real}\}\,.$$

Here the term *local* means that to every point $x \in \mathbb{R}^4$ of spacetime a copy of the group

$$\mathrm{U}(1) = \{g \in \mathbb{C} \,|\, |g|^2 = 1\,,\ \text{i.e. } g = \mathrm{e}^{\mathrm{i}\alpha}, \alpha \in [0, 2\pi]\}$$

is attached. This gauge group (which is Abelian) acts on the potentials A^μ by

$$A'^\mu(x) = A^\mu(x) - \mathrm{i}g(x)\partial^\mu g^{-1}(x)\,, \qquad (2.62)$$

but acts also on the source terms in Maxwell's equations. Equation (2.62) is a special case of a transformation for more general, non-Abelian groups that will be studied in Chap. 5. For the moment we just note that (2.62) with $\alpha(x) = \chi(x)$ reproduces the formula (2.59).

3. In case all charge and current densities are localized the continuity equation (2.56) implies that the time derivative of the integral of the charge density over the whole space vanishes,

$$\partial_0 \iiint \mathrm{d}^3x\, j^0(x) = \iiint \mathrm{d}^3x\, \partial_0 j^0(t, \boldsymbol{x})$$
$$= -\iiint \mathrm{d}^3x\, \partial_i j^i(t, \boldsymbol{x}) = 0\,.$$

This follows because the space integral, by Gauss' theorem (1.6), can be converted to a surface integral over a surface at infinity where the current density vanishes by assumption. The total charge contained in space

$$Q := \iiint \mathrm{d}^3x\, j^0(t, \boldsymbol{x}) \qquad (2.63)$$

is conserved. Although this is a beautiful and important result it seems to single out a certain class of Lorentz systems for which the division into space and time is fixed. Yet, the conservation law (2.63) of electric charge is Lorentz invariant. This becomes plausible by the following argument:
Let Σ_0 be the hypersurface $x^0 = \text{const.}$ in Minkowski space[4]. One says that this surface is *space*like and we understand by this that any two points on Σ_0 are spacelike relative to each other. This means that in each point $x \in \Sigma_0$ the positively oriented normal $n(x)$ (with respect to the time direction) to the surface is *time*like. In the special case of Σ_0 one has $n^\mu(x) = (1, \mathbf{0})^T$ for all x. In the general case, one has $n^2(x) \equiv n_\mu(x)n^\mu(x) = 1$ for every spacelike hypersurface. To be spacelike is a property of a hypersurface which is invariant under all $\Lambda \in L_+^\uparrow$. Thus, the assertion that the normal n is timelike remains true for all inertial observers who differ from one another by proper, orthochronous Lorentz transformations.
It is not difficult to guess the four-dimensional variant of Gauss' theorem (1.6)

$$\iiiint\limits_V \mathrm{d}^4x\, \partial^\mu F_\mu(x) = \iiint\limits_{\Delta(V)} \mathrm{d}\sigma^\mu F_\mu(x) . \tag{2.64}$$

In this formula $\Delta(V)$ is a piecewise smooth, closed hypersurface in Minkowski space, V denotes the (four-dimensional) volume enclosed by it, and $F^\mu(x)$ is a smooth vector field. The integration on the right-hand side contains $\mathrm{d}\sigma^\mu = n^\mu(x)\mathrm{d}\sigma$, where $n^\mu(x)$ is the positive normal to the surface, and $\mathrm{d}\sigma$ is the surface element on $\Delta(V)$.
The total charge (2.63) is given by the integral of the current density $j(x)$ over the space-like hypersurface Σ_0,

$$Q = \iiint \mathrm{d}^3x\, j^0(x) = \iiint\limits_{\Sigma_0} \mathrm{d}\sigma^\mu j_\mu(x) , \tag{2.63a}$$

$$\left(\mathrm{d}\sigma^\mu = n^\mu \mathrm{d}\sigma ,\ n^\mu = (1,0,0,0)^T ,\ \mathrm{d}\sigma = \mathrm{d}^3x\right) .$$

Consider now another space-like smooth hypersurface Σ which differs from Σ_0 only at finite distances in the way sketched in Fig. 2.2. Then the difference $\Sigma - \Sigma_0 =: \Delta(V)$ is a piecewise smooth, closed hypersurface which encloses a finite volume V. Gauss' theorem in the form of (2.64) applied to $\Delta(V)$ and V shows that the two integrals

$$Q' = \iiint\limits_{\Sigma} \mathrm{d}\sigma^\mu j_\mu(x) \quad \text{and} \quad Q = \iiint\limits_{\Sigma_0} \mathrm{d}\sigma^\mu j_\mu(x)$$

[4] Every smooth finite-dimensional surface that is embedded in a manifold with higher dimension is called a *hypersurface.*

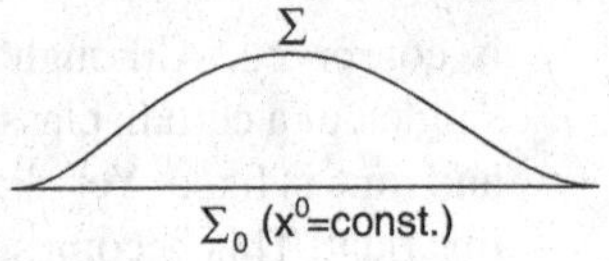

Fig. 2.2 The three-dimensional hypersurface $x^0 = \text{const.}$ is deformed locally and continuously to the spacelike hypersurface Σ such that $\Sigma - \Sigma_0$ encloses a finite volume

differ only by the integral of $\partial^\mu j_\mu(x)$ over the volume V. This difference vanishes if and only if $j^\mu(x)$ is a conserved current. In this case the charge (2.63a) is independent of the choice of the space-like hypersurface Σ. Therefore, in spite of the apparent dependence on the splitting into space and time the definition (2.63) is Lorentz invariant.

2.4 Fields of a Uniformly Moving Point Charge

A special Lorentz transformation sort of "tumbles" electric and induction fields. While the transformation behaviour of the tensor field $F^{\mu\nu}(x)$ under $\mathsf{L}(\boldsymbol{v}) \in L_+^\uparrow$ is straightforward and transparent, this is not obviously so for the fields $\boldsymbol{E}$ and $\boldsymbol{B}$. In order to clarify this matter and to work out its physical significance we study the example of a particle with charge q which moves at constant velocity $\boldsymbol{v}$ with respect to some inertial frame K (called a laboratory system, in short, in what follows). We calculate the fields $\boldsymbol{E}(t, \boldsymbol{x})$ and $\boldsymbol{B}(t, \boldsymbol{x})$, as measured by an observer who keeps a fixed position in the laboratory system.

Let the rest system of the particle be denoted by K′. It is chosen such that at $t = t' = 0$ it coincides with the laboratory system. The observer B is at rest relative to K, its spatial coordinates are $(0, b, 0)$; the charged particle sits at the origin of K′, and, as seen from the laboratory system, moves with the constant velocity $\boldsymbol{v} = v\hat{\boldsymbol{e}}_1$, i.e. along the 1-axis of K. This is sketched in Fig. 2.3. With $\beta = v/c$ and $\gamma = 1/\sqrt{1-\beta^2}$ the coordinates of B are

$$x'^0 = \gamma(x^0 - \beta x^1)\,, \qquad x'^2 = x^2\,,$$
$$x'^1 = \gamma(x^1 - \beta x^0)\,, \qquad x'^3 = x^3\,.$$

Inserting the laboratory coordinates $x^2 = b$ and $x^1 = 0 = x^3$ one has

$$B : \left(ct, 0, b, 0\right)\Big|_{\text{(rel. to K)}}\,,$$
$$B : \left(ct' = c\gamma t, x'^1 = -v\gamma t = -vt', x'^2 = b, x'^3 = 0\right)\Big|_{\text{(rel. to K}')}\,.$$

The special Lorentz transformation which links K to K′, and the formula which takes

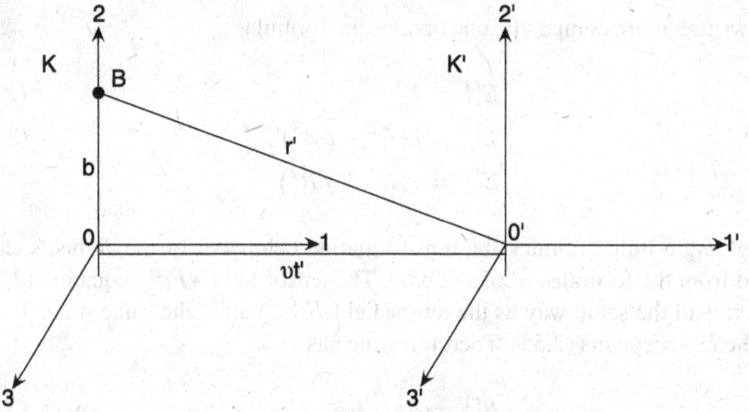

Fig. 2.3 An observer is assumed to be at rest relative to the inertial system K and has the position $(x^1 = 0, x^2 = b, x^3 = 0)$. He or she sees a charged particle moving at constant velocity which at the coordinate time $t = t' = 0$ passes through the origin of K

the field strength tensor from one system to the other are, respectively,

$$\mathsf{L}(-\boldsymbol{v}) = \begin{pmatrix} \gamma & -\gamma\beta & 0 & 0 \\ -\gamma\beta & \gamma & 0 & 0 \\ 0 & 0 & 1 & 0 \\ 0 & 0 & 0 & 1 \end{pmatrix}$$

$$F'^{\sigma\tau}(x') = \Lambda^{\sigma}{}_{\mu}\Lambda^{\tau}{}_{\nu}\, F^{\mu\nu}(\Lambda^{-1}x') .$$

As we will see in a moment it suffices to calculate the time-space components only. With $\sigma = 0$ they are for the three values of τ, respectively,

$$\begin{aligned} \tau = 1: \qquad F'^{01} &= -E'^1 = \Lambda^0{}_\mu \Lambda^1{}_\nu F^{\mu\nu} \\ &= \Lambda^0{}_0\Lambda^1{}_1 F^{01} + \Lambda^0{}_1\Lambda^1{}_0 F^{10} \\ &= (\gamma)^2(-E^1) + (\gamma\beta)^2 E^1 = -E^1 , \\ \tau = 2: \qquad F'^{02} &= -E'^2 = \Lambda^0{}_\mu \Lambda^2{}_\nu F^{\mu\nu} \\ &= \Lambda^0{}_0\Lambda^2{}_2 F^{02} + \Lambda^0{}_1\Lambda^2{}_2 F^{12} \\ &= \gamma(-E^2) + (-\gamma\beta)(-B^3) , \\ \tau = 3: \qquad F'^{03} &= -E'^3 = \Lambda^0{}_\mu \Lambda^3{}_\nu F^{\mu\nu} \\ &= \Lambda^0{}_0\Lambda^3{}_3 F^{03} + \Lambda^0{}_1\Lambda^3{}_3 F^{13} \\ &= \gamma(-E^3) - \gamma\beta B^2 . \end{aligned}$$

Here we made use of

$$\Lambda^1{}_2 = 0 = \Lambda^1{}_3 , \quad \Lambda^2{}_1 = 0 = \Lambda^2{}_3 , \quad \Lambda^3{}_1 = 0 = \Lambda^3{}_2 .$$

Thus, written more compactly, one obtains the formulae

$$E'^1 = E^1 , \tag{2.65a}$$

$$E'^2 = \gamma\left(E^2 - \beta B^3\right) , \tag{2.65b}$$

$$E'^3 = \gamma\left(E^3 + \beta B^2\right) . \tag{2.65c}$$

By applying a little gimmick the transformation behaviour of the $\boldsymbol{B}$ fields can be derived from the formulae (2.65a–2.65c). The tensor field $\star F^{\mu\nu}$, equation (2.54c), transforms in the same way as the tensor field $F^{\mu\nu}$, but, at the same time, it arises from the replacements (2.55). Therefore, one has

$$B'^1 = B^1 , \tag{2.65d}$$

$$B'^2 = \gamma\left(B^2 + \beta E^3\right) , \tag{2.65e}$$

$$B'^3 = \gamma\left(B^3 - \beta E^2\right) . \tag{2.65f}$$

Finally, it is not difficult to generalize these formulae to an arbitrary direction of the velocity $\boldsymbol{v}$: The results (2.65a–2.65f) show that the components which are parallel to $\boldsymbol{v}$ remain unchanged while the components perpendicular to $\boldsymbol{v}$ depend on the cross product of the velocity and the other field, respectively. Therefore, one has

$$E'_\parallel = E_\parallel , \quad B'_\parallel = B_\parallel , \tag{2.66a}$$

$$\boldsymbol{E}'_\perp = \gamma\left(\boldsymbol{E}_\perp + \frac{1}{c}\boldsymbol{v} \times \boldsymbol{B}\right) \tag{2.66b}$$

$$\boldsymbol{B}'_\perp = \gamma\left(\boldsymbol{B}_\perp - \frac{1}{c}\boldsymbol{v} \times \boldsymbol{E}\right) . \tag{2.66c}$$

Returning to the concrete example $\boldsymbol{v} = v\hat{\boldsymbol{e}}_1$ and referring to Fig. 2.3 one has $r' = \sqrt{b^2 + (vt')^2}$. In its own rest system K' the particle creates the spherically symmetric electric field

$$\boldsymbol{E} = \frac{q}{r'^3}\boldsymbol{r}' .$$

There is no induction field. At the position of the observer one has, in particular,

$$E'^1 = -\frac{q}{r'^2}\frac{(vt')}{r'} , \quad E'^2 = \frac{q}{r'^2}\frac{b}{r'} , \quad E'^3 = 0 ,$$

$$B'^1 = 0 = B'^2 = B'^3 .$$

As in the rest system $\boldsymbol{B}' = 0$, it follows from (2.65e) or (2.65f) that $B^2 = -\beta E^3$ and $B^3 = \beta E^2$. One inserts this into (2.65c) or (2.65b) to obtain $E'^2 = E^2/\gamma$ and $E'^3 = E^3/\gamma$, respectively. This is used to calculate the fields in the laboratory system K: At the position of B and with the given state of motion of the particle, the

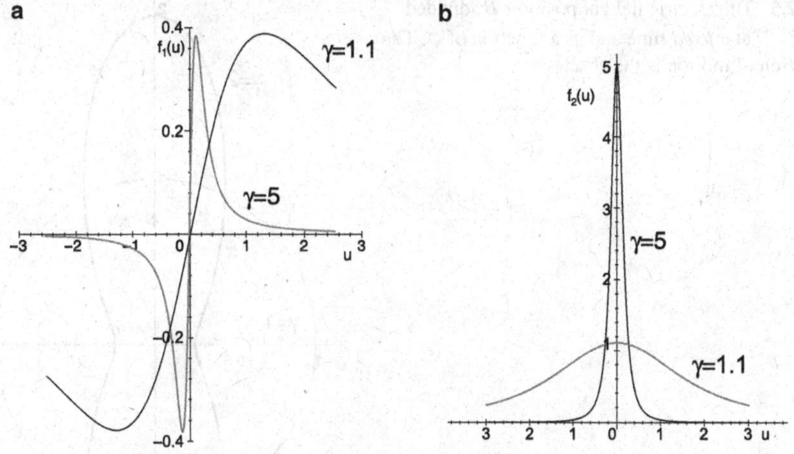

Fig. 2.4 **a** The component E^1 of the electric field in the direction of $\boldsymbol{v}$, and **b** the component E^2 perpendicular to $\boldsymbol{v}$ as a function of time, for two values of γ

field components are

$$E^1 = -q\frac{v\gamma t}{\left(b^2 + (v\gamma t)^2\right)^{\frac{3}{2}}}\,, \tag{2.67a}$$

$$E^2 = \gamma E'^2 = q\frac{\gamma b}{\left(b^2 + (v\gamma t)^2\right)^{\frac{3}{2}}}\,, \tag{2.67b}$$

$$\boldsymbol{B}_\perp = \frac{\gamma}{c}\boldsymbol{v}\times\boldsymbol{E}\,, \quad \text{i.e.} \quad B^3 = \frac{v}{c}\gamma E'^2 = \frac{v}{c}E^2\,, \tag{2.67c}$$

and, evidently, $E^3 = E'^3 = 0$. In order to illustrate this result it is useful to introduce the dimensionless variable

$$u := \frac{ct}{b}$$

and to express the fields in units of q/b^2. Equations (2.67a) and (2.67b) then become, with $vt/b = \beta u$ and with $\beta^2\gamma^2 = \gamma^2 - 1$,

$$f_1(u) := -\frac{E^1}{(\frac{q}{b^2})} = \frac{\sqrt{\gamma^2-1}\,u}{\left(1+(\gamma^2-1)u^2\right)^{\frac{3}{2}}}\,,$$

$$f_2(u) := \frac{E^2}{(\frac{q}{b^2})} = \frac{\gamma}{\left(1+(\gamma^2-1)u^2\right)^{\frac{3}{2}}}\,.$$

Figure 2.4a shows the function f_1 as a function of u, i.e. as a function of coordinate time t. This function is odd, its maximum and its minimum are at $u_{\text{max/min}} =$

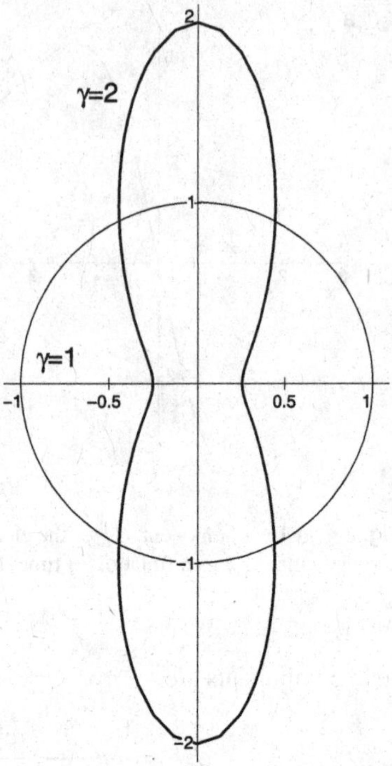

Fig. 2.5 The electric field at position $\boldsymbol{B}$, divided by $\boldsymbol{E}^{(0)}$, at a *fixed* time and as a function of φ. The direction of motion is the abscissa

$\mp 1/(\sqrt{2(\gamma^2-1)})$, respectively, its absolute value being $2/(3\sqrt{3})$ in both positions. Figure 2.4b shows the function $f_2(u)$. This function has its maximum at $u = 0$. Its value at time zero is $f_2(0) = \gamma$, the width of this curve, i.e. the distance between the two points at which it has decreased to half its value at $u = 0$, is found to be

$$\Delta u = \frac{2\sqrt{4^{\frac{1}{3}}-1}}{\sqrt{\gamma^2-1}} \simeq \frac{1.533}{\sqrt{\gamma^2-1}} .$$

The larger the value of γ, the more pronounced and narrow the "pulse" that the observer sees in the 2-direction. The phenomena seen *in* the direction of flight are the result of Lorentz contraction. This is seen most clearly if one calculates the electric field at the position of the observer B, in the laboratory system and at an arbitrary but fixed point in time. From the geometry of Fig. 2.5 one sees that

$$\frac{E^1}{E^2} = -\frac{vt}{b} ,$$

i.e. $\boldsymbol{E}$ has the same direction as the position vector $\boldsymbol{r}$. Regarding the denominator of the expressions (2.67a) and (2.67b) one calculates

$$\begin{aligned} b^2 + (v\gamma t)^2 &= \gamma^2 \left(b^2 + (vt)^2\right) + b^2 \left(1 - \gamma^2\right) \\ &= \gamma^2 r^2 \left[1 + \frac{1-\gamma^2}{\gamma^2} \frac{b^2}{r^2}\right] = \gamma^2 r^2 \left(1 - \beta^2 \sin^2 \varphi\right) . \end{aligned}$$

Thus, at the fixed time t the electric field at the position of B is

$$\boldsymbol{E}\left(t_{\text{fixed}}, \boldsymbol{r}\right) = \frac{q\boldsymbol{r}}{r^3 \gamma^2 \left(1 - \beta^2 \sin^2 \varphi\right)^{\frac{3}{2}}} . \tag{2.68}$$

The effect of Lorentz contraction can be read off this result: With $\boldsymbol{E}^{(0)} = q\boldsymbol{r}/r^3$ being the field of the particle at rest one has

$$\begin{aligned} &\text{for } \varphi = \pm\frac{\pi}{2}: && \boldsymbol{E}\left(t_{\text{fixed}}, \boldsymbol{r}\right) = \gamma \boldsymbol{E}^{(0)}\left(t_{\text{fixed}}, \boldsymbol{r}\right) , \\ &\text{for } \varphi = 0 \text{ and } \pi: && \boldsymbol{E}\left(t_{\text{fixed}}, \boldsymbol{r}\right) = \frac{1}{\gamma^2} \boldsymbol{E}^{(0)}\left(t_{\text{fixed}}, \boldsymbol{r}\right) . \end{aligned}$$

In the direction of motion ($\varphi = 0$ or π) the spherically symmetric field appears contracted when compared to the directions $\varphi = \pm\pi/2$ perpendicular to the motion.

2.5 Lorentz Invariant Exterior Forms and the Maxwell Equations

As was shown in Sect. 2.2.2 for the case of a fixed division of spacetime into time axis $\mathbb{R}_t$ and coordinate space $\mathbb{R}^3$, the association of simple exterior forms over $\mathbb{R}^3$ to the fields and potentials of Maxwell theory proved useful in reformulating Maxwell's equations in a concise and transparent manner. On the basis of this experience it is suggestive to interpret the observables and the potentials of Maxwell theory, written in covariant form, as geometric objects on Minkowski space $\mathbb{R}^4$. In this Section we show that the field strength tensor, the Lorentz force, and the external sources can be written as exterior forms which are even simpler than in the case of $\mathbb{R}^3$ and which satisfy simple and natural equations. By the same token we show that the apparent asymmetry between the electric field that was a *one*-form and the induction field that was a *two*-form on $\mathbb{R}^3$, disappears. Finally, we provide the basis for the generalization to non-Abelian gauge theories which are studied in Chap. 5.

2.5.1 Field Strength Tensor and Lorentz Force

The tensor field $F^{\mu\nu}(x)$ which in a given inertial system decomposes into observable $\boldsymbol{E}$-fields and $\boldsymbol{B}$-fields according to (2.46), is defined on Minkowski space $(\mathbb{R}^4, \mathrm{g} = \mathrm{diag}(1,-1,-1,-1))$. Denoting the base one-forms over this space by $\mathrm{d}x^\mu$, the base two-forms by $\mathrm{d}x^\mu \wedge \mathrm{d}x^\nu$ $(\mu < \nu)$ we define

$$\omega_F := \sum_{\mu<\nu} F_{\mu\nu}(x)\,\mathrm{d}x^\mu \wedge \mathrm{d}x^\nu\,. \tag{2.69}$$

The sums on μ and ν are written explicitly because of the condition $\mu < \nu$. If one prefers to also apply the summation convention here one must add the factor $1/2$. We have inserted the *co*variant tensor field in (2.69) which is obtained by calculating – now using the summation convention! –

$$F_{\mu\nu}(x) = g_{\mu\sigma}F^{\sigma\tau}(x)g_{\tau\nu}\,.$$

Like the coordinates the base one-forms

$$\mathrm{d}x^0 = c\,\mathrm{d}t\,,\ \mathrm{d}x^1\,,\ \mathrm{d}x^2\,,\ \mathrm{d}x^3$$

are ordered from 0 to 3, they refer to the chosen coordinate system. Note that we simplified the notation somewhat by omitting the degree of the form above the symbol. Obviously, the definition (2.69) indicates that one is dealing with a two-form.

Already at this point there is an important remark to be made: In contrast to the representation of the field strength tensor by $F^{\mu\nu}$, with its obvious recurrence to a given frame of reference, the definition of the two-form ω_F is Lorentz invariant. If, nevertheless, one sticks to the given frame in which, taking proper account of the signs from the two factors g,

$$F_{\mu\nu}(x) = \begin{pmatrix} 0 & +E_1 & +E_2 & +E_3 \\ -E_1 & 0 & -B_3 & +B_2 \\ -E_2 & +B_3 & 0 & -B_1 \\ -E_3 & -B_2 & +B_1 & 0 \end{pmatrix} \tag{2.69a}$$

there follows[5]

$$\begin{aligned} \omega_F &= \mathrm{d}x^0 \wedge \left[E_1\,\mathrm{d}x^1 + E_2\,\mathrm{d}x^2 + E_3\,\mathrm{d}x^3\right] \\ &\quad - \left[B_3\,\mathrm{d}x^1 \wedge \mathrm{d}x^2 + B_1\,\mathrm{d}x^2 \wedge \mathrm{d}x^3 + B_2\,\mathrm{d}x^3 \wedge \mathrm{d}x^1\right] \\ &= \mathrm{d}x^0 \wedge \overset{1}{\omega}_{\boldsymbol{E}} - \overset{2}{\omega}_{\boldsymbol{B}}\,, \end{aligned} \tag{2.70}$$

[5] One should notice that on $\mathbb{R}^3$, the Euclidean space, one has $E_i = E^i$ and $B_k = B^k$. On $\mathbb{R}^3$ and with cartesian coordinates covariant and contravariant indices can be identified and need not be distinguished.

which contains the exterior forms defined in (2.26a) and (2.26b). This analysis explains at once why on $\mathbb{R}^3$ the field $\boldsymbol{E}$ is associated to a one-form, while the field $\boldsymbol{B}$ is associated to a two-form. On Minkowski space, in turn, both kinds of fields are represented by two-forms. Therefore, it is appropriate to replace the definitions (2.26a) and (2.26b) by the following:

$$\omega_E \equiv \overset{2}{\omega}_E := \sum_{i=1}^{3} E_i(t, \boldsymbol{x})\, \mathrm{d}x^0 \wedge \mathrm{d}x^i\,, \tag{2.71a}$$

$$\omega_B \equiv \overset{2}{\omega}_B := \frac{1}{2} \sum_{i,j,k} \varepsilon_{ijk} B_i(t, \boldsymbol{x})\, \mathrm{d}x^j \wedge \mathrm{d}x^k\,. \tag{2.71b}$$

In what follows we shall make use of either notation, the one defined in (2.69) but also the ones of (2.71a) and (2.71b).

Being a two-form ω_F can act on up to two vector fields. Thus, one has with

$$u = u^\alpha \partial_\alpha \quad \text{and} \quad v = v^\beta \partial_\beta$$

$$\omega_F(u, v) = \sum_{\mu<\nu} F_{\mu\nu} \left(v^\mu u^\nu - u^\mu v^\nu\right) = 2 \sum_{\mu<\nu} F_{\mu\nu} u^\nu v^\mu$$
$$= 2 \frac{1}{2} \sum_{\mu,\nu} F_{\mu\nu} u^\nu v^\mu\,.$$

In turn, if one lets ω_F act on one vector field only, then one obtains a one-form

$$\omega_F(u, \bullet) = \sum_{\mu,\nu} F_{\mu\nu} u^\nu\, \mathrm{d}x^\mu\,.$$

Multiplication of this form with q/c yields the one-form which is to be associated to the Lorentz force (2.47) by way of the definition

$$\omega_{\text{Lor}} := \sum_{\mu=0}^{3} K_\mu(x)\, \mathrm{d}x^\mu\,, \quad \text{with} \quad K_\mu(x) = \frac{q}{c} \sum_{\nu=0}^{3} F_{\mu\nu}(x) u^\nu\,. \tag{2.72}$$

It is then not difficult to rewrite Maxwell's equations in their manifestly covariant form in terms of the two-form (2.69) and its Hodge dual. The Hodge star operation in Minkowski space is a bit more subtle to handle, as compared to the case of a Euclidean space and the definition (2.16), because of the characteristic signs of the metric. We list here the duals of all base forms over Minkowski space

$$\star\, \mathrm{d}x^\mu = \frac{1}{3!} g^{\mu\lambda} \varepsilon_{\lambda\nu\sigma\tau}\, \mathrm{d}x^\nu \wedge \mathrm{d}x^\sigma \wedge \mathrm{d}x^\tau\,, \tag{2.73a}$$

$$\star \left(\mathrm{d}x^\mu \wedge \mathrm{d}x^\nu\right) = \frac{1}{2!} g^{\mu\lambda} g^{\nu\varrho} \varepsilon_{\lambda\varrho\sigma\tau}\, \mathrm{d}x^\sigma \wedge \mathrm{d}x^\tau\,, \tag{2.73b}$$

$$\star \left(\mathrm{d}x^\mu \wedge \mathrm{d}x^\nu \wedge \mathrm{d}x^\sigma\right) = g^{\mu\lambda} g^{\nu\varrho} g^{\sigma\eta} \varepsilon_{\lambda\varrho\eta\tau}\, \mathrm{d}x^\tau\,, \tag{2.73c}$$

$$\star \left(\mathrm{d}x^0 \wedge \mathrm{d}x^1 \wedge \mathrm{d}x^2 \wedge \mathrm{d}x^3\right) = \det \mathrm{g} = -1\,. \tag{2.73d}$$

▶ **Remark**

In these formulae $\varepsilon_{\alpha\beta\gamma\delta}$ denotes the totally antisymmetric Levi-Civita symbol in dimension four that was introduced in (2.50). Note the convention $\varepsilon_{0123} = +1$. The (inverse) metric which appears in the formulae (2.73a–2.73d) gives rise to signs because while the time-time element is $g^{00} = +1$, the space-space elements are $g^{ii} = -1$. This also implies a sign change in the relation (2.18) between the doubly dualized $\star\star\omega$ and the original ω which is different from the case of Euclidean spaces. Here it reads

$$\star\star\omega = (-)^{k(n-k)+1}\,\omega\,, \tag{2.74}$$

where the 1 in the exponent stems from the signature of the semi-Euclidian space $\mathbb{R}^{(p,q)}$ (with p space coordinates and q time coordinates), with $p+q=n$, and with the metric $\mathrm{g} = \big(1,1,\ldots\ (q \text{ times}), -1,-1,\ldots\ (p \text{ times})\big)$. The signature s is the codimension of the biggest subspace on which the metric is *definite*. Consider the example of Minkowski space $\mathbb{R}^{(1,3)}$: The metric is $\mathrm{g} = \mathrm{diag}(1,-1,-1,-1)$. Its restriction to the space part $\mathrm{g}|_{\mathbb{R}^3}$ is negative definite, hence, $s = 4-3 = 1$.

Let us consider a few examples: The formula (2.73d) can be written, alternatively, as follows

$$\star\big(\mathrm{d}x^\mu \wedge \mathrm{d}x^\nu \wedge \mathrm{d}x^\sigma \wedge \mathrm{d}x^\tau\big) = g^{\mu\alpha} g^{\nu\beta} g^{\sigma\gamma} g^{\tau\delta} \varepsilon_{\alpha\beta\gamma\delta}\,.$$

Similarly, the dual of the constant function 1 is

$$\star 1 = \frac{1}{4!}\varepsilon_{\mu\nu\sigma\tau}\,\mathrm{d}x^\mu \wedge \mathrm{d}x^\nu \wedge \mathrm{d}x^\sigma \wedge \mathrm{d}x^\tau\,.$$

From (2.73b) one obtains

$$\star\mathrm{d}x^0 \wedge \mathrm{d}x^1 = g^{00} g^{11} \varepsilon_{0123}\,\mathrm{d}x^2 \wedge \mathrm{d}x^3 = -\mathrm{d}x^2 \wedge \mathrm{d}x^3\,,$$
$$\star\mathrm{d}x^2 \wedge \mathrm{d}x^3 = g^{22} g^{33} \varepsilon_{2301}\,\mathrm{d}x^0 \wedge \mathrm{d}x^1 = +\mathrm{d}x^0 \wedge \mathrm{d}x^1\,,$$

and from there

$$\star\star\,\mathrm{d}x^0 \wedge \mathrm{d}x^1 = -\mathrm{d}x^0 \wedge \mathrm{d}x^1\,,$$

in agreement with (2.74) where $k = 2$ and $s = 1$. Starting from the formula (2.73a) one has

$$\begin{aligned}
\star\mathrm{d}x^0 &= g^{00}\varepsilon_{0123}\,\mathrm{d}x^1 \wedge \mathrm{d}x^2 \wedge \mathrm{d}x^3 = \mathrm{d}x^1 \wedge \mathrm{d}x^2 \wedge \mathrm{d}x^3\,,\\
\star\mathrm{d}x^1 &= g^{11}\varepsilon_{1023}\,\mathrm{d}x^0 \wedge \mathrm{d}x^2 \wedge \mathrm{d}x^3 = -\varepsilon_{1023}\,\mathrm{d}x^0 \wedge \mathrm{d}x^2 \wedge \mathrm{d}x^3\\
&= \mathrm{d}x^0 \wedge \mathrm{d}x^2 \wedge \mathrm{d}x^3\,,\\
\star\mathrm{d}x^2 &= g^{22}\varepsilon_{2013}\,\mathrm{d}x^0 \wedge \mathrm{d}x^1 \wedge \mathrm{d}x^3 = -\mathrm{d}x^0 \wedge \mathrm{d}x^1 \wedge \mathrm{d}x^3\\
&= +\mathrm{d}x^0 \wedge \mathrm{d}x^3 \wedge \mathrm{d}x^1\,,\\
\star\mathrm{d}x^3 &= g^{33}\varepsilon_{3012}\,\mathrm{d}x^0 \wedge \mathrm{d}x^1 \wedge \mathrm{d}x^2 = \mathrm{d}x^0 \wedge \mathrm{d}x^1 \wedge \mathrm{d}x^2\,.
\end{aligned}$$

(Note that the last three formulae show the cyclic symmetry in the space indices.) Upon comparison with the formula (2.73c) one sees that the base three-forms as well as the base one-forms do not change under the double star operation, in agreement with (2.74) for $n = 4$ and $s = 1$, $k = 1$ and $k = 3$, respectively.

The star operation applied to ω_F, (2.69), using the rules (2.73b) derived above, yields

$$\star\omega_F = -E_1\,\mathrm{d}x^2 \wedge \mathrm{d}x^3 - E_2\,\mathrm{d}x^3 \wedge \mathrm{d}x^1 - E_3\,\mathrm{d}x^1 \wedge \mathrm{d}x^2$$
$$- B_1\,\mathrm{d}x^0 \wedge \mathrm{d}x^1 - B_2\,\mathrm{d}x^0 \wedge \mathrm{d}x^2 - B_3\,\mathrm{d}x^0 \wedge \mathrm{d}x^3 \,.$$

This is seen to be the two-form (2.70), with the replacements

$$\boldsymbol{E} \longmapsto -\boldsymbol{B}\,, \quad \boldsymbol{B} \longmapsto \boldsymbol{E}\,.$$

If one compares this with (2.55) it is clear that "dual" and "dual" are identical i.e. that

$$\star\,\omega_F = \omega_{(\star F)}\,. \tag{2.75}$$

The two-form (2.69), constructed from $\star F$, is the same as the Hodge dual of ω_F, (2.69).

Of course, the same construction can be applied to $\mathcal{F}^{\mu\nu}(x)$, the tensor field of the $\boldsymbol{D}$ and $\boldsymbol{H}$ fields, equation (2.52a). In analogy to (2.69) one defines

$$\omega_{\mathcal{F}} = \sum_{\mu<\nu} \mathcal{F}_{\mu\nu}(x)\,\mathrm{d}x^\mu \wedge \mathrm{d}x^\nu\,. \tag{2.76}$$

Both types of two-forms, ω_F and $\omega_{\mathcal{F}}$, appear in the Maxwell equations to which we now turn.

2.5.2 Differential Equations for the Two-Forms ω_F and $\omega_{\mathcal{F}}$

The homogeneous Maxwell equations (2.49) or (2.49a) become very simple in the language of exterior forms. They just say that ω_F is a *closed* form,

$$\mathrm{d}\omega_F = 0\,. \tag{2.77}$$

This is verified by applying the formula (2.15e) for exterior derivatives:

$$\mathrm{d}\omega_F = (\mu < \nu)\,\partial_\lambda F_{\mu\nu}(x)\,\mathrm{d}x^\lambda \wedge \mathrm{d}x^\mu \wedge \mathrm{d}x^\nu$$
$$= \frac{1}{2}\,\partial_\lambda F_{\mu\nu}(x)\,\mathrm{d}x^\lambda \wedge \mathrm{d}x^\mu \wedge \mathrm{d}x^\nu\,.$$

The three indices λ, μ, and ν must all be different. As the base three-forms are linearly independent, the coefficient

$$\partial_\lambda F_{\mu\nu} + \partial_\mu F_{\nu\lambda} + \partial_\nu F_{\lambda\mu} ,$$

which multiplies the base form $\mathrm{d}x^\lambda \wedge \mathrm{d}x^\mu \wedge \mathrm{d}x^\nu$ must vanish. This is the content of the equation (2.49). As an alternative, one may calculate the dual of $\mathrm{d}\omega_F$ by means of (2.73c):

$$\star\, \mathrm{d}\omega_F = \frac{1}{2}\, \partial^\nu F^{\sigma\tau}(x)\varepsilon_{\mu\nu\sigma\tau}\, \mathrm{d}x^\mu .$$

As the coefficient of every base one-form $\mathrm{d}x^\mu$ must vanish this yields the homogeneous Maxwell equations in the form (2.49a).

In order to obtain the inhomogeneous equations (2.53) one starts by calculating the exterior derivative of the dualized form $\star\omega_{\mathcal{F}}$,

$$\begin{aligned}
\mathrm{d}(\star\omega_{\mathcal{F}}) &= \frac{1}{4}\, \partial_\alpha \mathcal{F}_{\mu\nu}(x) g^{\mu\lambda} g^{\nu\varrho} \varepsilon_{\lambda\varrho\beta\gamma}\, \mathrm{d}x^\alpha \wedge \mathrm{d}x^\beta \wedge \mathrm{d}x^\gamma \\
&= \frac{1}{4}\, \partial_\alpha \mathcal{F}^{\lambda\varrho}(x) \varepsilon_{\lambda\varrho\beta\gamma}\, \mathrm{d}x^\alpha \wedge \mathrm{d}x^\beta \wedge \mathrm{d}x^\gamma .
\end{aligned}$$

The indices α, β, and γ of the base three-form must all be different from one another. In addition, as β and γ in the ε symbol must also differ from λ and from ϱ; one must have either $\lambda = \alpha$ or $\varrho = \alpha$. This becomes even more obvious if the last result is subject once more to the star operation. Making use of the tensor

$$\varepsilon^{\alpha\beta\gamma\delta} = g^{\alpha\mu} g^{\beta\nu} g^{\gamma\sigma} g^{\delta\tau} \varepsilon_{\mu\nu\sigma\tau}$$

and of the formula (2.73d), one has

$$\star\, \mathrm{d} \star \omega_{\mathcal{F}} = \frac{1}{4}\, \partial_\alpha \mathcal{F}^{\lambda\varrho}(x) \varepsilon_{\lambda\varrho\beta\gamma} \varepsilon^{\alpha\beta\gamma\delta}\, g_{\delta\eta}\, \mathrm{d}x^\eta .$$

Each coefficient multiplying a one-form $\mathrm{d}x^\eta$ must be considered separately. First one notes that the sums over β and γ can be evaluated by means of the formula

$$\varepsilon_{\lambda\varrho\beta\gamma} \varepsilon^{\alpha\beta\gamma\delta} = \varepsilon_{\beta\gamma\lambda\varrho} \varepsilon^{\beta\gamma\alpha\delta} = -2 \left\{ \delta^\alpha_\lambda \delta^\delta_\varrho - \delta^\alpha_\varrho \delta^\delta_\lambda \right\} . \tag{2.78}$$

Upon inserting this formula one obtains four times the same term so that

$$\star\, \mathrm{d} \star \omega_{\mathcal{F}} = -\partial_\lambda \mathcal{F}^{\lambda\varrho}(x) g_{\varrho\eta}\, \mathrm{d}x^\eta .$$

This equation contains the operator (2.19a) with $n = 4$, $k = 2$, $\delta = -\star\, \mathrm{d}\, \star$, so that

$$\delta\, \omega_{\mathcal{F}} = \partial_\lambda \mathcal{F}^{\lambda\varrho}(x) g_{\varrho\eta}\, \mathrm{d}x^\eta .$$

If, conversely, one compares the three-form (2.28a) of the charge density and the two-form (2.28b) of the current density one reckons that in the covariant formulation both will appear as three-forms over Minkowski space, viz.

$$\omega_j = \frac{1}{3!}\,\varepsilon_{\mu\alpha\beta\gamma}\, j^\mu(x)\, \mathrm{d}x^\alpha \wedge \mathrm{d}x^\beta \wedge \mathrm{d}x^\gamma\,. \tag{2.79}$$

Taking the dual of this expression and making use of the formula

$$\varepsilon_{\mu\alpha\beta\gamma}\varepsilon^{\alpha\beta\gamma\eta} = -\varepsilon_{\alpha\beta\gamma\mu}\varepsilon^{\alpha\beta\gamma\eta} = 3!\delta^\eta_\mu\,,$$

one has

$$\star\,\omega_j = j^\mu(x) g_{\mu\eta}\, \mathrm{d}x^\eta\,. \tag{2.79a}$$

From this and by comparison of coefficients of $\mathrm{d}x^\eta$ it is clear that the inhomogeneous Maxwell equations (2.53) in exterior forms must be

$$\delta\,\omega_{\mathcal{F}} = \frac{4\pi}{c}\star\omega_j\,. \tag{2.80}$$

Remarks

1. When one uses exclusively the geometric language in formulating electrodynamics in terms of exterior forms one simply writes F instead of ω_F, $\mathcal{F}$ instead of $\omega_{\mathcal{F}}$, etc. In this notation one has

$$F \equiv \frac{1}{2} F_{\mu\nu}(x)\, \mathrm{d}x^\mu \wedge \mathrm{d}x^\nu\,.$$

Covariance of Maxwell's equations in the form (2.77) and (2.80) is obvious because both are written in a coordinate-free way. Their independence of specific coordinates, in fact, means that Maxwell's equations hold in *all* inertial systems.

2. The homogeneous equations which are summarized in (2.77), do not make use yet of the metric on Minkowski space. In turn, the inhomogeneous equations which slumber in (2.80), depend on the Hodge star operation which assumes a metric. In their axiomatic approach to electrodynamics Hehl and Obukhov start from topological manifolds without presupposing the existence of a metric.

3. As one easily verifies the mapping δ, like the exterior derivative d, when applied twice, gives zero, $\delta \circ \delta = \star\mathrm{d}\star\circ\star\,\mathrm{d}\star = 0$. Therefore, by applying δ to the inhomogeneous equations (2.80), one concludes

$$\delta(\star\omega_j) = 0 \quad\text{or}\quad \partial_\mu j^\mu(x) = 0\,. \tag{2.81}$$

This is the result found earlier: Current conservation follows from the inhomogeneous Maxwell equations. In other terms, only a *conserved* current can be a source of Maxwell's equations.

2.5.3 Potentials and Gauge Transformations

The four-potential (2.57) can be written as an exterior form as well, by using the covariant components $A_\nu(x) = g_{\nu\lambda}A^\lambda(x)$ in the definition of the following one-form

$$\omega_A := A_\nu(x)\,\mathrm{d}x^\nu\,. \tag{2.82}$$

Taking the exterior derivative, one finds

$$\begin{aligned}\mathrm{d}\omega_A &= \mathrm{d}A_\nu(x)\wedge \mathrm{d}x^\nu = \partial_\mu A_\nu(x)\,\mathrm{d}x^\mu \wedge \mathrm{d}x^\nu \\ &= \sum_{\mu<\nu}\Big(\partial_\mu A_\nu(x) - \partial_\mu A_\nu(x)\Big)\,\mathrm{d}x^\mu \wedge \mathrm{d}x^\nu\,.\end{aligned}$$

Comparison with the definition (2.69) shows that ω_F is the exterior derivative of ω_A, viz.

$$\omega_F = \mathrm{d}\omega_A\,. \tag{2.83}$$

We rediscover here a known fact in a particularly simple form: If one introduces potentials the homogeneous Maxwell equations are trivially fulfilled. Indeed, from the property (2.15d) of the exterior derivative

$$\omega_F = \mathrm{d}\omega_A \Longrightarrow \mathrm{d}\omega_F = \mathrm{d}^2\omega_A = 0\,.$$

Likewise, the gauge transformations (2.59) fit well into the framework of exterior forms. Let $\Lambda(x)$ be a smooth function on Minkowski space. Its total differential $\mathrm{d}\Lambda$ is a one-form that may be added to ω_A without modifying the Maxwell equation (2.77):

$$\begin{aligned}\omega_A \mapsto \omega_{A'} = \omega_A + \mathrm{d}\Lambda \Longrightarrow \omega_{F'} &= \mathrm{d}\omega_{A'} = \mathrm{d}\omega_A + \mathrm{d}^2\Lambda \\ &= \mathrm{d}\omega_A \;= \omega_F\,.\end{aligned}$$

The exact form $\mathrm{d}\Lambda$ is closed. The gauge freedom in the choice of the potential $A_\mu(x)$ is equivalent to ω_A being determined only up to an arbitrary exact form. (Note, when comparing to (2.59), that $\Lambda(x)$ is the same function as $\chi(x)$, up to a sign.)

The action of the Laplace–de-Rham operator (2.19b) on a one-form of the kind of (2.82) is calculated by means of the relations (2.73a–2.73d) as follows:

$$\begin{aligned}\mathbf{\Delta}_{\mathrm{LdR}}\big(A_\mu\,\mathrm{d}x^\mu\big) &= \mathrm{d}\circ\delta\,\big(A_\mu\,\mathrm{d}x^\mu\big) + \delta\circ\mathrm{d}\,\big(A_\mu\,\mathrm{d}x^\mu\big) \\ &= -\big(\mathrm{d}\star\mathrm{d}\star + \star\,\mathrm{d}\star\mathrm{d}\big)\big(A_\mu\,\mathrm{d}x^\mu\big)\,.\end{aligned}$$

Using (2.73a) and (2.73d) the first term on the right-hand side leads to

$$-\frac{1}{3!}\partial_\varrho\partial_\lambda A_\mu g^{\mu\alpha}\varepsilon_{\alpha\nu\sigma\tau}\varepsilon^{\lambda\nu\sigma\tau}\,\mathrm{d}x^\varrho\,.$$

The contraction of the two ε-symbols is obtained from (2.78),

$$\varepsilon_{\alpha\nu\sigma\tau}\varepsilon^{\lambda\nu\sigma\tau} = \varepsilon_{\nu\sigma\tau\alpha}\varepsilon^{\nu\sigma\tau\lambda} = -2\left(\delta^{\beta}_{\tau}\delta^{\lambda}_{\alpha} - \delta^{\beta}_{\alpha}\delta^{\lambda}_{\tau}\right)\delta^{\tau}_{\beta}$$
$$= -2(4-1)\delta^{\lambda}_{\alpha} = -3!\,\delta^{\lambda}_{\alpha}\,,$$

so that the first term gives

$$\partial_{\varrho}\partial_{\lambda}A^{\lambda}(x)\,\mathrm{d}x^{\varrho}\,.$$

For the second term one needs the formulae (2.73b) and (2.73c) for calculating

$$-\frac{1}{2}\partial_{\eta}\partial_{\lambda}A_{\mu}g^{\lambda\bar{\lambda}}g^{\mu\bar{\mu}}g_{\bar{\gamma}\gamma}\varepsilon_{\bar{\lambda}\bar{\mu}\alpha\beta}\varepsilon^{\eta\alpha\beta\bar{\gamma}}\,\mathrm{d}x^{\gamma}\,.$$

The contraction of the two ε-symbols is given in (2.78), the second term yields two contributions,

$$\partial^{\lambda}\partial_{\lambda}A_{\mu}\,\mathrm{d}x^{\mu} - \partial^{\mu}\partial_{\lambda}A_{\mu}\,\mathrm{d}x^{\lambda}\,.$$

Taking the sum of the two terms one obtains

$$\mathbf{\Delta}_{\mathrm{LdR}}A_{\mu}(x)\,\mathrm{d}x^{\mu} = \partial^{\lambda}\partial_{\lambda}A_{\mu}(x)\,\mathrm{d}x^{\mu} \equiv \left(\Box A_{\mu}(x)\right)\mathrm{d}x^{\mu}\,, \tag{2.84}$$

where $\Box = \partial_0^2 - \mathbf{\Delta}$ with $\mathbf{\Delta}$ the well-known Laplace(–Beltrami) operator in $\mathbb{R}^3$.

From this calculation and from the inhomogeneous equation (2.80) follows the equation of motion for ω_A:

$$\delta\omega_F = \delta\circ\mathrm{d}\omega_A = \mathbf{\Delta}_{\mathrm{LdR}}\,\omega_A - \mathrm{d}\circ\delta\,\omega_A = \frac{4\pi}{c}\star\omega_j\,.$$

Inserting here the expansions (2.82) and (2.79a), respectively, for the one-forms ω_A and $\star\omega_j$ in terms of base one-forms and comparing the coefficients multiplying $\mathrm{d}x^{\mu}$, one finds the differential equation

$$\Box A_{\nu}(x) - \partial_{\nu}\left(\partial^{\mu}A_{\mu}(x)\right) = \frac{4\pi}{c}j_{\nu}(x)\,. \tag{2.85}$$

Finally, by means of the inverse metric one raises the covariant index ν in the three terms of this equation and recovers the equation (2.60).

2.5.4 Behaviour Under the Discrete Transformations

In this section we study the behaviour of the exterior forms of Maxwell theory under space reflection Π, time reversal T, and charge conjugation C. Note that these exterior forms are now defined over four-dimensional Minkowski space! Comparing ω_E, (2.71a), with $\omega_{\boldsymbol{E}}$, (2.26a), one notices at once that these two forms differ in their transformation behaviour under time reversal. In contrast, the two-form $\omega_{\boldsymbol{B}}$,

Table 2.2 In the covariant formalism the electromagnetic exterior forms behave under the three discrete transformations as shown in the table

	Π	T	C	ΠTC
ω_E	+	−	−	+
ω_B	+	−	−	+
ω_F	+	−	−	+
$\omega_{\mathcal{F}}$	+	−	−	+
ω_A	+	−	−	+
ω_j	−	+	−	+

(2.71b), does not differ from the two-form (2.26b) over the space $\mathbb{R}^3$. The transformation behaviour of ω_E and of ω_B under Π, T, and C is now the same. This holds also for the two-form ω_F, (2.69), and, of course, also for $\omega_{\mathcal{F}}$, (2.76). These observations are listed in the first four rows of Table 2.2.

Considering ω_A as defined in (2.82) and taking into account that the scalar potential $\Phi(t, \boldsymbol{x})$ is a genuine scalar, the vector potential $\boldsymbol{A}$ is a genuine vector field over $\mathbb{R}^3$, and that $\boldsymbol{B} = \nabla \times \boldsymbol{A}$, it is clear that the one-form ω_A has the same transformation properties as the first four two-forms. This is noted in the fifth row of Table 2.2.

The three-form ω_j, (2.79), which for a given partition of Minkowski space into time and space contains the charge density $\varrho(t, \boldsymbol{x})$ and the current density $\boldsymbol{j}(t, \boldsymbol{x})$, has the same transformation behaviour as $\overset{3}{\omega}_{\varrho}$, (2.28a), cf. Table 2.1. The result obtained there can be taken over directly to Table 2.2. One easily verifies that the one-form $\star\omega_j$ which is its Hodge dual, is *even* under Π, but *odd* under T.

A common feature of all forms considered here is that they are invariant under the combined transformation ΠTC. It is instructive to compare the results with Table 2.1: The invariance under the combined transformation ΠTC rests in an essential way on the fact that the exterior forms of Maxwell theory are defined on four-dimensional Minkowski space. Invariance of Maxwell theory, as well as of all other known theories of fundamental interactions, under the combined, so-called "PCT" symmetry touches upon a deeply significant result of quantum field theory.

2.5.5 * Covariant Derivative and Structure Equation

This section is merely a long remark which anticipates the more general framework of non-Abelian gauge theories. Therefore, it might not be fully understandable at this point and the reader might wish to come back to it at a later stage.

On the spaces $\Lambda^k(M)$ of exterior forms, $k = 1, \ldots, n$, define the following differential operator:

$$D_A := \mathrm{d} + \mathrm{i}\frac{q}{\hbar c}\omega_A \qquad (2.86)$$
$$= \mathrm{i}\left(-\mathrm{i}\,\mathrm{d} + \frac{q}{\hbar c}\omega_A\right).$$

The simple rewriting in the second line of (2.86) serves the purpose of preparing a first intuitive understanding of this definition. Recall that in quantum mechanics the spatial momentum $\boldsymbol{p}$ is replaced by the operator $-\mathrm{i}\hbar\nabla$. Thus, by multiplying D_A with $\hbar$, one sees that (2.86) is the natural generalization of the term

$$\boldsymbol{p} - \frac{q}{c}\boldsymbol{A}$$

whose square appears in the Hamiltonian function for a charged particle in external fields (see, e.g. [ME], Sect. 2.16), and which is prescribed by the principle of *minimal coupling*. In differential geometry as well as in quantum physics one calls this the *covariant derivative*.

When applied to an arbitrary exterior form ω the operator D_A shall act by

$$D_A\,\omega = \mathrm{d}\,\omega + \mathrm{i}\frac{q}{\hbar c}\omega_A \wedge \omega\,. \qquad (2.86a)$$

Its action on functions (i.e. on zero-forms), in particular, reads

$$D_A\,f = \left(\partial_\mu f + \mathrm{i}\frac{q}{\hbar c}A_\mu f\right)\,\mathrm{d}x^\mu\,. \qquad (2.86b)$$

The operator D_A is a linear combination of exterior derivative and exterior product with the one-form ω_A. In other terms, very much like d, D_A maps a k-form onto a $(k+1)$-form.

If one takes the square of the operator D_A, i.e. if one applies D_A twice successively on an arbitrary exterior form ω, one finds a remarkable result,

$$D_A \circ D_A\,\omega = \left(\mathrm{d} + \mathrm{i}\frac{q}{\hbar c}\omega_A\right) \circ \left(\mathrm{d} + \mathrm{i}\frac{q}{\hbar c}\omega_A\right)\,\omega$$
$$= \left\{\mathrm{d} \circ \mathrm{d} + \mathrm{i}\frac{q}{\hbar c}\left(\mathrm{d}\omega_A \wedge + \omega_A \wedge \mathrm{d}\right) + \left(\mathrm{i}\frac{q}{\hbar c}\right)^2 \omega_A \wedge \omega_A \wedge\right\}\,\omega\,.$$

The first term in the curly brackets gives zero because of $\mathrm{d}^2 = 0$ (cf. (2.15d)). The third term vanishes as well by the asymmetry of the wedge product. As for the middle term, using the graded Leibniz rule (2.15c), one has

$$\mathrm{d}\omega_A \wedge \omega + \omega_A \wedge \mathrm{d}\omega = \left(\mathrm{d}\omega_A\right) \wedge \omega - \omega_A \wedge \mathrm{d}\omega + \omega_A \wedge \mathrm{d}\omega$$
$$= \left(\mathrm{d}\omega_A\right) \wedge \omega\,.$$

From there and with (2.83) one obtains an important and most interesting result, viz.

$$D_A^2 = \mathrm{i}\frac{q}{\hbar c}(\mathrm{d}\omega_A) = \mathrm{i}\frac{q}{\hbar c}\omega_F \,. \tag{2.87}$$

It is particularly noteworthy that the operator $(\mathrm{d}\omega_A)$ saturates in itself, that is to say, the exterior derivative does not act further to the right – in contrast to the original operator D_A. The square of the covariant derivative D_A yields the two-form (2.69) of the field strengths, up to the factor $\mathrm{i}q/(\hbar c)$. In more mathematical terms (2.87) tells us that D_A^2 acts *linearly*. This does not hold for D_A.

These matters become a little more transparent if we replace the one-form ω_A and the two-form ω_F by a one-form A and a two-form F, respectively, which are defined as follows

$$A := \mathrm{i}\frac{q}{\hbar c}\omega_A \,, \quad F := \mathrm{i}\frac{q}{\hbar c}\omega_F \,. \tag{2.88a}$$

With these definitions the structure equations above simplify to

$$D_A = \mathrm{d} + A \,, \tag{2.88b}$$

$$D_A^2 = (\mathrm{d}A) + A \wedge A = (\mathrm{d}A) = F \,. \tag{2.88c}$$

Equations of this type are well known in differential geometry. The one-form A, defined here in a particularly simple case, is called the *connection form*. The operator $D_A = \mathrm{d} + A$ is the covariant derivative, while F is called the *curvature form* pertaining to the given connection. Indeed, one can show that $F = D_A^2$ may be interpreted as a "round trip" along a small closed loop, in analogy to what one would do in order to detect curvature in a hypersurface.

Maxwell Theory as a Classical Field Theory

3

3.1 Introduction

Hamilton's variational principle and the Lagrangian mechanics that rests on it are exceedingly successful in their application to mechanical systems with a *finite* number of degrees of freedom. Hamilton's principle characterizes the physically realizable orbits, among the set of all possible orbits, as being the critical elements of the action integral. The Lagrangian function, although not an observable on its own, is not only useful in deriving the equations of motion but is also an important tool for identifying symmetries of the theory and constructing the corresponding conserved quantities, via Noether's theorem.

The variational principle and Lagrangian mechanics can be generalized to dynamical systems with infinitely many, nondenumerable, degrees of freedom. The Lagrangian function is replaced by the Lagrangian density, the (generalized) *coordinates* are replaced by time- and space-dependent *fields.* The Euler–Lagrange equations are equations of motion for these fields, showing that Maxwell's equations are derivable from a variational principle. As in mechanical systems, the theorem of Noether yields the relationship between invariance of the Lagrangian density under transformations in space and time and the conservation laws.

3.2 Lagrangian Function and Symmetries in Finite Systems

To start with, we recall the notion of Lagrangian function in the mechanics of systems with a finite number of degrees of freedom with special emphasis on the description of symmetries of a given theory. Readers who are familiar with this topic may wish to skip this section.

Invariance of the Lagrangian function $L(q_1, q_2, \dots, q_f, \dot{q}_1, \dots \dot{q}_f, t)$ (up to gauge terms) under a symmetry transformation implies that the Euler–Lagrange equations pertaining to L are *covariant,* i.e. are form invariant with respect to that

F. Scheck, *Classical Field Theory*, Graduate Texts in Physics,
DOI 10.1007/978-3-642-27985-0_3,

symmetry. This fact illustrates the central importance of the notion of Lagrangian function: Indeed, it is often simpler to construct invariants which may be suitable candidates for possible Lagrangian functions rather than to magically invent covariant equations of motion. An elementary example may serve as an illustration for this assertion.

Example 3.1 Force-free particle

A point particle of mass m in $\mathbb{R}^3$ which is not subject to any forces may be described by the Lagrangian function

$$L(\boldsymbol{x}, \dot{\boldsymbol{x}}) = T_{\text{kin}} = \frac{1}{2} m \dot{\boldsymbol{x}}^2 .$$

Obviously, this very simple Lagrangian function is invariant under the Galilei transformations

$$t \mapsto t' = t + s \,, \quad s \in \mathbb{R} \,, \tag{3.1a}$$

$$\boldsymbol{x} \mapsto \boldsymbol{x}' = \mathrm{R}\boldsymbol{x} + \boldsymbol{a} \,, \quad \mathrm{R} \in \mathrm{SO}(3) \,, \quad \boldsymbol{w}, \boldsymbol{a} \quad \text{real} \,. \tag{3.1b}$$

The corresponding Euler–Lagrange equations

$$\frac{\mathrm{d}}{\mathrm{d}t} \frac{\partial L}{\partial \dot{\boldsymbol{x}}} - \frac{\partial L}{\partial \boldsymbol{x}} = m \ddot{\boldsymbol{x}} = 0$$

are covariant under transformations (3.1a) and (3.1b), meaning that if one of the following equations is fulfilled then so is the other:

$$\frac{\mathrm{d}^2 \boldsymbol{x}(t)}{\mathrm{d}t^2} = 0 \quad \Longleftrightarrow \quad \frac{\mathrm{d}^2 \boldsymbol{x}'(t')}{\mathrm{d}t'^2} = 0 \,.$$

The equations of motion are the same in all inertial systems. Joining to (3.1a) and (3.1b) also the special Galilei transformations

$$\boldsymbol{x} \mapsto \boldsymbol{x}' = \boldsymbol{x} + \boldsymbol{w} t \,, \quad \boldsymbol{w} \quad \text{real} \tag{3.1c}$$

the Lagrangian function no longer stays strictly invariant,

$$L'(\boldsymbol{x}', \dot{\boldsymbol{x}}') = L(\boldsymbol{x}, \dot{\boldsymbol{x}}) + m \dot{\boldsymbol{x}} \cdot \left(\mathrm{R}^{-1} \boldsymbol{w}\right) + \frac{m}{2} \boldsymbol{w}^2 \,,$$

but is modified by an additive term which is seen to be the time derivative of a function of $\boldsymbol{x}$ and t:

$$M(\boldsymbol{x}, t) = m \boldsymbol{x} \cdot \left(\mathrm{R}^{-1} \boldsymbol{w}\right) + t \frac{m}{2} \boldsymbol{w}^2 \,.$$

A gauge transformation of this kind (in the sense given to this notion in mechanics) leaves the equations of motion unchanged, or, more precisely, these equations stay covariant (cf. [ME], Sect. 2.10).

Invariance under time translations (3.1a) implies conservation of the energy, which in this example is pure kinetic energy $E = T_{\text{kin}}$; invariance under translations in space implies the conservation of momentum $\boldsymbol{p} = m\dot{\boldsymbol{x}}$, and from invariance under rotations follows the conservation of the angular momentum $\boldsymbol{\ell} = \boldsymbol{x} \times \boldsymbol{p}$. (The centre-of-mass theorem is trivial here because we are concerned with a single particle.)

3.2.1 Noether's Theorem with Strict Invariance

Here and in what follows a set of generalized coordinates of a mechanical system with f degrees of freedom is denoted by $q = (q_1, q_2, \dots, q_f)$. Geometrically speaking, these variables are coordinates of points in the manifold of motions Q. In its simplest form, the theorem of E. Noether applies to autonomous systems as follows.

If the Lagrangian function $L(q, \dot{q})$ is invariant under continuous transformations which can be deformed continuously into the identity, i.e.

$$q \longmapsto q' = h^s(q) \quad \text{with} \quad h^{s=0}(q) = q \ , \tag{3.2}$$

then the function $I : TQ \to \mathbb{R} : (q, \dot{q}) \mapsto I(q, \dot{q})$, given by

$$I(q, \dot{q}) = \sum_{i=1}^{f} \frac{\partial L(q, \dot{q})}{\partial \dot{q}^i} \frac{\mathrm{d}h^s(q^i)}{\mathrm{d}s}\bigg|_{s=0} \ , \tag{3.3}$$

is constant along solutions $q = \varphi(t)$ of the equations of motion.

This integral of the motion is a composition of the (generalized) momenta $p_i = \partial L / \partial \dot{q}^i$ and of the generating function in its action on the coordinates q^i. Note that the transformation h^s depends on one parameter only, i.e. in a group action such as with SO(3) in (3.1b), one must consider one-parameter subgroups. In the case of the rotation group, these are rotations about a fixed direction. In the simple example 3.1, consider the (active) rotation about the direction $\hat{\boldsymbol{n}}$:

$$\boldsymbol{x} \mapsto \boldsymbol{x}' = \boldsymbol{x} \cos s + \hat{\boldsymbol{n}} \times \boldsymbol{x} \sin s \equiv h^s(\boldsymbol{x}) \ .$$

Here one has $\mathrm{d}h^s(\boldsymbol{x})/\mathrm{d}s|_0 = \hat{\boldsymbol{n}} \times \boldsymbol{x}$, so that the integral (3.3) is

$$m\dot{\boldsymbol{x}} \cdot (\hat{\boldsymbol{n}} \times \boldsymbol{x}) = \hat{\boldsymbol{n}} \cdot (\boldsymbol{x} \times (m\dot{\boldsymbol{x}})) = \hat{\boldsymbol{n}} \cdot \boldsymbol{\ell} \ .$$

As was to be expected, it is the projection of the orbital angular momentum onto the given direction that is conserved.

A converse of the theorem of Noether is also well known from mechanics: Every smooth dynamical quantity $f(q, p)$ whose Poisson bracket with the Hamiltonian function vanishes generates a symmetry transformation of the system.

3.2.2 Generalized Theorem of Noether

As worked out in, e. g. [ME], Sect. 2.41, the theorem of Noether can be generalized in two ways. On the one hand, covariance of the equations of motion is also guaranteed if the Lagrangian function is modified by a general gauge term,

$$L(q, \dot{q}, t) \mapsto L'(q, \dot{q}, t) = L(q, \dot{q}, t) + \frac{\mathrm{d}}{\mathrm{d}t} M(t, q) , \tag{3.4}$$

where the function M depends only on q and t. On the other hand, and this is the genuine generalization, the gauge function may also depend on $\dot{q}$, provided one makes sure that any new acceleration terms $\ddot{q}^i$ generated by the symmetry transformation add up to zero identically. As this is developed in detail in [ME], we retain only the assumptions and the main result here.

One allows for general transformations of time and coordinates

$$t' = g(t, q, \dot{q}, s) , \tag{3.5}$$

$$q'^i = h^i(t, q, \dot{q}, s) , \tag{3.6}$$

with functions g and h^i being at least twice differentiable in their $2f + 2$ variables. The parameter s is taken from an interval of the real axis which contains the origin. For $s = 0$, one must have

$$g(t, q, \dot{q}, s = 0) = t , \tag{3.7}$$

$$h^i(t, q, \dot{q}, s = 0) = q^i . \tag{3.8}$$

As in the case of strict invariance, only the neighbourhood of $s = 0$ is relevant. One defines the derivatives at $s = 0$,

$$\tau(t, q, \dot{q}) = \left. \frac{\partial g(t, q, \dot{q}, s)}{\partial s} \right|_{s=0} , \tag{3.9}$$

$$\kappa^i(t, q, \dot{q}) = \left. \frac{\partial h^i(t, q, \dot{q}, s)}{\partial s} \right|_{s=0} , \tag{3.10}$$

and interprets these as the infinitesimal generators of transformations (3.5) and (3.6). One shows that the integral of the motion given by (3.3) is replaced by the more general function $I : \mathbb{R}_t \times TQ \to \mathbb{R}$:

$$I(t,q,\dot{q}) = L(t,q,\dot{q})\tau(t,q,\dot{q}) + \sum_i \frac{\partial L}{\partial \dot{q}^i}\left[\kappa^i(t,q,\dot{q}) - \dot{q}^i \tau(t,q,\dot{q})\right] - M(t,q,\dot{q})\,, \tag{3.11}$$

which is invariant along solutions $q = \varphi(t)$ of the equations of motion.

Examples for finite systems which illustrate the nature of this function are found in [ME], Sect. 2.41, and in Boccaletti and Pucacco (1999).

Remarks

1. Of course, in the case of strict invariance, the integral of motion given by (3.3) is contained in the general formula (3.11). It suffices to take the generating function τ and the gauge function M to be identically zero:
$$\tau(t,q,\dot{q}) \equiv 0\,, \quad M(t,q,\dot{q}) \equiv 0\,. \tag{3.12}$$
The functions κ^i, defined in (3.10), are identical to the second factor on the right-hand side of (3.3).
2. Whenever the gauge term vanishes identically, $M(t,q,\dot{q}) \equiv 0$, it is appropriate to talk about *strict invariance,* as in Sect. 3.2.1, because in this case the Lagrangian function stays unchanged. If there is a nonvanishing gauge term, then one may talk about a *quasi symmetry.* Examples may be found in [ME], Sect. 2.41.

3.3 Lagrangian Density and Equations of Motion for a Field Theory

Hamilton's variational principle can be generalized to systems with infinitely many, nondenumerable degrees of freedom. Instead of the generalized coordinates q of mechanics, such systems are described by classical fields $\psi^{(i)}(t,\boldsymbol{x})$, which depend on time and space (see, e. g., [ME] Chap. 7). To such a system, called a field theory, one associates a *Lagrangian density*, which is a function of the fields and of their time and space derivatives. The Lagrangian density may also have an explicit dependence on time t and position $\boldsymbol{x}$, and possibly on some external sources $j^{(k)}(x)$,

$$\mathcal{L}(t,\boldsymbol{x},j^{(k)},\psi^{(i)},\partial_\mu\psi^{(i)})\,, \quad i = 1,2,\ldots,N\,. \tag{3.13}$$

Its integral over the whole space,

$$L = \iiint \mathrm{d}^3x\,\mathcal{L}\,, \tag{3.14}$$

is the analogue of the Lagrangian function of mechanics with a finite number of degrees of freedom. This is the quantity which defines the action integral of Hamilton's principle. This action integral is a functional of the fields

$$I\left[\psi^{(1)},\ldots,\psi^{(N)}\right]=\int_{t_1}^{t_2}\mathrm{d}t\; L=\int_{t_1}^{t_2}\mathrm{d}t\iiint \mathrm{d}^3x\;\mathcal{L}\,. \tag{3.15a}$$

According to Hamilton's principle, suitably generalized to a field theory, functional (3.15a) is stationary for physical solutions. From this requirement follow the Euler–Lagrange equations for the variational problem,

$$\delta I\left[\psi^{(1)},\ldots,\psi^{(N)}\right]=0 \quad (\text{fixed boundaries } t_1 \text{ and } t_2)\,, \tag{3.15b}$$

with the condition that the variations $\delta\psi^{(i)}$ of the fields vanish on the hypersurfaces $t = t_1$ and $t = t_2$. The equations read (using the summation convention in the second term)

$$\frac{\partial\mathcal{L}}{\partial\psi^{(i)}}-\partial_\mu\frac{\partial\mathcal{L}}{\partial(\partial_\mu\psi^{(i)})}=0\,,\quad i=1,2,\ldots,N\,. \tag{3.16}$$

The derivation of these equations is formally the same as for the Euler–Lagrange equations in mechanics. We sketch it for the case of a single field ψ: One has

$$\begin{aligned}
\delta I[\psi] &= \int_{t_1}^{t_2}\mathrm{d}t\iiint \mathrm{d}^3x\left\{\frac{\partial\mathcal{L}}{\partial\psi}\delta\psi+\frac{\partial\mathcal{L}}{\partial(\partial_\mu\psi)}\delta(\partial_\mu\psi)\right\}\\
&= \int_{t_1}^{t_2}\mathrm{d}t\iiint \mathrm{d}^3x\left\{\frac{\partial\mathcal{L}}{\partial\psi}\delta\psi+\frac{\partial\mathcal{L}}{\partial(\partial_\mu\psi)}\partial_\mu(\delta\psi)\right\}\\
&= \int_{t_1}^{t_2}\mathrm{d}t\iiint \mathrm{d}^3x\left\{\frac{\partial\mathcal{L}}{\partial\psi}-\partial_\mu\frac{\partial\mathcal{L}}{\partial(\partial_\mu\psi)}\right\}\delta\psi\,.
\end{aligned}$$

In the first step, it is assumed that the variation of a derivative of the fields $\delta(\partial_\mu\phi)$ is equal to the derivative $\partial_\mu(\delta\phi)$ of the variation. This is adequate if one assumes that the fields are smooth functions and the variations are smooth as well. Going from the second to the third step involves a partial integration – hence the minus sign – and makes use of the assumption that $\delta\psi$ vanishes on the hypersurfaces $t = t_1$ and $t = t_2$. If this integral is required to be zero for all admissible variations $\delta\psi$, then the integrand in the curly brackets of the third line must vanish. This is the Euler–Lagrange equation for the single field ψ. If there are several fields, say N kinds, then one obtains the equations of (3.16).

Of course, with regard to Maxwell's equations, we have in mind Minkowski space M^4 and covariance under the proper, orthochronous Lorentz group $L_+^\uparrow$, possibly also the discrete transformations Π, T and C. In this case, the fields are defined as functions of $x \in M^4 = \mathbb{R}^4$. The representation of point x by a time coordinate t and by space coordinates $\boldsymbol{x}$ refers to a class of frames of reference in which the splitting into time axis and space $\mathbb{R}^3$ is fixed. The Lorentz *covariance* of the equations of motion given by (3.16) is guaranteed if the Lagrangian density (3.13) is *invariant* under Lorentz transformations.

Remarks

1. In many cases, the $\psi^{(i)}(x)$ fields are independent of each other. Index i serves merely to distinguish different types of fields, such as electromagnetic fields, from their external sources. There are cases, however, where the fields are subject to subsidiary conditions such as the requirement that the set $\{\psi^{(i)}\}$ must describe a particle with spin. As an example, consider a massive particle with spin 1. As the spin vector (this is the expectation value of spin) has three components, we will need three different fields for its description. Yet these fields are correlated by their properties under rotations in space.[1] As a natural candidate, one chooses a Lorentz vector field $V(x) = (V^0(x), V^1(x), V^2(x), V^3(x))^T$ which, however, has four components, not three. In this example, one would have

$$\mathcal{L} = \mathcal{L}\big(x, V^\alpha(x), \partial_\mu V^\alpha(x)\big)\,, \quad \text{i.e.}$$
$$\psi^{(1)}(x) \equiv V^0(x)\,,\ \psi^{(2)}(x) \equiv V^1(x)\,,\ \psi^{(3)}(x) \equiv V^2(x)\,,$$
$$\psi^{(4)}(x) \equiv V^3(x)\,.$$

In this case, the fields $V^\alpha(x)$ cannot be independent because otherwise they would describe both spin 1, as required, and a spin 0 object. The second, unwanted, part is eliminated by a subsidiary condition, viz.

$$\partial_\alpha V^\alpha(x) = 0\,.$$

This divergence is a Lorentz scalar, and it seems plausible that it contains the spin 0 part of the field which one wishes to exclude.

2. Functional (3.15a) and the variational principle given by (3.15b) seem to single out the time axis as opposed to three-dimensional space. This does not break the Lorentz invariance of the procedure, though. One can always replace the two surfaces $t = t_1$ and $t = t_2$ and their complement at infinity by a closed, smooth three-manifold $\partial\Sigma$ and take the integral in (3.15a) over the volume Σ enclosed by it. (Note that we again use a geometric notation: $\partial\Sigma$ is

[1] Although spin is a quantum property and, hence, is described by self-adjoint operators $\hat{\boldsymbol{S}} = \{\hat{S}_i\}$, $i = 1, 2, 3$, of quantum mechanics, its expectation values $\langle\hat{\boldsymbol{S}}\rangle$ in quantum states are classical observables.

the boundary of Σ.) It is sufficient to require that the variations $\delta\psi^{(i)}$ vanish on $\partial\Sigma$ to obtain the same equations of motion, (3.16), as before. Therefore, these equations are indeed covariant.

Example 3.2 Real scalar field

As an example, consider a single real Lorentz scalar field $\phi(x)$ coupled to an external scalar density $\varrho(x)$.[1] With these assumptions, the following Lagrangian density is a scalar with respect to Lorentz transformations as well:

$$\frac{1}{\hbar c}\mathcal{L}(\phi, \partial_\mu\phi, \varrho) = \frac{1}{2}\left[\partial_\mu\phi(x)\partial^\mu\phi(x) - \kappa^2\phi^2(x)\right] - \varrho(x)\phi(x) . \qquad (3.17a)$$

In this ansatz one should choose $\kappa = mc^2/(\hbar c)$, for dimensional reasons – up to a factor 2π this is the inverse of the Compton wavelength $\lambda_C = h/mc$ of a particle with mass m. The factor which multiplies $\mathcal{L}$ on the left-hand side is introduced to provide $\mathcal{L}$ with the correct physical dimension.[2] Note, however, that this factor is irrelevant for the equation of motion because the latter is linear in $\mathcal{L}$.

The Euler–Lagrange equation (3.16) (there is only one in this example) reads as follows. One calculates

$$\frac{1}{\hbar c}\frac{\partial L}{\partial\phi} = -\kappa^2\phi(x) - \varrho(x) ,$$

$$\frac{1}{\hbar c}\frac{\partial L}{\partial(\partial_\mu\phi)} = \frac{1}{2}\frac{\partial}{\partial(\partial_\mu\phi)}\left(\partial_\mu\phi(x)\right)g^{\mu\nu}\left(\partial_\nu\phi(x)\right) = \partial^\mu\phi(x)$$

(noting that $\partial_\mu\phi(x)$ appears twice), then takes the four-divergence of the second expression and inserts the two terms in (3.16). This gives the equation of motion

$$\left(\partial_\mu\partial^\mu + \kappa^2\right)\phi(x) = \left(\Box + \kappa^2\right)\phi(x) = -\varrho(x) . \qquad (3.17b)$$

There is a close analogy to the wave equation (2.60) if the Lorenz condition, (2.61), $\partial_\mu A^\mu(x) = 0$, is chosen: The left-hand side of (3.17b) contains the field $\phi(x)$ only; the right-hand side is the source of the field.

By comparing the Lagrangian density given by (3.17a) with the Lagrangian function of point mechanics in its natural form, $L = T_{\text{kin}} - U$, one is led to a physical interpretation of the three terms. The first term,

$$\frac{1}{2}\partial_\mu\phi(x)\partial^\mu\phi(x) ,$$

[1] More on this example is found in [QP], Sect. 7.1.

[2] In quantum field theory, it is useful to provide the (quantized) scalar field ϕ with the dimension (length) $^{-1}$. As one verifies, the density $\mathcal{L}$ then has dimension E/L^3, i.e. energy/volume.

or rather its integral over all space, is the analogue of the kinetic energy in the mechanics of a point particle. The remaining two terms,

$$\frac{1}{2}\kappa^2\phi^2(x) + \varrho(x)\phi(x) ,$$

describe a mass term (which is new) and a coupling $\varrho(x)\phi(x)$ whose integral is the analogue of the potential energy in mechanics.

Example 3.3 shows that the formal analogy to mechanics with a finite number of degrees of freedom extends even further. As one can talk about the energy content of a field, or of a set of fields, there must be an analogue of the Hamiltonian function and, by the same token, a generalization of the momentum canonically conjugate to a field. With reference to the definition $p := \partial L/\partial \dot{q}$ of mechanics, one defines the momentum field $\pi^{(i)}(x)$ canonically conjugate to the field $\psi^{(i)}(x)$ by

$$\pi^{(i)}(x) := \frac{\partial \mathcal{L}}{\partial(\partial_0 \psi^{(i)})} . \tag{3.18}$$

The Hamilton density is obtained from the function

$$\widetilde{\mathcal{H}} := \sum_{i=1}^{N} \pi^{(i)}(x)\partial_0\psi^{(i)}(x) - \mathcal{L}\big(t, \boldsymbol{x}, j^{(k)}, \psi^{(i)}, \partial_\mu \psi^{(i)}\big) \tag{3.19}$$

by a Legendre transform with respect to the variables $\partial_0\psi^{(i)}$. This means that (3.18) must be solved for $\partial_0\psi^{(i)}$ and that one must replace these variables in $\tilde{\mathcal{H}}$. In the simplest case, the density $\mathcal{L}$ contains a kinetic term of the kind

$$\mathcal{L}_{\text{kin}} = \frac{1}{2}\sum_{i=1}^{N}\Big(\partial_\mu\psi^{(i)}(x)\partial^\mu\psi^{(i)}(x)\Big) ,$$

so that the canonically conjugate momentum is found to be

$$\pi^{(i)}(x) = \partial_0\psi^{(i)}(x) .$$

From the Lagrangian density (3.17a) of Example 3.3 the Hamilton density is obtained from $\widetilde{\mathcal{H}}$, (3.19), as follows:

$$\mathcal{H} = \frac{1}{2}\Big\{\pi^2(x) + \big(\nabla\phi(x)\big)^2 + \kappa^2\phi^2(x)\Big\} + \varrho(x)\phi(x) . \tag{3.20}$$

Suppose the Lagrangian density given by (3.13) is autonomous, i.e. depends only on the fields and their derivatives, but has no explicit dependence on time and position or on external sources. We take the derivative of $\mathcal{L}$ by x_α,

$$\partial^\alpha \mathcal{L}\big(\psi^{(i)}, \partial_\mu\psi^{(i)}\big) = g^{\alpha\beta}\partial_\beta\mathcal{L}\big(\psi^{(i)}, \partial_\mu\psi^{(i)}\big) ,$$

and compare it with the same derivative as obtained by the chain rule and by using the equations of motion (3.16):

$$\begin{aligned}
\partial^\alpha \mathcal{L}(\psi^{(i)}, \partial_\mu \psi^{(i)}) &= \sum_i \left\{ \frac{\partial \mathcal{L}}{\partial \psi^{(i)}} \partial^\alpha \psi^{(i)} + \frac{\partial \mathcal{L}}{\partial(\partial_\beta \psi^{(i)})} \partial^\alpha (\partial_\beta \psi^{(i)}) \right\} \\
&= \sum_i \left\{ \left[\partial_\beta \left(\frac{\partial \mathcal{L}}{\partial(\partial_\beta \psi^{(i)})} \right) + \frac{\partial \mathcal{L}}{\partial(\partial_\beta \psi^{(i)})} \partial_\beta \right] \partial^\alpha \psi^{(i)} \right\} \\
&= \partial_\beta \left(\sum_i \frac{\partial \mathcal{L}}{\partial(\partial_\beta \psi^{(i)})} \partial^\alpha \psi^{(i)} \right) .
\end{aligned}$$

(Note that the index α is fixed while β is a summation index.)
Defining the tensor field

$$\mathcal{T}^{\mu\nu} := \sum_i \left(\frac{\partial \mathcal{L}}{\partial(\partial_\mu \psi^{(i)})} \right) \partial^\nu \psi^{(i)} - g^{\mu\nu} \mathcal{L}(\psi^{(i)}, \partial_0 \psi^{(i)}, \nabla \psi^{(i)}) , \tag{3.21}$$

the preceding calculation tells us that this tensor field satisfies a continuity equation for all solutions of the equations of motion (3.16):

$$\partial_\mu \mathcal{T}^{\mu\nu} = 0 . \tag{3.22}$$

In Sect. 3.5.3, we will analyse in detail the analogue of this tensor field, (3.21), for the case of Maxwell theory showing that the conservation law, (3.22), describes, among other features, the conservation of the energy and momentum content of the fields. For the moment we return to Example 3.3. Starting from the Lagrangian density given by (3.17a), one finds

$$\mathcal{T}^{\mu\nu} = \partial^\mu \phi(x) \partial^\nu \phi(x) - g^{\mu\nu} \left(\tfrac{1}{2} \partial_\lambda \phi g^{\lambda\eta} \partial_\eta \phi - \tfrac{1}{2} \kappa^2 \phi^2(x) - \varrho(x) \phi(x) \right) . \tag{3.23a}$$

Calculating the components ($\mu = 0, \nu = 0$) and ($\mu = 0, \nu = i$), one finds

$$\begin{aligned}
\mathcal{T}^{00} &= (\partial^0 \phi(x))^2 - \frac{1}{2} \partial_\lambda \phi(x) \partial^\lambda \phi(x) + \frac{1}{2} \kappa^2 \phi^2(x) + \varrho(x) \phi(x) \\
&= \frac{1}{2} \{ \pi^2(x) + (\nabla \phi(x))^2 + \kappa^2 \phi^2(x) \} + \varrho(x) \phi(x) , && (3.23b) \\
\mathcal{T}^{0i} &= \pi(x) \partial^i \phi(x) . && (3.23c)
\end{aligned}$$

The component $\mathcal{T}^{00}$, (3.23c), is indeed identical to (3.20). The components $\mathcal{T}^{0i}$ describe the momentum density whose divergence is related to the time derivative of $\mathcal{T}^{00}$ by the continuity equation (3.22). As a further important property we note

$$T^{\mu\nu}(x) = T^{\nu\mu}(x) ; \tag{3.24}$$

the energy-momentum tensor field is symmetric in μ and ν.

3.4 Lagrangian Density for Maxwell Fields with Sources

The tools we have at our disposal for the construction of a Lagrangian density which is invariant under the proper orthochronous Lorentz group $L_+^\uparrow$ are as follows: the tensor field $F^{\mu\nu}(x)$, (2.46), or its companion $\mathcal{F}^{\mu\nu}(x)$, equation (2.52a), the dual tensor field $\star F^{\mu\nu}(x)$, equation (2.54c), the four-vector potential $A^\mu(x)$, equation (2.57), and the current density $j^\mu(x)$, (2.51). Obvious invariants are

$$F_{\mu\nu}(x)F^{\mu\nu}(x)\,, \quad (\star F)_{\mu\nu}(x)F^{\mu\nu}(x)\,, \quad \text{and} \quad j^\mu(x)A_\mu(x)\,.$$

To analyse the physical content of these invariants, we calculate them in terms of electric and magnetic fields within a fixed partition of spacetime into coordinate time $\mathbb{R}_t$ and space part $\mathbb{R}^3$. In calculating, e. g. the contraction $F_{\mu\nu}F^{\mu\nu}$, one takes $F_{\mu\nu}$ from formula (2.69a) and multiplies by the transpose of expression (2.46), $F^{\nu\mu} = -F^{\mu\nu}$. The product $F_{\mu\nu}F^{\mu\nu}$ is then seen to be the trace of a product of two 4×4 matrices. In more detail, one has

$$\begin{aligned} F_{\mu\nu}F^{\mu\nu} &= \operatorname{tr}\left\{\begin{pmatrix} 0 & E^1 & E^2 & E^3 \\ -E^1 & 0 & -B^3 & B^2 \\ -E^2 & B^3 & 0 & -B^1 \\ -E^3 & -B^2 & B^1 & 0 \end{pmatrix}\begin{pmatrix} 0 & E^1 & E^2 & E^3 \\ -E^1 & 0 & B^3 & -B^2 \\ -E^2 & -B^3 & 0 & B^1 \\ -E^3 & B^2 & -B^1 & 0 \end{pmatrix}\right\} \\ &= -2\left(\boldsymbol{E}^2 - \boldsymbol{B}^2\right)\,, \qquad (3.25\text{a}) \\ \star F_{\mu\nu}F^{\mu\nu} &= \operatorname{tr}\left\{\begin{pmatrix} 0 & -B^1 & -B^2 & -B^3 \\ B^1 & 0 & -E^3 & E^2 \\ B^2 & E^3 & 0 & -E^1 \\ -B^3 & -E^2 & E^1 & 0 \end{pmatrix}\begin{pmatrix} 0 & E^1 & E^2 & E^3 \\ -E^1 & 0 & B^3 & -B^2 \\ -E^2 & -B^3 & 0 & B^1 \\ -E^3 & B^2 & -B^1 & 0 \end{pmatrix}\right\} \\ &= 4\boldsymbol{B}\cdot\boldsymbol{E}\,, \qquad (3.25\text{b}) \\ j^\mu A_\mu &= c\varrho\Phi - \boldsymbol{j}\cdot\boldsymbol{A}\,. \qquad (3.25\text{c}) \end{aligned}$$

Clearly, the products $\mathcal{F}_{\mu\nu}\mathcal{F}^{\mu\nu}$ etc. yield very similar formulae in which $\boldsymbol{E}$ is replaced by $\boldsymbol{D}$ and $\boldsymbol{B}$ is replaced by $\boldsymbol{H}$.

Going back to Sect. 2.2.1, one sees that the $L_+^\uparrow$-invariant (3.25a) is also invariant under the three discrete transformations Π (space reflection), T (time reversal) and C (charge conjugation). This is also true for the third term, (3.25c). The term (3.25b), however, is *odd* under Π, being a pseudoscalar. It is odd under time reversal T, too, but since it is even under charge conjugation, it stays invariant under the combined transformation ΠCT, much like the two other terms.

Practical experience and a great number of experiments tell us that no parity-violating effects are observed in electromagnetic processes. Therefore, the interaction which describes these processes must have a well-defined behaviour under space reflection. A parity-violating effect, e. g. would be signalled by an observable proportional to the scalar product of a momentum and an angular momentum, $\mathcal{O} =$

$f(E)\,\boldsymbol{p}\cdot\boldsymbol{\ell}$, with f being a function of, for example, the energy.[3] Parity-violating terms of this kind, well known from *weak* interactions (β-decay, neutrino reactions, etc.), as illustrated by Fig. 3.1, have never been observed in purely *electromagnetic* processes. When applied to Maxwell theory, this means that the Lagrangian density cannot contain simultaneously the terms given by (3.25a) and (3.25c) on the one hand and the term given by (3.25b) on the other.

On the basis of these considerations and with the expectation that the Hamilton density that derives from $\mathcal{L}$ will definitely contain the interaction $j^\mu(x)A_\mu(x)$ between matter and radiation, it seems plausible that the Lagrangian density should be a linear combination of the first and third terms, viz.

$$\mathcal{L} = a_1 F_{\mu\nu}(x)F^{\mu\nu}(x) + a_2 j^\mu(x)A_\mu(x)\,. \tag{3.26a}$$

The coefficients a_1 and a_2 are fixed by two requirements:
(i) The Lagrangian density, (3.26a), must yield Maxwell's equations in the form of the homogeneous equations of (2.49a) and the inhomogeneous equations of (2.53), with the correct factor on the right-hand side of (2.53).
(ii) After transformation to the Hamilton density, the energy density must be compatible with the previously known expression (1.130), which applies to stationary situations.
However, prior to this construction, one must identify the degrees of freedom of Maxwell theory whose variations are used in Hamilton's principle. One possible choice could be the tensor field $F_{\mu\nu}(x)$, which contains the observable fields; another could be the set of potentials $A_\mu(x)$, which, being defined only up to gauge transformations, are not observables. If one decides to choose the fields $A_\mu(x)$ as the field variables, then the observables $F_{\mu\nu}(x)$ must be expressed in terms of potentials, as given in (2.58).

Remark

The choice of the potentials as the dynamical degrees of freedom is somewhat problematic because the four fields $A^\mu(x)$ cannot be independent. Consider, as an example, the class of gauges which guarantee the Lorenz condition $\partial_\mu A^\mu(x) = 0$. In quantum theory, one learns that the photon carries spin 1, the Lorenz condition is found to be the condition that eliminates the spin 0 part of the vector field A^μ. Furthermore, as a consequence of the linear dependency of the fields A^μ, one of them has a vanishing canonically conjugate momentum. This, in turn, causes difficulties in the quantization of Maxwell theory, which must be studied carefully (for more on this see, e. g. [QP]).

[3] The precise statement is this: If an initial state which has a well-defined behaviour under space reflection went over into another state which exhibits such a momentum-angular momentum correlation, by the effect of electromagnetic interaction, the interaction would contain both parity-even and parity-odd terms.

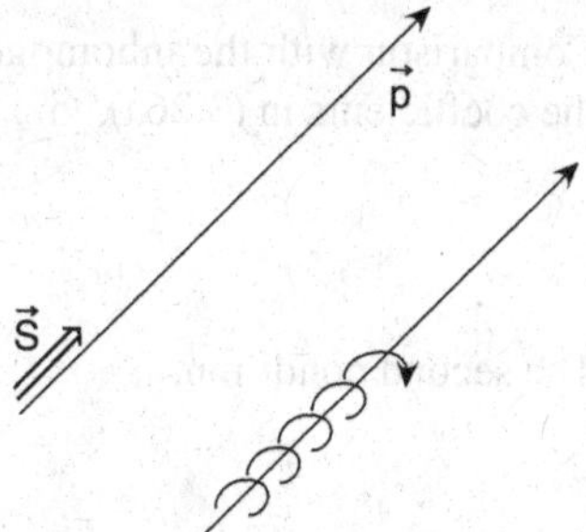

Fig. 3.1 In β decays of many nuclei, one observes positrons whose spin is oriented almost completely along the direction of their spatial momentum. The initial state, being an atomic nucleus, is an eigenstate of parity. The occurrence of a spin-momentum correlation of this kind signals that weak interactions are not invariant under space reflection

For the sake of simplicity, we treat Maxwell fields in a vacuum, outside the sources which are assumed to be localized. The Lagrangian density is a function of the four fields $A_\tau(x)$, $\tau = 0, 1, 2, 3$, and of their derivatives $\partial_\sigma A_\tau(x)$,

$$\mathcal{L}(A_\tau, \partial_\sigma A_\tau) = a_1 F_{\mu\nu} g^{\mu\lambda} g^{\nu\eta} F_{\lambda\eta} + a_2 j^\mu A_\mu \qquad (3.26b)$$
$$= a_1 (\partial_\mu A_\nu - \partial_\nu A_\mu) g^{\mu\lambda} g^{\nu\eta} (\partial_\lambda A_\eta - \partial_\eta A_\lambda) + a_2 j^\mu A_\mu \,.$$

(Note that the Lorentz index τ attached to the arguments of $\mathcal{L}$ counts the fields in analogy to the superscript (i) of the fields $\psi^{(i)}$ in the general form of the Lagrangian density given by (3.13), whereas on the right-hand side, μ and ν are summation indices.) The partial derivative of $\mathcal{L}$ by A_τ equals $a_2 j^\tau(x)$. The partial derivatives by the first derivatives $\partial_\sigma A_\tau$ are best calculated by means of the chain rule:

$$\frac{\partial}{\partial(\partial_\sigma A_\tau)} = \frac{\partial F_{\mu\nu}}{\partial(\partial_\sigma A_\tau)} \frac{\partial}{\partial F_{\mu\nu}} = (\delta_{\mu\sigma}\delta_{\nu\tau} - \delta_{\mu\tau}\delta_{\nu\sigma}) \frac{\partial}{\partial F_{\mu\nu}}$$

(where σ and τ have fixed values and μ and ν are to be summed over). The partial derivatives which are contained in the Euler–Lagrange equations (3.16) and the equations of motion pertaining to the Lagrangian density (3.26b) are given by, respectively,

$$\frac{\partial \mathcal{L}}{\partial A_\tau} = a_2 j^\tau(x) \,,$$
$$\frac{\partial \mathcal{L}}{\partial(\partial_\sigma A_\tau)} = (\delta_{\mu\sigma}\delta_{\nu\tau} - \delta_{\mu\tau}\delta_{\nu\sigma}) \, 2a_1 F^{\mu\nu}(x) = 4a_1 F^{\sigma\tau}(x) \,,$$
$$\frac{\partial \mathcal{L}}{\partial A_\tau} - \partial_\sigma \frac{\partial \mathcal{L}}{\partial(\partial_\sigma A_\tau)} = a_2 j^\tau(x) - 4a_1 \partial_\sigma F^{\sigma\tau}(x) = 0 \,.$$

In the second line use was made of the antisymmetry of the field strength tensor, and the third line was obtained by inserting the first two lines into the equations of motion, (3.16). In the vacuum and using Gaussian units, one has $\mathcal{F}^{\mu\nu} = F^{\mu\nu}$.

Comparison with the inhomogeneous equations (2.53) yields the first condition on the coefficients in (3.26a), viz.

$$\frac{a_2}{4a_1} = \frac{4\pi}{c} . \tag{3.27a}$$

The second condition,

$$a_2 = -\frac{1}{c} , \tag{3.27b}$$

from which one obtains the value of a_1, follows from inspection of physical dimensions and from the comparison with static or stationary situations. The product of the current density j, (2.51), and the potential A, (2.57), has a dimension of $c\varrho\Phi$, i.e.

$$\left[c\varrho(x)\Phi(x)\right] = \frac{\text{length} \times \text{energy}}{\text{time} \times (\text{length})^3} .$$

This yields the desired dimension for $\mathcal{L}$, which should be (energy)/(length)^3, only if one divides by c. By the same token, the term $-a_2 j_\mu A^\mu$, which will appear in the Hamilton density, (3.19), after integration over the whole space, yields the correct expression (charge) × (potential) of electrostatics. Thus, in the system of units we are using here, we find that the constants a_1 and a_2 have the values

$$a_1 = -\frac{1}{16\pi} , \qquad a_2 = -\frac{1}{c} . \tag{3.27c}$$

The Lagrangian density which follows from there,

$$\mathcal{L}(A_\tau, \partial_\sigma A_\tau) = -\frac{1}{16\pi} F_{\mu\nu}(x) F^{\mu\nu}(x) - \frac{1}{c} j^\mu(x) A_\mu(x) , \tag{3.28}$$

yields the inhomogeneous Maxwell equations as its Euler–Lagrange equations. Thus, Maxwell's equations can indeed be derived by means of Hamilton's variational principle.

Remarks

1. We assumed that the tensor field of field strengths was expressed in terms of potentials,

$$F_{\mu\nu}(x) = \partial_\mu A_\nu(x) - \partial_\nu A_\mu(x) .$$

Therefore, the homogeneous Maxwell equations are fulfilled automatically.

2. Although in this book we do not make use of this possibility, we point out that the choice of units can be simplified even further by introducing *natural units*. In these units, the speed of light takes the value 1, $c = 1$, and fields and sources are defined with specific factors, as compared to Gaussian units, as follows. One chooses

$$F^{(\text{nat})}_{\mu\nu}(x) := \frac{1}{\sqrt{4\pi}} F^{(\text{Gauss})}_{\mu\nu}(x) , \quad A^{(\text{nat})}_\mu(x) := \frac{1}{\sqrt{4\pi}} A^{(\text{Gauss})}_\mu(x)$$

and

$$j^{\mu}_{(\text{nat})}(x) := \sqrt{4\pi}\, j^{\mu}_{(\text{Gauß})}(x)\ .$$

In these units the Lagrangian density (3.28) becomes

$$\mathcal{L}(A_\tau, \partial_\sigma A_\tau) = -\frac{1}{4} F^{(\text{nat})}_{\mu\nu}(x) F^{\mu\nu}_{(\text{nat})}(x) - j^{\mu}_{(\text{nat})}(x) A^{(\text{nat})}_{\mu}(x)\ .$$

In the literature, the kinetic term $F_{\mu\nu}F^{\mu\nu}$ of the Lagrangian density is often given with the factor $-1/(16\pi)$ or, alternatively, with the factor $-1/4$. This is always the same term, only the units chosen are different.

3. The Lagrangian density, (3.28), is not only invariant under the group $L_+^\uparrow$, it is also invariant under space reflection (parity) Π and under time reversal T. Maxwell's equations follow from (3.28) by an invariant variational principle, and therefore they are covariant with respect to these transformations, meaning they are *form invariant.* This is a nice illustration of the general principle discussed above.

"invariant Lagrangian density" $\Longrightarrow$ "covariant equations of motion"

In a next step, we calculate the canonically conjugate field momenta and the Hamilton density, which follow from the Lagrangian density given by (3.28). First, one obtains

$$\pi^0(x) = \frac{\partial \mathcal{L}}{\partial(\partial_0 A_0)} = 0\ , \tag{3.29a}$$

$$\pi^i(x) = \frac{\partial \mathcal{L}}{\partial(\partial_0 A_i)} = -\frac{1}{4\pi} F^{0i}(x) = \frac{1}{4\pi} E^i(x)\ . \tag{3.29b}$$

Here the first surprise appears: The field A_0 has no associated conjugate momentum. Leaving this feature as is, one calculates the Hamilton density following the model of (3.19) and inserts the formula $E^i = -\nabla^i A_0 - \partial A^i/\partial x^0$ to find

$$\begin{aligned}
\mathcal{H} &= -\pi^i \frac{\partial A^i}{\partial x^0} - \mathcal{L} \\
&= \frac{1}{4\pi} \boldsymbol{E} \cdot (\boldsymbol{E} + \nabla A_0) - \frac{1}{8\pi}(\boldsymbol{E}^2 - \boldsymbol{B}^2) + \frac{1}{c} j^\mu(x) A_\mu(x) \\
&= \frac{1}{8\pi}(\boldsymbol{E}^2 + \boldsymbol{B}^2) + \frac{1}{c} j^\mu(x) A_\mu(x) + \frac{1}{4\pi} \boldsymbol{E} \cdot \nabla A_0\ .
\end{aligned} \tag{3.30a}$$

The first of the three terms on the right-hand side (last line) is the electric and magnetic energy density known from Sect. 1.9.4, (1.130), with $\boldsymbol{D} = \boldsymbol{E}$ and $\boldsymbol{H} = \boldsymbol{B}$. The sum of the second and third terms in (3.30a) has a physical interpretation, too.

This can be seen if one derives the Hamiltonian function by taking the integral of the density over the whole space. Using partial integration one finds

$$\frac{1}{4\pi}\iiint \mathrm{d}^3x\, \boldsymbol{E}\cdot\nabla A_0 = -\frac{1}{4\pi}\iiint \mathrm{d}^3x\, (\nabla\cdot\boldsymbol{E})A_0$$
$$= -\iiint \mathrm{d}^3x\, \varrho(x)A_0(x)\,,$$

where the Maxwell equation (2.1c) was used. From this result and with $j^0(x) = c\varrho(x)$ one has

$$H = \iiint \mathrm{d}^3x\, \mathcal{H} \tag{3.30b}$$
$$= \iiint \mathrm{d}^3x \left\{\frac{1}{8\pi}(\boldsymbol{E}^2(x)+\boldsymbol{B}^2(x)) - \frac{1}{c}\boldsymbol{j}(x)\cdot\boldsymbol{A}(x)\right\}.$$

Besides the pure field energy, this expression contains the interaction between the fields and the electric current in the source. If there is no external source at all, $\boldsymbol{j}(t,\boldsymbol{x}) \equiv 0$, the total energy contained in the Maxwell fields is given by

$$H^{(0)} = \frac{1}{8\pi}\iiint \mathrm{d}^3x\,\{\boldsymbol{E}^2(x)+\boldsymbol{B}^2(x)\}\,. \tag{3.30c}$$

► **Remark**

In Sect. 1.9.4, we started from (1.129a), i.e. from the integral of $\boldsymbol{j}\cdot\boldsymbol{A}$, to express the energy content of magnetic fields by the integral of $\boldsymbol{B}\cdot\boldsymbol{H}$. In (3.30b), in turn, the two kinds of contribution appear on the same footing. This is not in contradiction to the static situations considered in Sect. 1.9.4. The assumption there was that the current density $\boldsymbol{j}(\boldsymbol{x})$ is the source of the magnetic field. Here the idea is that certain Maxwell fields due to some other sources interact with matter in which there flows an (external) current density.

3.5 Symmetries and Noether Invariants

This section is devoted to gauge transformations and spacetime transformations, applied to the fields and to their sources, which are such that the Lagrangian density is either strictly invariant or changed by gauge terms only. This analysis exhibits a close analogy to the results of Sects. 3.2.1 and 3.2.2, which applied to mechanical systems. The invariance of the Lagrangian density, up to gauge terms, under one-parameter groups of transformations implies the existence of constants of motion. This is the central proposition of the theorem of E. Noether.

The variational principle of Hamilton utilizes the functional of (3.15a), or, written somewhat differently, the functional

$$I\left[\psi^{(i)}\right] = \int\limits_{\Sigma} \mathrm{d}^4 x\, \mathcal{L}(\psi^{(i)}, \partial_\mu \psi^{(i)}, \ldots) , \tag{3.31}$$

where the integral is to be taken over a four-dimensional volume Σ in spacetime whose surface $\partial\Sigma$ is (at least piecewise) smooth. This functional must assume an extremum under variations $\delta\psi^{(i)}(x)$ of the fields which vanish on $\partial\Sigma$. A given transformation representing a symmetry of the theory described by $\mathcal{L}$ either leaves $\mathcal{L}$ strictly invariant or replaces $\mathcal{L}$ by $\mathcal{L}' = \mathcal{L} + \partial_\mu \Lambda^\mu(x)$, i.e. adds a (four-)divergence to it. An additive term of this kind, by Stokes' theorem, is transformed into an integral over the surface $\partial\Sigma$, where the variations of the fields are assumed to vanish. This means, in other terms, that $\mathcal{L}'$ leads to the same equations of motion as $\mathcal{L}$. We illustrate these assertions by some specific examples which are relevant for Maxwell theory.

3.5.1 Invariance Under One-Parameter Groups

Assume the Lagrangian density $\mathcal{L}(\psi^{(i)}(x), \partial_\mu \psi^{(i)}(x))$, which describes a set of fields $\psi^{(i)}(x)$, $i = 1, 2, \ldots, N$, to be invariant under a one-parameter group of transformations:

$$\psi^{(i)}(x) \longmapsto \psi^{(i)\prime} = h^\alpha\left(\psi^{(1)}(x), \ldots, \psi^{(N)}(x)\right) , \tag{3.32a}$$
$$\text{with} \quad h^{\alpha=0}\left(\psi^{(1)}(x), \ldots, \psi^{(N)}(x)\right) = \psi^{(i)}(x) .$$

For α close to zero, the variations of the fields are linear in α:

$$\delta\psi^{(i)} = \psi^{(i)\prime} - \psi^{(i)}(x) = \frac{\mathrm{d}}{\mathrm{d}\alpha} h^\alpha\left(\psi^{(1)}(x), \ldots, \psi^{(N)}(x)\right)\bigg|_{\alpha=0} \alpha . \tag{3.32b}$$

Note the similarity to the expansions of (3.5) and (3.6) around $s = 0$. As in (3.9) and (3.10), the derivative of h^α by α, taken at zero, is the generator of a one-parameter group of transformations.

The variation $\delta\mathcal{L}$ resulting from the variation $\delta\psi^{(i)}$ of each of the fields is calculated as follows:

$$\begin{aligned} \delta\mathcal{L} &= \sum_{i=1}^{N} \left\{ \frac{\partial \mathcal{L}}{\partial \psi^{(i)}} \delta\psi^{(i)} + \frac{\partial \mathcal{L}}{\partial(\partial_\mu \psi^{(i)})} \partial_\mu \delta\psi^{(i)} \right\} \\ &= \sum_{i=1}^{N} \left\{ \frac{\partial \mathcal{L}}{\partial \psi^{(i)}} - \partial_\mu \left(\frac{\partial \mathcal{L}}{\partial(\partial_\mu \psi^{(i)})} \right) \right\} \delta\psi^{(i)} + \partial_\mu \left(\sum_{i=1}^{N} \frac{\partial \mathcal{L}}{\partial(\partial_\mu \psi^{(i)})} \delta\psi^{(i)} \right) . \end{aligned} \tag{3.33}$$

In the first step, we used again $\delta(\partial_\mu \psi^{(i)}) = \partial_\mu \delta\psi^{(i)}$. In the second term, we subtracted and added a new term so as to introduce the left-hand side of the equations of motion, (3.16), in the curly brackets. The argument then continues as follows: If the Lagrangian density is strictly invariant under (3.32a) or (3.32b), and if $\psi^{(i)}$ are solutions of the Euler–Lagrange equations (3.16), then the following four-vector is conserved:

$$J^\mu(x) = \sum_{i=1}^{N} \frac{\partial \mathcal{L}}{\partial(\partial_\mu \psi^{(i)})} \delta\psi^{(i)} , \quad \partial_\mu J^\mu(x) = 0 . \tag{3.34}$$

The examples which follow show that it is justified to call the dynamical quantity (3.34) a "current" since either it is proportional to the electric current or it is a physically meaningful generalization thereof.

One sees already at this point that the assumptions and the conservation law given by (3.34) generalize to field theory the case of strict invariance of a Lagrangian function.

Example 3.3 Complex scalar field

We take up Example 3.2 (but without the external source $\varrho(x)$) and generalize to the case where the scalar field is a complex field:

$$\Phi(x) = \phi_1(x) + \mathrm{i}\phi_2(x) , \quad \text{with } \phi_1(x), \phi_2(x) \text{ real} .$$

The Lagrangian density which describes this field, still without interaction, is constructed in close analogy to (3.17a):

$$\frac{1}{\hbar c}\mathcal{L}^{(0)}(\Phi, \partial_\mu \Phi, \varrho) = \frac{1}{2}\left[\partial_\mu \Phi^*(x)\partial^\mu \Phi(x) - \kappa^2 \Phi^*(x)\Phi(x)\right] . \tag{3.35a}$$

It contains the field Φ and its complex conjugate Φ^* in such a way as to make the Lagrangian density a real function. In contrast to Example 3.3, this model has two degrees of freedom, ϕ_1 and ϕ_2. Instead of these two real fields, one might as well vary the fields Φ and Φ^* independently to obtain the equations of motion. In doing so, the present example of an interaction-free Lagrangian density, (3.35a), yields the Klein–Gordon equation (3.17b) without external source,

$$(\Box + \kappa^2)\Phi(x) = 0$$

and, likewise, its complex conjugate.

The Lagrangian density given by (3.35a) is strictly invariant under what is called a *gauge transformation of the first kind* or, in somewhat more geometric terms, a *global gauge transformation*:

$$\Phi(x) \longmapsto \mathrm{e}^{\mathrm{i}\alpha}\Phi(x) , \quad \Phi^*(x) \longmapsto \mathrm{e}^{-\mathrm{i}\alpha}\Phi^*(x) , \quad \alpha \in \mathbb{R} . \tag{3.35b}$$

These transformation formulae are needed only in the neighbourhood of the identity. Thus, by (3.32b),

$$\delta\Phi(x) = \mathrm{i}\alpha\Phi(x)\,, \quad \delta\Phi^*(x) = -\mathrm{i}\alpha\Phi^*(x)\,. \tag{3.35c}$$

The corresponding conserved current as obtained from (3.34) is found to be

$$\begin{aligned} J^\mu(x) &= \sum_{i=1}^{N} \frac{\partial\mathcal{L}^{(0)}}{\partial(\partial_\mu\psi^{(i)})}\delta\psi^{(i)} \\ &= \hbar c\Big\{\big(\partial^\mu\Phi^*(x)\big)\Phi(x) - \Phi^*(x)\big(\partial^\mu\Phi(x)\big)\Big\}\,. \end{aligned} \tag{3.35d}$$

Clearly, the current $J^\mu(x)$ is defined only up to a multiplicative constant. Therefore, one can always transform it to a four-current density $j^\mu(x)$, which has the same physical dimension as the electric current density. If the field Φ has the dimension (1/length), this could be

$$j^\mu(x) = ecJ^\mu(x)\,.$$

The invariance of the Lagrangian density, (3.35a), with respect to the transformations of (3.35b) yields the continuity equation $\partial_\mu j^\mu(x) = 0$ and, therefore, the conservation of electric charge.

3.5.2 Gauge Transformations and Lagrangian Density

In general, a gauge transformation (2.59), $A'_\mu = A_\mu - \partial_\mu\Lambda(x)$, changes the Lagrangian density given by (3.28) as follows:

$$\mathcal{L}'\big(A'_\tau, \partial_\sigma A'_\tau\big) = \mathcal{L}\big(A_\tau, \partial_\sigma A_\tau\big) + \frac{1}{c} j^\mu(x)\partial_\mu\Lambda(x)\,.$$

However, there is no harm in replacing the additional term

$$j^\mu(x)\partial_\mu\Lambda(x) \quad \text{by} \quad \partial_\mu\Big(j^\mu(x)\Lambda(x)\Big)$$

simply because the current $j^\mu(x)$ is conserved, $\partial_\mu j^\mu(x) = 0$. The additional term is seen to be a (four-)divergence which, in the action integral (3.31), gives a surface term on $\partial\Sigma$. As the variations of the fields, by assumption, vanish on this surface, the equations of motion do not change. This is an important result:
The coupling term $j^\mu(x)A_\mu(x)$ of the radiation field to matter is gauge invariant only if the electromagnetic current density $j^\mu(x)$ satisfies the continuity equation $\partial_\mu j^\mu(x) = 0$.

When, besides the Lagrangian density (3.28) for the Maxwell fields, one also has a Lagrangian density that describes charged matter at one's disposal, the invariance under gauge transformations can be analysed even further. The following example illustrates what we have in mind.

Example 3.4 Atoms in external fields

In atomic physics, one describes the electron in the framework of nonrelativistic quantum mechanics and by means of the Schrödinger equation. We analyse its interactions with Maxwell fields in a semiclassical approach, i.e. in experimental situations where the electron is subject to external, classical (i.e. unquantized) electric and magnetic fields. This is an appropriate description, for instance, when one studies the orbital dynamics of electrons in particle accelerators and in beam lines.

We assign to the Maxwell fields and to the electron each a Lagrangian density with the choice that the interaction of the electron with the Maxwell fields is contained in the Lagrangian density, which yields the Schrödinger equation with external fields, $\mathcal{L} = \mathcal{L}_\mathrm{M} + \mathcal{L}_\mathrm{E}$,

$$\mathcal{L}_\mathrm{M} = -\frac{1}{16\pi} F_{\mu\nu}(x) F^{\mu\nu}(x) \,, \tag{3.36a}$$

$$\begin{aligned} \mathcal{L}_\mathrm{E} &= \frac{1}{2}\mathrm{i}\hbar\left[\psi^* \partial_t \psi - (\partial_t \psi^*)\psi\right] - e\Phi(t,\boldsymbol{x})\psi^*\psi \\ &\quad - \frac{1}{2m}\left(\left[\mathrm{i}\hbar\nabla - \frac{e}{c}\boldsymbol{A}(t,\boldsymbol{x})\right]\psi^*\right)\left(\left[-\mathrm{i}\hbar\nabla - \frac{e}{c}\boldsymbol{A}(t,\boldsymbol{x})\right]\psi\right) . \end{aligned} \tag{3.36b}$$

To familiarize ourselves with the Lagrangian density given by (3.36b), it is useful to set the vector potential $\boldsymbol{A}$ equal to zero in a first step. With $e\Phi(t,\boldsymbol{x}) = U(t,\boldsymbol{x})$, the Lagrangian density then reads

$$\mathcal{L}_\mathrm{E} = \frac{1}{2}\mathrm{i}\hbar\left[\psi^* \partial_t \psi - (\partial_t \psi^*)\psi\right] - U(t,\boldsymbol{x})\psi^*\psi - \frac{\hbar^2}{2m}(\nabla\psi^*)(\nabla\psi) .$$

Varying, for example, the complex conjugate field ψ^*, one must take the derivatives of $\mathcal{L}_\mathrm{E}$ by ψ^*, by $(\partial_t\psi^*)$ and by $(\nabla\psi^*)$:

$$\frac{\partial \mathcal{L}_\mathrm{E}}{\partial \psi^*} = \frac{1}{2}\mathrm{i}\hbar\partial_t\psi - U(t,\boldsymbol{x})\psi \,,$$

$$\frac{\partial \mathcal{L}_\mathrm{E}}{\partial(\partial_t\psi^*)} = -\frac{1}{2}\mathrm{i}\hbar\psi \,, \qquad \frac{\partial \mathcal{L}_\mathrm{E}}{\partial(\nabla\psi^*)} = -\frac{\hbar^2}{2m}\nabla\psi \,.$$

Inserting these expressions into the equation of motion (3.16) (with $\psi^{(i)} = \psi^*$), one obtains

$$\frac{\partial \mathcal{L}_E}{\partial \psi^*} - \partial_t \left(\frac{\partial \mathcal{L}_E}{\partial(\partial_t \psi^*)} \right) - \nabla \left(\frac{\partial \mathcal{L}_E}{\partial(\nabla \psi^*)} \right) = 0 \,, \quad \text{giving}$$

$$i\hbar \partial_t \psi(t, \boldsymbol{x}) - U(t, \boldsymbol{x})\psi(t, \boldsymbol{x}) + \frac{\hbar^2}{2m} \Delta \psi(t, \boldsymbol{x}) = 0 \,.$$

This is the Schrödinger equation $i\hbar\partial_t \psi(t, \boldsymbol{x}) = [(\boldsymbol{p}^2/2m) + U(t, \boldsymbol{x})] \psi(t, \boldsymbol{x})$, with $\boldsymbol{p} = -i\hbar\nabla$. (Note that if one varies ψ instead of ψ^*, the same procedure yields the complex conjugate Schrödinger equation for ψ^*.) Performing the same calculation for the more general Lagrangian density (3.36b), with arbitrary potentials $\boldsymbol{A}$ and Φ, one obtains the equation of motion

$$i\hbar\partial_t \psi(t, \boldsymbol{x}) = \frac{1}{2m} \left[-i\hbar\nabla - \frac{e}{c} \boldsymbol{A}(t, \boldsymbol{x}) \right]^2 \psi(t, \boldsymbol{x}) + e\Phi(t, \boldsymbol{x})\psi(t, \boldsymbol{x}) \,. \tag{3.37}$$

Replacing again $-i\hbar\nabla$ by $\boldsymbol{p}$, the right-hand side is seen to be the Hamiltonian function of a charged particle in external fields, familiar from classical mechanics; cf., e. g. [ME], Sect. 2.16 (ii):

$$H = \frac{1}{2m} \left[\boldsymbol{p} - \frac{e}{c} \boldsymbol{A}(t, \boldsymbol{x}) \right]^2 + e\Phi(t, \boldsymbol{x}) \,.$$

The substitution

$$\boldsymbol{p} \mapsto \boldsymbol{p} - \frac{e}{c} \boldsymbol{A} \quad \text{or} \quad -i\hbar\nabla \mapsto -i\hbar\nabla - \frac{e}{c} \boldsymbol{A} \tag{3.38}$$

is called the rule of *minimal substitution,* and the specific coupling to the radiation field generated by it is called *minimal coupling.*

A general gauge transformation, (2.59), when applied to the potentials,

$$\Phi \mapsto \Phi' = \Phi - \frac{1}{c} \partial_t \chi(t, \boldsymbol{x}) \,, \quad \boldsymbol{A} \mapsto \boldsymbol{A}' = \boldsymbol{A} + \nabla\chi(t, \boldsymbol{x}) \,, \tag{3.39a}$$

modifies $\mathcal{L}_E$ and hence $\mathcal{L}$ by terms which can be treated as was previously done, at the start of this section. This would not be new compared to purely classical situations. However, when one performs a space- and time-dependent phase transformation of the Schrödinger wave function, viz.

$$\psi(t, \boldsymbol{x}) \mapsto \psi'(t, \boldsymbol{x}) = e^{i\alpha(t, \boldsymbol{x})} \psi(t, \boldsymbol{x}) \,, \tag{3.39b}$$

simultaneously with the gauge transformation given by (3.39a), then a very interesting observation emerges. A phase transformation with a *constant* real phase α

(see Example (3.35b)) leaves invariant all quantum mechanical expectation values and, therefore, all observables. This is the assertion, known from quantum mechanics, that physical states are not described by a single wave function ψ but by what is called a unit ray $\{e^{i\alpha}\psi\}$, with $\alpha \in \mathbb{R}$. These phase factors are elements of a Lie group U(1), which is parametrized by $\alpha \pmod{2\pi}$ and whose only generator is the identity. In turn, (3.39b) shows something new: It contains a phase factor which depends on a smooth function $\alpha(t, \boldsymbol{x})$, i.e. a transformation which at every point $(t, \boldsymbol{x})$ of spacetime provides a copy $\mathrm{U}(1)|_x$ of the group U(1). The difference between the two cases is quite obvious. Using a constant phase, one transforms by $e^{i\alpha}$ the wave function independently of time and over the whole universe. In contrast, if α is a function of time and space, one may perfectly choose this function to be localized in the sense that it is different from zero, in an essential way, only in a given time interval and in a finite spatial domain.

For the time derivatives and the gradients in the Schrödinger equation the transformation given by (3.39b) implies the replacements

$$\mathrm{i}\partial_t \psi \mapsto \mathrm{i}\partial_t \psi' = e^{\mathrm{i}\alpha(t,\boldsymbol{x})}\Big\{\mathrm{i}\partial_t - \big(\partial_t \alpha\big)\Big\}\psi \,, \tag{3.40a}$$

$$\nabla\psi \mapsto \nabla\psi' = e^{\mathrm{i}\alpha(t,\boldsymbol{x})}\Big\{\nabla + \mathrm{i}\big(\nabla\alpha\big)\Big\}\psi \tag{3.40b}$$

(the parentheses emphasize the range of the action to the right of the operators ∂_t and ∇). Applying simultaneously the transformations given by (3.39a) and (3.39b) in $\mathcal{L}_\mathrm{E}$, (3.36b), the first two terms are substituted as follows:

$$\begin{aligned} &\frac{1}{2}\mathrm{i}\hbar\big[\psi^{*\prime}\partial_t\psi' - (\partial_t\psi^{*\prime})\psi'\big] - e\Phi(t,\boldsymbol{x})\psi^{*\prime}\psi' \\ &= \frac{1}{2}\mathrm{i}\hbar\big[\psi^{*}\partial_t\psi - (\partial_t\psi^{*})\psi\big] - \hbar\big(\partial_t\alpha\big)\psi^*\psi \\ &\qquad - e\Phi(t,\boldsymbol{x})\psi^*\psi + \frac{e}{c}\big(\partial_t\chi\big)\psi^*\psi \,, \end{aligned} \tag{3.40c}$$

whereas in the third term one has

$$\begin{aligned} &\Big[-\mathrm{i}\hbar\nabla - \frac{e}{c}\boldsymbol{A}'\Big]\psi' \\ &= e^{\mathrm{i}\alpha(t,\boldsymbol{x})}\Big[-\mathrm{i}\hbar\nabla - \frac{e}{c}\boldsymbol{A} + \hbar\big(\nabla\alpha\big) - \frac{e}{c}\big(\nabla\chi\big)\Big]\psi \end{aligned} \tag{3.40d}$$

and an analogous expression for the complex conjugate factor. The exponential $\exp\{\mathrm{i}\alpha(t,\boldsymbol{x})\}$ cancels in all contributions to $\mathcal{L}_\mathrm{E}$. Furthermore, if one chooses the phase function $\alpha(t,\boldsymbol{x})$ proportional to the gauge function $\chi(t,\boldsymbol{x})$, or more precisely, if one takes

$$\alpha(t,\boldsymbol{x}) = \frac{e}{\hbar c}\chi(t,\boldsymbol{x}) \,, \tag{3.41}$$

then the second and fourth terms of the right-hand side of (3.40cc) cancel each other, as do the last two terms on the right-hand side of (3.40d). In conclusion, the simultaneous transformations given by (3.39a) and (3.39b) and supplemented by relation (3.41) leave $\mathcal{L}_E$ and, hence, $\mathcal{L} = \mathcal{L}_M + \mathcal{L}_E$ completely unchanged!

The example illustrates quite well why one talks about *local* gauge transformation. The transformations given by (3.39a) and (3.39b), which act simultaneously on the Schrödinger wave function and on the potentials of the Maxwell fields, are elements of the local gauge group U(1). In that sense, U(1) is the gauge group of Maxwell theory.

Remark

In the older literature, the two types of gauge transformations were called gauge transformations of the first kind (referring to global transformations) and of the second kind (referring to local transformations). Nowadays, one rather uses the nomenclature customary from differential geometry, even though this is not standardized. The group of global, rigid, U(1) transformations

$$G = \left\{ \mathrm{e}^{\mathrm{i}\alpha} \middle| \alpha \in [0, 2\pi) \right\}$$

is often called the *structure group*; the group of local gauge transformations

$$\mathfrak{G} = \left\{ \mathrm{e}^{\mathrm{i}\alpha(x)} \middle| x \in \mathbb{R}^4, \alpha \in \mathfrak{F}(\mathbb{R}^4) \pmod{2\pi} \right\} ,$$

with α a smooth function on Minkowski space, is called the *gauge group*. As explained previously, the gauge group provides a copy of the structure group at every point x of spacetime. It does so in a smooth, i.e. differentiable, manner.

3.5.3 Invariance Under Translations

In this section, we return to the original form of Noether's theorem. We study the action of a translation in space and in time, first for a general Lagrangian density without explicit dependence on x

$$\mathcal{L}(\psi^{(i)}(x), \partial_\mu \psi^{(i)}(x)) , \quad i = 1, 2, \ldots, N , \tag{3.42}$$

and then for the Lagrangian density given by (3.28) of Maxwell theory. Given the transformation

$$x^\nu \longmapsto x'^\nu = x^\nu + \varepsilon a^\nu , \tag{3.43}$$

one obtains up to the first order in ε (the symbol $+\mathcal{O}(\varepsilon^2)$ is suppressed)

$$\psi^{(i)}(x) \mapsto \psi'^{(i)}(x') = \psi^{(i)}(x + \varepsilon a) = \psi^{(i)}(x) + \delta\psi^{(i)}(x) , \quad \text{with}$$

$$\delta\psi^{(i)} = \left. \frac{\partial \psi^{(i)}}{\partial \varepsilon} \right|_{\varepsilon=0} \varepsilon = \varepsilon a_\nu \partial^\nu \psi^{(i)} . \tag{3.43a}$$

The corresponding variation of the Lagrangian density is

$$\begin{aligned}\delta\mathcal{L} &= \mathcal{L}(\psi'^{(i)}(x'), \partial_\mu\psi'^{(i)}(x')) - \mathcal{L}(\psi^{(i)}(x), \partial_\mu\psi^{(i)}(x)) \\ &= \left.\frac{\partial\mathcal{L}}{\partial\varepsilon}\right|_{\varepsilon=0} \varepsilon = \varepsilon a_\nu \partial^\nu \mathcal{L}(\psi^{(i)}(x), \partial_\mu\psi^{(i)}(x)) .\end{aligned} \tag{3.43b}$$

If the fields $\psi^{(i)}$ are solutions of the equations of motion, one obtains according to (3.33)

$$\delta\mathcal{L} = \partial_\mu \left(\sum_{i=1}^{N} \frac{\partial\mathcal{L}}{\partial(\partial_\mu\psi^{(i)})} \delta\psi^{(i)} \right) . \tag{3.43c}$$

From expression (3.43b) for $\delta\mathcal{L}$ and with (3.43a) for the variations $\delta\psi^{(i)}$, one obtains the equation

$$\varepsilon a_\nu \left\{ \partial^\nu \mathcal{L} - \partial_\mu \left(\sum_{i=1}^{N} \frac{\partial\mathcal{L}}{\partial(\partial_\mu\psi^{(i)})} \partial^\nu \psi^{(i)} \right) \right\} = 0 , \quad \text{or}$$

$$\varepsilon a_\nu \partial_\mu \left\{ -g^{\mu\nu}\mathcal{L} + \left(\sum_{i=1}^{N} \frac{\partial\mathcal{L}}{\partial(\partial_\mu\psi^{(i)})} \partial^\nu \psi^{(i)} \right) \right\} = 0 .$$

As the components of translation vector a can be chosen arbitrarily, one concludes that the tensor field

$$T^{\mu\nu} := \left(\sum_{i=1}^{N} \frac{\partial\mathcal{L}}{\partial(\partial_\mu\psi^{(i)})} \partial^\nu \psi^{(i)} \right) - g^{\mu\nu}\mathcal{L}(\psi^{(i)}, \partial_\alpha\psi^{(i)}) \tag{3.44}$$

satisfies the four conservation laws

$$\partial_\mu T^{\mu\nu}(\psi^{(i)}, \partial_\alpha\psi^{(i)}) = 0 . \tag{3.45}$$

We rediscover here the same tensor field as in (3.21) and the same conservation laws as in (3.22). However, we derived it here from a deeper principle: The conservation laws, (3.45), follow from the invariance under translations, (3.43), of the theory defined by the Lagrangian density, (3.42). Before working out the physical content of these conservation laws we repeat the same analysis for the specific case of Maxwell theory. Let the Lagrangian density given by (3.28), without coupling to external sources

$$\mathcal{L} = -\frac{1}{16\pi} F_{\mu\nu}(x) F^{\mu\nu}(x) , \tag{3.46a}$$

be invariant under the translations given by (3.43). We vary the fields $A_\sigma(x)$ in the following way:

$$\delta A_\sigma = \varepsilon a_\nu F^\nu{}_\sigma(x) = \varepsilon a_\nu (\partial^\nu A_\sigma - \partial_\sigma A^\nu) . \tag{3.46b}$$

In this ansatz, the first term on the right-hand side is of the same nature as in the general case of the transformation given by (3.43a). In the second term, a gauge transformation was added with the gauge function $\chi = (\varepsilon a_\nu A^\nu)$. This is a permissible operation! The special feature, and the advantage of this ansatz, is that the variation of the field A_σ remains unchanged under any further gauge transformation,

$$A_\sigma \longmapsto A'_\sigma = A_\sigma - \partial_\sigma \Lambda(x) .$$

All potentials of a class of gauge-equivalent potentials are varied by the same amount, (3.46b).

The variation of $\mathcal{L}$ is calculated as follows:

$$\delta\mathcal{L} = \varepsilon a_\nu \partial_\mu \left(\frac{\partial \mathcal{L}}{\partial(\partial_\mu A_\sigma)} F^\nu{}_\sigma \right) .$$

Following the same arguments as previously, one obtains the equation

$$\varepsilon a_\nu \partial_\mu \left\{ -g^{\mu\nu}\mathcal{L} + \frac{\partial \mathcal{L}}{\partial(\partial_\mu A_\sigma)} F^\nu{}_\sigma \right\} = 0 .$$

The partial derivative of $\mathcal{L}$ by the derivatives of the fields is given by

$$\frac{\partial \mathcal{L}}{\partial(\partial_\mu A_\sigma)} = -\frac{1}{4\pi} F^{\mu\sigma} .$$

Inserting $\mathcal{L}$ from (3.46a) yields the *Maxwell tensor field*:

$$T_{\mathrm{M}}^{\mu\nu}(x) = \frac{1}{4\pi} \left\{ F^{\mu\sigma}(x) F_\sigma{}^\nu(x) + \frac{1}{4} g^{\mu\nu} F_{\alpha\beta}(x) F^{\alpha\beta}(x) \right\} , \qquad (3.47)$$

where $F_\sigma{}^\nu = g_{\sigma\tau} F^{\tau\nu} = -F^\nu{}_\sigma$. The Maxwell tensor field obeys the conservation laws

$$\partial_\mu T_{\mathrm{M}}^{\mu\nu}(x) = 0 , \qquad (j^\alpha(x) \equiv 0) , \qquad (3.47a)$$

as long as there are no external sources.

If the fields contained in the field strength tensor $F^{\mu\nu}$ are assumed to interact with some external sources j^μ, then one calculates the divergence $\partial_\mu T^{\mu\nu}$ by means of Maxwell's equations as follows:

$$\begin{aligned}
\partial_\mu T_{\mathrm{M}}^{\mu\nu}(x) &= \frac{1}{4\pi} \left\{ \partial_\mu \left(F^{\mu\alpha} F_\alpha{}^\nu \right) + \frac{1}{4} \partial^\nu \left(F_{\alpha\beta} F^{\alpha\beta} \right) \right\} \\
&= \frac{1}{4\pi} \left\{ \left(\partial^\mu F_{\mu\alpha} \right) F^{\alpha\nu} + F_{\mu\alpha} \partial^\mu \left(F^{\alpha\nu} \right) + \frac{1}{2} F_{\alpha\beta} \partial^\nu F^{\alpha\beta} \right\} \\
&= \frac{1}{c} j_\alpha F^{\alpha\nu} + \frac{1}{8\pi} F_{\mu\alpha} \left\{ \left[\partial^\mu F^{\alpha\nu} + \partial^\nu F^{\mu\alpha} \right] + \partial^\mu F^{\alpha\nu} \right\} .
\end{aligned}$$

In the first step, some indices were moved consistently from lower to upper, or upper to lower, positions, and derivatives were worked out in accordance with the chain rule. In particular, one should note that

$$\begin{aligned}\partial^\nu\big(F_{\alpha\beta}F^{\alpha\beta}\big) &= -\partial^\nu\big(F_{\alpha\beta}F^{\beta\alpha}\big) = -\partial^\nu\operatorname{tr}\big(F^2\big)\\ &= -2\operatorname{tr}\big(F(\partial^\nu F)\big) = 2F_{\alpha\beta}\partial^\nu F^{\alpha\beta}\,.\end{aligned}$$

due to the cyclic property of the trace, i.e. $\operatorname{tr}(F\partial^\nu F) = \operatorname{tr}((\partial^\nu F)F)$. In going from the second to the third line, the inhomogeneous Maxwell equations were inserted, and the term $\partial^\mu F^{\alpha\nu}$ was duplicated (with a factor $1/2$). The two terms in square brackets can be transformed by means of the homogeneous Maxwell equations so that they are replaced by $-\partial^\alpha F^{\nu\mu} = \partial^\alpha F^{\mu\nu}$. The source term $j_\alpha F^{\alpha\nu} = -j_\alpha F^{\nu\alpha}$ is taken to the left-hand side. On the right-hand side there remains

$$F_{\mu\alpha}\Big\{\partial^\mu F^{\alpha\nu} + \partial^\alpha F^{\mu\nu}\Big\}\,,$$

a term that vanishes because the expression in curly brackets is symmetric under $\mu \leftrightarrow \alpha$ and is contracted with the antisymmetric $F_{\mu\alpha}$. As a result, one obtains the important equation

$$\partial_\mu T_{\rm M}^{\mu\nu}(x) + \frac{1}{c}F^{\nu\alpha}(x)j_\alpha(x) = 0\,. \tag{3.47b}$$

When there is no external current density, $j_\alpha(x) \equiv 0$, the Maxwell tensor field $T_{\rm M}^{\mu\nu}$ is conserved. If this is not so, then the second term on the left-hand side of (3.47b) describes the exchange of energy and momentum between the radiation field and matter. In that sense, (3.47b) is a balance equation. Further questions of interpretation are dealt with in the next subsection.

Before turning to its interpretation, one reads off the following properties of the Maxwell tensor field (3.47):

(i) The tensor field $T_{\rm M}^{\mu\nu}(x)$ is symmetric. When expressing the field strengths in terms of potentials, the Maxwell tensor field is seen to be invariant under gauge transformations.
(ii) With $g_{\mu\nu}g^{\mu\nu} = 4$, the trace of (3.47) is seen to be zero:

$$\operatorname{tr} T_{\rm M}^{\mu\nu}(x) = g_{\mu\nu}T_{\rm M}^{\mu\nu}(x) = \frac{1}{4\pi}\left\{F^{\mu\sigma}F_{\sigma\mu} + \frac{1}{4}4F_{\alpha\beta}F^{\alpha\beta}\right\} = 0\,. \tag{3.48}$$

In the absence of external sources, the Maxwell tensor field is a symmetric, traceless tensor field of degree two.

3.5.4 Interpretation of the Conservation Laws

As emphasized earlier, the tensor field given by (3.47) pertains to Maxwell theory in the vacuum where there is no essential, physical, difference between $\boldsymbol{E}$ and $\boldsymbol{D}$, and between $\boldsymbol{B}$ and $\boldsymbol{H}$, respectively. (Recall that with Gaussian units, they are pairwise equal.) When we stick to this case for a while and, yet, utilize balance equation (3.47b), we have in mind a perturbative framework: The given radiation field in the vacuum hits (external) charges and currents due to single particles or to localized charge and current densities, which are such that their interaction with the Maxwell fields can be treated like a perturbation. Maxwell fields and matter exchange energy and momentum as described by balance equation (3.47b), but the back-reaction onto the given fields remains small. This picture provides the basis for the quantized version of the theory, i.e. quantum electrodynamics, which describes the interaction of free quantized Maxwell fields with electrons and other charged elementary particles.

Electromagnetic fields in matter will be considered later by a suitable generalization of the expression (3.47).

Maxwell Tensor Field in the Vacuum

The physical interpretation of the tensor field, given by (3.47), becomes more transparent if we follow an experimentalist in his laboratory who measures electric and magnetic fields and, in doing so, singles out a class of systems of reference in which the partition of $\mathbb{R}^4$ into time axis $\mathbb{R}_t$ and three-dimensional space $\mathbb{R}^3$ is fixed. With $g^{\mu\eta} g_{\eta\nu} = \delta^\mu{}_\nu$ we have

$$(T_\mathrm{M}(x))^\mu{}_\nu = \frac{1}{4\pi}\left\{F^{\mu\sigma}(x)F_{\sigma\nu}(x) + \frac{1}{4}\delta^\mu{}_\nu F_{\alpha\beta}(x)F^{\alpha\beta}(x)\right\},$$

and when the explicit representations (2.46) and (2.69a) are inserted, the first term on the right-hand side is the product of the two 4×4 matrices

$$F^{\mu\sigma} = \begin{pmatrix} 0 & -E^1 & -E^2 & -E^3 \\ E^1 & 0 & -B^3 & B^2 \\ E^2 & B^3 & 0 & -B^1 \\ E^3 & -B^2 & B^1 & 0 \end{pmatrix}, \quad F_{\sigma\nu} = \begin{pmatrix} 0 & E^1 & E^2 & E^3 \\ -E^1 & 0 & -B^3 & B^2 \\ -E^2 & B^3 & 0 & -B^1 \\ -E^3 & -B^2 & B^1 & 0 \end{pmatrix}.$$

The second term is the unit matrix multiplied by $F_{\alpha\beta}F^{\alpha\beta}/4$, a factor that was calculated previously in (3.25a). As a result, the components $(T_\mathrm{M})^\mu{}_\nu$ are found to be the following functions of the electric and magnetic fields:

$$(T_\mathrm{M})^0{}_0 = \frac{1}{8\pi}\{\boldsymbol{E}^2 + \boldsymbol{B}^2\} =: u(t, \boldsymbol{x}), \tag{3.49a}$$

$$(T_\mathrm{M})^0{}_i = -\frac{1}{4\pi}(\boldsymbol{E} \times \boldsymbol{B})^i =: -cP^i(t, \boldsymbol{x}), \tag{3.49b}$$

$$(T_{\mathrm{M}})^i{}_0 = +\frac{1}{4\pi}(\boldsymbol{E}\times\boldsymbol{B})^i =: \frac{1}{c}S^i(t,\boldsymbol{x})\,, \tag{3.49c}$$

$$(T_{\mathrm{M}})^k{}_i = \frac{1}{4\pi}\Big[E^k E^i + B^k B^i - \frac{1}{2}\delta^{ki}(\boldsymbol{E}^2+\boldsymbol{B}^2)\Big]\,. \tag{3.49d}$$

The notations to the right of these equations hint at their interpretation as follows. If the expressions obtained in Sect. 1.9.4 are also correct for time-dependent fields (and this is indeed the case), then $u(t,\boldsymbol{x})$ is the *energy density.* The vector $\boldsymbol{P} = (P^1, P^2, P^3)^T$ is interpreted to be the *momentum density.* The vector $\boldsymbol{S}$ will be found to be the flux density of energy, the energy flow. It is called the *Poynting vector.* The space–space components $(T_{\mathrm{M}})^k{}_i$ in the last line, (3.49d), are the entries of the *Maxwell stress tensor.*

Slightly rewritten, balance equation (3.47b) reads

$$\partial_\mu (T_{\mathrm{M}}(x))^\mu{}_\nu + \frac{1}{c}F_{\nu\alpha}(x)j^\alpha(x) = 0\,.$$

With the following decompositions into space and time coordinates

$$\partial_\mu = \left\{\frac{1}{c}\partial_t, \nabla\right\} \quad \text{and} \quad j^\alpha = \{c\varrho(t,\boldsymbol{x}), \boldsymbol{j}(t,\boldsymbol{x})\}$$

and taking $\nu = 0$, one obtains the equation

$$\frac{\partial}{\partial t}u(t,\boldsymbol{x}) + \nabla\cdot\boldsymbol{S}(t,\boldsymbol{x}) + \boldsymbol{E}(t,\boldsymbol{x})\cdot\boldsymbol{j}(t,\boldsymbol{x}) = 0\,. \tag{3.50}$$

This equation relates the change in time of the energy density and the divergence of the flux field $\boldsymbol{S}(t,\boldsymbol{x})$, on the one hand, to the work per unit time and volume done by the electromagnetic fields on the sources, on the other. The vector field

$$\boldsymbol{S}(t,\boldsymbol{x}) = \frac{c}{4\pi}\boldsymbol{E}(t,\boldsymbol{x})\times\boldsymbol{H}(t,\boldsymbol{x}) \tag{3.51}$$

is called the *Poynting vector field,* and equation (3.50) is often paraphrased as *Poynting's theorem.*

Integrate now over a finite volume V whose surface ∂V is piecewise smooth. Then the term

$$\iiint_V \mathrm{d}^3x\,\frac{\partial}{\partial t}u(t,\boldsymbol{x}) = \frac{\mathrm{d}}{\mathrm{d}t}\iiint_V \mathrm{d}^3x\,u(t,\boldsymbol{x}) = \frac{\mathrm{d}}{\mathrm{d}t}W_{\text{field}}$$

is the change in time of the energy content of the fields, whereas

$$\iiint_V \mathrm{d}^3x\,\boldsymbol{E}(t,\boldsymbol{x})\cdot\boldsymbol{j}(t,\boldsymbol{x}) = \frac{\mathrm{d}}{\mathrm{d}t}W_{\text{mech}}$$

is the change in mechanical energy of the particles enclosed by the volume. The volume integral of the divergence of $\boldsymbol{S}(t,\boldsymbol{x})$ becomes a surface integral over ∂V, so that the integral form of (3.50) reads as follows:

$$\frac{\mathrm{d}}{\mathrm{d}t}\left(W_{\text{field}}+W_{\text{mech}}\right)=-\iint\limits_{\partial V}\mathrm{d}\sigma\,\hat{\boldsymbol{n}}\cdot\boldsymbol{S}(t,\boldsymbol{x})\,,$$

with $\mathrm{d}\sigma$ the surface element and $\hat{\boldsymbol{n}}$ the outward normal of the surface. This relation between the change in time of the total energy and the surface integral on the right-hand side shows that, indeed, the vector field $\boldsymbol{S}(t,\boldsymbol{x})$ describes the energy flow.

The space components ($\nu=i$) of balance equation (3.47b) yield

$$-\frac{\partial}{\partial t}P^i+\frac{1}{4\pi}\sum_{k=1}^{3}\nabla^k\left[E^kE^i+B^kB^i-\frac{1}{2}\delta^{ki}\left(\boldsymbol{E}^2+\boldsymbol{B}^2\right)\right]$$
$$-E^i\varrho-\frac{1}{c}(\boldsymbol{j}\times\boldsymbol{B})^i=0\,.$$

One rewrites this slightly and introduces the space–space components of the Maxwell tensor field to obtain an equation which is easier to interpret:

$$\varrho(t,\boldsymbol{x})E^i(t,\boldsymbol{x})+\frac{1}{c}\left(\boldsymbol{j}(t,\boldsymbol{x})\times\boldsymbol{B}(t,\boldsymbol{x})\right)^i+\frac{\partial P^i}{\partial t}-\sum_{k=1}^{3}\partial_k(T_{\mathrm{M}})^k{}_i=0\,. \tag{3.52}$$

The first two terms remind one of the Lorentz force, and it seems appropriate to call it the *Lorentz force density.* Together with the third term, one has here

$$\frac{\partial}{\partial t}\left(P^i_{\text{mech}}+P^i_{\text{field}}\right).$$

The significance of the last term in (3.52) is better understood when one considers the integral form of the balance equation: One integrates again over a finite volume V whose surface ∂V is (piecewise) smooth so that, by Gauss' theorem,

$$\sum_{k=1}^{3}\iiint\limits_{V}\mathrm{d}^3x\,\partial_k(T_{\mathrm{M}})^k{}_i=\iint\limits_{\partial V}\mathrm{d}\sigma\,\sum_{k=1}^{3}(T_{\mathrm{M}})^k{}_i n^k\,.$$

The integrand represents the flux of the ith component of the momentum per unit surface on ∂V, i.e. the force per unit of surface that acts on particles and fields in V. This is what is called the *radiation pressure.*

Maxwell Tensor Field in Matter

The study of Maxwell fields inside electrically and magnetically polarizable matter is considerably more complicated. Strictly speaking, the question of how to calculate the momentum density, the energy flow and the stress tensor in these cases can only be answered if one has a concrete model describing the block of matter that one investigates. In the simplest case with *linear* relationships between the fields and with constant coefficients,

$$\boldsymbol{D} = \varepsilon \boldsymbol{E} , \qquad \boldsymbol{B} = \mu \boldsymbol{H} ,$$

(cf. Sects. 1.7.2 and 1.7.3), one may try to guess the shape of the tensor field, (3.47).

For this purpose, one repeats the individual steps of the derivation of balance equation (3.47b), noting carefully where one used the inhomogeneous Maxwell equations and where one used the homogeneous ones. Furthermore, one confirms that the reasoning used in calculating the trace still holds, provided the relations between fields $\boldsymbol{D}$ and $\boldsymbol{E}$ and between $\boldsymbol{B}$ and $\boldsymbol{H}$, respectively, are linear and have constant coefficients. (This was the assumption.) Then

$$\partial^\nu \left(\mathcal{F}_{\alpha\beta} F^{\alpha\beta}\right) = -\partial^\nu \left(\mathcal{F}_{\alpha\beta} F^{\beta\alpha}\right) = -\partial^\nu \left(\boldsymbol{E} \cdot \boldsymbol{D} + \boldsymbol{B} \cdot \boldsymbol{H}\right) = \mathcal{F}_{\alpha\beta} \partial^\nu F^{\alpha\beta} .$$

Thus, the tensor field given by (3.47) can be replaced by the following composition of the two types of tensor fields:

$$T^{\mu\nu}(x) = \frac{1}{4\pi} \left\{ \mathcal{F}^{\mu\sigma}(x) F_\sigma{}^\nu(x) + \frac{1}{4} g^{\mu\nu} \mathcal{F}_{\alpha\beta}(x) F^{\alpha\beta}(x) \right\} . \tag{3.53}$$

Working out the individual components as in (3.49a–3.49d), one obtains

$$u(t, \boldsymbol{x}) = \frac{1}{8\pi} \{\boldsymbol{E} \cdot \boldsymbol{D} + \boldsymbol{H} \cdot \boldsymbol{B}\} = \frac{1}{8\pi} \{\varepsilon \boldsymbol{E}^2 + \mu \boldsymbol{H}^2\} , \tag{3.54a}$$

$$\boldsymbol{P}(t, \boldsymbol{x}) = \frac{1}{4\pi c} \boldsymbol{D} \times \boldsymbol{B} = \frac{1}{4\pi c} \varepsilon\mu \boldsymbol{E} \times \boldsymbol{H} , \tag{3.54b}$$

$$\boldsymbol{S}(t, \boldsymbol{x}) = \frac{c}{4\pi} \boldsymbol{E} \times \boldsymbol{H} , \tag{3.54c}$$

$$T^k{}_j(t, \boldsymbol{x}) = \frac{1}{4\pi} \left\{ \varepsilon \left(E^k E^j - \frac{1}{2} \delta^{kj} \boldsymbol{E}^2 \right) + \mu \left(H^k H^j - \frac{1}{2} \delta^{kj} \boldsymbol{H}^2 \right) \right\} . \tag{3.54d}$$

Equations (3.50) and (3.52) as well as their interpretation remain unchanged.

Remark

Given the interpretation of $\boldsymbol{P}$ in (3.54b) as the momentum density, it makes sense to interpret

$$\boldsymbol{\ell} := \frac{1}{4\pi c} \boldsymbol{x} \times (\boldsymbol{D} \times \boldsymbol{B}) = \frac{\varepsilon\mu}{4\pi c} \boldsymbol{x} \times (\boldsymbol{E} \times \boldsymbol{H}) \tag{3.55}$$

as the *angular momentum density* of the radiation field.

3.6 Wave Equation and Green Functions

In this section and in the next, we work out some general methods of solving the wave equation and investigate how the solutions reflect the causal structure of Minkowski spacetime. Concrete examples of applications to optics and selected oscillatory solutions are deferred to the next chapter.

Introducing potentials for the description of the tensor field $F^{\mu\nu}(x)$,

$$F^{\mu\nu}(x) = \partial^\mu A^\nu(x) - \partial^\nu A^\mu(x) ,$$

and imposing the Lorenz condition (2.61), differential equation (2.60) is reduced to the inhomogeneous wave equation

$$\Box A^\mu(x) = \frac{4\pi}{c} j^\mu(x) , \tag{3.56}$$

in which the current density j (as a manifestation of matter!) is the source term, up to a factor. With a given partition of time and space, i.e. in a special class of systems of reference, one has

$$j = \left(c\varrho(t,\boldsymbol{x}), \boldsymbol{j}(t,\boldsymbol{x})\right)^T , \quad A = \left(\Phi(t,\boldsymbol{x}), \boldsymbol{A}(t,\boldsymbol{x})\right)^T ,$$

$$\Box = \partial_\mu \partial^\mu = \partial_0^2 - \boldsymbol{\Delta} = \frac{1}{c^2}\frac{\partial^2}{\partial t^2} - \boldsymbol{\Delta} .$$

Thus, wave equation (3.56) has the general form

$$\Box \Psi(x) = 4\pi F(x) , \tag{3.57a}$$

where $\Psi(x)$ is a field quantity, i.e. in the case of (3.56), a component of A or a component of one of the physical fields, whereas $F(x)$ is a source term.

3.6.1 Solutions in Noncovariant Form

In a noncovariant notation, the inhomogeneous wave equation reads

$$\left(\boldsymbol{\Delta}_x - \frac{1}{c^2}\frac{\partial^2}{\partial t^2}\right)\Psi(t,\boldsymbol{x}) = -4\pi F(t,\boldsymbol{x}) . \tag{3.57b}$$

A standard method of solving this equation goes as follows: One subjects the differential equation in x-space to a Fourier transformation such as to transform it into an algebraic equation which can be solved in an elementary way. One starts from the ansatz[4]

$$\Psi(t,\boldsymbol{x}) = \frac{1}{\sqrt{2\pi}} \int_{-\infty}^{+\infty} \mathrm{d}\omega\, \widetilde{\Psi}(\omega,\boldsymbol{x})\, \mathrm{e}^{-\mathrm{i}\omega t}\,, \tag{3.58a}$$

$$F(t,\boldsymbol{x}) = \frac{1}{\sqrt{2\pi}} \int_{-\infty}^{+\infty} \mathrm{d}\omega\, \widetilde{F}(\omega,\boldsymbol{x})\, \mathrm{e}^{-\mathrm{i}\omega t}\,. \tag{3.58b}$$

Thus, the Fourier transformation is performed in the time variable t and one expands in terms of the base functions

$$\varphi(\omega,t) = \frac{1}{\sqrt{2\pi}} \mathrm{e}^{-\mathrm{i}\omega t}\,,$$

which are orthogonal and complete in the sense of distributions:

$$\int_{-\infty}^{+\infty} \mathrm{d}t\, \varphi^*(\omega,t)\varphi(\omega',t) = \delta(\omega-\omega') \quad \text{(orthogonality)}\,, \tag{3.59a}$$

$$\int_{-\infty}^{+\infty} \mathrm{d}\omega\, \varphi^*(\omega,t)\varphi(\omega,t') = \delta(t-t') \quad \text{(completeness)}\,. \tag{3.59b}$$

By means of these formulae, one derives the inverse formulae of (3.58a) and (3.58b). These are, respectively,

$$\widetilde{\Psi}(\omega,\boldsymbol{x}) = \frac{1}{\sqrt{2\pi}} \int_{-\infty}^{+\infty} \mathrm{d}t\, \Psi(t,\boldsymbol{x})\, \mathrm{e}^{\mathrm{i}\omega t}\,, \tag{3.60a}$$

$$\widetilde{F}(\omega,\boldsymbol{x}) = \frac{1}{\sqrt{2\pi}} \int_{-\infty}^{+\infty} \mathrm{d}t\, F(t,\boldsymbol{x})\, \mathrm{e}^{\mathrm{i}\omega t}\,. \tag{3.60b}$$

The Fourier transform of differential equation (3.57b) is an algebraic equation; the action of the second derivative by the time variable turns into the factor $(\omega/c)^2$, so that one has

$$\{\boldsymbol{\Delta} + k^2\}\widetilde{\Psi}(\omega,\boldsymbol{x}) = -4\pi\widetilde{F}(\omega,\boldsymbol{x})\,, \quad \text{with} \quad k := \frac{\omega}{c}\,. \tag{3.61}$$

The differential equation (3.61), called an *inhomogeneous Helmholtz equation,* is solved by the method of Green functions, in direct analogy to Sect. 1.8.1, (1.81)

[4] We follow the generally accepted convention of denoting the Fourier components by a "tilde".

and (1.83). In the present context, this means that one should find a distribution $G_k(\boldsymbol{x}, \boldsymbol{x}')$ which is a solution of

$$(\boldsymbol{\Delta}_x + k^2) G_k(\boldsymbol{x}, \boldsymbol{x}') = \delta(\boldsymbol{x} - \boldsymbol{x}') . \tag{3.62a}$$

Once this is achieved, one derives from it a Green function $G(t, \boldsymbol{x}; t', \boldsymbol{x}')$ which is a solution of the original differential equation

$$\left(\boldsymbol{\Delta}_x - \frac{1}{c^2} \frac{\partial^2}{\partial t^2} \right) G(t, \boldsymbol{x}; t', \boldsymbol{x}') = \delta(\boldsymbol{x} - \boldsymbol{x}') \delta(t - t') . \tag{3.62b}$$

This program is carried out in two steps.

Green Function of the Helmholtz Equation

If there are no special boundary conditions which may break the translational and rotational invariance, the Green function $G_k(\boldsymbol{x}, \boldsymbol{x}')$ can depend only on the *difference* $\boldsymbol{r} := (\boldsymbol{x} - \boldsymbol{x}')$. Furthermore, it must be *spherically symmetric* in the variable $\boldsymbol{r}$,

$$G_k(\boldsymbol{x}, \boldsymbol{x}') = G_k(r) \quad \text{with} \quad r := \left| \boldsymbol{x} - \boldsymbol{x}' \right| .$$

Inserting this into (3.61) and making use of the Laplace operator in spherical polar coordinates, one obtains a differential equation in the variable r:

$$\frac{1}{r^2} \frac{\mathrm{d}}{\mathrm{d}r} \left(r^2 \frac{\mathrm{d}G_k}{\mathrm{d}r} \right) + k^2 G_k = \delta(\boldsymbol{r}) , \quad \text{where} \tag{3.61a}$$

$$\frac{1}{r^2} \frac{\mathrm{d}}{\mathrm{d}r} \left(r^2 \frac{\mathrm{d}f(r)}{\mathrm{d}r} \right) = \frac{\mathrm{d}^2 f(r)}{\mathrm{d}r^2} + \frac{2}{r} \frac{\mathrm{d}f(r)}{\mathrm{d}r} = \frac{1}{r} \frac{\mathrm{d}^2}{\mathrm{d}r^2} (r f(r)) .$$

As long as $r \neq 0$, we are left with the homogeneous differential equation

$$\frac{\mathrm{d}^2}{\mathrm{d}r^2} (r G_k(r)) + k^2 (r G_k(r)) = 0 ,$$

whose general solution is easily seen to be

$$r G_k(r) = a_+ \mathrm{e}^{\mathrm{i}kr} + a_- \mathrm{e}^{-\mathrm{i}kr} , \quad \text{with} \quad a_\pm \in \mathbb{R} .$$

For $r \to 0$ and for a fixed value of k, the product (kr) becomes small compared to 1, $kr \ll 1$. In this limit, radial equation (3.61a) goes over into the radial part of the Poisson equation whose solution is known from (1.83):

$$\lim_{kr \ll 1} G_k(r) = -\frac{1}{4\pi} \frac{1}{r} .$$

By combining the two results and replacing r by $|\boldsymbol{x} - \boldsymbol{x}'|$ one finds the desired solution of (3.61a):

$$G_k\left(|\boldsymbol{x} - \boldsymbol{x}'|\right) = a_+ G_k^{(+)}\left(|\boldsymbol{x} - \boldsymbol{x}'|\right) + a_- G_k^{(-)}\left(|\boldsymbol{x} - \boldsymbol{x}'|\right), \quad \text{with} \quad a_+ + a_- = 1, \tag{3.63a}$$

$$G_k^{(\pm)}\left(|\boldsymbol{x} - \boldsymbol{x}'|\right) = -\frac{1}{4\pi}\frac{\mathrm{e}^{\pm \mathrm{i}k|\boldsymbol{x}-\boldsymbol{x}'|}}{|\boldsymbol{x} - \boldsymbol{x}'|}. \tag{3.63b}$$

Green Function of the Wave Equation

From the results given by (3.63a) and (3.63b), and by a twofold Fourier transformation, one obtains the corresponding Green functions for (3.62b). Transforming with respect to the variable t, i.e. taking

$$G(t, \boldsymbol{x}, t', \boldsymbol{x}') = \frac{1}{\sqrt{2\pi}} \int_{-\infty}^{+\infty} \mathrm{d}\omega\, \widetilde{G}(\omega, \boldsymbol{x}; t', \boldsymbol{x}')\mathrm{e}^{-\mathrm{i}\omega t}, \tag{3.64a}$$

the left-hand side of (3.62b) becomes

$$\left(\boldsymbol{\Delta}_x - \frac{1}{c^2}\frac{\partial^2}{\partial t^2}\right) G(t, \boldsymbol{x}; t', \boldsymbol{x}') = \frac{1}{\sqrt{2\pi}} \int_{-\infty}^{+\infty} \mathrm{d}\omega\, [\boldsymbol{\Delta}_x + k^2]\widetilde{G}(\omega, \boldsymbol{x}; t', \boldsymbol{x}')\mathrm{e}^{-\mathrm{i}\omega t},$$

while on the right-hand side of (3.62b) the δ-distribution in $(t - t')$ can be replaced by an integral representation,

$$\delta(\boldsymbol{x} - \boldsymbol{x}')\delta(t - t') = \delta(\boldsymbol{x} - \boldsymbol{x}')\frac{1}{2\pi}\int_{-\infty}^{+\infty} \mathrm{d}\omega\ \mathrm{e}^{-\mathrm{i}\omega(t-t')}. \tag{3.64b}$$

From this one obtains a differential equation for $\widetilde{G}$:

$$(\boldsymbol{\Delta}_x + k^2)\widetilde{G}(\omega, \boldsymbol{x}; t', \boldsymbol{x}') = \frac{1}{\sqrt{2\pi}}\delta(\boldsymbol{x} - \boldsymbol{x}')\mathrm{e}^{\mathrm{i}\omega t'}.$$

Inserting the base solutions given by (3.63b) of (3.62a), one has

$$\widetilde{G}^{(\pm)}(\omega, \boldsymbol{x}; t', \boldsymbol{x}') = \frac{1}{\sqrt{2\pi}}\mathrm{e}^{\mathrm{i}\omega t'} G_k^{(\pm)}\left(|\boldsymbol{x} - \boldsymbol{x}'|\right), \quad (\omega = kc).$$

In a last step, one uses the inverse Fourier formula of (3.64a) and finds

$$G^{(\pm)}(t, \boldsymbol{x}, t', \boldsymbol{x}') = \frac{1}{2\pi}\int_{-\infty}^{+\infty} \mathrm{d}\omega\, \frac{-1}{4\pi|\boldsymbol{x} - \boldsymbol{x}'|}\exp\left\{\mathrm{i}\left[-\omega(t - t') \pm \frac{\omega}{c}|\boldsymbol{x} - \boldsymbol{x}'|\right]\right\}.$$

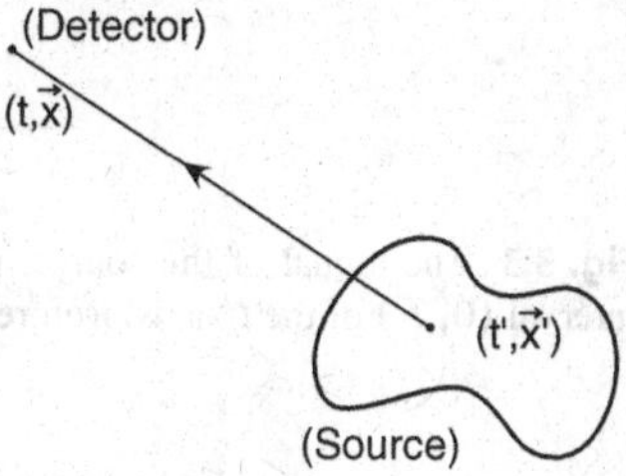

Fig. 3.2 A signal emitted at time t' from the position $\boldsymbol{x}'$ causes an observable effect at time t in the detector situated in $\boldsymbol{x}$. Causality is respected by the use of the retarded Green function $G^{(+)}$: t is later than t', and $t - t'$ is the correct time of flight from the source to the detector

Here k was replaced by ω/c. The integral over ω is

$$\frac{1}{2\pi}\int_{-\infty}^{+\infty} \mathrm{d}\omega \, \exp\left\{-\mathrm{i}\omega\left[(t-t') \mp \frac{1}{c}\left|\boldsymbol{x}-\boldsymbol{x}'\right|\right]\right\} = \delta\left(t - \left[t' \pm \frac{1}{c}\left|\boldsymbol{x}-\boldsymbol{x}'\right|\right]\right).$$

Thus, the Green functions for (3.62b) are given by

$$G^{(\pm)}(t,\boldsymbol{x};t',\boldsymbol{x}') = -\frac{1}{4\pi|\boldsymbol{x}-\boldsymbol{x}'|}\delta\left(t - \left[t' \pm \tfrac{1}{c}\left|\boldsymbol{x}-\boldsymbol{x}'\right|\right]\right). \tag{3.65}$$

As an application, we construct solutions of the original inhomogeneous wave equation (3.57b) as follows:

$$\Psi^{(\pm)}(t,\boldsymbol{x}) = -4\pi \int_{-\infty}^{+\infty} \mathrm{d}t' \iiint \mathrm{d}^3x' \, G^{(\pm)}(t,\boldsymbol{x};t',\boldsymbol{x}')\, F(t',\boldsymbol{x}'). \tag{3.66}$$

These formulae are well suited for an analysis of their physical interpretation. We distinguish the two cases of interest.

The Retarded Green Function $G^{(+)}$

In this case, the δ-distribution enforces the relation

$$t = t' + \frac{1}{c}\left|\boldsymbol{x}-\boldsymbol{x}'\right|, \quad \text{i.e.} \quad t > t'.$$

The signal which emanates from the source at the position $\boldsymbol{x}'$ at time t' propagates with the speed of light towards the observer at the position $\boldsymbol{x}$ and reaches that observer at time

$$t = t' + (\text{time-of-flight from } \boldsymbol{x}' \text{ to } \boldsymbol{x}).$$

This Green function describes the intuitive causal relationship between cause and effect, the signal reaches the observer after the correct retardation, as sketched in Fig. 3.2. The distribution $G^{(+)}$ is called the *retarded Green function*.

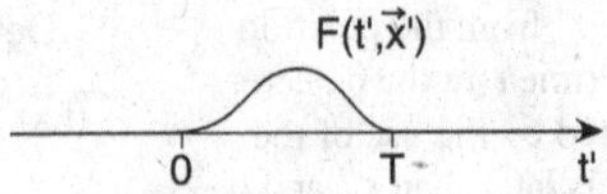

Fig. 3.3 The signal of the source, plotted as a function of time, is active only in a finite time interval $(0, T)$ of the t'-axis. Before and after that interval the source remains silent

Let $F(t', \boldsymbol{x}')$ describe a source which is localized in space and in time. This means, specifically for the time variable, that the source depicted in Fig. 3.2 is absent for all $t' < 0$ and for all $t' \geqslant T > 0$, as sketched in Fig. 3.3. If at time $t = -\infty$ a certain initial state $\Psi_{\text{in}}(-\infty, \boldsymbol{x})$ was present ("in" for *incoming*), i.e. a solution of the homogeneous wave equation corresponding to (3.57b), the complete solution of (3.57b) reads

$$\Psi(t, \boldsymbol{x}) = \Psi_{\text{in}}(t, \boldsymbol{x}) - 4\pi \int_{-\infty}^{+\infty} \mathrm{d}t' \iiint \mathrm{d}^3x' \, G^{(+)}(t, \boldsymbol{x}; t', \boldsymbol{x}') F(t', \boldsymbol{x}') \, . \quad (3.67)$$

Long before the source starts emitting there is an initial signal. The source emits additional contributions only when t is equal to $t' + |\boldsymbol{x} - \boldsymbol{x}'|/c$. It contributes to the total field Ψ causally, i.e. in a properly retarded fashion.

The Advanced Green Function $G^{(-)}$
In this case, $t = t' - |\boldsymbol{x} - \boldsymbol{x}'|/c$, the time at which the observer sees an effect is earlier than the time at which the source emits. The source time t' is causally related to the observer time t, and not t to t'. Nevertheless, there is no reason to discard the advanced distribution $G^{(-)}$. The solution given by (3.67) provides the clue for the explanation of this apparent contradiction, too: Indeed, it may happen that it is not the *in*coming field that is given but the *out*going field Ψ_{out}, which will build up as $t \to +\infty$. In this case, the complete solution of (3.57b) must have the form

$$\Psi(t, \boldsymbol{x}) = \Psi_{\text{out}}(t, \boldsymbol{x}) - 4\pi \int_{-\infty}^{+\infty} \mathrm{d}t' \iiint \mathrm{d}^3x' \, G^{(-)}(t, \boldsymbol{x}; t', \boldsymbol{x}') F(t', \boldsymbol{x}') \, . \quad (3.68)$$

The distribution $G^{(-)}$ ensures that whenever $t \neq t' - |\boldsymbol{x} - \boldsymbol{x}'|/c$, no contribution comes from the source. This means, in particular, that the field Ψ receives no further contributions after the source has been shut down. All such contributions are already contained in Ψ_{out}.

3.6.2 Solutions of the Wave Equation in Covariant Form

In the preceding section, we chose to analyse the relevant equations in noncovariant form to work out the causal structure of the connection between source and observation. Of course, one can treat the differential equation (3.57a) in a Lorentz covariant form from the start. That is the subject matter of this section.

We wish to construct distributions $G(x, x')$ which satisfy the differential equation[5]

$$\Box_x G(x, x') = \delta^{(4)}(x - x') . \tag{3.69}$$

Without any special boundary conditions, the distribution $G(x, x')$ depends on the variable $z := x - x'$ only, so that one must solve the equation

$$\Box G(z) = \delta^{(4)}(z) . \tag{3.69a}$$

Also in this case, it is advisable to consider the Fourier transforms, but now in all four variables so that the differential equation (3.69a) is mapped to a purely algebraic equation in a single step.

We denote the Fourier transform of $G(z)$ by $\tilde{G}(k)$, but normalize it in a way somewhat different from formulae (3.58a) and (3.58b):

$$G(z) = \frac{1}{(2\pi)^4} \int \mathrm{d}^4 k \, \widetilde{G}(k) \mathrm{e}^{-\mathrm{i}(kz)} . \tag{3.70}$$

Here $(kz) = k^0 z^0 - \boldsymbol{k} \cdot \boldsymbol{z}$, where $\boldsymbol{k}$ is a wave number vector whose four independent components are the integration variables (for the sake of clarity we write only a single integral symbol). With this ansatz in (3.69a) and using the representation

$$\delta^{(4)}(z) = \frac{1}{(2\pi)^4} \int \mathrm{d}^4 k \ \mathrm{e}^{-\mathrm{i}(kz)}$$

for the δ-distribution, one concludes

$$(2\pi)^4 \Box G(z) = \int \mathrm{d}^4 k \ \mathrm{e}^{-\mathrm{i}(kz)} \left(-k^2\right) \widetilde{G}(k) = \int \mathrm{d}^4 k \ \mathrm{e}^{-\mathrm{i}(kz)} \quad \text{and}$$

$$\widetilde{G}(k) = -\frac{1}{k^2} . \tag{3.71}$$

The Green function as a function of z is calculated from (3.70) as follows:

$$\begin{aligned} G(z) &= -\frac{1}{(2\pi)^4} \int \mathrm{d}^4 k \ \frac{1}{k^2} \mathrm{e}^{-\mathrm{i}(kz)} \\ &= -\frac{1}{(2\pi)^4} \int \mathrm{d}^3 k \int \mathrm{d}k^0 \ \mathrm{e}^{\mathrm{i}\boldsymbol{k}\cdot\boldsymbol{z}} \frac{\mathrm{e}^{-\mathrm{i}k^0 z^0}}{(k^0)^2 - \boldsymbol{k}^2} . \end{aligned}$$

Let $\kappa := |\boldsymbol{k}|$. The path of integration of the variable k^0 runs through the two poles of the integrand at $k^0 = \kappa$ and at $k^0 = -\kappa$. To assign a well-defined value to

[5] Compared to (3.62b), the other sign was chosen on the right-hand side. This choice, which entails a sign change of the corresponding Green functions, does not limit the generality of the method.

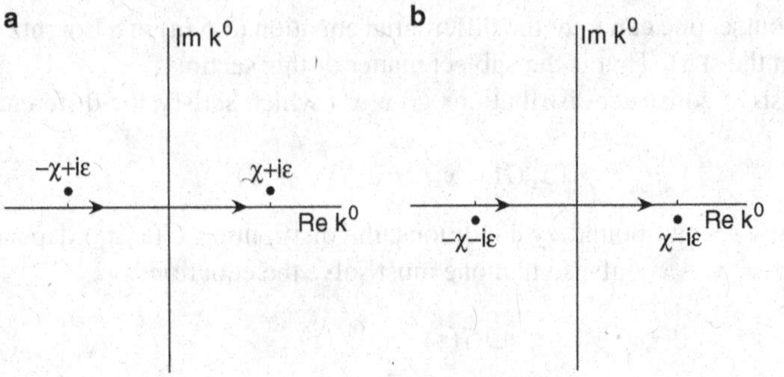

Fig. 3.4 **a** Position of poles of integrand of $G(z)$ for the choice of the upper sign in (3.72). **b** Position of poles for the choice of the lower sign in (3.72)

the integral, one must deform this path in such a way that the poles lie above or below the path. Depending on how one deforms the path, one obtains different Green functions. Equivalently, one might as well leave the path as it is but move the poles away from the real axis slightly. For example, one may choose

$$\frac{1}{(k^0)^2 - \kappa^2} \longrightarrow \frac{1}{(k^0 \mp \mathrm{i}\varepsilon)^2 - \kappa^2} = \frac{1}{\big(k^0 - (\kappa \pm \mathrm{i}\varepsilon)\big)\big(k^0 + (\kappa \mp \mathrm{i}\varepsilon)\big)}$$
$$\text{with} \quad \varepsilon \to 0^+ \,. \qquad (3.72)$$

Depending on the choice of signs in (3.72) and Fig. 3.4, one obtains different Green functions. Choosing the upper sign in (3.72) yields the advanced Green function, whereas choosing the lower sign yields the retarded function. We verify this for the case of the retarded function.

Obviously, as ε is simply an arbitrarily small quantity which is sent to zero from above, the integrand $((k^0+\mathrm{i}\varepsilon)^2-\kappa^2)^{-1}$ has the same effect as the integrand $((k^0)^2-\kappa^2+\mathrm{i}\varepsilon)^{-1}$, so that, specifically, one must calculate the integral

$$\int_{-\infty}^{+\infty} \mathrm{d}k^0 \, \frac{\mathrm{e}^{-\mathrm{i}k^0 z^0}}{(k^0)^2 - \kappa^2 + \mathrm{i}\varepsilon} = \lim_{R\to\infty} \int_{-R}^{+R} \mathrm{d}k^0 \, \frac{\mathrm{e}^{-\mathrm{i}k^0 z^0}}{(k^0)^2 - \kappa^2 + \mathrm{i}\varepsilon} \,.$$

For negative values of z^0, one completes the interval $[-R, +R]$ on the real axis by the semicircle in the upper half-plane $\mathrm{Im}\, k^0 > 0$, which is drawn in Fig. 3.5a. As $R \to \infty$, the integrand goes to zero due to the exponential $\mathrm{e}^{-\mathrm{Im}\, k^0 z^0}$, so that the integral we wish to calculate is equal to the integral over the closed path of Fig. 3.5a. As there are no poles inside the enclosed domain, by Cauchy's theorem, the integral is equal to zero.

For positive values of z^0, the interval is completed by the semicircle shown in Fig. 3.5b in the lower half-plane. The integral we wish to calculate is the integral over the closed path and, hence, equals $-2\pi\mathrm{i}$ times the sum of the residua of

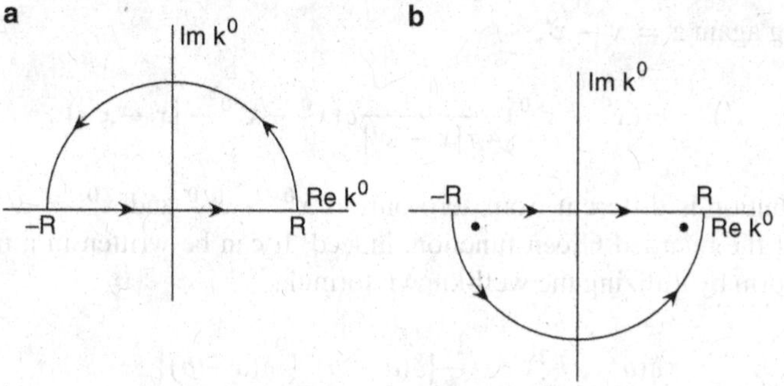

Fig. 3.5 **a** If $z^0 < 0$, then the segment $[-R, +R]$ of the real k^0-axis is supplemented by a semicircle in the upper half-plane so as to obtain a closed path. **b** If $z^0 > 0$, then one adds a semicircle in the lower half-plane

the enclosed poles of the first order (the minus sign comes from the orientation of the path). Therefore, in the limit $\varepsilon \to 0^+$, one has

$$\int_{-\infty}^{+\infty} dk^0 \, \frac{e^{-ik^0z^0}}{(k^0)^2 - \kappa^2 + i\varepsilon} = -2\pi i \Big\{ \frac{e^{-i\kappa z^0}}{2\kappa} + \frac{e^{i\kappa z^0}}{-2\kappa} \Big\} = -\frac{2\pi}{\kappa} \sin(\kappa z^0) \,,$$

$$(z^0 > 0) \,.$$

The condition that the integral is only different from zero if z^0 is positive is handled by a Heaviside step function $\Theta(z^0)$. Introducing spherical polar coordinates in the space $\mathbb{R}^3_\kappa$, one has

$$G_{\text{ret}}(z) = \frac{1}{(2\pi)^2}\Theta(z^0) \int_{-1}^{+1} d(\cos\theta) \int_0^\infty \kappa^2 \, d\kappa \; e^{i\kappa r\cos\theta} \, \frac{\sin(\kappa z^0)}{\kappa}$$

$$= \frac{2}{r(2\pi)^2}\Theta(z^0) \int_0^\infty d\kappa \; \sin(\kappa r) \sin(\kappa z^0) \,.$$

As the integral over κ concerns an integrand which is even in this variable, it can be extended to the interval $(-\infty, +\infty)$. It is equal to

$$\frac{1}{2} \int_{-\infty}^{+\infty} d\kappa \; \sin(\kappa r) \sin(\kappa z^0) = \frac{1}{2} \frac{2}{(2i)^2} \int_{-\infty}^{+\infty} d\kappa \, \left[e^{i\kappa(r+z^0)} - e^{-i\kappa(r-z^0)} \right] .$$

However, as one must have $z^0 > 0$ and as r is positive, only the second term of the integrand contributes; it is equal to $(\pi/2)\delta(z^0 - r)$. As a result,

$$G_{\text{ret}}(z) = \Theta(z^0) \frac{1}{4\pi r} \delta(z^0 - r) \,, \tag{3.73a}$$

or, inserting again $z = x - x'$,

$$G_{\text{ret}}(x, x') = \Theta(x^0 - x'^0)\frac{1}{4\pi|\boldsymbol{x} - \boldsymbol{x}'|}\delta\left(x^0 - x'^0 - |\boldsymbol{x} - \boldsymbol{x}'|\right). \tag{3.73b}$$

This distribution is different from zero only if $x^0 > x'^0$ and $x^0 = x'^0 + |\boldsymbol{x} - \boldsymbol{x}'|$. This is the retarded Green function, indeed. It can be written in a manifestly covariant form by utilizing the well-known formula

$$\delta(a^2 - b^2) = \frac{1}{2b}\{\delta(a + b) + \delta(a - b)\},$$

which implies, with $z = x - x'$,

$$\delta(z^2) = \delta((z^0)^2 - \boldsymbol{z}^2) = \delta((z^0)^2 - r^2) = \frac{1}{2r}\{\delta(z^0 - r) + \delta(z^0 + r)\}.$$

As the step function contained in $G_{\text{ret}}(x, x')$ allows for positive values of $z^0 = x^0 - x'^0$ only, only the first term of this formula contributes to the Green function, so that one obtains

$$G_{\text{ret}}(x, x') = \frac{1}{2\pi}\Theta(x^0 - x'^0)\delta\left((x - x')^2\right). \tag{3.74a}$$

In an analogous manner, one confirms that the *advanced* Green function is given by the same formula, provided one interchanges x^0 and x'^0,

$$G_{\text{adv}}(x, x') = \frac{1}{2\pi}\Theta(x'^0 - x^0)\delta\left((x - x')^2\right). \tag{3.74b}$$

Remarks

1. The retarded Green function $G_{\text{ret}}(z)$ is different from zero only on the forward light cone in the space of variables z^0 and $\boldsymbol{z}$ (Fig. 3.6a). As the proper, orthochronous Lorentz transformations $\Lambda \in L_+^\uparrow$ leave the light cone invariant as a whole, the distinction between the cases $z^0 \geqslant 0$ and $z^0 \leqslant 0$, when multiplied by $\delta(z^2)$, is independent of the frame of reference one has chosen. Despite the fact that the step function depends on the frame of reference, the retarded Green function given by (3.74a) is invariant. For example, the action of a special Lorentz transformation with $z^1 = z^0$ gives

$$z'^0 = \gamma z^0 - \beta\gamma z^1 = \gamma(1 - \beta)z^0.$$

   However, as $1 - \beta \geqslant 0$, one has always sign $z'^0 = $ sign z^0.
2. The support of the advanced Green function, (3.74b), obtained by the choice of the upper sign in (3.72) is the backward light cone of Fig. 3.6b. This Green function is invariant under all $\Lambda \in L_+^\uparrow$ as well.

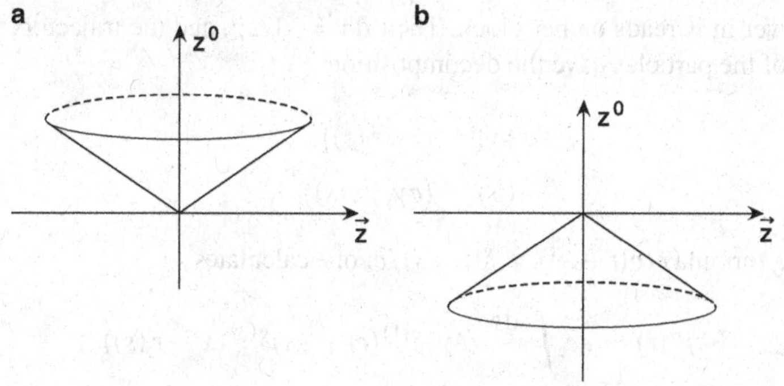

Fig. 3.6 **a** The retarded Green function is different from zero only on the forward light cone. **b** The advanced Green function is different from zero only on the backward light cone

3.7 Radiation of an Accelerated Charge

An electrically charged particle, which in a reference system K moves along the trajectory $\boldsymbol{r}(t)$ in space, generates a pointlike charge density $\varrho(t, \boldsymbol{x})$ as well as a current density $\boldsymbol{j}(t, \boldsymbol{x})$ which is proportional to its spatial velocity. Thus, these quantities are given by the following coordinate expressions:

$$\varrho(t, \boldsymbol{x}) = e\,\delta^{(3)}(\boldsymbol{x} - \boldsymbol{r}(t)), \tag{3.75a}$$

$$\boldsymbol{j}(t, \boldsymbol{x}) = e\,\boldsymbol{v}(t)\delta^{(3)}(\boldsymbol{x} - \boldsymbol{r}(t)). \tag{3.75b}$$

Here e denotes the (obviously Lorentz invariant) charge of the particle and $\boldsymbol{v}(t) = \dot{\boldsymbol{r}}(t)$ denotes its instantaneous spatial velocity.

The choice of a fixed frame of reference does not mean that the Lorentz covariance is broken. Rather, it represents what an observer sees or measures in his frame of reference. Expressions (3.75a) and (3.75b) can equally well be written in a covariant manner, and hence independently of the frame of reference, viz.

$$j^\alpha(x) = e\,c \int \mathrm{d}\sigma\; u^\alpha(\sigma)\delta^{(4)}\big(x - r(\sigma)\big), \tag{3.76}$$

with c the speed of light, σ the (Lorentz invariant) proper time, x a spacetime point and $r(\sigma) \equiv \{r^\alpha(\sigma)\}$ the world line of the particle. The Lorentz vector $u(\sigma) = \mathrm{d}r/\mathrm{d}\sigma$ is the four-velocity of the particle. As on the right-hand side of (3.76) one integrates over proper time, and as the δ-distribution is a Lorentz scalar, the vector nature of $j(x)$ follows from the fact that the velocity u is a four-vector.

As a first step, we show that the covariant formula (3.76) does indeed represent the physical quantities given by (3.75a) and (3.75b) when expressed in the frame K. Let s be the coordinate time corresponding to the proper time σ, i.e. the time which the observer in K reads on her clock. Then $d\sigma = ds/\gamma$, and the trajectory and the velocity of the particles have the decomposition

$$r(\sigma) = \left(cs, \boldsymbol{r}(s)\right)^T ,$$
$$u(\sigma) = \left(c\gamma, \gamma\boldsymbol{v}(s)\right)^T .$$

Using the formula $\delta(c(t-s)) = \delta(t-s)/c$, one calculates

$$\begin{aligned} j^0(t) &= ec \int \frac{ds}{\gamma}\,(c\gamma)\delta^{(1)}(ct-cs)\delta^{(3)}(\boldsymbol{x}-\boldsymbol{r}(s)) \\ &= ec\,\delta^{(3)}(\boldsymbol{x}-\boldsymbol{r}(t)) \equiv c\varrho(t,\boldsymbol{x}) . \end{aligned}$$

The spatial components of the four-component current density are calculated in the same way:

$$\begin{aligned} j^k(t) &= ec \int \frac{ds}{\gamma}\,(\gamma v^k(s))\delta^{(1)}(ct-cs)\delta^{(3)}(\boldsymbol{x}-\boldsymbol{r}(s)) \\ &= e\,v^k(t)\delta^{(3)}(\boldsymbol{x}-\boldsymbol{r}(t)) , \quad k=1,2,3 . \end{aligned}$$

These are expressions (3.75a) and (3.75b), respectively, whose physical interpretation is more transparent than that of formula (3.76): The pointlike charge is located where the particle sits at time t; the current density that it creates at point $\boldsymbol{r}$ at time t is proportional to the velocity of the particle in this position and at this time.

We insert the current density, (3.76), into the wave equation for the four-potential $A^\alpha(x)$, at the place of the source term, and we assume that there is no incident field. Under this assumption and taking account of (3.69), (3.67) yields the solution for the potential created by the particle:

$$A^\alpha(x) = \frac{4\pi}{c} \int d^4x'\; G_{\text{ret}}(x-x')\, j^\alpha(x')\,. \tag{3.77}$$

Upon insertion of expression (3.76) for the current density and formula (3.74a) for the Green function, one has

$$A^\alpha(x) = 2e\int d\sigma \int d^4x'\; \Theta(x^0-x'^0)\,\delta^{(1)}((x-x')^2)u^\alpha(\sigma)\,\delta^{(4)}\left(x'-r(\sigma)\right) .$$

The integration $\int d^4x'$ over the variable x' is obvious and shows that x' is replaced by $r(\sigma)$. Thus, this integration gives

$$A^\alpha(x) = 2e \int d\sigma\; \Theta\left(x^0-r^0(\sigma)\right)\delta^{(1)}((x-r(\sigma)^2)u^\alpha(\sigma) . \tag{3.78}$$

Already this intermediate result, (3.78), can be analysed and interpreted in terms of causality. For this purpose, we fix the point x where the potential is determined.

From the integral on the right-hand side of (3.78) we analyse where and when the particle contributes to the potential. The one-dimensional δ-distribution enforces the condition $(x-r(\sigma))^2=0$, i.e. the source point $r(\tau)$ and the observer's position x must lie on a light cone relative to each other. Furthermore, the step function makes sure that the time t^0 at which the particle influences the potential $A^\alpha(x)$ is earlier than the time x^0. Only if these two conditions are fulfilled are the source (the particle flying by) and the effect (the potential $A^\alpha(x)$) related in the correct causal way. Expressed with respect to an arbitrarily chosen inertial system and as illustrated by Fig. 3.7, the measuring point x with its coordinate time $x^0 = ct$ always lies on the forward light cone of the world point where the particle was at the earlier coordinate time t^0.

For the calculation of the integral over σ with $\delta^{(1)}(x-r(\sigma)^2)$ in the integrand, one makes use of the auxiliary formula

$$\delta(f(u)) = \sum_i \frac{1}{|f'(u_i)|}\delta(u-u_i) ,$$

where the values u_i are the *simple* zeros of $f(u)$. Obviously, nonvanishing contributions come only from where the integration variable σ is equal to the proper time τ^0. Furthermore, one has

$$\begin{aligned}\frac{\mathrm{d}}{\mathrm{d}\sigma}\left(x-r(\sigma)\right)^2\bigg|_{\tau=\tau^0} &= -2\left(x-r(\tau^0)\right)_\mu \frac{\mathrm{d}}{\mathrm{d}\sigma} r^\mu(\sigma)\bigg|_{\tau=\tau^0} \\ &= -2\left(x-r(\tau^0)\right)_\mu u^\mu(\tau^0) .\end{aligned}$$

This is inserted in (3.78), the result being what is called the *Liénard–Wiechert potential*:

$$A^\alpha(x) = e\,\frac{u^\alpha(\tau)}{u(\tau)\cdot\left(x-r(\tau)\right)}\bigg|_{\tau=\tau^0} . \tag{3.79}$$

As expected, this potential is proportional to the four-velocity $u(\tau)$ of the particle. The denominator contains the Lorentz scalar product of u and $(x-r(\tau))$; the whole expression is to be evaluated at $\tau=\tau^0$ such that x lies on the forward light cone of $r(\tau^0)$. To work this out even more clearly, one may compute the space and time components of the potential in the reference system K starting from (3.79). In such a system, one has

$$u\cdot(x-r) = u^0(x^0-r^0) - \boldsymbol{u}\cdot(\boldsymbol{x}-\boldsymbol{r}) = \gamma c^2(t_x-t_r) - \gamma\boldsymbol{v}\cdot(\boldsymbol{x}-\boldsymbol{r}) .$$

The conditions $(x-r(\tau^0))^2 = 0$ and $x^0 > r^0(\tau^0)$ yield

$$x^0 - r^0(\tau^0) = c(t-t^0) = \left|\boldsymbol{x}-\boldsymbol{r}(\tau^0)\right| =: R ,$$

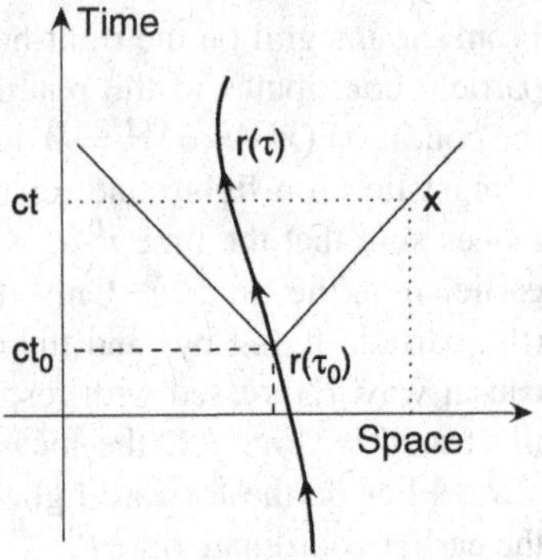

Fig. 3.7 Seen from an inertial system of reference K, the charged particle moves along the world line $r(\tau)$. At coordinate time t^0, its position is the world point $r(\tau^0)$. In this position, it can cause actions only which lie on the forward light cone of $r(\tau^0)$

where the spatial distance between $r(\tau^0)$ and x (as it is defined in the system of reference K) is denoted by R. With the definition $\hat{\boldsymbol{n}} = (\boldsymbol{x} - \boldsymbol{r}(\tau^0))/|\boldsymbol{x} - \boldsymbol{r}(\tau^0)|$, there follows the coordinate expression in K

$$u \cdot (x - r) = \gamma c R - \gamma \boldsymbol{v} \cdot \hat{\boldsymbol{n}}\, R = \gamma c R \left[1 - \frac{1}{c} \boldsymbol{v} \cdot \hat{\boldsymbol{n}} \right] .$$

Finally, inserting the decomposition of the four-velocity $u = (c\gamma, \gamma \boldsymbol{v})^T$, the time and space components of $A^\alpha = (\Phi, \boldsymbol{A})^T$ are, respectively,

$$\Phi(t, \boldsymbol{x}) = \frac{e}{R} \frac{1}{1 - \frac{1}{c} \boldsymbol{v} \cdot \hat{\boldsymbol{n}}} \bigg|_{\text{ret}} , \tag{3.80a}$$

$$\boldsymbol{A}(t, \boldsymbol{x}) = \frac{e}{R} \frac{\frac{\boldsymbol{v}}{c}}{1 - \frac{1}{c} \boldsymbol{v} \cdot \hat{\boldsymbol{n}}} \bigg|_{\text{ret}} . \tag{3.80b}$$

The subscript "ret" indicates that the potentials Φ and $\boldsymbol{A}$ are retarded, i.e. that these formulae apply at the distance R from the particle and at the time

$$t = t^0 + \frac{R}{c} . \tag{3.81}$$

This guarantees that the time-of-flight from where the particle passes by (the cause) to the measuring point at which the potentials are computed (the effect) is correctly implemented.

In a second step, one calculates the field strengths $F^{\mu\nu} = \partial^\mu A^\nu - \partial^\nu A^\mu$ from $A^\alpha(x)$ in (3.79), either by calculating the first derivatives by means of formula (3.79) or by returning to the integral representation (3.78). We choose the second alternative. The variables x^μ appear in two places: once in the step function, once in the δ-distribution.

(a) The derivative by x of the step function in the integrand gives a δ-distribution, $\delta(x^0 - r^0(\sigma))$. This means that from the original distribution $\delta^{(1)}((x - r(\sigma))^2)$ there remains only a spatial part, $\delta^{(1)}(-\boldsymbol{R}^2)$, which does not contribute as long as one excludes the origin $\boldsymbol{R} = 0$.

(b) The derivative of the δ-distribution in the integrand of (3.78) is calculated by means of the following auxiliary calculations:

$$\partial^\mu \delta\big(f(x,\sigma)\big) = (\partial^\mu f)\frac{\mathrm{d}}{\mathrm{d}f}\delta(f) = (\partial^\mu f)\frac{\mathrm{d}\sigma}{\mathrm{d}f}\frac{\mathrm{d}}{\mathrm{d}\sigma}\delta\big(f(x,\sigma)\big)\,,$$

where $f(x,\sigma) = ((x - r(\sigma))^2)$. One has

$$\partial^\mu f = 2\big(x - r(\sigma)\big)^\mu\,, \quad \frac{\mathrm{d}f}{\mathrm{d}\sigma} = -2(x - r)\cdot\frac{\mathrm{d}r}{\mathrm{d}\sigma} = -2(x - r)\cdot u$$

and, hence,

$$\partial^\mu \delta\big(f(x,\sigma)\big) = -\frac{(x - r)^\mu}{(x - r)\cdot u}\frac{\mathrm{d}}{\mathrm{d}\sigma}\delta\big(f(x,\sigma)\big)\,.$$

Inserting these auxiliary formulae into (3.78) and performing a partial integration, one obtains

$$\partial^\mu A^\nu(x) = 2e\int \mathrm{d}\sigma\,\frac{\mathrm{d}}{\mathrm{d}\sigma}\left(\frac{(x - r)^\mu u^\nu}{(x - r)\cdot u}\right)\Theta\big(x^0 - r^0(\sigma)\big)\,\delta^{(1)}\big((x - r(\sigma)^2)\big)\,. \tag{3.82}$$

The integration over σ is carried out in the same way as in the transition from (3.78) to (3.79). The result is

$$F^{\mu\nu}(x) = \frac{e}{u(\tau)\cdot\big(x - r(\tau)\big)} \times \frac{\mathrm{d}}{\mathrm{d}\tau}\left\{\frac{\big(x - r(\tau)\big)^\mu u(\tau)^\nu - \big(x - r(\tau)\big)^\nu u(\tau)^\mu}{u(\tau)\cdot\big(x - r(\tau)\big)}\right\}\Bigg|_{\tau=\tau^0}\,. \tag{3.83}$$

This is the tensor field of the electromagnetic fields which are created by a charge in motion.

To gain some "feeling" for this important result, it is useful to calculate the electric field and the magnetic induction in a frame of reference K. (The reader is invited to do this exercise!) One finds

$$\boldsymbol{E}(t,\boldsymbol{x}) = \boldsymbol{E}_{\text{stat}}(t,\boldsymbol{x}) + \boldsymbol{E}_{\text{acc}}(t,\boldsymbol{x})\,, \tag{3.84}$$

where the two terms are given by the expressions, respectively,

$$\boldsymbol{E}_{\text{stat}}(t,\boldsymbol{x}) = \frac{e}{R^2}\,\frac{\hat{\boldsymbol{n}} - \frac{\boldsymbol{v}}{c}}{\gamma^2\big(1 - \frac{1}{c}\boldsymbol{v}\cdot\hat{\boldsymbol{n}}\big)^3}\Bigg|_{\text{ret}}\,, \tag{3.84a}$$

$$\boldsymbol{E}_{\text{acc}}(t,\boldsymbol{x}) = \frac{e}{R}\,\frac{\hat{\boldsymbol{n}}\times\big[\big(\hat{\boldsymbol{n}} - \frac{\boldsymbol{v}}{c}\big)\times\dot{\boldsymbol{v}}\big]}{c^2\big(1 - \frac{1}{c}\boldsymbol{v}\cdot\hat{\boldsymbol{n}}\big)^3}\Bigg|_{\text{ret}}\,. \tag{3.84b}$$

The distance R and the unit vector $\hat{\boldsymbol{n}}$ are defined as in (3.80a) and (3.80b). The interpretation of the two parts of the electric field will be explained in a moment, but first one notes the result for the magnetic induction:

$$\boldsymbol{B}(t, \boldsymbol{x}) = \hat{\boldsymbol{n}} \times \boldsymbol{E}(t, \boldsymbol{x}) . \tag{3.85}$$

The first part $\boldsymbol{E}_{\text{stat}}$ in (3.84) is called a *static field* or *velocity field* because it is present also when the particle moves uniformly along a straight line. This field, in essence, is a static field because, although it contains the velocity, it does not depend on the acceleration. The second term, $\boldsymbol{E}_{\text{acc}}(t, \boldsymbol{x})$, is called the *acceleration field*; it is a function of $\dot{\boldsymbol{v}}$, the instantaneous acceleration, and it vanishes whenever the velocity of the particle is constant.

For velocities which are small compared to the speed of light, $|\boldsymbol{v}| \ll c$, the Poynting vector field can be approximated as follows (for an application see [QP], Example 1.2):

$$\boldsymbol{S}(t, \boldsymbol{x}) = \frac{c}{4\pi} \boldsymbol{E}(t, \boldsymbol{x}) \times \boldsymbol{B}(t, \boldsymbol{x}) = \frac{c}{4\pi} \boldsymbol{E} \times \left(\hat{\boldsymbol{n}} \times \boldsymbol{E} \right) \simeq \frac{c}{4\pi} \boldsymbol{E}^2(t, \boldsymbol{x})\, \hat{\boldsymbol{n}} . \tag{3.86}$$

In this approximation, only the acceleration field

$$\boldsymbol{E}(t, \boldsymbol{x}) \simeq \frac{e}{R} \frac{1}{c^2} \hat{\boldsymbol{n}} \times \left(\hat{\boldsymbol{n}} \times \dot{\boldsymbol{v}} \right) \Big|_{\text{ret}}$$

contributes to the Poynting vector. These formulae are derived from the results of this section and from expression (3.49c) for the Poynting vector field. For example, the power radiated to the element of solid angle $\mathrm{d}\Omega$, with θ being the angle between $\dot{\boldsymbol{v}}$ and $\hat{\boldsymbol{n}}$, is found to be

$$\frac{\mathrm{d}P}{\mathrm{d}\Omega} \simeq R^2 \frac{c}{4\pi} \boldsymbol{E}^2(t, \boldsymbol{x}) = \frac{e^2 \dot{\boldsymbol{v}}^2}{4\pi c^3} \sin^2\theta . \tag{3.87}$$

Integration of this expression over the complete solid angle yields the total radiated power

$$P \simeq \frac{2e^2}{3c^3} \dot{\boldsymbol{v}}^2 . \tag{3.88}$$

If the velocity is no longer small as compared to c, one finds a rather similar result, viz.

$$P = \frac{2e^2}{3c^3} \gamma^6 \left\{ \dot{\boldsymbol{v}}^2 - \frac{1}{c^2} \left(\boldsymbol{v} \times \dot{\boldsymbol{v}} \right)^2 \right\} . \tag{3.89}$$

In either case, when $|\boldsymbol{v}| \ll c$, but also when this is no longer true, the radiated power is equal to zero if the particle is neither accelerated nor decelerated. *Only an accelerated charged particle radiates energy. If it moves with constant velocity, then it does not radiate.*

Simple Applications of Maxwell Theory

4

4.1 Introduction

In this chapter, we select some characteristic examples from the great wealth of electromagnetic and optical phenomena which are described by Maxwell's equations. These case studies are restricted to the classical, nonquantized version of the theory. The field of semiclassical interactions of quantum matter and classical radiation field, as well as the full quantum field-theoretic treatment of Maxwell theory is described in many monographs or textbooks, such as, for example, [QP].

4.2 Plane Waves in a Vacuum and in Homogeneous Insulating Media

In the simplest case, nonconducting media, as far as their electromagnetic properties are concerned, are homogeneous and isotropic. This means that they can be described by scalar material constants, a dielectric constant ε and a magnetic permeability μ so that the relationships (2.2) between displacement field $\boldsymbol{D}$ and electric field $\boldsymbol{E}$, and, similarly, between magnetic induction $\boldsymbol{B}$ and magnetic field $\boldsymbol{H}$ are linear. In this section, we study harmonic solutions of the wave equation in simple media of this kind, including the vacuum, and we analyse the polarization of electromagnetic waves.

4.2.1 Dispersion Relation and Harmonic Solutions

With the assumptions just mentioned and in the absence of any charge or current densities, every component of the electric field $\boldsymbol{E}(t, \boldsymbol{x})$ and every component of the

F. Scheck, *Classical Field Theory*, Graduate Texts in Physics,
DOI 10.1007/978-3-642-27985-0_4, © Springer-Verlag Berlin Heidelberg 2012

induction field $\boldsymbol{B}(t, \boldsymbol{x})$ satisfies a wave equation of the type of (1.45):

$$\frac{1}{v^2}\frac{\partial^2}{\partial t^2} f(t, \boldsymbol{x}) - \boldsymbol{\Delta} f(t, \boldsymbol{x}) = 0 \,, \tag{4.1}$$

with the velocity modified by the properties of the medium,

$$v = \frac{c}{\sqrt{\varepsilon\mu}} \,. \tag{4.2}$$

The function $f(t, \boldsymbol{x})$ stands generically for an arbitrary component of either $\boldsymbol{E}$ or $\boldsymbol{B}$. The reason that we have chosen a specific class of reference systems and that we did not use the covariant form of the wave equation is that we are dealing here with the observable fields. As was explained earlier, the mere fact that one *measures* electric and magnetic fields entails a partition of the $\mathbb{R}^4$ of Maxwell theory into time axis and space parts. On the other hand, as we are dealing with observables only, no gauge condition must be taken into account. Wave equation (4.1) applies without subsidiary conditions.

We demonstrate (4.1) for the example of the electric field. Take the curl of the vector field $\boldsymbol{\nabla} \times \boldsymbol{E} + (1/c)\partial\boldsymbol{B}/\partial t$, which, by (1.44b), should be zero. For the first term one uses the identity (1.47c):

$$\boldsymbol{\nabla} \times (\boldsymbol{\nabla} \times \boldsymbol{E}) = \boldsymbol{\nabla}(\boldsymbol{\nabla} \cdot \boldsymbol{E}) - \boldsymbol{\Delta}\boldsymbol{E}$$

and notes that the first term on the right-hand side is equal to zero by (1.44c). Indeed, as there is no free charge density, one has

$$\boldsymbol{\nabla} \cdot \boldsymbol{D} = \varepsilon \boldsymbol{\nabla} \cdot \boldsymbol{E} = 0 \,.$$

The curl of the induction field is calculated from (1.44d) with $\boldsymbol{j}(t, \boldsymbol{x}) \equiv 0$, yielding

$$\boldsymbol{\nabla} \times \boldsymbol{B} = \mu \boldsymbol{\nabla} \times \boldsymbol{H} = \frac{\mu}{c}\frac{\partial \boldsymbol{D}}{\partial t} = \frac{\mu\varepsilon}{c}\frac{\partial \boldsymbol{E}}{\partial t} \,.$$

Inserting these results one obtains

$$\left(-\boldsymbol{\Delta} + \frac{\mu\varepsilon}{c^2}\frac{\partial^2}{\partial t^2}\right) \boldsymbol{E}(t, \boldsymbol{x}) = 0 \,.$$

This is wave equation (4.1), which applies to every component of $\boldsymbol{E}$. The same differential equation is obtained for $\boldsymbol{B}(t, \boldsymbol{x})$ by taking the curl of (1.44d).

The wave equation (4.1), very much like the full set of Maxwell's equations without external sources, is a *linear* equation in the function $f(t, \boldsymbol{x})$. Therefore, the superposition principle applies: With any two solutions $f_1(t, \boldsymbol{x})$ and $f_2(t, \boldsymbol{x})$ every linear combination

$$c_1 f_1(t, \boldsymbol{x}) + c_2 f_2(t, \boldsymbol{x}) \quad \text{with} \quad c_1, c_2 \in \mathbb{R} \quad \text{or} \quad c_1, c_2 \in \mathbb{C} \tag{4.3}$$

is also a solution of (4.1). The significance and the usefulness of complex coefficients instead of real ones will become evident later.

Even in the case of an isotropic and homogeneous medium, it may happen that the permeability μ and the dielectric constant ε, and hence also the speed of propagation v, are functions of the circular frequency ω of the radiation in focus. Therefore, one distinguishes two cases as follows.

μ and ε Independent of the Frequency

Obviously, the vacuum belongs to this class of media: Using Gaussian units, both constants are equal to 1, independently of the frequency of the harmonic oscillation.

A harmonic solution (a so-called pure "sine oscillation"), for example, has the form

$$f_k(t, \boldsymbol{x}) = \mathrm{e}^{-\mathrm{i}\omega t}\,\mathrm{e}^{\pm \mathrm{i}\boldsymbol{k}\cdot\boldsymbol{x}} \quad \text{with} \tag{4.4a}$$

$$k \equiv |\boldsymbol{k}| = \frac{\omega}{v} = \sqrt{\mu\varepsilon}\,\frac{\omega}{c}\,. \tag{4.4b}$$

The vector $\boldsymbol{k}$ is the *wave vector*; its modulus k is called the *wave number*. Replacing the circular frequency by $\omega = 2\pi\nu$, with ν the frequency, and taking $k = 2\pi/\lambda$, with λ the wavelength, one recovers the well-known relation $v = \lambda\nu$. As an example, choose the wave vector along the 3-axis, $\boldsymbol{k} = k\hat{\boldsymbol{e}}_3$. The general solution with this wave number then reads

$$f_k(t, \boldsymbol{x}) = c_+\,\mathrm{e}^{\mathrm{i}k(x^3 - vt)} + c_-\,\mathrm{e}^{\mathrm{i}k(x^3 + vt)}\,.$$

Fundamental solutions of this kind can be combined linearly in a rather arbitrary way such as, for instance,

$$g(x^3 - vt) = \frac{1}{\sqrt{2\pi}} \int_{-\infty}^{+\infty} \mathrm{d}k\; \tilde{g}(k)\,\mathrm{e}^{\mathrm{i}k(x^3 - vt)}\,, \tag{4.5a}$$

$$h(x^3 + vt) = \frac{1}{\sqrt{2\pi}} \int_{-\infty}^{+\infty} \mathrm{d}k\; \tilde{h}(k)\,\mathrm{e}^{\mathrm{i}k(x^3 + vt)}\,, \tag{4.5b}$$

so that the general differentiable solution reads

$$f(t, x^3) = g(x^3 - vt) + h(x^3 + vt)\,. \tag{4.6}$$

Obviously, the two types of solutions, (4.5a) and (4.5b), are distinguished by their direction of motion – they propagate either along the 3-axis or in the opposite direction.

Media with Dispersion

In dispersive media, the product $\mu\varepsilon$ and, hence, the speed of propagation v are functions of the circular frequency ω. Every field function f such as the Cartesian components E^i or B^k, respectively, may be written in terms of its Fourier components in the circular frequency ω, viz.

$$f(t, \boldsymbol{x}) = \int_{-\infty}^{+\infty} \mathrm{d}\omega\, \widetilde{f}(\omega, \boldsymbol{x})\mathrm{e}^{-\mathrm{i}\omega t} . \tag{4.7}$$

The wave equation (4.1), which is a partial differential equation in space and time coordinates, becomes a differential equation in the space components only:

$$\left[\boldsymbol{\Delta} + \mu\varepsilon\frac{\omega^2}{c^2}\right]\widetilde{f}(\omega, \boldsymbol{x}) = 0 . \tag{4.8}$$

This differential equation is the homogeneous version of the *Helmholtz equation* (3.61). In general, the wave number is no longer a linear function of the circular frequency, so that one should write

$$k(\omega) = \sqrt{\mu(\omega)\varepsilon(\omega)}\,\frac{\omega}{c} . \tag{4.9}$$

The relation between the circular frequency and the wave number is called the *dispersion relation*. The detailed nature of this function is not a problem of Maxwell theory but rather a question about the composition of the medium, i.e. about matter with which the Maxwell fields interact.

Considering plane wave solutions of the kind of (4.4a), also called *harmonic solutions*, the electric field and the induction field are taken to be

$$\boldsymbol{E}_{\mathrm{c}}(t, \boldsymbol{x}) = \boldsymbol{e}\ \mathrm{e}^{\mathrm{i}(k\hat{\boldsymbol{n}}\cdot\boldsymbol{x} - \omega t)} , \tag{4.10a}$$

$$\boldsymbol{B}_{\mathrm{c}}(t, \boldsymbol{x}) = \boldsymbol{b}\ \mathrm{e}^{\mathrm{i}(k\hat{\boldsymbol{n}}\cdot\boldsymbol{x} - \omega t)} . \tag{4.10b}$$

The index "c" is meant to indicate that the fields have been continued to the complex plane whose real parts are the physical, i.e. observable, fields. In the ansatz (4.10a), or (4.10b), the variable $k = |\boldsymbol{k}|$ with $\boldsymbol{k} = k\hat{\boldsymbol{n}}$ is the wave number which satisfies the dispersion relation (4.9), while $\hat{\boldsymbol{n}}$ is the direction of propagation of the plane wave. Vectors $\boldsymbol{e}$ and $\boldsymbol{b}$ are constant vectors which may be real or complex and whose properties and physical interpretation need to be clarified. Before turning to this question, a remark about calculus with complex fields seems in order.

▶ **Remark: Complex Maxwell fields**

Maxwell's equations without external sources are linear in the observables they describe, i.e. in the electric and magnetic fields. All coefficients are real. This is seen most clearly in the covariant formulations (2.49a) and (2.53), in a vacuum,

$$\varepsilon_{\mu\nu\sigma\tau}\partial^{\nu}F^{\sigma\tau}(x) = 0\,,$$
$$\partial_{\mu}F^{\mu\nu}(x) = 0$$

or equally well in the original formulation of equations (1.44a)–(1.44d) with $\varrho = 0$ and $\boldsymbol{j} = 0$. Therefore, it is perfectly admissible to search for complex solutions, $\boldsymbol{E}_{\mathrm{c}}(t,\boldsymbol{x})$, $\boldsymbol{B}_{\mathrm{c}}(t,\boldsymbol{x})$, and to interpret the real parts of the complex fields as the physically realized fields. Note that the method of complex fields is often applied as a useful technique in electrical engineering, too.

There are applications such as optical oscillations dealt with in this chapter where the method of complex fields is not only a technical trick but also useful from a physical point of view. Two examples may illustrate this point.

Introducing complex fields, formula (3.54a) for the energy density and formula (3.54c) for the energy flow are replaced respectively by

$$u(t,\boldsymbol{x}) = \frac{1}{8\pi}\left\{\varepsilon\,(\mathrm{Re}\,\boldsymbol{E}_{\mathrm{c}})^2 + \mu\,(\mathrm{Re}\,\boldsymbol{H}_{\mathrm{c}})^2\right\},$$
$$\boldsymbol{S}(t,\boldsymbol{x}) = \frac{c}{4\pi}\,(\mathrm{Re}\,\boldsymbol{E}_{\mathrm{c}}) \times (\mathrm{Re}\,\boldsymbol{H}_{\mathrm{c}})\ .$$

Indeed, straightforward calculation of $u(t,\boldsymbol{x})$, for example, gives

$$\begin{aligned} u(t,\boldsymbol{x}) &= \frac{1}{32\pi}\left\{\varepsilon\,(\boldsymbol{E}_{\mathrm{c}} + \boldsymbol{E}_{\mathrm{c}}^{*})^2 + \mu\,(\boldsymbol{H}_{\mathrm{c}} + \boldsymbol{H}_{\mathrm{c}}^{*})^2\right\} \\ &= \frac{1}{16\pi}\left\{\varepsilon\,(\boldsymbol{E}_{\mathrm{c}}^{*}\cdot\boldsymbol{E}_{\mathrm{c}}) + \mu\,(\boldsymbol{H}_{\mathrm{c}}^{*}\cdot\boldsymbol{H}_{\mathrm{c}})\right\} \\ &\quad + \frac{1}{32\pi}\left\{\varepsilon\,(\boldsymbol{E}_{\mathrm{c}}^{2} + \boldsymbol{E}_{\mathrm{c}}^{*2}) + \mu\,(\boldsymbol{H}_{\mathrm{c}}^{2} + \boldsymbol{H}_{\mathrm{c}}^{*2})\right\}. \end{aligned}$$

If the fields have the harmonic time dependence of equations (4.10a) and (4.10b), then the first term on the right-hand side is independent of time, whereas the second term is proportional to $\mathrm{e}^{\pm 2\mathrm{i}\omega t}$. In typical applications to optics, the circular frequency ω is large compared to 1. Furthermore, in many situations, only time averages matter. In those cases, the second term does not contribute, and one has

$$\langle u\rangle = \frac{1}{16\pi}\left\{\varepsilon\,(\boldsymbol{E}_{\mathrm{c}}^{*}\cdot\boldsymbol{E}_{\mathrm{c}}) + \mu\,(\boldsymbol{H}_{\mathrm{c}}^{*}\cdot\boldsymbol{H}_{\mathrm{c}})\right\}, \qquad (4.11\mathrm{a})$$

where $\langle\ \rangle$ denotes the time averages.

An analogous formula applies to the energy flow, viz.

$$\begin{aligned} \boldsymbol{S}(t,\boldsymbol{x}) &= \frac{c}{4\pi}\mathrm{Re}\,\boldsymbol{E}_{\mathrm{c}} \times \mathrm{Re}\,\boldsymbol{H}_{\mathrm{c}} = \frac{c}{16\pi}\left\{(\boldsymbol{E}_{\mathrm{c}} + \boldsymbol{E}_{\mathrm{c}}^{*}) \times (\boldsymbol{H}_{\mathrm{c}} + \boldsymbol{H}_{\mathrm{c}}^{*})\right\} \\ &= \frac{c}{16\pi}\left\{\boldsymbol{E}_{\mathrm{c}} \times \boldsymbol{H}_{\mathrm{c}}^{*} + \boldsymbol{E}_{\mathrm{c}}^{*} \times \boldsymbol{H}_{\mathrm{c}}\right\} + \text{ terms in } \mathrm{e}^{\pm 2\mathrm{i}\omega t}\ , \end{aligned}$$

so that here, too, the time average is given by the first term:

$$\langle \boldsymbol{S}(t,\boldsymbol{x})\rangle = \frac{c}{16\pi}\left\{\boldsymbol{E}_{\rm c}\times\boldsymbol{H}_{\rm c}^{*}+\boldsymbol{E}_{\rm c}^{*}\times\boldsymbol{H}_{\rm c}\right\}. \tag{4.11b}$$

Both expressions (4.11a) and (4.11b) have an intuitive form that is easy to interpret.

We return to the ansatz (4.10a), (4.10b), and clarify the conditions on vectors $\boldsymbol{e}$ and $\boldsymbol{b}$ imposed by Maxwell's equations: As the induction field is always divergence free, $\nabla\cdot\boldsymbol{B}_{\rm c}=0$, one concludes with (4.10b):

$$\nabla\cdot\boldsymbol{B}_{\rm c}(t,\boldsymbol{x}) = \boldsymbol{b}\cdot\nabla {\rm e}^{{\rm i}(\boldsymbol{k}\cdot\boldsymbol{x}-\omega t)} = {\rm i}\,(\boldsymbol{b}\cdot\boldsymbol{k})\,{\rm e}^{{\rm i}(\boldsymbol{k}\cdot\boldsymbol{x}-\omega t)} = 0\,.$$

When there are no external sources, one has also $\nabla\cdot\boldsymbol{E}_{\rm c}(t,\boldsymbol{x}) = 0$. These two equations imply the conditions

$$\hat{\boldsymbol{n}}\cdot\boldsymbol{e}=0\,,\quad\text{and}\quad \hat{\boldsymbol{n}}\cdot\boldsymbol{b}=0\,. \tag{4.12a}$$

Both the electric field and the induction field are transversal fields: Both are perpendicular to the direction of propagation.

It remains to determine the relative orientation of the two fields. For that, one returns to the homogeneous Maxwell equation (1.44b):

$$\nabla\times\boldsymbol{E}_{\rm c}+\frac{1}{c}\frac{\partial\boldsymbol{B}_{\rm c}}{\partial t}=0\,.$$

Inserting harmonic functions (4.10a) and (4.10b), one concludes

$$k\,\hat{\boldsymbol{n}}\times\boldsymbol{e}-\frac{\omega}{c}\boldsymbol{b}=0\,,$$

or, making use of dispersion relation (4.9),

$$\boldsymbol{b}=\sqrt{\mu\varepsilon}\;\hat{\boldsymbol{n}}\times\boldsymbol{e}\,. \tag{4.12b}$$

As $\hat{\boldsymbol{n}}$ is a real unit vector, the vectors $\boldsymbol{e}$ and $\boldsymbol{b}$ must have the same phase. Thus, they can also be chosen to be real without loss of generality. According to relation (4.12b), the electric field and the induction field are perpendicular to each other, the direction of propagation $\hat{\boldsymbol{n}}$, field $\boldsymbol{E}$ and field $\boldsymbol{B}$ form a right-handed system of vectors, as sketched in Fig. 4.1.

We give two examples, the 3-axis being chosen along the direction of propagation, i.e. $\hat{\boldsymbol{n}}=\hat{\boldsymbol{e}}_3$.

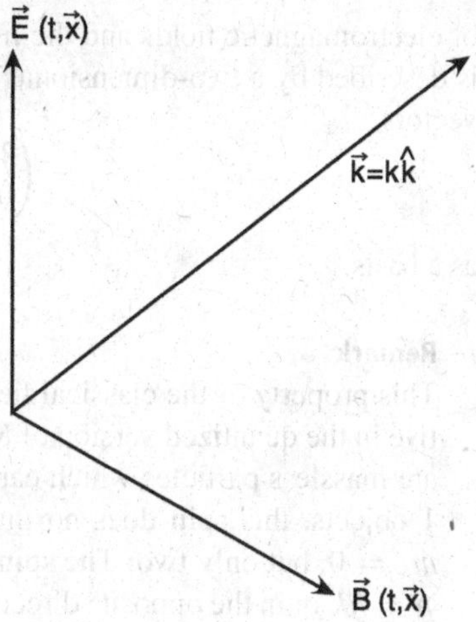

Fig. 4.1 Vector $\boldsymbol{k}$ defines the direction of propagation of the plane wave. Electric field $\boldsymbol{E}$ and induction field $\boldsymbol{B}$ are perpendicular to $\hat{\boldsymbol{k}}$. Unit vectors $\boldsymbol{n}$, $\boldsymbol{E}$ and $\boldsymbol{B}$ span a right-handed frame

(i) Let $\boldsymbol{e} = c_1\hat{\boldsymbol{e}}_1$, with c_1 a complex constant number. Then one has $\boldsymbol{b} = \sqrt{\mu\varepsilon}\, c_1\hat{\boldsymbol{e}}_2$.
(ii) Let $\boldsymbol{e} = c_2\hat{\boldsymbol{e}}_2$. There then follows $\boldsymbol{b} = -\sqrt{\mu\varepsilon}\, c_2\hat{\boldsymbol{e}}_1$.

In example (i), the electric field oscillates along the 1-direction; in example (ii) it oscillates along the 2-direction. In these cases, one talks about *linear polarization*.

The most general case is obtained by linear superposition, i.e.

$$\boldsymbol{E}_{\mathrm{c}}(t,\boldsymbol{x}) = \left(c_1\hat{\boldsymbol{e}}_1 + c_2\hat{\boldsymbol{e}}_2\right) \mathrm{e}^{\mathrm{i}(k\hat{\boldsymbol{n}}\cdot\boldsymbol{x}-\omega t)}\,, \tag{4.13a}$$

$$\boldsymbol{B}_{\mathrm{c}}(t,\boldsymbol{x}) = \frac{1}{k}\sqrt{\mu\varepsilon}\,\boldsymbol{k}\times\boldsymbol{E}_{\mathrm{c}}(t,\boldsymbol{x})\,, \tag{4.13b}$$

$$\text{with} \quad c_1,\ c_2 \in \mathbb{C}\,. \tag{4.13c}$$

If the two complex numbers (4.13c) have the same phase, i.e. if the ratio c_1/c_2 is real, then (4.13a) is again a *linearly polarized* wave whose direction of polarization is given by

$$\hat{\boldsymbol{e}} = \hat{\boldsymbol{e}}_1\cos\varphi + \hat{\boldsymbol{e}}_2\sin\varphi \qquad \text{with} \quad \tan\varphi = \frac{c_2}{c_1}\,.$$

In turn, if the phases of c_1 and c_2 differ from each other, then the polarization is no longer linear. This more general case, as well as the case where the light beam is only partially polarized, must be studied in more detail.

4.2.2 Completely Polarized Electromagnetic Waves

The essential result of the preceding section is that the electric field has only *two* possible orientations relative to the direction of propagation – in spite of the fact that it is a vector field over $\mathbb{R}^3$ – so that one would expect three possible orientations. (The same observation applies to the induction field.) Because of the transversality of electromagnetic fields and the linearity of Maxwell's equations, the polarization is described by a two-dimensional real vector space for which one may choose the vectors

$$\hat{\boldsymbol{e}}_1 = \begin{pmatrix} 1 \\ 0 \end{pmatrix} \quad \text{and} \quad \hat{\boldsymbol{e}}_2 = \begin{pmatrix} 0 \\ 1 \end{pmatrix}$$

as a basis.

Remark

This property of the classical field theory of light appears in a different perspective in the quantized version of Maxwell theory. There one finds that light quanta are massless particles which carry spin 1. However, in contrast to *massive* spin-1 objects, this spin does not have three possible orientations, $m_s = \pm 1$ and $m_s = 0$, but only two: The spin points either in the direction of the momentum $\boldsymbol{p} = \hbar \boldsymbol{k}$ or in the opposite direction. One says that the photon possesses a *helicity* $h = \boldsymbol{S} \cdot \boldsymbol{p}/|\boldsymbol{p}|$ which takes only the two values ± 1. Helicity is the projection of spin in the direction of the momentum. The two admissible values correspond to left-circular and right-circular polarization in the classical theory. They will be defined more precisely below.

As before and without loss of generality, let the 3-axis be chosen along $\boldsymbol{k}$. As we are interested in oscillations in time only, and in relative strength and relative phase of the transversal components, it is sufficient to study the situation in a fixed point, say $\boldsymbol{x} = \boldsymbol{0}$. Expressed in terms of real, physical fields, the general case can be written in the form

$$\boldsymbol{E}(t, \boldsymbol{x} = \boldsymbol{0}) = \varepsilon_1 \cos(\omega t) \begin{pmatrix} 1 \\ 0 \end{pmatrix} + \varepsilon_2 \cos(\omega t + \alpha) \begin{pmatrix} 0 \\ 1 \end{pmatrix}, \tag{4.14}$$

which depends on the (real) amplitudes ε_i and the phase shift α. We show that, in general, this corresponds to an *elliptic* polarization where the tips of the fields move along an ellipse, and we calculate the position and the parameters of this ellipse as functions of ε_1, ε_2 and α. When continued to the complex plane, one has $\boldsymbol{E} = \operatorname{Re} \boldsymbol{E}_{\mathrm{c}}$, where

$$\boldsymbol{E}_{\mathrm{c}}(t, \boldsymbol{x} = \boldsymbol{0}) = \mathrm{e}^{\mathrm{i}\omega t} \begin{pmatrix} \varepsilon_1 \\ \varepsilon_2 \mathrm{e}^{\mathrm{i}\alpha} \end{pmatrix}. \tag{4.15}$$

This is now a two-component vector with complex-valued coefficients. Therefore, it may always be written as a sum:

$$\boldsymbol{E}_{\mathrm{c}} = \boldsymbol{u} + \mathrm{i}\boldsymbol{v}, \tag{4.16a}$$

where $\boldsymbol{u}$ and $\boldsymbol{v}$ are vectors with real coefficients. By choosing these two vectors perpendicular to each other, $\boldsymbol{u} \cdot \boldsymbol{v} = 0$, one has already found the principal axes of the ellipse hidden in (4.14).

Define $u := |\boldsymbol{u}|$ and $v := |\boldsymbol{v}|$. One calculates

$$\boldsymbol{E}_c^* \cdot \boldsymbol{E}_c = u^2 + v^2 = \varepsilon_1^2 + \varepsilon_2^2 , \tag{4.16b}$$
$$\boldsymbol{E}_c^* \times \boldsymbol{E}_c = 2\mathrm{i}\,\boldsymbol{u} \times \boldsymbol{v} = 2\mathrm{i}\,uv\,\hat{\boldsymbol{e}}_3 = 2\mathrm{i}\,\varepsilon_1\varepsilon_2 \sin\alpha\,\hat{\boldsymbol{e}}_3 \tag{4.16c}$$

to obtain the moduli of u and v:

$$\left.\begin{matrix} u \\ v \end{matrix}\right\} = \frac{1}{2}\left\{\sqrt{\varepsilon_1^2 + \varepsilon_2^2 + 2\varepsilon_1\varepsilon_2 \sin\alpha} \pm \sqrt{\varepsilon_1^2 + \varepsilon_2^2 - 2\varepsilon_1\varepsilon_2 \sin\alpha}\right\} . \tag{4.16d}$$

These are the two semiaxes of the ellipse in the (real) (1, 2)-plane. One determines from them the angle ψ, shown in Fig. 4.2, whose tangent equals v/u:

$$\sin(2\psi) = \frac{2\tan\psi}{1+\tan^2\psi} = \frac{2uv}{u^2+v^2} = \frac{2\varepsilon_1\varepsilon_2 \sin\alpha}{\varepsilon_1^2+\varepsilon_2^2} . \tag{4.17}$$

The orientation of the semimajor axis in the (1, 2)-plane is found as follows. From (4.16a) one has $\boldsymbol{E}_c \cdot \boldsymbol{u} = u^2$ and $\boldsymbol{E}_c^* \cdot \boldsymbol{v} = -\mathrm{i}v^2$. On the other hand, these scalar products are calculated by means of the components of the three vectors involved, viz.

$$\boldsymbol{E}_c \cdot \boldsymbol{u} = u\,\mathrm{e}^{\mathrm{i}\omega t}\left\{\varepsilon_1 \cos\varphi + \varepsilon_2 \sin\varphi\,\mathrm{e}^{\mathrm{i}\alpha}\right\} ,$$
$$\boldsymbol{E}_c^* \cdot \boldsymbol{v} = v\,\mathrm{e}^{-\mathrm{i}\omega t}\left\{-\varepsilon_1 \sin\varphi + \varepsilon_2 \cos\varphi\,\mathrm{e}^{-\mathrm{i}\alpha}\right\} .$$

The product of these two scalar products is equal to $-\mathrm{i}u^2v^2$ and, thus, is pure imaginary. Therefore, calculating the real part of this product from the representation in terms of components one should find zero:

$$-\left(\varepsilon_1^2 - \varepsilon_2^2\right)\sin\varphi\cos\varphi + \varepsilon_1\varepsilon_2\left(\cos^2\varphi - \sin^2\varphi\right)\cos\alpha = 0 .$$

This yields the desired formula for the angle enclosed by the semimajor axis and the 1-direction:

$$\tan(2\varphi) = \frac{2\varepsilon_1\varepsilon_2 \cos\alpha}{\varepsilon_1^2 - \varepsilon_2^2} . \tag{4.18}$$

Various special cases can be identified in the results (4.16d), (4.17) and (4.18).

(i) If $\alpha = 0$, then $v = 0$; the polarization is linear and points along the φ direction with $\tan\varphi = \varepsilon_2/\varepsilon_1$:

$$\alpha = 0 : \quad \tan(2\varphi) = \frac{2(\varepsilon_2/\varepsilon_1)}{1-(\varepsilon_2/\varepsilon_1)^2} = \frac{2\tan\varphi}{1-\tan^2\varphi} .$$

This confirms the result of the preceding section.

Fig. 4.2 General case of elliptic polarization of a harmonic solution of the wave equation. One has $\tan\psi = v/u$; the angle ϕ defines the orientation of the semimajor axis relative to the 1-direction

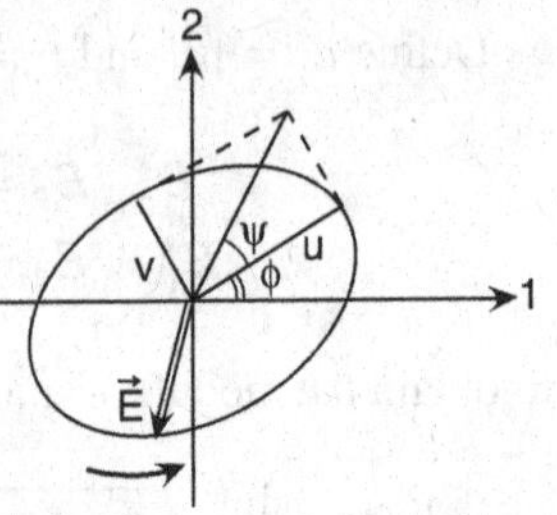

(ii) If $\alpha = \pi/2$ and if the amplitudes are equal, $\varepsilon_1 = \varepsilon_2$, then $v = u$, $\psi = \pi/4$, while φ remains undetermined. In this case, the tip of $\boldsymbol{E}_c$ moves along a circle in the (1, 2)-plane of Fig. 4.2, whose radius is $\varepsilon_1 = \varepsilon_2$, in a clockwise direction. In this case, one says there is *right-circular polarization.*

(iii) The case $\alpha = -\pi/2$, $\varepsilon_1 = \varepsilon_2$, is closely analogous to case (ii): As before, one has $v = u$, but $\psi = -\pi/4$. The tip of $\boldsymbol{E}_c$ moves in an anti-clockwise direction along a circle with radius $\varepsilon_1 = \varepsilon_2$ in the (1, 2)-plane of Fig. 4.2. One calls this case *left-circular polarization.* The rotating electric field, together with the oriented 3-axis (which in Fig. 4.2 points towards the observer), defines a sense of rotation – analogous to the motion of a standard corkscrew – which is called *positive helicity.*
The opposite sense of rotation of the previous example (ii) is described as *negative helicity.* In either case, this terminology refers to the quantized version of electrodynamics and the properties of photons.

Remark

In the quantum version of Maxwell theory, the two helicities ± 1 are the allowed states of a photon's spin $s = 1$. They replace the projection quantum number s_3 of quantum mechanics of massive particles. There is an essential difference, though: In the case of a *massive* particle which carries spin s, the spin projection onto the 3-axis has a range of $(2s + 1)$ possible values, viz. $m_s = -s, -s + 1, \ldots, s$, in accord with the quantum mechanics of angular momentum. In the case of a particle which has no rest mass and which has spin s, these projection quantum numbers are replaced by only two helicity states, $h = +s$ and $h = -s$. This is a characteristic property of *massless* particles.

In the complex two-dimensional vector space that was introduced in the special cases discussed above, these states correspond to the basis

$$\hat{\boldsymbol{e}}_+ = -\frac{1}{\sqrt{2}}\begin{pmatrix}1\\ \mathrm{i}\end{pmatrix} \quad \text{and} \quad \hat{\boldsymbol{e}}_- = \frac{1}{\sqrt{2}}\begin{pmatrix}1\\ -\mathrm{i}\end{pmatrix}, \tag{4.19}$$

which replaces the basis introduced in (4.14),

$$\hat{e}_1 = \begin{pmatrix} 1 \\ 0 \end{pmatrix} \quad \text{and} \quad \hat{e}_2 = \begin{pmatrix} 0 \\ 1 \end{pmatrix} . \tag{4.20}$$

Of course, there is no compelling reason to choose the elements of these bases normalized to 1. Also, the choice of signs is not fixed a priori. I have chosen the signs in (4.19) such that they correspond to the convention in the choice of the spherical basis in $\mathbb{R}^3$. Note, however, that the general case of elliptic polarization, using complexified fields, can be described equally well by linear combinations of the base vectors, (4.20), or of the base vectors, (4.19).

4.2.3 Description of Polarization

Laser beams, as a rule, are polarized. Likewise, the electromagnetic waves emitted by simple antennas are polarized, but this is not so for many light sources of daily life, including the light of the Sun. So how can we describe electromagnetic radiation, on the basis of the results of the preceding section, which is either partially polarized or not polarized at all?

In answering this question, the method of complex fields is very useful. Although complete information about a fully polarized monochromatic wave is already contained in the real formula (4.14), matters become more transparent if one complexifies the electric field and hence embeds this field in the vector space $V^2(\mathbb{C})$. In this space, for which one may use the "spherical" basis (4.19) or the "linear" basis (4.20), *hermitian* 2×2 matrices

$$\{\mathsf{H} \in M_2(\mathbb{C}) | \mathsf{H}^\dagger = \mathsf{H}\}$$

play a special role. This is so because they are diagonalizable and have real eigenvalues. They can be used as representations of observables in physics.

Remarks

1. An $n \times n$ matrix with complex entries is said to be hermitian if it satisfies the condition

$$\mathsf{H}^\dagger = \mathsf{H} , \quad \text{i.e.} \quad H^*_{ik} = H_{ki} , \quad i,k = 1,2,\ldots,n . \tag{4.21}$$

The "dagger", or cross, symbol denotes the transposed matrix whose entries are replaced by their complex conjugates. The notion of a hermitian matrix is the generalization of the notion of a *symmetric* matrix (whose entries are real) to complex-valued matrices. A hermitian matrix is diagonalized by means of a unitary transformation

$$\overset{0}{\mathsf{H}} = \mathsf{U}\mathsf{H}\mathsf{U}^\dagger \quad \text{with} \quad \mathsf{U}\mathsf{U}^\dagger = \mathbb{1}_n . \tag{4.22}$$

In dimension $n = 2$, this procedure is especially simple. Let the (necessarily real) eigenvalues of H be denoted by λ_1 and λ_2. A unitary transformation leaves the trace and the determinant of H unchanged, i.e.

$$\det(\mathsf{U}\mathsf{H}\mathsf{U}^\dagger) = \det \mathsf{H} = H_{11}H_{22} - |H_{12}|^2 = \lambda_1\lambda_2 \equiv P ,$$
$$\mathrm{tr}(\mathsf{U}\mathsf{H}\mathsf{U}^\dagger) = \mathrm{tr}\,\mathsf{H} = H_{11} + H_{22} = \lambda_1 + \lambda_2 \equiv S .$$

The two eigenvalues to be determined and whose product and sum are known are the roots of the quadratic equation $x^2 - Sx + P = 0$. Hence, they are given by

$$\lambda_1 = \frac{1}{2}\{H_{11} + H_{22} + \sqrt{(H_{11} - H_{22})^2 + 4|H_{12}|^2}\} , \tag{4.23a}$$
$$\lambda_2 = \frac{1}{2}\{H_{11} + H_{22} - \sqrt{(H_{11} - H_{22})^2 + 4|H_{12}|^2}\} . \tag{4.23b}$$

2. In quantum mechanics, such a two-dimensional vector space over the complex numbers $\mathbb{C}$ is used in the description of the spin of electrons or other fermions with spin 1/2. This space may be visualized as the space spanned by the eigenvectors of the 3-component s_3 corresponding to the spin orientations $s_3 = +1/2$ and $s_3 = -1/2$. Observables which act on the spin degrees of freedom only are represented by two-dimensional hermitian matrices. The reader who is familiar with these matters will find the following development a rather natural one.

Every hermitian 2×2 matrix can be represented in a unique manner as a linear combination of the unit matrix $\sigma_0 = \mathbb{1}_2$ and the three Pauli matrices

$$\sigma_1 = \begin{pmatrix} 0 & 1 \\ 1 & 0 \end{pmatrix} , \quad \sigma_2 = \begin{pmatrix} 0 & -\mathrm{i} \\ \mathrm{i} & 0 \end{pmatrix} , \quad \sigma_3 = \begin{pmatrix} 1 & 0 \\ 0 & -1 \end{pmatrix} , \tag{4.24}$$

that is to say,

$$\mathsf{H} = \sum_{\mu=0}^{3} a_\mu \sigma_\mu = \begin{pmatrix} a_0 + a_3 & a_1 - \mathrm{i}a_2 \\ a_1 + \mathrm{i}a_2 & a_0 - a_3 \end{pmatrix} , \qquad (\mathsf{H} = \mathsf{H}^\dagger) .$$

One easily verifies that the vectors given by (4.20) are eigenvectors of σ_3 and belong to the eigenvalues $+1$ and -1, respectively. The vectors of (4.19) are the eigenvectors of σ_2 and correspond to the eigenvalues ± 1 as well:

$$\begin{pmatrix} 0 & -\mathrm{i} \\ \mathrm{i} & 0 \end{pmatrix} \begin{pmatrix} 1 \\ \mathrm{i} \end{pmatrix} = \begin{pmatrix} 1 \\ \mathrm{i} \end{pmatrix} , \quad \begin{pmatrix} 0 & -\mathrm{i} \\ \mathrm{i} & 0 \end{pmatrix} \begin{pmatrix} 1 \\ -\mathrm{i} \end{pmatrix} = -\begin{pmatrix} 1 \\ -\mathrm{i} \end{pmatrix} .$$

In a similar way, one verifies that the vectors

$$\frac{1}{\sqrt{2}}\begin{pmatrix}1\\1\end{pmatrix} \quad \text{and} \quad \frac{1}{\sqrt{2}}\begin{pmatrix}1\\-1\end{pmatrix}$$

are eigenvectors of σ_1 and belong to the eigenvalues $+1$ and -1, respectively.

To relate the polarization of the wave to quantities which one can measure, it is suggestive to calculate the expectation values of the matrices σ_μ with vector (4.15). Using the shorthand

$$\mathcal{E} := \boldsymbol{E}_{\mathrm{c}}(t, \boldsymbol{0}) = \mathrm{e}^{\mathrm{i}\omega t}\begin{pmatrix}\varepsilon_1\\ \varepsilon_2\,\mathrm{e}^{\mathrm{i}\alpha}\end{pmatrix}$$

the scalar products of $\sigma_\mu \mathcal{E}$ with the complex conjugate $\mathcal{E}^*$ of (4.15) are found to be

$$s_0 := \left(\mathcal{E}^*, \sigma_0 \mathcal{E}\right) = \varepsilon_1^2 + \varepsilon_2^2\,, \tag{4.25a}$$

$$s_1 := \left(\mathcal{E}^*, \sigma_1 \mathcal{E}\right) = 2\varepsilon_1\varepsilon_2 \cos\alpha\,, \tag{4.25b}$$

$$s_2 := \left(\mathcal{E}^*, \sigma_2 \mathcal{E}\right) = 2\varepsilon_1\varepsilon_2 \sin\alpha\,, \tag{4.25c}$$

$$s_3 := \left(\mathcal{E}^*, \sigma_3 \mathcal{E}\right) = \varepsilon_1^2 - \varepsilon_2^2\,. \tag{4.25d}$$

As an example, we verify equation (4.25c):

$$\begin{aligned}\left(\mathcal{E}^*, \sigma_2 \mathcal{E}\right) &= \mathrm{e}^{-\mathrm{i}\omega t}\begin{pmatrix}\varepsilon_1 & \varepsilon_2\,\mathrm{e}^{-\mathrm{i}\alpha}\end{pmatrix}\begin{pmatrix}0 & -\mathrm{i}\\ \mathrm{i} & 0\end{pmatrix}\begin{pmatrix}\varepsilon_1\\ \varepsilon_2\,\mathrm{e}^{\mathrm{i}\alpha}\end{pmatrix}\mathrm{e}^{-\mathrm{i}\omega t}\\ &= \begin{pmatrix}\varepsilon_1 & \varepsilon_2\,\mathrm{e}^{-\mathrm{i}\alpha}\end{pmatrix}\begin{pmatrix}-\mathrm{i}\varepsilon_2\,\mathrm{e}^{\mathrm{i}\alpha}\\ \mathrm{i}\varepsilon_1\end{pmatrix}\\ &= \varepsilon_1\varepsilon_2(-\mathrm{i}\mathrm{e}^{\mathrm{i}\alpha} + \mathrm{i}\mathrm{e}^{-\mathrm{i}\alpha}) = 2\varepsilon_1\varepsilon_2 \sin\alpha\,.\end{aligned}$$

The four real parameters s_0, s_1, s_2 and s_3, which are called *Stokes parameters,* provide a complete description of polarization: Three of them are independent. For example, s_0 is a measure of the wave intensity and is related to the other parameters by

$$s_0 = \sqrt{s_1^2 + s_2^2 + s_3^2}\,. \tag{4.26}$$

The remaining three Stokes parameters may serve to extract the requested quantities ε_1, ε_2 and α. Also, the angles ψ and φ defined in Fig. 4.2 can be expressed in terms of Stokes parameters. Indeed, by (4.17) and according to (4.18), one has

$$\sin(2\psi) = \frac{s_2}{s_0}\,, \qquad \tan(2\varphi) = \frac{s_1}{s_3}\,. \tag{4.27}$$

A few special cases follow.

(i) If $\alpha = 0$, then $s_2 = 0$ and $\psi = 0$. This is the case of linear polarization in the direction given by φ.

(ii) If $\alpha = \pm\pi/2$ and if the amplitudes are equal, $\varepsilon_1 = \varepsilon_2$, then one has $s_1 = 0 = s_3$. The angle φ remains undetermined while ψ is equal to $\pm\pi/4$. This is circular polarization which is left- or right-moving, depending on the sign of α. Regarding parameter s_2, one has

$$\frac{s_2^{(+)}}{s_0} = +1 \text{ for right-circular,} \quad \frac{s_2^{(-)}}{s_0} = -1 \text{ for left-circular} \tag{4.28}$$

polarization. The Stokes parameter s_2, when normalized to s_0, corresponds to the two helicity states of the quantized theory.

The usefulness and significance of the Stokes parameters (4.25a)–(4.25d) become obvious in the description of unpolarized or partially polarized light. Indeed, one can talk about polarized electromagnetic waves only if there exist fixed phase relations between, e. g., harmonic solutions such as (4.4a). In turn, unpolarized radiation of a given frequency may be thought of as the result of adding nearly monochromatic wave trains whose phases are completely uncorrelated.[1]

Obviously, two waves whose Stokes parameters take opposite values have opposite polarizations. Equation (4.28) provides an example. The states ($s_1 = s_3 = 0$, $s_2 = \pm 1$) have opposite circular polarizations. Therefore, an unpolarized or partially polarized state will be represented by an *incoherent* mixture of two opposite polarizations, where the relative weights of the two components fix the degree of polarization. An example may help to illustrate this.

The incoherent mixture of a beam with right-circular polarization with weight w_+ and a beam with left-circular polarization with weight w_-, both having the same intensity, gives

$$s_1 = 0 = s_3\,, \quad s_2 = w_+ s_2^{(+)} + w_- s_2^{(-)} = (w_+ - w_-)s_0\,.$$

Referring to the direction of propagation, the degree of polarization is given by

$$P = \frac{w_+ - w_-}{w_+ + w_-}\,.$$

Thus, a beam with $w_+ = 0.7$ and $w_- = 0.3$ has a right-circular polarization of 40%. Another beam with $w_+ = w_- = 1/2$ is completely unpolarized.

Remark

The formal similarity to the description of the polarization of spin-1/2 particles in quantum mechanics is striking. In both cases, optics and quantum theory, one can introduce a density matrix which in the example just described is given by $\varrho = \operatorname{diag}(w_+, w_-)$, under the assumption that the sum of the weights is normalized to 1, $w_+ + w_- = 1$. If, as before, left-circular and right-circular polarization

[1] Such wave trains are modelled by wave packages, i.e. by superposition of plane waves with a k-dependent weight function, arranged in such a way that the contributing wave numbers k' lie in the immediate neighbourhood of k.

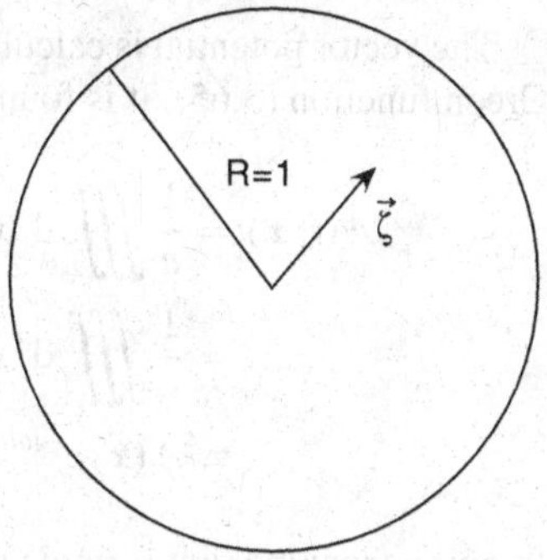

Fig. 4.3 Polarization vector $\boldsymbol{\zeta}$ lies inside a ball with radius 1. Its direction is the diretion of propagation, its modulus the degree of circular polarization

is defined with respect to the direction of propagation $\hat{\boldsymbol{n}}$ but the 3-axis is not chosen along that direction, then the density matrix reads

$$\varrho = \frac{1}{2}\Big[\,\mathbb{1}_2 + (w_+ - w_-)\,\hat{\boldsymbol{n}}\cdot\boldsymbol{\sigma}\Big] \equiv \frac{1}{2}\Big[\,\mathbb{1}_2 + \boldsymbol{\zeta}\cdot\boldsymbol{\sigma}\Big]\,.$$

The symbol $\boldsymbol{\sigma}$ stands for the three Pauli matrices; the scalar product with vector $\boldsymbol{a}$ is to be understood as $\boldsymbol{a}\cdot\boldsymbol{\sigma} = a_1\sigma_1 + a_2\sigma_2 + a_3\sigma_3$. Given the vector $\boldsymbol{\zeta} = (w_+ - w_-)\,\hat{\boldsymbol{n}}$, the direction of propagation is fixed at $\hat{\boldsymbol{n}}$, and its modulus $|\boldsymbol{\zeta}|$ is the degree of circular polarization. This vector is located in the centre of the ball with radius 1 (Fig. 4.3). If its tip touches the surface of the ball, then the beam is fully polarized. If it is inside the ball, then the beam is partially polarized or not polarized at all.

4.3 Simple Radiating Sources

Localized oscillating charge and current densities provide the simplest models for radiating sources. Localization means that the sources occupy a finite domain in space. One then is interested mostly in the waves emitted by them in the outer space, outside the sources. Because of the linearity of Maxwell's equations, it is sufficient to study *harmonic* solutions:

$$\varrho_c(t, \boldsymbol{x}) = \varrho(\boldsymbol{x})\,\mathrm{e}^{-\mathrm{i}\omega t}\,, \tag{4.29a}$$

$$\boldsymbol{j}_c(t, \boldsymbol{x}) = \boldsymbol{j}(\boldsymbol{x})\,\mathrm{e}^{-\mathrm{i}\omega t}\,, \tag{4.29b}$$

though in complexified form. Realistic source distributions are obtained from them by Fourier analysis in the variable t.

The vector potential is calculated within the Lorenz gauge and using the retarded Green function (3.65). It is found to be a time-harmonic function, too, viz.

$$\begin{aligned} \boldsymbol{A}(t,\boldsymbol{x}) &= \frac{1}{c}\iiint \mathrm{d}^3x' \int \mathrm{d}t' \frac{\boldsymbol{j}(t',\boldsymbol{x}')}{|\boldsymbol{x}-\boldsymbol{x}'|}\delta\left(t'-t+\frac{1}{c}|\boldsymbol{x}-\boldsymbol{x}'|\right) \\ &= \frac{1}{c}\iiint \mathrm{d}^3x' \frac{\boldsymbol{j}(\boldsymbol{x}')}{|\boldsymbol{x}-\boldsymbol{x}'|}\mathrm{e}^{\mathrm{i}(\frac{\omega}{c})|\boldsymbol{x}-\boldsymbol{x}'|}\mathrm{e}^{-\mathrm{i}\omega t} \\ &\equiv \boldsymbol{A}(\boldsymbol{x})\mathrm{e}^{-\mathrm{i}\omega t}\,. \end{aligned} \tag{4.30}$$

In going from the first formula to the second, the integral over t' was carried out:

$$\int_{-\infty}^{+\infty} \mathrm{d}t'\, \mathrm{e}^{-\mathrm{i}\omega t'}\delta\left(t'-t+\frac{1}{c}|\boldsymbol{x}-\boldsymbol{x}'|\right) = \mathrm{e}^{-\mathrm{i}\omega t}\,\mathrm{e}^{\mathrm{i}(\frac{\omega}{c})|\boldsymbol{x}-\boldsymbol{x}'|}\,.$$

As the radiation is to be calculated outside the source, i.e. in vacuum where there are no effects of any media, the dispersion relation is $(\omega/c) = k$. Furthermore, $\nabla\times\boldsymbol{H}-(1/c)\dot{\boldsymbol{E}} = 0$, and therefore

$$\dot{\boldsymbol{E}}(t,\boldsymbol{x}) = c\nabla\times\boldsymbol{B}(t,\boldsymbol{x}) = \frac{\mathrm{i}}{k}\nabla\times\dot{\boldsymbol{B}}(t,\boldsymbol{x})\,.$$

Regarding the factors in the Fourier components of the fields which depend on $\boldsymbol{x}$ only,

$$\boldsymbol{E}_{\mathrm{c}}(t,\boldsymbol{x}) = \boldsymbol{E}(\boldsymbol{x})\mathrm{e}^{-\mathrm{i}\omega t}\,,\quad \boldsymbol{B}_{\mathrm{c}}(t,\boldsymbol{x}) = \boldsymbol{B}(\boldsymbol{x})\mathrm{e}^{-\mathrm{i}\omega t}$$

one must solve equations which depend on $\boldsymbol{x}$ only:

$$\boldsymbol{E}(\boldsymbol{x}) = \frac{\mathrm{i}}{k}\nabla\times\boldsymbol{B}(\boldsymbol{x})\,, \tag{4.31a}$$

$$\boldsymbol{B}(\boldsymbol{x}) = \nabla\times\boldsymbol{A}(\boldsymbol{x})\,. \tag{4.31b}$$

The individual steps in the calculation are now clearly defined: Starting from the given current density $\boldsymbol{j}(\boldsymbol{x})$, one calculates the vector field $\boldsymbol{A}(\boldsymbol{x})$ by means of equation (4.30). The physical fields are obtained from (4.31a) and (4.31b).

4.3.1 Typical Dimensions of Radiating Sources

Let the source be localized and let its typical extension be d. In the description and the mathematical treatment of sources of this kind, the decisive parameter is the ratio of the wavelength $\lambda = 2\pi/k$ to the dimension d. In particular, one must distinguish the case where λ is much greater than d from a situation where the two parameters are of the same order of magnitude.

Ordinary Atoms

The atoms we find in nature have spatial extensions of the order of magnitude of the Bohr radius

$$a_B = \frac{\hbar^2}{e^2 m_e} \equiv \frac{\hbar c}{(\frac{e^2}{\hbar c}) m_e c^2} \simeq 5.3 \cdot 10^{-11}\ \text{m}\,,$$

$$\text{with} \quad \frac{e^2}{\hbar c} \simeq \frac{1}{137}\,, \quad m_e c^2 \simeq 0.511\ \text{MeV}\,, \quad \hbar c \simeq 197.3 \cdot 10^{-15}\ \text{MeVm}\,.$$

The typical radiation emitted by atoms lies in the visible range, and a good orientation is provided by a wavelength on the order of 1000 Å $= 10^{-7}$ m. Thus, in this case $\lambda \gg d$, the emitted wavelength is much larger than the typical dimension of the source. As a consequence, electric dipole transitions dominate in atoms and are found with much higher intensity than those with higher multiplicities.

Muonic Atoms

The Bohr radius for muons is smaller than that of electrons by the ratio m_e/m_μ. The binding energies in hydrogen-like atoms in which the electron is replaced by a muon, and likewise in turn the transition energies, are larger by the reciprocal of this value. The wavelengths of the emitted radiation are correspondingly smaller. In general, the wavelength λ will be barely larger than, or even comparable to, the dimension d of the source. What held in electronic atoms is only marginally true in muonic atoms: Though electric dipole transitions are important, higher multipoles such as, e. g., electric quadrupoles are seen with sizeable intensities.

Atomic Nuclei

Atomic nuclei have spatial extensions on the order of magnitude $d = 10^{-15}$ to 10^{-14} m, and transitions between different states of nuclei in which γ-rays are emitted correspond to energy differences of a few megaelectron volts. Therefore, typical wavelengths are $\lambda = 2\pi(\hbar c/E) \gtrsim 10^{-12}$ m, which is no longer significantly larger than the typical dimension of the source. As in muonic atoms, besides the dipoles, one observes higher multipolarities.

Classical Macroscopic Sources

In the case of macroscopic senders and their antennas, the wavelength, in general, is large compared to their physical extension $\lambda \gg d$. As we are talking here about macroscopic scales where practical measurements may well fall short of these, one must compare the distance r of the observer from the source, both with the typical extension d and with the wavelength λ. This is in contrast to the microscopic systems discussed previously where the observation is always done at distances which are larger than the extension of the source. Therefore, we distinguish three different situations:

(A) $d \ll r \ll \lambda$: The so-called *near-field zone*, or *static zone*, where an observer sits at a distance at which the source looks pointlike to him yet close enough to be still far from the first oscillation node;
(B) $d \ll \lambda \ll r$: The *far-field zone*, or *radiation zone*, where the observer sees the source as pointlike and is far enough to see the fully developed wave;
(C) $d \ll r \simeq \lambda$: The intermediate domain where the distance from the source is comparable with the wavelength, also called the *induction zone.*

A radio receiver, in general, will be located in the far-field zone of a given radio station. In turn, if one makes measurements in the vicinity of a long-wave radio station, then one is located rather in the near-field zone or the intermediate domain between near-field and far-field zones.

4.3.2 Description by Means of Multipole Radiation

A very useful framework for calculating the radiation emitted by a given source is provided by the method of multipole moments. In Sect. 1.8.4, this method was developed for electrostatics. For the static problems discussed there we constructed a set of base solutions of the Laplace equation using spherical harmonics. More general solutions were obtained by expansion in terms of base solutions, as indicated by (1.103). In the problems to be discussed in this section, the aim is to apply this method to the Helmholtz equation (4.8), that is, to the differential equation

$$[\boldsymbol{\Delta} + k^2]\widetilde{f}(k, \boldsymbol{x}) = 0 \,. \tag{4.32}$$

The procedure is similar to the case of electrostatics. One starts from a factorizing ansatz of the type of (1.95), with spherical harmonics $Y_{\ell m}(\theta, \phi)$ in the angular variables and with radial functions $f_\ell(r)$, such that

$$\widetilde{f}(k, \boldsymbol{x}) = \sum_{\ell=0}^{\infty} \sum_{m=-\ell}^{+\ell} f_\ell(k, r) Y_{\ell m}(\theta, \phi) \,.$$

One shows that the radial functions satisfy the ordinary differential equation

$$\left\{ \frac{1}{r^2} \frac{\mathrm{d}}{\mathrm{d}r} \left(r^2 \frac{\mathrm{d}}{\mathrm{d}r} \right) - \frac{\ell(\ell+1)}{r^2} + k^2 \right\} f_\ell(k, r) = 0 \,. \tag{4.33}$$

As one verifies immediately, the previously known static solutions $f_\ell^{\text{stat}} = r^\ell$ and $f_\ell^{\text{stat}} = r^{-\ell-1}$ satisfy this differential equation if $k^2 = 0$.

For $k \neq 0$ one introduces the dimensionless argument $z := kr$ and derives a differential equation which replaces (4.33), viz.

$$\left\{\frac{d^2}{z^2} + \frac{2}{z}\frac{d}{dz} - \frac{\ell(\ell+1)}{z^2} + 1\right\} f_\ell(z) = 0 . \tag{4.34}$$

This is a well-known differential equation of the theory of Bessel functions. A system of fundamental solutions which is well adapted to the problem posed in this section is given by

$$f_\ell^{(1)}(kr) = j_\ell(kr) , \tag{4.35a}$$

$$f_\ell^{(2)}(kr) = h_\ell^{(1)}(kr) = j_\ell(kr) + i n_\ell(kr) . \tag{4.35b}$$

Here $j_\ell(kr)$ are the spherical Bessel functions, $n_\ell(kr)$ the spherical Neumann functions, for which the formulae

$$j_\ell(z) = (-z)^\ell \left(\frac{1}{z}\frac{d}{dz}\right)^\ell \frac{\sin z}{z} , \tag{4.36a}$$

$$n_\ell(z) = -(-z)^\ell \left(\frac{1}{z}\frac{d}{dz}\right)^\ell \frac{\cos z}{z} \tag{4.36b}$$

provide a useful representation. The function $h_\ell^{(1)}$ denotes one of the two so-called spherical Hankel functions. This class of special functions is used in treating problems with central fields in quantum mechanics (cf., e. g., [QP]). The functions $h_\ell^{(1)}(z)$ are called spherical Hankel functions of the first kind.[2]

Here are a few examples:

$$j_0(z) = \frac{\sin z}{z} , \tag{4.37a}$$

$$j_1(z) = -\frac{d}{dz}\frac{\sin z}{z} = \frac{\sin z}{z^2} - \frac{\cos z}{z} , \tag{4.37b}$$

$$j_2(z) = z\frac{d}{dz}\frac{1}{z}\frac{d}{dz}\frac{\sin z}{z} = \frac{\sin z}{z}\left(\frac{3}{z^2} - 1\right) - \frac{3\cos z}{z^2} . \tag{4.37c}$$

$$n_0(z) = -\frac{\cos z}{z} , \tag{4.38a}$$

$$n_1(z) = -\frac{\cos z}{z^2} - \frac{\sin z}{z} , \tag{4.38b}$$

$$n_2(z) = -\frac{\cos z}{z}\left(\frac{3}{z^2} - 1\right) - \frac{3\sin z}{z^2} . \tag{4.38c}$$

---

[2] Another notation for spherical Hankel functions is $h_\ell^{(+)}$, which differs from the one in the main text by a factor i: $h_\ell^{(+)}(z) = i h_\ell^{(1)}(z)$.

Of special importance for what follows is the behaviour of these functions at $r \to 0$ and for $r \to \infty$, which we list here:

$$(kr) \ll 1 : j_\ell(kr) \sim \frac{(kr)^\ell}{(2\ell+1)!!}, \quad n_\ell^{(1)} \sim -\frac{(2\ell-1)!!}{(kr)^{\ell+1}}, \tag{4.39a}$$

$$h_\ell^{(1)} \sim -\mathrm{i}\frac{(2\ell-1)!!}{(kr)^{\ell+1}}, \tag{4.39b}$$

where $(2\ell+1)!! = 1 \cdot 3 \cdot 5 \cdots (2\ell+1)$ is the double factorial, and

$$kr \gg 1 : j_\ell(kr) \sim \frac{1}{kr}\sin\left(kr - \ell\frac{\pi}{2}\right), \quad n_\ell(kr) \sim -\frac{1}{kr}\cos\left(kr - \ell\frac{\pi}{2}\right), \tag{4.40a}$$

$$h_\ell^{(1)}(kr) \sim (-\mathrm{i})^{\ell+1}\frac{\mathrm{e}^{\mathrm{i}kr}}{kr}. \tag{4.40b}$$

The generalization of expansion (1.105) to Green functions with $k \neq 0$ reads as follows:

$$\frac{\mathrm{e}^{\mathrm{i}k|\boldsymbol{x}-\boldsymbol{x}'|}}{4\pi|\boldsymbol{x}-\boldsymbol{x}'|} = \mathrm{i}k\sum_{\ell=0}^{\infty} j_\ell(kr_<)h_\ell^{(1)}(kr_>)\sum_{m=-\ell}^{+\ell} Y_{\ell m}^*(\hat{\boldsymbol{x}}')Y_{\ell m}(\hat{\boldsymbol{x}}). \tag{4.41}$$

As before, the notations $r_<$ and $r_>$ indicate that of the two radial variables $r = |\boldsymbol{x}|$ and $r' = |\boldsymbol{x}'|$, the *smaller* of them must be inserted into the Bessel function, and the *larger* of them must be inserted into the Hankel function. As a shorthand, the angular coordinates (θ, ϕ) of $\boldsymbol{x}$, and (θ', ϕ') of $\boldsymbol{x}'$ are denoted by the unit vectors $\hat{\boldsymbol{x}}$ and $\hat{\boldsymbol{x}}'$, respectively.

One confirms that for $k = 0$ expansion (4.41) goes over into formula (1.105). In this limit, one has $(kr) \ll 1$, so that one can insert estimates (4.39a) and (4.39b):

$$\begin{aligned}
(kr) \to 0 : &\frac{\mathrm{e}^{\mathrm{i}k|\boldsymbol{x}-\boldsymbol{x}'|}}{4\pi|\boldsymbol{x}-\boldsymbol{x}'|} \\
&\sim \mathrm{i}k\sum_{\ell=0}^{\infty}\frac{(kr_<)^\ell}{(2\ell+1)!!}(-\mathrm{i})\frac{(2\ell-1)!!}{(kr_>)^{\ell+1}}\sum_{m=-\ell}^{+\ell} Y_{\ell m}^*(\hat{\boldsymbol{x}}')Y_{\ell m}(\hat{\boldsymbol{x}}) \\
&= \sum_{\ell=0}^{\infty}\frac{r_<^\ell}{r_>^{\ell+1}}\frac{1}{2\ell+1}\sum_{m=-\ell}^{+\ell} Y_{\ell m}^*(\hat{\boldsymbol{x}}')Y_{\ell m}(\hat{\boldsymbol{x}}).
\end{aligned}$$

This agrees with formula (1.105) of electrostatics.

We comment on the proof of (4.41) as follows. The left-hand side is a Green function of the Helmholtz equation (4.32). Therefore, the radial functions contained in (4.41) must be solutions of the differential equation (4.34), i.e. they must be linear combinations of spherical Bessel and Neumann functions. The limit $k \to 0$ tells us which are the ones with the argument $kr_<$ and which are the ones with the argument $kr_>$. As the expansion in terms of spherical harmonics is unique, expression (4.41) holds true.

Expansion (4.41) can be inserted into expression (4.30) for the vector potential. For the domain *outside* the sources one obtains in this way

$$A(x) = \frac{4\pi i k}{c} \sum_{\ell=0}^{\infty} h_\ell^{(1)}(kr) \sum_{m=-\ell}^{+\ell} Y_{\ell m}(\hat{x}) \cdot \int_0^\infty r'^2 \mathrm{d}r' \int \mathrm{d}\Omega' \, j(x') j_\ell(kr') Y^*_{\ell m}(\hat{x}') \,. \tag{4.42}$$

If the measuring point x is also allowed to lie inside the source, then the two cases $r > r'$ and $r < r'$ must be distinguished, and the integral over r' must be split in analogy to (1.106b). For instance, in expression (4.42), which applies to values of x outside the source, the assignments

$$r_> = r = |x| \quad \text{and} \quad r_< = r' = |x'|$$

must be chosen. The vector potential (4.42) is a sum of products of two factors each, one of which depends on the measuring point x only, while the other factor,

$$\int_0^\infty r'^2 \mathrm{d}r' \iint \mathrm{d}\Omega' \, j(x') j_\ell(kr') Y^*_{\ell m}(\hat{x}') \,, \tag{4.43}$$

depends on the source distribution only. This shows that the factors given by (4.43) are generalizations of the multipole moments (1.106d).

Remarks

1. In the near-field zone $d \ll r \ll \lambda$, the product kr is small compared to 1. Therefore, in (4.30), one can approximate

$$\mathrm{e}^{ik|x-x'|} \simeq 1$$

or, as regards the multipole expansion (4.42), replace the spherical Bessel and Hankel functions by their approximations (4.39a) and (4.39b), respectively, so that

$$A(x) \simeq \frac{4\pi}{c} \sum_{\ell,m} \frac{1}{2\ell+1} \frac{Y_{\ell m}(\hat{x})}{r^{\ell+1}} \int_0^\infty r'^2 \mathrm{d}r' \int \mathrm{d}\Omega' \, r'^\ell j(x') Y^*_{\ell m}(\hat{x}') \,. \tag{4.44}$$

The interpretation of these formulae in terms of physics is that retardation effects are negligible in the near-field zone and that the conditions are nearly *static* ones. Except for the harmonic time dependence, fields $\boldsymbol{E}$ and $\boldsymbol{B}$ are static.

2. Also, in the far-field zone $r \gg \lambda$, the conditions are simpler than in the intermediate zone. For example, returning to expression (4.30), one has $r \gg r'$, and one can expand in terms of r'/r

$$\left|\boldsymbol{x} - \boldsymbol{x}'\right| \simeq r - \hat{\boldsymbol{n}} \cdot \boldsymbol{x}' , \quad \hat{\boldsymbol{n}} = \frac{\boldsymbol{x}}{r} ,$$

so that in the limit $(kr) \to \infty$ one obtains

$$\begin{aligned} \boldsymbol{A}(\boldsymbol{x}) &\simeq \frac{\mathrm{e}^{ikr}}{cr} \iiint \mathrm{d}^3x' \, \boldsymbol{j}(\boldsymbol{x}') \mathrm{e}^{-ik\hat{\boldsymbol{n}} \cdot \boldsymbol{x}'} \\ &= \frac{\mathrm{e}^{ikr}}{cr} \sum_{\mu=0}^{\infty} \frac{(-\mathrm{i}k)^\mu}{\mu!} \iiint \mathrm{d}^3x' \, \boldsymbol{j}(\boldsymbol{x}')(\hat{\boldsymbol{n}} \cdot \boldsymbol{x}')^\mu . \end{aligned} \tag{4.45}$$

As the wavelength is large compared to the spatial extension of the source, $\lambda \gg d$, the product kd is small compared to 1. Therefore, the series in expression (4.45) above converges rapidly and is dominated by the first nonvanishing term.

4.3.3 The Hertzian Dipole

As an application of the general decomposition (4.42), consider the term with $\ell = 0$, noticing that

$$h_0^{(1)}(kr) = \frac{\mathrm{e}^{ikr}}{\mathrm{i}kr} , \quad j_0(kr) = \frac{\sin(kr)}{kr} .$$

If the source is nearly pointlike, then the integral in (4.43) is dominated by values of r' for which $kr' \ll 1$. Therefore, one can set $j_0(kr') \simeq 1$ so that one obtains

$$\boldsymbol{A}(\boldsymbol{x}) \simeq \frac{4\pi}{c} \frac{\mathrm{e}^{ikr}}{r} \iiint \mathrm{d}^3x' \, \boldsymbol{j}(\boldsymbol{x}') \frac{1}{4\pi} .$$

(Note that $Y_{00} = 1/\sqrt{4\pi}$ is inserted here.) By means of the continuity equation, the integral over the current density is replaced by an integral over the charge density. With harmonic time dependence the continuity equation yields

$$\boldsymbol{\nabla} \cdot \boldsymbol{j}(t, \boldsymbol{x}) + \frac{\partial \varrho(t, \boldsymbol{x})}{\partial t} = 0 \quad \Longrightarrow \quad \boldsymbol{\nabla} \cdot \boldsymbol{j}(\boldsymbol{x}) - \mathrm{i}\omega\varrho(\boldsymbol{x}) = 0 .$$

As the current density vanishes at infinity sufficiently fast, partial integration and insertion of the last equation yield

$$\iiint \mathrm{d}^3x' \, \boldsymbol{j}(\boldsymbol{x}') = -\iiint \mathrm{d}^3x' \, \boldsymbol{x}'(\nabla \cdot \boldsymbol{j}(\boldsymbol{x}'))$$

$$= -\mathrm{i}\omega \iiint \mathrm{d}^3x' \, \boldsymbol{x}' \varrho(\boldsymbol{x}') \, .$$

The right-hand side contains the dipole moment (1.109c)

$$\boldsymbol{d} = \iiint \mathrm{d}^3x' \, \boldsymbol{x}' \varrho(\boldsymbol{x}') \, , \tag{4.46}$$

which is known from electrostatics. Thus, one obtains a particularly simple expression for the $\boldsymbol{x}$ dependent factor in the vector potential, viz.

$$\boldsymbol{A}(\boldsymbol{x}) \simeq -\mathrm{i}k \frac{\mathrm{e}^{\mathrm{i}kr}}{r} \boldsymbol{d} \, , \qquad \left(k = \frac{\omega}{c} = \frac{2\pi}{\lambda} \right) , \tag{4.47}$$

from which one then derives the magnetic induction $\boldsymbol{B}_{\mathrm{c}}(\boldsymbol{x}) = \nabla \times \boldsymbol{A}(\boldsymbol{x})$ and the electric field $\boldsymbol{E}_{\mathrm{c}}(\boldsymbol{x}) = (\mathrm{i}/k) \nabla \times \boldsymbol{B}_{\mathrm{c}}(\boldsymbol{x})$. After a short calculation, one obtains

$$\boldsymbol{B}_{\mathrm{c}}(\boldsymbol{x}) = k^2 \frac{\mathrm{e}^{\mathrm{i}kr}}{r} \left(1 - \frac{1}{\mathrm{i}kr} \right) \hat{\boldsymbol{x}} \times \boldsymbol{d} \, , \tag{4.48}$$

$$\boldsymbol{E}_{\mathrm{c}}(\boldsymbol{x}) = k^2 \frac{\mathrm{e}^{\mathrm{i}kr}}{r} (\hat{\boldsymbol{x}} \times \boldsymbol{d}) \times \hat{\boldsymbol{x}} + \frac{\mathrm{e}^{\mathrm{i}kr}}{r^3} (1 - \mathrm{i}kr) \left[3\hat{\boldsymbol{x}}(\hat{\boldsymbol{x}} \cdot \boldsymbol{d}) - \boldsymbol{d} \right] . \tag{4.49}$$

Equipped with their harmonic time dependence the fields are

$$\boldsymbol{B}_{\mathrm{c}}(t, \boldsymbol{x}) = \mathrm{e}^{-\mathrm{i}\omega t} \boldsymbol{B}_{\mathrm{c}}(\boldsymbol{x}) \, , \quad \boldsymbol{E}_{\mathrm{c}}(t, \boldsymbol{x}) = \mathrm{e}^{-\mathrm{i}\omega t} \boldsymbol{E}_{\mathrm{c}}(\boldsymbol{x}) \, .$$

Finally, the physically realized fields are the real parts thereof,

$$\boldsymbol{B}(t, \boldsymbol{x}) = \mathrm{Re}\, \boldsymbol{B}_{\mathrm{c}}(t, \boldsymbol{x}) \, , \qquad \boldsymbol{E}(t, \boldsymbol{x}) = \mathrm{Re}\, \boldsymbol{E}_{\mathrm{c}}(t, \boldsymbol{x}) \, .$$

Before turning to an interpretation of these solutions, we give here a few steps of the calculation that leads from (4.47) to the complex fields of (4.48) and (4.49). First one must calculate

$$\boldsymbol{B}_{\mathrm{c}} = \nabla \times \boldsymbol{A} = -\mathrm{i}k \left(\nabla \times \boldsymbol{d} \, \frac{\mathrm{e}^{\mathrm{i}kr}}{r} \right) .$$

With $\frac{\partial f(r)}{\partial x^i} = \left(\frac{\partial f(r)}{\partial r} \right) (x^i/r)$ and with $\hat{\boldsymbol{x}} = \boldsymbol{x}/r$ one derives

$$\boldsymbol{B}_{\mathrm{c}} = -\mathrm{i}k \hat{\boldsymbol{x}} \times \boldsymbol{d} \left(-\frac{\mathrm{e}^{\mathrm{i}kr}}{r^2} + \mathrm{i}k \frac{\mathrm{e}^{\mathrm{i}kr}}{r} \right) \equiv -\mathrm{i}k \hat{\boldsymbol{x}} \times \boldsymbol{d} \, g(r) \, ,$$

where the following abbreviation is introduced:

$$g(r) = -\frac{e^{ikr}}{r^2} + ik\frac{e^{ikr}}{r} .$$

Note that this is precisely the result given by (4.48).

The electric field follows from this by applying once more the curl operator, $\boldsymbol{E}_c = (i/k)\nabla \times \boldsymbol{B}_c$. This is calculated by means of the formula

$$\left(\nabla \times (\boldsymbol{a} \times \boldsymbol{b})\right)_i = \varepsilon_{ikm}\varepsilon_{mnp}\partial_k a_n b_p$$

– summed over all repeated indices – and of the identity

$$\varepsilon_{ikm}\varepsilon_{mnp} = \delta_{in}\delta_{kp} - \delta_{ip}\delta_{kn} .$$

If the vector fields $\boldsymbol{a}$ and $\boldsymbol{b}$ depend on $\boldsymbol{x}$, this gives

$$\nabla \times (\boldsymbol{a} \times \boldsymbol{b}) = (\partial_k \boldsymbol{a})b_k - (\partial_n a_n)\boldsymbol{b} + \boldsymbol{a}(\partial_k b_k) - a_n(\partial_n \boldsymbol{b}) .$$

Abbreviating the r-dependent factor by $g(r)$, as previously, one has in the case at hand

$$\nabla \times (\hat{\boldsymbol{x}} \times \boldsymbol{d}\, g(r)) = \left(\boldsymbol{d}\, g(r) \cdot \nabla\right)\hat{\boldsymbol{x}} - \boldsymbol{d}\, g(r)\left(\nabla \cdot \hat{\boldsymbol{x}}\right) + \hat{\boldsymbol{x}}\left(\nabla \cdot \boldsymbol{d}\, g(r)\right) - \left(\hat{\boldsymbol{x}} \cdot \nabla g(r)\right)\boldsymbol{d} .$$

Finally, using the formulae

$$\frac{\partial}{x^i}\frac{x^j}{r} = \frac{1}{\partial r}\delta_{ij} - \frac{x^i x^j}{r^3} , \qquad \nabla \cdot \frac{\boldsymbol{x}}{r} = \frac{2}{r} ,$$

which are known to the reader or else are easy to derive, the first two terms on the right-hand side yield

$$\left(\boldsymbol{d}\, g(r) \cdot \nabla\right)\hat{\boldsymbol{x}} - \boldsymbol{d}\, g(r)\left(\nabla \cdot \hat{\boldsymbol{x}}\right) = -\frac{g(r)}{r}(\boldsymbol{d} \cdot \hat{\boldsymbol{x}})\hat{\boldsymbol{x}} - \frac{g(r)}{r}\boldsymbol{d} .$$

The third and fourth terms, taken together, give

$$\hat{\boldsymbol{x}}\left(\nabla \cdot \boldsymbol{d}\, g(r)\right) - \left(\hat{\boldsymbol{x}} \cdot \nabla g(r)\right)\boldsymbol{d} = (\boldsymbol{d} \cdot \hat{\boldsymbol{x}})\hat{\boldsymbol{x}}\frac{\partial g(r)}{\partial r} - \boldsymbol{d}\frac{\partial g(r)}{\partial r} = -\hat{\boldsymbol{x}} \times (\boldsymbol{d} \times \hat{\boldsymbol{x}})\frac{\partial g(r)}{\partial r} ,$$

where the identity $\hat{\boldsymbol{x}} \times (\boldsymbol{d} \times \hat{\boldsymbol{x}}) = \boldsymbol{d} - \hat{\boldsymbol{x}}(\hat{\boldsymbol{x}} \cdot \boldsymbol{p})$ was inserted. Finally, using this identity once more in the preceding terms, and inserting the explicit expression for $g(r)$ and its derivative, the result for electric field (4.49) follows.

Hertzian Dipole in the Far Zone

Far outside the source and at distances which are large compared to the wavelength emitted by the Hertz dipole, only the first term on the right-hand side of (4.48) contributes. One finds

$$\boldsymbol{B}_{\mathrm{c}}(\boldsymbol{x}) \simeq k^2 \frac{\mathrm{e}^{ikr}}{r} \hat{\boldsymbol{x}} \times \boldsymbol{d} \,, \tag{4.50}$$

$$\boldsymbol{E}_{\mathrm{c}}(\boldsymbol{x}) \simeq \boldsymbol{B}_{\mathrm{c}}(\boldsymbol{x}) \times \hat{\boldsymbol{x}} \,. \tag{4.51}$$

The electric field and the magnetic induction field, which are given by the real parts of these expressions, respectively, oscillate *in phase* and their magnitudes are of the same order. Both fields are perpendicular to the direction of propagation $\hat{\boldsymbol{x}}$ and both decrease like $1/r$.

Hertzian Dipole in the Near Zone

Because we assumed the dipole to be pointlike, the variable r is still large compared to the dimension d of the emitter, but at the same time it is small compared to the wavelength. Thus, as for formula (4.44) the approximation $j_0(kr) \simeq 1$ is justified. Hence, the fields are

$$\boldsymbol{B}_{\mathrm{c}}(\boldsymbol{x}) \simeq \mathrm{i}k \frac{1}{r^2} \hat{\boldsymbol{x}} \times \boldsymbol{d} \,, \tag{4.52}$$

$$\boldsymbol{E}_{\mathrm{c}}(\boldsymbol{x}) \simeq \frac{1}{r^3} \left[3\hat{\boldsymbol{x}}(\hat{\boldsymbol{x}} \cdot \boldsymbol{d}) - \boldsymbol{d} \right] . \tag{4.53}$$

As expected, the electric field is static, but for the harmonic time dependence. It equals the field of a static electric dipole. The magnitude of the magnetic induction is smaller by a factor of (kr) than that of the electric field. Furthermore, the two physical fields have a relative phase shift of $\pi/2$. Indeed,

$$\boldsymbol{E}(t, \boldsymbol{x}) = \mathrm{Re}\left(\mathrm{e}^{\mathrm{i}\omega t} \boldsymbol{E}_{\mathrm{c}}(\boldsymbol{x})\right) \propto \cos(\omega t) \,,$$

$$\boldsymbol{B}(t, \boldsymbol{x}) = \mathrm{Re}\left(\mathrm{e}^{\mathrm{i}\omega t} \boldsymbol{B}_{\mathrm{c}}(\boldsymbol{x})\right) \propto -\sin(\omega t) \,.$$

It is instructive to calculate the power of a Hertzian sender. Enclosing the dipole by a sphere with radius r (it is sufficient to choose this radius in the far zone), the power radiated into the solid angle $\mathrm{d}\Omega$ is equal to the surface element $r^2\,\mathrm{d}\Omega$ multiplied by the mean value of the radial component of the Poynting vector

$$\begin{aligned}\frac{\mathrm{d}\overline{W}}{\mathrm{d}\Omega} &= r^2\,(\hat{\boldsymbol{r}} \cdot \langle \boldsymbol{S} \rangle) = \frac{r^2 c}{16\pi} \hat{\boldsymbol{r}} \{ \boldsymbol{E}_{\mathrm{c}}^* \times \boldsymbol{B}_{\mathrm{c}} + \boldsymbol{E}_{\mathrm{c}} \times \boldsymbol{B}_{\mathrm{c}}^* \} \\ &= \frac{c}{8\pi} k^4 \left| (\hat{\boldsymbol{r}} \times \boldsymbol{d}) \times \hat{\boldsymbol{r}} \right|^2 = \frac{c}{8\pi} k^4 d^2 \sin^2\theta \,.\end{aligned} \tag{4.54}$$

The total time-averaged power is given by the integral of this expression:

$$\overline{W} = \iint \mathrm{d}\Omega \, \frac{\mathrm{d}\overline{W}}{\mathrm{d}\Omega} = \frac{c}{8\pi} k^4 d^2 2\pi \left(2 - \frac{2}{3} \right) = \frac{c}{3} k^4 d^2 \,. \tag{4.55a}$$

The power radiated during one period of oscillation is given by

$$T\overline{W} = \frac{2\pi}{ck}\overline{W} = \frac{16\pi^4}{3}\frac{1}{\lambda^3}\boldsymbol{d}^2 , \tag{4.55b}$$

where we inserted the relation $k = 2\pi/\lambda$ between wavelength and wave number.

► **Remarks**

1. Harmonically oscillating sources cannot produce an electric monopole field. The scalar potential

$$\Phi(t, \boldsymbol{x}) = \int \mathrm{d}t' \iiint \mathrm{d}^3x' \, \frac{\varrho(t', \boldsymbol{x}')}{|\boldsymbol{x} - \boldsymbol{x}'|} \delta\left(t' - t + \frac{1}{c}|\boldsymbol{x} - \boldsymbol{x}'|\right)$$

contains the term with $\ell = 0$

$$\Phi_{\text{monopole}}(t, \boldsymbol{x}) = \frac{q(t' = t - r/c)}{r} = \frac{q}{r} ,$$

which is independent of time because the electric charge is conserved.

2. Similarly, the term with $\ell = 1$ of (4.42) with a pointlike source is easy to interpret. One uses the formula

$$h_1^{(1)}(kr) = (-\mathrm{i})\frac{\mathrm{e}^{\mathrm{i}kr}}{kr}\left(\frac{1}{kr} - \mathrm{i}\right)$$

for the Hankel function. In integrating over the source, one can use the approximation

$$j_1(kr') \simeq \frac{kr'}{3} ,$$

valid for $(kr') \ll 1$. Then one has

$$\boldsymbol{A}(\boldsymbol{x}) \simeq \frac{4\pi}{3cr}\left(\frac{1}{r} - \mathrm{i}k\right)\sum_m Y_{1m}(\hat{\boldsymbol{x}}) \int_0^\infty r'^2 \,\mathrm{d}r' \iint \mathrm{d}\Omega' \, \boldsymbol{j}(\boldsymbol{x}')r'Y_{1m}^*(\hat{\boldsymbol{x}}') . \tag{4.56}$$

Except for a factor which contains some constants and the radial variable r, this is seen to be the scalar product of $\hat{\boldsymbol{x}}$, the unit vector in the direction of propagation, and $\boldsymbol{x}'$,

$$\frac{4\pi}{3}\sum_m Y_{1m}(\hat{\boldsymbol{x}})r'Y_{1m}^*(\hat{\boldsymbol{x}}') = \hat{\boldsymbol{x}} \cdot \boldsymbol{x}' ,$$

expressed once in spherical coordinates, once in Cartesian coordinates. Using the identity $(\boldsymbol{a} \times \boldsymbol{b}) \times \boldsymbol{c} = \boldsymbol{b}(\boldsymbol{a} \cdot \boldsymbol{c}) - \boldsymbol{a}(\boldsymbol{b} \cdot \boldsymbol{c})$, one obtains the relation

$$\frac{1}{c}(\hat{\boldsymbol{x}} \times \boldsymbol{x}')\boldsymbol{j} = \frac{1}{2c}[(\hat{\boldsymbol{x}} \cdot \boldsymbol{x}')\boldsymbol{j} + (\hat{\boldsymbol{x}} \cdot \boldsymbol{j})\boldsymbol{x}'] + \frac{1}{2c}(\boldsymbol{x}' \times \boldsymbol{j}) \times \hat{\boldsymbol{x}} . \tag{4.57}$$

The first term on the right-hand side of (4.57) contains an electric quadrupole density, and the second term contains the magnetic dipole density, by now well known from (1.120a) whose integral over the source yields the magnetic moment $\boldsymbol{\mu}$, (1.120b). Isolating this term, one has

$$\boldsymbol{A}_{\text{magn}}(\boldsymbol{x}) = \mathrm{i}k\,\hat{\boldsymbol{x}} \times \boldsymbol{\mu}\frac{\mathrm{e}^{\mathrm{i}kr}}{r}\left(1 - \frac{1}{\mathrm{i}kr}\right). \tag{4.58}$$

This *magnetic* dipole potential is the analogue of the Hertzian dipole, which is an oscillating *electric* dipole. It is interesting to note that the potential has the same form as the induction field, (4.48), of the electric dipole. The corresponding physical fields formally resemble the fields given by (4.49) and (4.48), respectively:

$$\boldsymbol{B}_{\text{c}}^{\text{magn}}(\boldsymbol{x}) = k^2\frac{\mathrm{e}^{\mathrm{i}kr}}{r}(\hat{\boldsymbol{x}} \times \boldsymbol{\mu}) \times \hat{\boldsymbol{x}} + \frac{\mathrm{e}^{\mathrm{i}kr}}{r^3}(1 - \mathrm{i}kr)\left[3\hat{\boldsymbol{x}}(\hat{\boldsymbol{x}} \cdot \boldsymbol{\mu}) - \boldsymbol{\mu}\right], \tag{4.59}$$

$$\boldsymbol{E}_{\text{c}}^{\text{magn}}(\boldsymbol{x}) = -k^2\frac{\mathrm{e}^{\mathrm{i}kr}}{r}\left(1 - \frac{1}{\mathrm{i}kr}\right)\hat{\boldsymbol{x}} \times \boldsymbol{\mu}. \tag{4.60}$$

These expressions illustrate well the close relationship of the electric and the magnetic dipole sender.

4.4 Refraction of Harmonic Waves

In this section, we study the refraction of a harmonic wave at the interface between two isolating homogeneous media with dielectric constants ε and ε' and with magnetic permeabilities μ and μ', respectively. The 3-axis in $\mathbb{R}^3$ is chosen perpendicular to the boundary plane. A plane wave with wave vector $\boldsymbol{k}$ incident on this plane splits into a refracted beam in the neighbouring medium and a reflected beam in the first medium.

4.4.1 Index of Refraction and Angular Relations

Without loss of generality we choose the 1- and 2-axes in the boundary plane such that $\boldsymbol{k}$ lies in the (1, 3)-plane. With the well-known relation between wave number $k = |\boldsymbol{k}|$ and wavelength λ, and using (4.9), one has

$$k = \frac{2\pi}{\lambda}, \qquad k = \sqrt{\mu\varepsilon}\,\frac{\omega}{c}.$$

Thus, in the case of the vacuum,

$$\frac{\omega}{c} = \frac{2\pi}{\lambda_0} \,. \tag{4.61}$$

In a homogeneous medium, the same wave propagates with a wavelength λ which is related to the wavelength λ_0 in the vacuum by

$$\lambda = \lambda_0 \frac{1}{\sqrt{\mu\varepsilon}} \,. \tag{4.62a}$$

The *index of refraction* is inversely proportional to the velocity c_{M} at which the harmonic oscillation propagates in the medium:

$$n = \frac{c}{c_{\mathrm{M}}} \,. \tag{4.62b}$$

Taking the index of refraction of the vacuum as being equal to 1 and taking account of the relations $\nu\lambda_0 = c$ and $\nu\lambda = c_{\mathrm{M}}$, one derives *Maxwell's relation*

$$n = \sqrt{\mu\varepsilon} \,. \tag{4.62c}$$

Comparing the two media whose common boundary is the plane $x^3 = 0$, one has

$$n = \sqrt{\mu\varepsilon} \,, \qquad n' = \sqrt{\mu'\varepsilon'} \,.$$

At the interface with the neighbouring medium, the incoming wave splits into a *refracted wave* and a *reflected wave*, which propagates in the original medium. Denoting the wave vector of the refracted wave by $\boldsymbol{k}'$ and that of the reflected wave by $\boldsymbol{k}''$, the electric field and the magnetic induction are, in complex form,

$$\boldsymbol{E}_{\mathrm{c}}(t,\boldsymbol{x}) = \boldsymbol{e}_0 \mathrm{e}^{-\mathrm{i}(\omega t - \boldsymbol{k}\cdot\boldsymbol{x})} \,, \quad \boldsymbol{B}_{\mathrm{c}}(t,\boldsymbol{x}) = n\,\hat{\boldsymbol{k}} \times \boldsymbol{E}_{\mathrm{c}}(t,\boldsymbol{x}) \,, \tag{4.63a}$$

$$\boldsymbol{E}'_{\mathrm{c}}(t,\boldsymbol{x}) = \boldsymbol{e}'_0 \mathrm{e}^{-\mathrm{i}(\omega' t - \boldsymbol{k}'\cdot\boldsymbol{x})} \,, \quad \boldsymbol{B}'_{\mathrm{c}}(t,\boldsymbol{x}) = n'\,\hat{\boldsymbol{k}}' \times \boldsymbol{E}'_{\mathrm{c}}(t,\boldsymbol{x}) \,, \tag{4.63b}$$

$$\boldsymbol{E}''_{\mathrm{c}}(t,\boldsymbol{x}) = \boldsymbol{e}''_0 \mathrm{e}^{-\mathrm{i}(\omega'' t - \boldsymbol{k}''\cdot\boldsymbol{x})} \,, \quad \boldsymbol{B}''_{\mathrm{c}}(t,\boldsymbol{x}) = n''\,\hat{\boldsymbol{k}}'' \times \boldsymbol{E}''_{\mathrm{c}}(t,\boldsymbol{x}) \,. \tag{4.63c}$$

At this point, one needs to know the boundary conditions which must be fulfilled at the interface. Exercise 4.1 shows the following: At the boundary, the tangential components of fields $\boldsymbol{E}$ and $\boldsymbol{H}$, as well as the normal components of fields $\boldsymbol{D}$ and $\boldsymbol{B}$, are continuous. This is possible only if the three phases in (4.63a)–(4.63c) at $x^3 = 0$ coincide for all times t and all values of x^1 and x^2. This implies an equality of the frequencies, $\omega = \omega' = \omega''$ (note that we already made use of this in (4.62c)), as well as the equality of the scalar products at the boundary

$$(\boldsymbol{k}\cdot\boldsymbol{x})\big|_{x^3=0} = (\boldsymbol{k}'\cdot\boldsymbol{x})\big|_{x^3=0} = (\boldsymbol{k}''\cdot\boldsymbol{x})\big|_{x^3=0} \tag{4.64a}$$

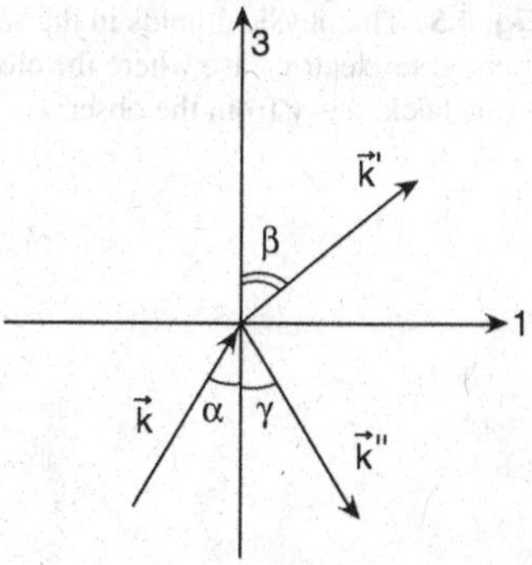

Fig. 4.4 A given wave in the $(1, 3)$-plane which comes in under the angle α relative to the 3-axis. The wave vectors of both the refracted and the reflected waves lie in the same plane

and the relations

$$|\boldsymbol{k}| = |\boldsymbol{k}''| = k = n\frac{\omega}{c} , \tag{4.64b}$$

$$|\boldsymbol{k}'| = k' = n'\frac{\omega}{c} . \tag{4.64c}$$

Having chosen the 1-axis such that $\boldsymbol{k}$ lies in the $(1, 3)$-plane, one concludes from (4.64a) that $\boldsymbol{k}'$ and $\boldsymbol{k}''$ lie in that plane, too. With the notations of Fig. 4.4 one has

$$\boldsymbol{k} = \frac{n\omega}{c}(\sin\alpha, 0, \cos\alpha)^T ,$$
$$\boldsymbol{k}' = \frac{n'\omega}{c}(\sin\beta, 0, \cos\beta)^T ,$$
$$\boldsymbol{k}'' = \frac{n\omega}{c}(\sin\gamma, 0, -\cos\gamma)^T .$$

A comparison with condition (4.64a) yields the following relations between the angles α, β and γ:

$$\alpha = \gamma \quad \text{and} \quad n\sin\alpha = n'\sin\beta . \tag{4.65}$$

The reflected beam propagates under the same angle relative to the 3-axis as the incoming beam but is oriented in the negative 3 direction. The second equation is called *Snellius' law.* The refracted beam deviates from the direction of the incoming beam whenever the indices of refraction are not equal.

4.4.2 Dynamics of Refraction and Reflection

In detail, the boundary conditions for the fields are

$$\boldsymbol{D}_\mathrm{n} = \boldsymbol{D}'_\mathrm{n} , \quad \boldsymbol{B}_\mathrm{n} = \boldsymbol{B}'_\mathrm{n} , \quad \boldsymbol{E}_\mathrm{t} = \boldsymbol{E}'_\mathrm{t} , \quad \boldsymbol{H}_\mathrm{t} = \boldsymbol{H}'_\mathrm{t} . \tag{4.66}$$

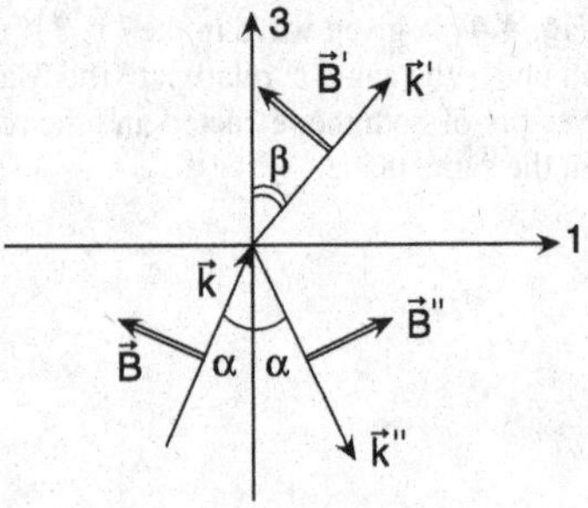

Fig. 4.5 The physical fields in the so-called transverse-electric case where the electric field points to the back, away from the observer

Denoting the positive normal to the boundary plane by $\hat{\boldsymbol{n}} = \hat{\boldsymbol{e}}_3$, one obtains the following system of equations (following the same order as in (4.66)):

$$\left[\varepsilon(\boldsymbol{e}_0 + \boldsymbol{e}''_0) - \varepsilon'\boldsymbol{e}'_0\right] \cdot \hat{\boldsymbol{n}} = 0\,, \tag{4.67a}$$

$$\left[\boldsymbol{k} \times \boldsymbol{e}_0 + \boldsymbol{k}'' \times \boldsymbol{e}''_0 - \boldsymbol{k}' \times \boldsymbol{e}'_0\right] \cdot \hat{\boldsymbol{n}} = 0\,, \tag{4.67b}$$

$$\left[\boldsymbol{e}_0 + \boldsymbol{e}''_0 - \boldsymbol{e}'_0\right] \times \hat{\boldsymbol{n}} = 0\,, \tag{4.67c}$$

$$\left[\frac{1}{\mu}(\boldsymbol{k} \times \boldsymbol{e}_0 + \boldsymbol{k}'' \times \boldsymbol{e}''_0) - \frac{1}{\mu'}(\boldsymbol{k}' \times \boldsymbol{e}'_0)\right] \times \hat{\boldsymbol{n}} = 0\,. \tag{4.67d}$$

As the fields are transversal to their respective directions of propagation, one needs to distinguish only two basic situations in a complete analysis of these equations: the case where $\boldsymbol{E}$ is *perpendicular* to the $(1,3)$-plane, i.e. to the plane spanned by $\boldsymbol{k}$ and $\hat{\boldsymbol{n}}$, and the case where $\boldsymbol{E}$ lies *in* this plane.

In what follows, let the absolute values of vectors $\boldsymbol{e}_0$, $\boldsymbol{e}'_0$ and $\boldsymbol{e}''_0$ be denoted by e_0, e'_0 and e''_0, respectively.

The Transverse Electric Case

In a situation where the electric field is perpendicular to $\boldsymbol{k}$ and to $\hat{\boldsymbol{n}}$, i.e. points "away" from the observer, i.e. to the back of Fig. 4.5, field $\boldsymbol{B}$ lies in the $(1,3)$-plane. By assumption, $\boldsymbol{E} = e_0\hat{\boldsymbol{e}}_2$. For the amplitudes, (4.67c) gives $e_0 + e''_0 - e'_0 = 0$, whereas (4.67d) yields the relation

$$\sqrt{\frac{\varepsilon}{\mu}}(e_0 - e''_0)\cos\alpha - \sqrt{\frac{\varepsilon'}{\mu'}}e'_0\cos\beta = 0\,.$$

Boundary conditions (4.67a)–(4.67d) and the transversality condition

$$\boldsymbol{k}' \cdot \boldsymbol{E}' = 0 = \boldsymbol{k}'' \cdot \boldsymbol{E}''$$

show that $\boldsymbol{E}' = e_0'\hat{\boldsymbol{e}}_2$ and $\boldsymbol{E}'' = e_0''\hat{\boldsymbol{e}}_2$, i.e. that the electric fields of the refracted and the reflected waves point in the same direction as the electric field of the incoming wave.

One rewrites the second relation of (4.65) as

$$n' \cos\beta = \sqrt{n'^2 - n^2 \sin^2\alpha}$$

and solves these equations for e_0'/e_0 and for e_0''/e_0. This gives

$$\frac{e_0'}{e_0} = \frac{2n\cos\alpha}{n\cos\alpha + (\frac{\mu}{\mu'})n'\cos\beta} = \frac{2n\cos\alpha}{n\cos\alpha + (\frac{\mu}{\mu'})\sqrt{n'^2 - n^2\sin^2\alpha}} , \tag{4.68a}$$

$$\frac{e_0''}{e_0} = \frac{n\cos\alpha - (\frac{\mu}{\mu'})n'\cos\beta}{n\cos\alpha + (\frac{\mu}{\mu'})n'\cos\beta} = \frac{n\cos\alpha - (\frac{\mu}{\mu'})\sqrt{n'^2 - n^2\sin^2\alpha}}{n\cos\alpha + (\frac{\mu}{\mu'})\sqrt{n'^2 - n^2\sin^2\alpha}}$$
$$=: R_{\text{para}} . \tag{4.68b}$$

The second result simplifies when the two media have the same magnetic permeability, $\mu' = \mu$. Using once more the second relation of (4.65), one has[3]

$$R_{\text{para}}(\mu' = \mu) = \frac{e_0''}{e_0} = -\frac{\sin(\alpha - \beta)}{\sin(\alpha + \beta)} . \tag{4.69}$$

This is one of *Fresnel's formulae.* Fresnel derived this and other formulae for refraction and reflection in 1821.

The Transverse Magnetic Case

In this case, the electric fields lie in the (1, 3)-plane, as shown in Fig. 4.6. In particular, one has

$$\boldsymbol{E} = e_0(-\hat{\boldsymbol{e}}_1 \cos\alpha + \hat{\boldsymbol{e}}_2 \sin\alpha) ,$$
$$\boldsymbol{E}' = e_0'(-\hat{\boldsymbol{e}}_1 \cos\beta + \hat{\boldsymbol{e}}_2 \sin\beta) ,$$
$$\boldsymbol{E}'' = e_0''(\hat{\boldsymbol{e}}_1 \cos\alpha + \hat{\boldsymbol{e}}_2 \sin\alpha) .$$

The boundary conditions (4.67c) and (4.67d) yield the equations

$$(e_0 - e_0'')\cos\alpha - e_0'\cos\beta = 0 ,$$
$$\sqrt{\frac{\varepsilon}{\mu}}(e_0 + e_0'') - \sqrt{\frac{\varepsilon'}{\mu'}}e_0' = 0 .$$

[3] The subscript "para" (parallel) refers to the three electric fields of the incoming, the refracted and the reflected waves.

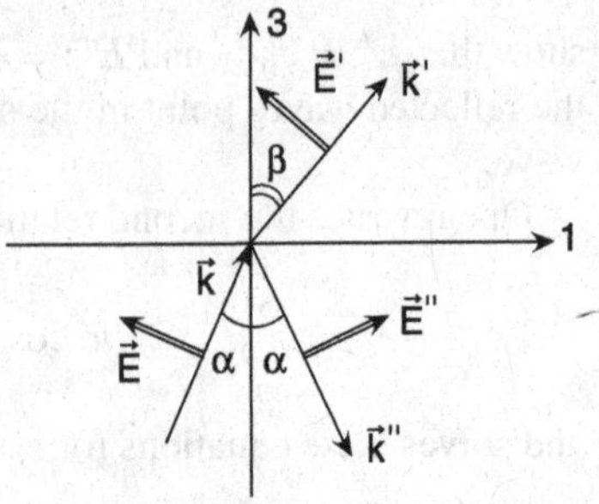

Fig. 4.6 The physical fields in the so-called transverse-magnetic case where the electric fields of the incoming, refracted and reflected beams lie in the (1, 3)-plane

These, in turn, as well as (4.62c) and (4.65), determine the ratio of the amplitudes:

$$R_{\text{trans}} := \frac{e_0''}{e_0} = \frac{(\frac{\mu}{\mu'})n'^2 \cos\alpha - n\sqrt{n'^2 - n^2 \sin^2\alpha}}{(\frac{\mu}{\mu'})n'^2 \cos\alpha + n\sqrt{n'^2 - n^2 \sin^2\alpha}}\,, \tag{4.70a}$$

$$\frac{e_0'}{e_0} = \frac{2nn'\cos\alpha}{(\frac{\mu}{\mu'})n'^2 \cos\alpha + n\sqrt{n'^2 - n^2 \sin^2\alpha}}\,. \tag{4.70b}$$

(The notation "trans" stands for transversal.)

In either case, the transverse-electric or the transverse-magnetic, the square of the ratio e_0''/e_0 is a measure of the intensity of the reflected wave as compared to the incoming wave. To visualize these ratios, we consider a situation where $\mu' = \mu$ but distinguish the cases $n' > n$ and $n' < n$. As a shorthand we denote the ratio of the indices of refraction by

$$r := \frac{n'}{n}\,. \tag{4.71}$$

The squares of the ratios are

$$R^2_{\text{para}} = \left(\frac{\cos\alpha - \sqrt{r^2 - \sin^2\alpha}}{\cos\alpha + \sqrt{r^2 - \sin^2\alpha}}\right)^2\,, \tag{4.72a}$$

$$R^2_{\text{trans}} = \left(\frac{r^2\cos\alpha - \sqrt{r^2 - \sin^2\alpha}}{r^2\cos\alpha + \sqrt{r^2 - \sin^2\alpha}}\right)^2\,. \tag{4.72b}$$

The case $r > 1$:

First, one identifies the following special cases from the results given by (4.68b) and (4.70a)

$$\alpha = 0: \qquad R^2_{\text{para}} = \left(\frac{1-r}{1+r}\right)^2\,, \qquad R^2_{\text{trans}} = \left(\frac{1-r}{1+r}\right)^2\,,$$

$$\alpha = \frac{\pi}{2}: \qquad R^2_{\text{para}} = 1\,, \qquad R^2_{\text{trans}} = 1\,.$$

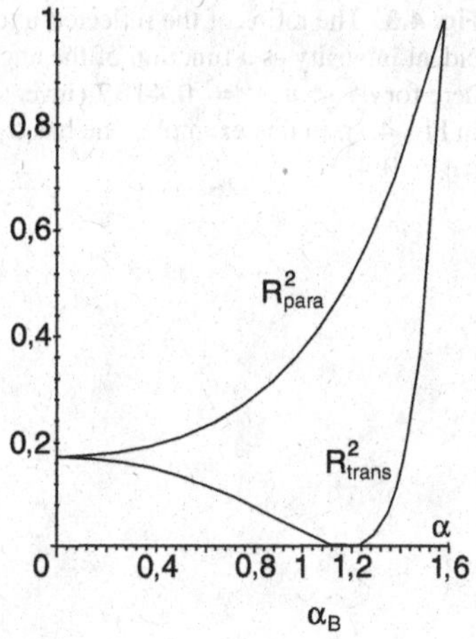

Fig. 4.7 The ratios of the intensity of the reflected beam to the intensity of the incident beam as a function of the angle of incidence, for the example $n' > n$: $r = 2.4173$ (diamond and sodium light at room temperature). In this example, one has $\alpha_B = 1.178$

While R^2_{para} grows monotonously in the interval $(0, \pi/2)$, the ratio R^2_{trans} has a zero at an angle α_B which is obtained from the equation

$$r^2 \cos\alpha - \sqrt{r^2 - \sin^2\alpha} = 0\,.$$

The angle α_B is called the *Brewster angle.* One verifies that it is obtained from

$$\alpha_B = \arctan(r)\,, \quad \text{or } \sin^2\alpha_B = \frac{r^2}{1+r^2}\,, \quad \cos^2\alpha_B = \frac{1}{1+r^2}\,. \tag{4.73}$$

As we assumed $r = n'/n > 1$, the Brewster angle α_B lies between $\pi/4$ and $\pi/2$. From Snellius' law (4.65) one concludes for $\alpha = \alpha_B$

$$\sin\beta|_{\alpha=\alpha_B} = \frac{1}{\sqrt{1+r^2}} = \cos\alpha_B$$

and, as all angles are in the interval $(0, \pi/2)$, one obtains the condition

$$\beta = \frac{\pi}{2} - \alpha_B\,. \tag{4.74}$$

These results have the following interpretation:

(i) In the transverse-magnetic case in which the electric field oscillates in the $(1, 3)$-plane, there is no reflected wave if the angle of incidence is α_B.
(ii) If the electric field of the incoming wave has components both in the $(1, 3)$-plane and in directions perpendicular to it, that is to say, if its (partial or full) polarization is elliptic, then only the component of the reflected beam perpendicular to the $(1, 3)$-plane survives. In this case, the reflected beam is

Fig. 4.8 The ratios of the reflected intensity to the incident intensity as a function of the angle of incidence, here for $n' < n$: $r = 0.4137$ (inverse of the value in Fig. 4.7). In this example, one has $\alpha_B = 0.392$ and $\alpha_G = 0.426$

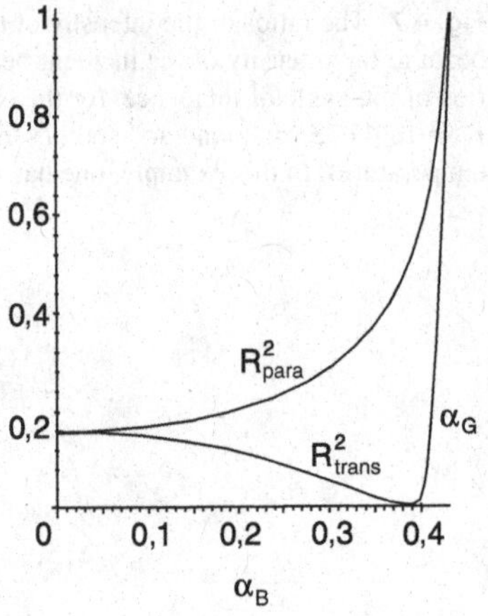

fully and linearly polarized. The behaviour of R^2_{para} and of R^2_{trans} as functions of α is shown in Fig. 4.7.

The case $r < 1$:
In this case, Snellius' law (4.65) says that the angle β is larger than the angle of incidence α. Therefore, there must exist a limiting angle α_G for which β is equal to $\pi/2$, viz.

$$\alpha_G = \arcsin r = \arcsin\left(\frac{n'}{n}\right) . \tag{4.75}$$

This is the case of total reflection. As before, there still exists a Brewster angle which, however, is below $\pi/4$. The squares of the ratios of (4.68b) and (4.70a) (for $\mu' = \mu$) are shown in Fig. 4.8 as functions of the angle of incidence.

4.5 Geometric Optics, Lenses and Negative Index of Refraction

This section is devoted to additional examples of the application of Maxwell's equations to optics: the limiting transition to *geometric optics,* a few formulae for thin optical lenses, needed in subsequent sections, and new and surprising phenomena which occur when the index of refraction takes negative values.

4.5.1 Optical Signals in Coordinate and in Momentum Space

As explained in Sect. 4.2, the homogeneous Maxwell equations are linear in all four types of fields, $\boldsymbol{E}(t, \boldsymbol{x})$, $\boldsymbol{D}(t, \boldsymbol{x})$, $\boldsymbol{B}(t, \boldsymbol{x})$ and $\boldsymbol{H}(t, \boldsymbol{x})$. This means that the superposition principle applies: With any two solutions, every linear combination of them is also a solution. Furthermore, the fields may be assumed to be *complex*, provided one respects the rules of how to extract from them the *observable* fields and their properties. As the wave equation (4.1) or (1.45) holds for every component of a field, it will be sufficient in what follows to study this differential equation for generic functions $g(t, \boldsymbol{x})$, irrespective of whether this is a scalar function or a component of one of the vector fields or the vector potential.

In this analysis, an important tool is provided by Fourier transformation which allows one to expand arbitrary signals localized in position or momentum space, in terms of their harmonic components. In other terms, one expands such signals in terms of solutions with fixed wave number, fixed frequency and a given direction of propagation. In Sect. 3.6.1, we used Fourier transformation in the variables

$$t \in \mathbb{R}_t \ \text{(time)} \longleftrightarrow \omega \in \mathbb{R}_\omega \ \text{(circular frequency)}\,,$$

i.e. the time t was replaced by the circular frequency ω, or, conversely, ω by t. In the considerations that follow, one needs a Fourier transformation in the variables $\boldsymbol{x}$ (position vector in $\mathbb{R}^3$) and the wave vector $\boldsymbol{k}$:

$$\boldsymbol{x} \in \mathbb{R}^3_x \ \text{(space)} \longleftrightarrow \boldsymbol{k} \in \mathbb{R}^3_k \ \text{(wave vector)}\,.$$

In other terms, one expands the measurable functions $g(t, \boldsymbol{x}) \in L^1(\mathbb{R}^3)$ in terms of the base system of harmonic functions

$$f_{\boldsymbol{k}}(t, \boldsymbol{x}) = \frac{1}{(2\pi)^{3/2}} \mathrm{e}^{\mathrm{i}(\boldsymbol{k}\cdot\boldsymbol{x} - \omega(k)t)}\,, \quad \boldsymbol{x} \in \mathbb{R}^3_x\,,\ \boldsymbol{k} \in \mathbb{R}^3_k\,. \tag{4.76}$$

Here $\boldsymbol{k}$ is the vector whose direction indicates the direction of propagation of the plane wave and whose modulus $k = |\boldsymbol{k}|$ specifies the wave number. For electromagnetic waves in vacuum, the dispersion relation reads

$$\omega(k) = kc \quad \text{with} \quad k = \frac{2\pi}{\lambda} \quad \text{and} \quad \omega = \frac{2\pi}{T} = 2\pi\nu\,. \tag{4.77}$$

The dispersion relation yields the well-known relation between wavelength λ and frequency ν.

The simplest case is a Fourier transformation in *one* dimension, in which case the base system given by (4.76) is replaced by

$$f_k(t, x) = \frac{1}{(2\pi)^{1/2}} \mathrm{e}^{\mathrm{i}(kx - \omega(k)t)}\,, \quad x \in \mathbb{R}_x\,,\ k \in \mathbb{R}_k\,. \tag{4.78a}$$

It has the following properties:

$$\int_{-\infty}^{+\infty} \mathrm{d}x \; f_{k'}^*(t,x) f_k(t,x) = \delta(k'-k) \quad \text{(orthogonality)}\,, \tag{4.78b}$$

$$\int_{-\infty}^{+\infty} \mathrm{d}k \; f_k^*(t,x') f_k(t,x) = \delta(x'-x) \quad \text{(completeness)}\,. \tag{4.78c}$$

Remark

Despite the fact that these formulae are completely symmetric in the space variable x and the wave number k, so that one could adjust the notation by writing $f(t,k,x)$ instead of $f_k(t,x)$, it seems fully justified, from the perspective of position space, to talk about orthogonality in (4.78b). Indeed, if one imagines the wave to be enclosed in a box and imposes periodic boundary conditions, then k is no longer a continuous variable but an element of a discrete spectrum. Equation (4.78b) then becomes

$$\int_{-\infty}^{+\infty} \mathrm{d}x \; f_{k_m}^*(t,x) f_{k_n}(t,x) = \delta_{mn}\,.$$

Whereas in (4.78c) the integral on the left-hand side is replaced by the sum over this spectrum, the right-hand side keeps its form, viz.

$$\sum_n f_{k_n}^*(t,x) f_{k_n}(t,x') = \delta(x'-x)\,.$$

The restriction to a box may be compared to the fixation of a vibrating string between two bridges which are affixed at $x = r$ on the sounding board and at $x = s$ on the wrest plank of a monochord: The variable $x \in [r,s]$ remains continuous, but the frequencies belong to a discrete spectrum.

A measurable function $g : \mathbb{R} \to \mathbb{C} : x \mapsto g(t,x)$, i.e. a function for which the norm $\|g\|_1 = \int_{-\infty}^{+\infty} \mathrm{d}x \; |g|$ exists, can be expanded in terms of the basis given by (4.78a):

$$g(t,x) = \frac{1}{\sqrt{2\pi}} \int_{-\infty}^{+\infty} \mathrm{d}k \; \tilde{g}(k)\, \mathrm{e}^{\mathrm{i}(kx-\omega(k)t)}\,. \tag{4.79a}$$

As an example, consider the case $\omega(k) = kc$ and a given amplitude in k-space chosen to be

$$\tilde{g}(k) = \alpha b\, \mathrm{e}^{-(k-k_0)^2 b^2/2}\,, \quad \alpha \in \mathbb{C}\,. \tag{4.79b}$$

Parameter b has a physical dimension length. It is contained in the factor in front to yield a dimensionless number in (4.79a) – when multiplied by dk. The function $g(t,x)$ then has the dimension of the (complex) amplitude α, whose dimension may be what it may. Using the well-known integral

$$\int_{-\infty}^{+\infty} \mathrm{d}x\ \mathrm{e}^{-(px^2+2qx+r)} = \sqrt{\frac{\pi}{p}}\,\mathrm{e}^{(q^2-pr)/p}\,,$$

in which p must have a positive real part, and where q and r may be arbitrary real or complex numbers, one obtains

$$g(t,x) = \alpha\,\mathrm{e}^{-(x-ct)^2/(2b^2)}\,\mathrm{e}^{\mathrm{i}k_0(x-ct)}\,. \tag{4.79c}$$

As expected, this is a function of $x - ct$ only, $g(t,x) \equiv g(x - ct)$, and, hence, is a solution of the wave equation in one dimension:

$$\left(\frac{1}{c^2}\frac{\partial^2}{\partial t^2} - \frac{\partial^2}{\partial x^2}\right) g(t,x) = 0\,.$$

A comparison of the amplitude given by (4.79b), which is defined on the space $\mathbb{R}_k$, with the function $g(t,x)$ of (4.79c), defined on $\mathbb{R}_x$, shows an important property of Fourier transformation. Both amplitudes have the shape of a *Gauss curve* whose width is determined by the length b and by its inverse, respectively. The modulus of the amplitude given by (4.79b) has its maximum at $k = k_0$ and decreases to half its maximum value at

$$k = k_0 \pm \frac{\sqrt{2\ln 2}}{b} = k_0 \pm \frac{1.177}{b}\,. \tag{4.80a}$$

The modulus of the solution given by (4.79c) has its maximum at $x = ct$ and falls to half its maximal value when

$$x - ct = \pm\sqrt{2\ln 2}\,b = \pm 1.177\,b\,. \tag{4.80b}$$

The amplitude given by (4.79c) represents a signal which is more localized the smaller length b is chosen to be. Expression (4.79b) for its Fourier transformed function shows that in this case the signal contains a broad spectrum of wave numbers. The narrower a signal is localized in x-space, the broader it is in k-space. Of course, this correlation holds also for its converse: If one wishes to construct a nearly monochromatic wave localized around the value k_0, then it is necessarily very broad in position space. A strictly monochromatic wave is not localizable at all in x-space.

4.5.2 Geometric (Ray) Optics and Thin Lenses

Geometric, or *ray*, *optics* neglects all diffraction phenomena and constructs light paths through an optical arrangement of mirrors, lenses, prisms, etc. consisting of segments of straight lines, following the simple rules of reflection and refraction. We return for a while to the Helmholtz equation (3.61) or (4.8), i.e. in the terminology of the preceding section, to a $t \leftrightarrow \omega$ Fourier transformation. In the quest for solutions of the Helmholtz equation, consider a complex signal

$$u(\boldsymbol{x}) = a(\boldsymbol{x})\mathrm{e}^{\mathrm{i}k_0 S(\boldsymbol{x})} , \tag{4.81}$$

where $k_0 = 2\pi/\lambda_0$ is a given wave number, $a(\boldsymbol{x})$ is the amplitude and $S(\boldsymbol{x})$ is a real function, still to be determined, which fixes the phase $k_0 S(\boldsymbol{x})$ of the solution given by (4.81). The amplitude $a(\boldsymbol{x})$ is assumed to be slowly varying such that it may be taken to be constant over the range of one wavelength λ_0. The two-dimensional surfaces $S(\boldsymbol{x}) = \text{const.}$ in (4.81) are the wave fronts, and their orthogonal trajectories follow the gradient field $\nabla S(\boldsymbol{x})$. Locally, in the neighbourhood of a point $\boldsymbol{x}_0$, the signal is approximately a plane wave with wave number and direction of propagation, respectively,

$$k = n(\boldsymbol{x}_0)k_0 \quad \text{and} \quad \hat{\boldsymbol{k}} \propto \nabla S(\boldsymbol{x})|_{\boldsymbol{x}=\boldsymbol{x}_0} .$$

The function $S(\boldsymbol{x})$, which is called the *eikonal*, plays an important role in geometric optics. The orthogonal trajectories of the surfaces of constant value of this function determine the local wave vectors and, hence, are identical to the optical paths of geometric optics. This is seen explicitly if one inserts the ansatz (4.81) into the Helmholtz equation

$$(\boldsymbol{\Delta} + k^2)\, u(\boldsymbol{x}) = 0 \tag{4.82}$$

and makes use of the approximation just described. One calculates first

$$\nabla\left(a\,\mathrm{e}^{\mathrm{i}k_0 S}\right) = \mathrm{e}^{\mathrm{i}k_0 S}\left[(\nabla a) + \mathrm{i}k_0 a(\nabla S)\right] ,$$

$$\boldsymbol{\Delta}\left(a\,\mathrm{e}^{\mathrm{i}k_0 S}\right) = \mathrm{e}^{\mathrm{i}k_0 S}\left[(\boldsymbol{\Delta} a) + 2\mathrm{i}k_0(\nabla a)\cdot(\nabla S) + \mathrm{i}k_0 a(\boldsymbol{\Delta} S) - k_0^2 a(\nabla S)^2\right] .$$

Insertion into the Helmholtz equation (4.82) with $k = nk_0$ gives

$$\begin{aligned} &k_0^2 a\left[n^2 - (\nabla S)^2\right] + \boldsymbol{\Delta} a \\ &+ \mathrm{i}k_0\left[2(\nabla a)\cdot(\nabla S) + a(\boldsymbol{\Delta} S)\right] = 0 . \end{aligned} \tag{4.83a}$$

With $1/k_0^2 = (\lambda_0/(2\pi))^2$ the real part of this equation yields the equation

$$(\nabla S)^2 = n^2 + \left(\frac{\lambda_0}{2\pi}\right)^2 \frac{1}{a}(\Delta a) . \tag{4.83b}$$

The assumption that the amplitude varies slowly on the scale of the wavelength means that its second derivatives are small or, more precisely, that

$$\frac{\lambda_0^2 \Delta a(\boldsymbol{x})}{a(\boldsymbol{x})} \ll 1 .$$

In this situation, the second term on the right-hand side of (4.83b) can be neglected so that equation (4.83b) becomes the eikonal equation.

Eikonal equation

$$(\nabla S(\boldsymbol{x}))^2 = n^2(\boldsymbol{x}) . \tag{4.83c}$$

The phase of the solution given by (4.81) of the Helmholtz equation is determined by the slowly varying index of refraction only.

In a somewhat different interpretation, one may say that this equation holds in the limit $\lambda_0 \to 0$ and that it provides the basis for geometric optics. The eikonal equation (4.83c) expresses in a formula Fermat's principle of optics. This also justifies the somewhat qualitative definition of geometric optics as being the limit $\lambda_0 \to 0$ of wave optics. As a more precise definition, one says that geometric or ray optics is applicable whenever the wavelength of diffracted light is small compared to typical dimensions of the objects on which it scatters.

In geometric optics, with prisms and lenses, no more information about such optical components is needed than their geometric shape and the index of refraction n of the material they are made of. A simple example is provided by the *plano-convex lens.* Assume its spherically curved side to have radius R. If this lens is used in a vacuum, then its focal distance f and the curvature radius r are related by the equation

$$\frac{1}{f} = \frac{n-1}{r} . \tag{4.84}$$

Besides simple relations of this kind and other properties of lenses (aberration, astigmatism, field of curvature, etc.), all of which can be obtained by construction of straight light rays, there are also properties for which wave optics is essential. This is what we deal with next, preparing at the same time the ground for the following sections.

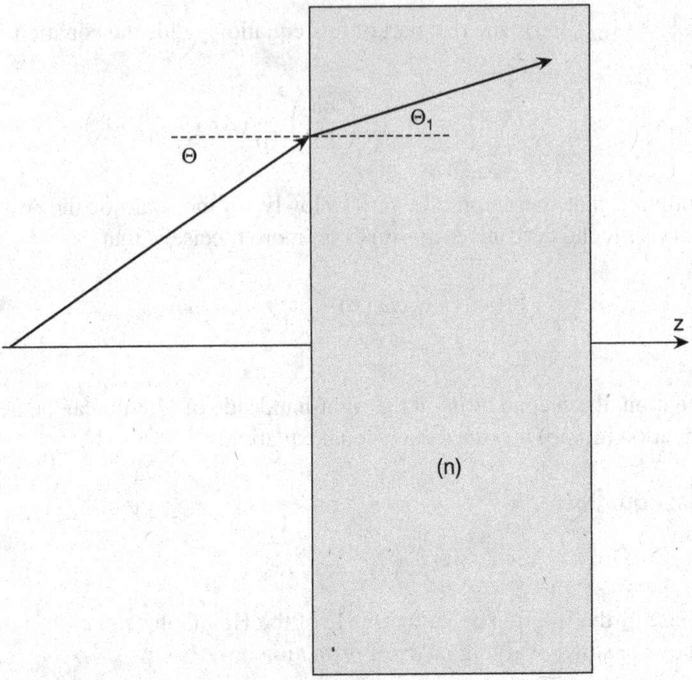

Fig. 4.9 Refraction of a light beam on a plane parallel block with (positive) index of refraction n

Figure 4.9 shows a plane parallel block of transparent material whose thickness is d and which is made of a homogeneous material with index of refraction n. The block is perpendicular to the z-axis. A light beam in the (x, z)-plane hits the block under an angle θ. Snellius' law, given by (4.65), yields the angle under which the refracted beam emerges, viz.

$$\sin\theta_1 = \frac{1}{n}\sin\theta \, .$$

An amplitude which is characteristic for the incoming beam has the form

$$u(\boldsymbol{x}) = a\,\mathrm{e}^{\mathrm{i}\boldsymbol{k}\cdot\boldsymbol{x}} = a\,\mathrm{e}^{\mathrm{i}k_0(x\sin\theta + z\cos\theta)} \, , \quad z < 0 \, .$$

This could be, for instance, one of the components of the electric field $\boldsymbol{E}(t,\boldsymbol{x}) = \boldsymbol{\varepsilon}(\boldsymbol{k})\exp\{\mathrm{i}\boldsymbol{k}\cdot\boldsymbol{x}\}$ of a monochromatic wave. In the medium of the block, this becomes

$$u(\boldsymbol{x}) = a\,\mathrm{e}^{\mathrm{i}\boldsymbol{k}_1\cdot\boldsymbol{x}} = a\,\mathrm{e}^{\mathrm{i}nk_0(x\sin\theta_1 + z\cos\theta_1)} \, , \quad 0 \leqslant z \leqslant d \, ,$$

Fig. 4.10 Plano-convex lens made of a material with index of refraction n

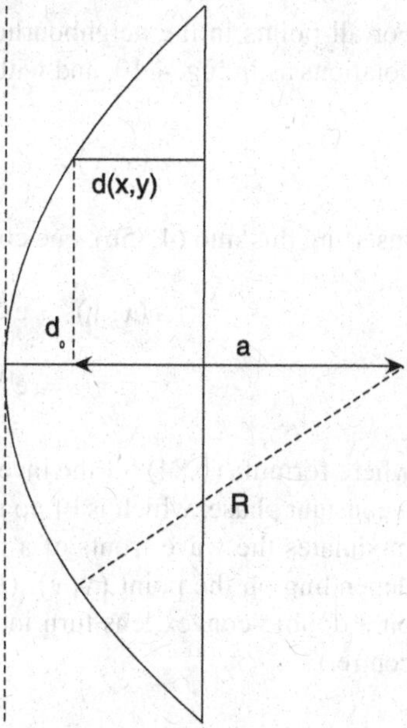

so that the transmission coefficient for the complex amplitude is given by

$$t(x, y) := \frac{u(x, y, z = d)}{u(x, y, z = 0)} = \mathrm{e}^{\mathrm{i}nk_0(x \sin\theta_1 + d \cos\theta_1)} \,. \qquad (4.85a)$$

In the case of *paraxial beams,* i.e. beams which are incident under a small angle θ_0, one has

$$\theta_1 \simeq \frac{\theta_0}{n}\,, \quad \sin\theta_1 \simeq \theta_1 \simeq \frac{\theta_0}{n}\,, \quad \cos\theta_1 \simeq 1 - \frac{1}{2}\theta_1^2 \simeq 1 - \frac{\theta_0^2}{2n^2}\,.$$

Expression (4.85a) is then replaced by the approximation

$$t(x, y) \simeq \mathrm{e}^{\mathrm{i}nk_0 d} \exp\left\{-\mathrm{i}\left(\frac{k_0 d}{2n}\theta_0^2 - k_0 x\theta_0\right)\right\} \,. \qquad (4.85b)$$

This is called the *paraxial approximation*. The result, given by (4.85a), may be applied to the example of the plano-convex lens sketched in Fig. 4.10. Let d_0 denote the thickness of the lens at the origin $d_0 = d(0, 0)$, and let $d(x, y)$ be its thickness at x and y different from zero. The transmission coefficient given by (4.85a) is equal to the product of its value for the horizontal layer $d_0 - d(x, y)$ and its value for the local thickness $d(x, y)$,

$$t(x, y) = \mathrm{e}^{\mathrm{i}k_0(d_0 - d(x,y))}\, \mathrm{e}^{\mathrm{i}nk_0 d(x,y))} = \mathrm{e}^{\mathrm{i}k_0 d_0}\, \mathrm{e}^{\mathrm{i}(n-1)k_0 d(x,y)} \,.$$

For all points in the neighbourhood of the axis, one has $(x^2 + y^2) \ll R^2$. With notations as in Fig. 4.10, and with $a = \sqrt{R^2 - (x^2 + y^2)}$, one has

$$d(x, y) = d_0 - (R - a) \simeq d_0 - \frac{x^2 + y^2}{2R} .$$

Inserting this into (4.85b), one concludes

$$\begin{aligned} t(x, y) &\simeq e^{ik_0 d_0} e^{i(n-1)k_0[d_0-(x^2+y^2)/(2R)]} \\ &= e^{ink_0 d_0} \exp\left\{-ik_0 \frac{x^2 + y^2}{2f}\right\} , \end{aligned} \tag{4.86}$$

where formula (4.84) for the inverse of the focal length was used. The first factor is a constant phase, which is of no relevance for most applications. The second factor modulates the wave fronts of a plane wave incident along the z-axis in a manner depending on the point (x, y). (Indeed, the planar wave fronts of a beam incident on a double-convex lens turn into paraboloid waves, with one of the foci as their centre.)

4.5.3 Media with Negative Index of Refraction

We return to Fig. 4.4 and consider once more the refraction of a light ray at the plane interface that separates two media with indices of refraction n and n'. In Fig. 4.11, a light ray shines from point A in the medium with index of refraction n to point B in the neighbouring medium whose index of refraction is n'. In the initial medium, the ray is incident under the angle α, relative to the normal to the interface. In the second medium, it encloses the angle β with that normal. The relative position of the points $A = (x_A = -x, 0, z_A = -a)$ and $B = (x_B = x_A + d, 0, z_B = b)$ is assumed to be given; their horizontal distance is d, and their vertical distance is $a + b$. The domain of definition of the angles α, β, etc. is chosen to be the interval $[-\pi/2, +\pi/2]$. The point S where the beam hits the boundary plane is taken to be the origin of the frame of reference. Only the $(1, 3)$-plane of that system is relevant. One has

$$\sin\alpha = \frac{x}{\sqrt{x^2 + a^2}} , \quad \sin\beta = \frac{d - x}{\sqrt{(d - x)^2 + b^2}} . \tag{4.87}$$

In a medium with index of refraction n, light propagates at velocity c/n. Thus, the time that light from A needs to reach B, via S, is equal to

$$\tau = t_{AS} + t_{SB} = \frac{n}{c}\sqrt{x^2 + a^2} + \frac{n'}{c}\sqrt{(d - x)^2 + b^2} . \tag{4.88a}$$

Multiplying both sides by the speed of light c yields the *optical path length*:

$$\lambda = n(AS) + n'(SB) = n\sqrt{x^2 + a^2} + n'\sqrt{(d - x)^2 + b^2} \equiv \lambda(x) . \tag{4.88b}$$

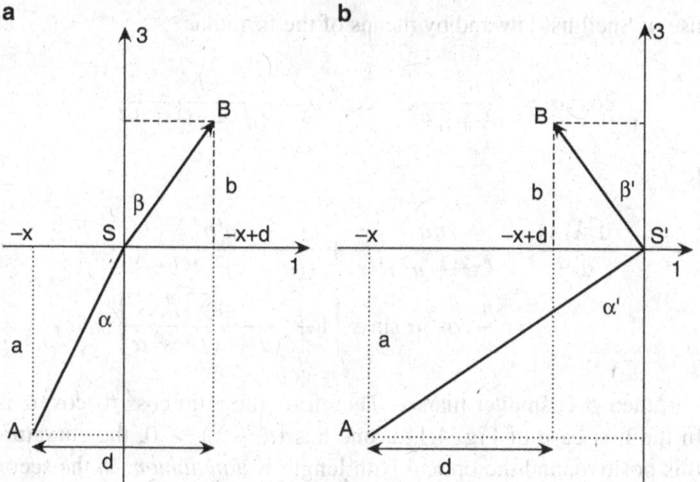

Fig. 4.11 Paths of minimal and maximal optical path length between the given points A in a medium with index of refraction n and B in a medium with index of refraction n'. The axis denoted by 1 is the x-axis, that denoted by 3 is the z-axis in the text

In the case of a more general path in a medium with a variable index of refraction, this example is replaced by a path integral:

$$\lambda := \int_A^B \mathrm{d}s\, n(s) . \tag{4.89}$$

The following principle is applicable to the optical path length of ray optics:

Fermat's principle
Within the domain of validity of ray optics, light takes its path through optical components in such a way that the optical path length (4.89) assumes an extremum.

Applying this principle to example (4.88b) means that the derivative

$$\frac{\mathrm{d}\lambda(x)}{\mathrm{d}x} = n\frac{x}{\sqrt{x^2+a^2}} - n'\frac{d-x}{\sqrt{(d-x)^2+b^2}} = n\sin\alpha - n'\sin\beta$$

must vanish. This is seen to be Snellius' relation (4.65). Whether this extremum is a minimum or a maximum follows from the second derivative, which is calculated making use of Snellius' law and by means of the formulae

$$\cos^2\alpha = \frac{a^2}{x^2+a^2}\,, \quad \cos^2\beta = \frac{b^2}{(d-x)^2+b^2}\,.$$

One finds

$$\begin{aligned}\frac{\mathrm{d}^2\lambda(x)}{\mathrm{d}x^2} &= \frac{na^2}{(x^2+a^2)^{3/2}} + \frac{n'b^2}{((d-x)^2+b^2)^{3/2}}\\ &= \frac{n}{x}\cos^2\alpha\sin\alpha\left\{1+\frac{x\cos^2\beta}{(d-x)\cos^2\alpha}\right\}\,. \end{aligned} \tag{4.90}$$

If $|n'| > n$, then β is smaller than α. Therefore, the ratio $\cos^2\beta/\cos^2\alpha$ is larger than 1. In the first case of Fig. 4.11a, one has $(d-x) > 0$, the curvature given by (4.90) is positive, and the optical path length is a *minimum*. In the second case of Fig. 4.11b, one has $(d-x) < 0$, the modulus of the second term in the curly brackets in (4.90) is larger than 1, and the curvature given by (4.90) becomes negative. The optical path length is a *maximum*. A glance at (4.87) shows that the angle β lies in the interval $-\pi/2 \le \beta \le 0$, and the refracted beam is now on the same side of the normal to the boundary plane as the incoming beam. On the other hand, comparison with Snellius' law shows that this is possible only if the index of refraction n' takes a *negative* value!

In what follows, we will see that there are indeed "metamaterials" which exhibit the phenomenon of negative index of refraction in specific ranges of frequencies. We will also give a qualitative explanation of how this comes about.[4] For the moment we simply note that the optical path length assumes a maximum if $n > 0$ and $n' < 0$, and we study the optical properties of plane parallel blocks made of a metamaterial with a negative index of refraction.

For the sake of comparison, let us return for a moment to the example of the plano-convex lens of Fig. 4.10 and investigate the optical properties of the lens for a wave with given circular frequency ω. Imagine a small dipole source on the optical axis fixed in front of the lens. It emits a wave whose electric field has the form

$$\boldsymbol{E}(t,\boldsymbol{x}) = \mathrm{e}^{\mathrm{i}k_z z}\sum_\sigma \boldsymbol{\varepsilon}_\sigma \frac{1}{2\pi}\iint \mathrm{d}^2k\ \exp\left\{\mathrm{i}k_x x + k_y y - \omega t)\right\}\,. \tag{4.91}$$

The z-axis is chosen along the optical axis, and the polarization $\boldsymbol{\varepsilon}_\sigma$ is contained in the (x, y)-plane transversal to the wave vector. In terms of a somewhat simplified

[4] The basic idea that a simultaneous change of sign of the dielectric constant ε and of the magnetic permeability μ would lead to new phenomena for electromagnetic waves was put forward in 1968 by V.G. Veselago (cf. V.G. Veselago, Soviet Physics USPEKxHI **10**, 509). But it took until the year 2000 before it became clear that this was not only a theoretical speculation but also a realistic experimental option (see e. g. J.B. Pendry, Phys. Rev. Lett. 85 (2000) 3966; D.W. Ward, K. Nelson, and K.J. Webb, Physics/0409083.

description, the purpose of the lens, in the framework of wave optics, is to modify the phases of the individual components of (4.91) by means of a transmission coefficient, (4.86) or (4.85a), such that the field components behind the lens combine to a focal point as an image of the pointlike source. Without going into the mechanism of this reconstruction, one realizes immediately that there must be a restriction of principle. The components of the wave vector must fulfill the dispersion relation

$$\left[(k_x^2 + k_y^2) + k_z^2\right] c^2 = \omega^2 . \tag{4.92a}$$

The waves for which (4.92a) holds with *real* values of $k_z = \sqrt{(\omega/c)^2 - (k_x^2 + k_y^2)}$ are said to be *propagating* waves. They are the only ones to which the qualitative argument developed previously applies. Waves, in turn, for which $(k_x^2+k_y^2)$ becomes larger than $(\omega/c)^2$ are called *evanescent* (i.e. "disappearing") waves. For waves of this kind, k_z becomes pure imaginary:

$$k_z = \mathrm{i}\sqrt{(k_x^2 + k_y^2) - (\omega/c)^2} . \tag{4.92b}$$

Insertion into (4.91) shows that these waves decay exponentially with increasing z and, hence, that they no longer contribute to the construction of the image. This is why in optical imaging by lenses the resolution of the image is restricted to a maximal value of

$$\delta\ell_{\text{max}} \simeq \frac{2\pi}{k_{\text{max}}} = \frac{2\pi c}{\omega} = \lambda , \tag{4.93}$$

even if one uses perfect lenses with maximal aperture.

The situation changes when one uses metamaterial with a negative index of refraction, making use of the special features of the *maximal* optical path length of Fig. 4.11. For the sake of simplicity, we assume that we have a cuboid of thickness d made of a metamaterial with $n = -1$, as drawn in Fig. 4.12. In the vacuum ($n = 1$), we place a light source Q at distance a in front of the cuboid. If the distance a is smaller than the thickness d (this is assumed in Fig. 4.12), then the construction with the tools of geometric optics shows that rays emanating from source Q are focused twice. The source has two images, one in B_1 and another in B_2. The cuboid with a negative index of refraction acts like a lens. The real surprise, however, comes from the fact that this "lens" is not subject to the restriction, given by (4.93), of ordinary lenses and, from the point of view of wave optics, is a genuine *perfect* lens. The following argument sketches the proof of this important property.

Metamaterials, in fact, are microstructured objects whose dielectric function ε and permeability μ can take complex values and, thus, in special cases can even be $\varepsilon = -1$ and $\mu = -1$. By Maxwell's formula (4.62c) the index of refraction is given by

$$n = \pm\sqrt{\varepsilon\mu} , \tag{4.94a}$$

whereby for "normal" matter we took the positive square root. For negative values of ε or μ, one must indeed insert the negative root. The impedance of the medium,

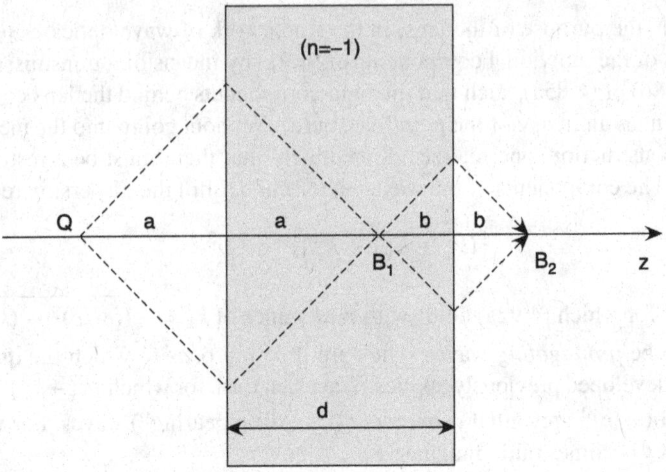

Fig. 4.12 A block of metamaterial with negative index of refraction $n = -1$ focuses the beams emanating from source Q at point B_1 inside and at point B_2 behind the block

defined by

$$Z := \sqrt{\frac{\mu}{\varepsilon}}\,, \tag{4.94b}$$

remains unchanged when ε is replaced by $-\varepsilon$ and μ is replaced by $-\mu$. Neither on the boundary facing the source nor on the opposite boundary of the cuboid is there any reflection. The light from the vacuum in front of the cuboid is transmitted to the vacuum behind it, without any losses. These assertions can be further justified. One starts by verifying by means of the formulae in Sects. 3.5.4 and 4.2.1 that in the example of the cuboid, the energy flows in the positive direction if for propagating waves (4.92a) is solved by the negative square root:[5]

$$k'_z = -\sqrt{(\omega/c)^2 - (k_x^2 + k_y^2)}\,. \tag{4.95a}$$

The transmission coefficient for propagating waves is then

$$t = \exp\left\{ik'_z d\right\} = \exp\left\{-id\sqrt{(\omega/c)^2 - (k_x^2 + k_y^2)}\right\}. \tag{4.95b}$$

It is this change of sign in the phase which is responsible for the focusing of the light in the cuboid and behind it.

But what happens to (formerly) evanescent waves whose amplitudes decreased with increasing distance? The following analysis shows that such waves, in fact, are

[5] As shown in the work of D.W. Ward et al., the phase velocity points in the negative z-direction, but the group velocity points in the positive direction.

amplified such that they are focused, too, and that there is no loss in the resolution of the image.

Transverse-Electric Case

As a first example we consider the transverse-electric case as in Sect. 4.4.2 and calculate the ratio, (4.68a), of the refracted amplitude, as well as the ratio, (4.68b), of the reflected amplitude to the incoming amplitude for a wave in the vacuum incident on the cuboid. This means that we must now set $n = 1$ and $\mu = 1$ in these formulae. From equations (4.64b) and (4.64c)

$$\left|\boldsymbol{k}'\right| \equiv k' = \left|n'\right| k = \left|n'\right| \frac{\omega}{c} \equiv \left|n'\right| \left|\boldsymbol{k}\right| ,$$

and the factors $\cos\alpha$ and $\cos\beta$ can be replaced by k_z and k'_z, respectively:

$$t_\mathrm{a} := \frac{e'_0}{e_0} = \frac{2\mu'\cos\alpha}{\mu'\cos\alpha + n'\cos\beta} = \frac{2\mu' k_z}{\mu' k_z + k'_z} ,$$
$$r_\mathrm{a} := \frac{e''_0}{e_0} = \frac{\mu'\cos\alpha - n'\cos\beta}{\mu'\cos\alpha + n'\cos\beta} = \frac{\mu' k_z - k'_z}{\mu' k_z + k'_z} .$$

For simplicity, the permeability in the interior is denoted by μ instead of μ' (in the outer space the permeability is now equal to 1):

$$t_\mathrm{a} = \frac{2\mu k_z}{\mu k_z + k'_z} , \tag{4.96a}$$
$$r_\mathrm{a} = \frac{\mu k_z - k'_z}{\mu k_z + k'_z} . \tag{4.96b}$$

Furthermore, it is necessary to let μ (and likewise the dielectric function ε) tend to -1 in a limit.

When the wave from the interior hits the interface to the vacuum, one must set $\mu' = 1$ and $n' = 1$ in (4.68a) and (4.68b). Furthermore, k and k' are interchanged so that one has

$$t_\mathrm{i} = \frac{2k'_z}{k'_z + \mu k_z} , \tag{4.97a}$$
$$r_\mathrm{i} = \frac{k'_z - \mu k_z}{k'_z + \mu k_z} . \tag{4.97b}$$

In calculating the transmission across the cuboid, one must add to the direct passage all those scattering processes in which the light is reflected from the inner boundaries twice, four times, etc. This series of processes yields a geometric series:

From expression (4.85a) and using the abbreviation $\phi = x\,k'_x + d\,k'_z$, one has

$$T^{(\mathrm{TE})}(x,y) = t_\mathrm{a}t_\mathrm{i}\,\mathrm{e}^{\mathrm{i}\phi} + t_\mathrm{a}t_\mathrm{i}r_\mathrm{i}^2\,\mathrm{e}^{3\mathrm{i}\phi} + t_\mathrm{a}t_\mathrm{i}r_\mathrm{i}^4\,\mathrm{e}^{5\mathrm{i}\phi} + \ldots$$
$$= \frac{t_\mathrm{a}t_\mathrm{i}\,\mathrm{e}^{\mathrm{i}\phi}}{1 - r_\mathrm{i}^2\,\mathrm{e}^{2\mathrm{i}\phi}}\,. \tag{4.98a}$$

Inserting formulae (4.96a), (4.97a) and (4.97b) and taking the limit ($\mu \to -1, \varepsilon \to -1$), this gives

$$\lim_{\mu,\varepsilon\to-1} T^{(\mathrm{TE})}(x,y) = \lim_{\mu,\varepsilon\to-1} \frac{4\mu k_z k'_z}{(k'_z + \mu k_z)^2}$$
$$\times \frac{1}{1 - [(k'_z - \mu k_z)/(k'_z + \mu k_z)]^2\,\mathrm{e}^{2\mathrm{i}\phi}}\,\mathrm{e}^{\mathrm{i}\phi}$$
$$= \lim_{\mu,\varepsilon\to-1} \frac{4\mu k_z k'_z}{(k'_z + \mu k_z)^2 - (k'_z - \mu k_z)^2\,\mathrm{e}^{2\mathrm{i}\phi}}\,\mathrm{e}^{\mathrm{i}\phi}$$
$$= \mathrm{e}^{-\mathrm{i}\phi}\,, \quad (\phi = x\,k'_x + d\,k'_z)\,. \tag{4.98b}$$

Note that we make use of the fact that k_z and k'_z coincide in this limit.

In a similar way, one calculates the reflection coefficient which results from the multiple scattering inside the cuboid:

$$\lim_{\mu,\varepsilon\to-1} R^{(\mathrm{TE})}(x,y) = \lim_{\mu,\varepsilon\to-1} \left\{ r_\mathrm{a} + \frac{t_\mathrm{a}t_\mathrm{i}\,\mathrm{e}^{\mathrm{i}\phi}}{1 - r_\mathrm{i}^2\,\mathrm{e}^{2\mathrm{i}\phi}} r_\mathrm{i}\,\mathrm{e}^{\mathrm{i}\phi} \right\}$$
$$= \lim_{\mu,\varepsilon\to-1} \left\{ r_\mathrm{a} + T^{(\mathrm{TM})} r_\mathrm{i}\,\mathrm{e}^{\mathrm{i}\phi} \right\}$$
$$= \lim_{\mu,\varepsilon\to-1} \{ r_\mathrm{a} + r_\mathrm{i} \} = 0\,. \tag{4.99}$$

Thus, nothing is being reflected at all! Evanescent waves which, according to (4.92b), become pure imaginary are *amplified* – in contrast to the case of ordinary lenses treated earlier where they were exponentially damped. In the case of an idealized cuboid with index of refraction $n = -1$, both the propagating and the formerly evanescent waves contribute to the resolution. Except for effects of the aperture and of possible impurities of the surfaces of the cuboid, there is no obstacle of principle against complete reconstruction of the image.

Transverse-Magnetic Case

The transverse-magnetic case is treated in an analogous manner. In equations (4.70a) and (4.70b), n is replaced with 1, μ with 1, and for simplicity $\mu \equiv \mu'$, $\varepsilon \equiv \varepsilon'$,

$n \equiv n'$ are written again. Then one obtains

$$t_{\mathrm{a}} = \frac{e_0'}{e_0} = \frac{2n\cos\alpha}{\varepsilon\cos\alpha + n\cos\beta} = \frac{2nk_z}{\varepsilon k_z + k_z'}\,, \tag{4.100a}$$

$$r_{\mathrm{a}} = \frac{e_0''}{e_0} = \frac{\varepsilon\cos\alpha - n\cos\beta}{\varepsilon\cos\alpha + n\cos\beta} = \frac{\varepsilon k_z - k_z'}{\varepsilon k_z + k_z'} \tag{4.100b}$$

$$t_{\mathrm{i}} = \frac{2nk_z'}{\mu(k_z' + \varepsilon k_z)}\,, \tag{4.100c}$$

$$r_{\mathrm{i}} = \frac{k_z' - \varepsilon k_z}{k_z' + \varepsilon k_z}\,. \tag{4.100d}$$

In this derivation, we inserted the square of Maxwell's relation (4.62c), i.e. $n^2 = \varepsilon\mu$. The product $t_{\mathrm{a}}t_{\mathrm{i}}$ which appears in the multiple scattering series is given by

$$t_{\mathrm{a}}t_{\mathrm{i}} = \frac{n^2}{\mu}\frac{4k_zk_z'}{(k_z' + \varepsilon k_z)^2} = \varepsilon\frac{4k_zk_z'}{(k_z' + \varepsilon k_z)^2}\,.$$

Thus, also in this case one obtains with $n^2 = \mu\varepsilon$

$$\begin{aligned}\lim_{\mu,\varepsilon\to -1} T^{(\mathrm{TM})}(x,y) &= \lim_{\mu,\varepsilon\to -1} \frac{4\varepsilon k_zk_z'}{(k_z' + \varepsilon k_z)^2 - (k_z' - \varepsilon k_z)^2\,\mathrm{e}^{2\mathrm{i}\phi}}\,\mathrm{e}^{\mathrm{i}\phi}\\ &= \mathrm{e}^{-\mathrm{i}\phi}\,, \quad (\phi = x\,k_x' + d\,k_z')\,.\end{aligned} \tag{4.101}$$

One confirms that here, too, the coefficient of reflection vanishes.

Clearly, the analytic solution given here applies to an idealized situation. Nevertheless, this case is realistic enough to exhibit the salient optical properties. Further arguments and illustrations in the optics of metamaterials with a negative index of refraction can be found, e. g., in a nicely illustrated article in *Physics Today*.[6]

4.5.4 Metamaterials with Negative Index of Refraction

Metamaterials, which are also called *left-handed media*, are microstructured materials composed of an array of wires and *split ring resonators.* To take an example, there are metamaterials which react to microwaves with frequencies on the order of 10 GHz by a negative index of refraction. The first reports on the production of such materials and on a proof of their optical properties were published in 2000.[7] Without going into the technical aspects of experiments of this kind, let us describe in a qualitative manner how one may understand the occurrence of complex-valued

[6] John B. Pendry, David R. Smith, *Physics Today*, December 2003

[7] D.R. Smith, W.J. Padilla, D.C. Vier, S.C. Nemat-Nasser, S. Schultz, Phys. Rev. Lett. **84** (2000) 4184.

material parameters ε and μ and, hence, of the possibility of negative indices of refraction.

The index of refraction is a macroscopic property of a medium, whereas the electric and magnetic susceptibilities follow from the medium's microscopic properties. Relations between these quantities were obtained in Chap. 1. For example, according to (1.73b), we have

$$\varepsilon(\boldsymbol{x}) = 1 + 4\pi\chi_e(\boldsymbol{x}) .$$

Similarly, μ, according to (1.78d), is given by

$$\mu(\boldsymbol{x}) = 1 + 4\pi\chi_m(\boldsymbol{x}) .$$

If the incident light contains frequencies ω in the immediate neighbourhood of a resonance ω_0 in the medium, the function χ_e depends on the frequency as follows:[8]

$$\chi_e = \frac{\chi_0\omega_0^2}{(\omega-\omega_0)^2+\Gamma^2\omega^2}\left[(\omega-\omega_0)^2+\mathrm{i}\Gamma\omega\right] . \tag{4.102}$$

The frequency-dependent function ε, and possibly also the magnetic permeability μ, is shifted to the complex plane. Expressing them in polar decomposition,

$$\varepsilon = |\varepsilon|\,\mathrm{e}^{\mathrm{i}\varphi_\varepsilon} , \qquad \mu = |\mu|\,\mathrm{e}^{\mathrm{i}\varphi_\mu} ,$$

the index of refraction, (4.62c), and the impedance, (4.94b), take the form

$$n = \sqrt{|\varepsilon|\,|\mu|}\,\mathrm{e}^{\mathrm{i}(\varphi_\varepsilon+\varphi_\mu)/2} \equiv |n|\,\mathrm{e}^{\mathrm{i}\phi_n} , \tag{4.103a}$$

$$Z = \sqrt{\frac{|\mu|}{|\varepsilon|}}\,\mathrm{e}^{\mathrm{i}(\varphi_\mu-\varphi_\varepsilon)/2} \equiv |Z|\,\mathrm{e}^{\mathrm{i}\phi_Z} . \tag{4.103b}$$

If the susceptibilities χ_e and χ_m exhibit resonances of the type given by (4.102), then these quantities lie in the upper half of the complex plane. Therefore, the two phases φ_ε and φ_μ are always in the interval $[0,\pi]$. As a consequence, the phase of the complex index of refraction also lies in the interval $[0,\pi]$, whereas the phase of the impedance lies in the interval $[-\pi/2,\pi/2]$:

$$0 \leqslant \phi_n \leqslant \pi , \qquad -\frac{\pi}{2} \leqslant \phi_Z \leqslant \frac{\pi}{2} .$$

The phase ϕ_n is larger than $\pi/2$, and the real part of n is negative only if both susceptibilities are complex. In contrast, if only one of the susceptibilities shows a resonance, for instance if $\varphi_\mu = 0$, the phases ϕ_n and ϕ_Z both take values in the interval $[0,\pi/2]$, and the index of refraction has always a positive-semidefinite real part. As it seems unlikely that in nature one would find materials for which χ_e and χ_m have one or more resonances in the same range of frequencies, one understands the need to manufacture composite metamaterials to satisfy this condition.

[8] M. Born, K. Huang, *Dynamical Theory of Crystal Lattices*, Oxford University Press, 1954.

4.6 The Approximation of Paraxial Beams

Laser light beams are nearly monochromatic and are strongly collimated. The direction of propagation of a laser beam defines an optical axis from which no partial beam deviates appreciably. The first property implies that the Helmholtz equation (3.61), or (4.82), which describes harmonic functions with fixed wave number $k = 2\pi/\lambda$, may be used for the analysis of laser beams. The second property is made use of by constructing approximate solutions of this equation which describe *paraxial beams.*

4.6.1 Helmholtz Equation in Paraxial Approximation

As in Sect. 4.5.2, we consider a typical harmonic function which propagates predominantly along the z-direction,

$$u(\boldsymbol{x}) = a(\boldsymbol{x})\mathrm{e}^{ikz} , \tag{4.104}$$

and assume that its amplitude $a(\boldsymbol{x})$ varies little in the z-direction on the scale of wavelength λ. In this situation, we deal locally with a plane wave whose components remain nearly parallel to the optical axis, i.e. to the z-direction. More technically speaking, this assumption means that the second derivative of the amplitude by z can be neglected:

$$\begin{aligned} \partial_z^2 u(\boldsymbol{x}) &= \partial_z^2 \left(a(\boldsymbol{x})\mathrm{e}^{ikz}\right) \\ &= \left\{-k^2 a(\boldsymbol{x}) + 2ik\,\partial_z a(\boldsymbol{x}) + \partial_z^2 a(\boldsymbol{x})\right\} \mathrm{e}^{ikz} \\ &\simeq \left\{-k^2 a(\boldsymbol{x}) + 2ik\,\partial_z a(\boldsymbol{x})\right\} \mathrm{e}^{ikz} . \end{aligned}$$

In this approximation, the Helmholtz equation $(\boldsymbol{\Delta} + k^2)u(\boldsymbol{x}) = 0$ simplifies to

$$\left(\partial_x^2 + \partial_y^2 + 2ik\,\partial_z\right) a(\boldsymbol{x}) \simeq 0 , \tag{4.105a}$$

$$\left(\partial_\varrho^2 + \frac{1}{\varrho}\partial_\varrho + 2ik\,\partial_z\right) a(\boldsymbol{x}) \simeq 0 , \tag{4.105b}$$

in Cartesian and in cylindrical coordinates, respectively. In the conversion to cylindrical coordinates, we have made use of the definition $\varrho = \sqrt{x^2 + y^2}$ from which follow the formulae

$$\partial_x = \frac{x}{\varrho}\partial_\varrho , \qquad \partial_y = \frac{y}{\varrho}\partial_\varrho$$

for the first derivatives and

$$\begin{aligned} \partial_x^2 &= \frac{1}{\varrho}\left(1 - \frac{x^2}{\varrho^2}\right)\partial_\varrho + \frac{x^2}{\varrho^2}\partial_\varrho^2 , \\ \partial_y^2 &= \frac{1}{\varrho}\left(1 - \frac{y^2}{\varrho^2}\right)\partial_\varrho + \frac{y^2}{\varrho^2}\partial_\varrho^2 \end{aligned}$$

for the second derivatives. The approximate forms (4.105a) and (4.105b) of the Helmholtz equation (4.82) hold for light bunches which contain primarily beams remaining close to the beam axis. As will be shown in the next section, one can construct physical solutions analytically which describe strongly collimated (laser) beams.

4.6.2 The Gaussian Solution

A special solution of differential equation (4.105b), which is useful in the context described above, is obtained from an outgoing (or incoming) spherical wave with constant, possibly complex, amplitude a,

$$u^{(\mathrm{K})} = a\frac{1}{r}\,\mathrm{e}^{ikr} , \tag{4.106a}$$

by restricting the argument r to values close to the optical axis. As before, the optical axis is the z-axis, whereas x and y are coordinates in planes perpendicular to the z-axis. For values of $|x|$ and $|y|$ small compared to $|z|$, one has

$$r = \sqrt{x^2+y^2+z^2} = z\sqrt{1+\frac{\varrho^2}{z^2}} \simeq z + \frac{\varrho^2}{2z} .$$

The spherical wave then turns into a solution of the Helmholtz equation in paraxial approximation, viz.

$$a\frac{1}{r}\,\mathrm{e}^{ikr} \simeq a\frac{1}{z}\,\mathrm{e}^{ikz}\,\mathrm{e}^{ik\varrho^2/(2z)} =: \mathrm{e}^{ikz}\,a^{(0)}(\boldsymbol{x}) . \tag{4.106b}$$

Indeed, one easily verifies that

$$a^{(0)}(\boldsymbol{x}) = a\frac{1}{z}\,\mathrm{e}^{ik\varrho^2/(2z)} , \quad a \in \mathbb{C} \tag{4.107}$$

is a solution of the approximate differential equation (4.105b): The derivatives are found to be

$$\begin{aligned}
\partial_\varrho a^{(0)} &= \mathrm{i}ka\frac{\varrho}{z^2}\,\mathrm{e}^{ik\varrho^2/(2z)} = \mathrm{i}k\frac{\varrho}{z}a^{(0)} , \\
\partial_\varrho^2 a^{(0)} &= \mathrm{i}k\frac{1}{z}a^{(0)} - k^2\frac{\varrho^2}{z^2}a^{(0)} , \\
2\mathrm{i}k\,\partial_z a^{(0)} &= \left\{-\frac{2\mathrm{i}k}{z} + k^2\frac{\varrho^2}{z^2}\right\}a^{(0)} ,
\end{aligned}$$

and one verifies that (4.105b) holds.

Thus, in (4.107) we found a first, in fact rather simple, solution of the approximate Helmholtz equation. From this, further solutions can be derived by means of the following reasoning. Equation (4.105a) resembles the force-free Schrödinger equation in two space dimensions if z is interpreted as the time variable and x and y as the space variables. Much like classical autonomous systems, the Schrödinger equation is invariant under time translations. To time translations in the Schrödinger equation correspond translations in the variable z in (4.105a). Therefore, starting from (4.107), new solutions are generated by the replacement

$$z \longmapsto z - \zeta \,.$$

Choosing ζ to be a real number means shifting the origin of the z-axis from 0 to ζ. Note, however, that ζ must by no means be restricted to the reals. For instance, if one assigns to it a pure imaginary value,

$$\zeta = \mathrm{i}z_0 \quad \text{with} \quad z_0 \in \mathbb{R} \,,$$

this generates a particularly interesting solution, viz.

$$u^{(1)}(\boldsymbol{x}) = a^{(1)}(\boldsymbol{x}) \,, \quad \text{with} \quad a^{(1)}(\boldsymbol{x}) = \frac{a}{z - \mathrm{i}z_0} \mathrm{e}^{\mathrm{i}k\varrho^2/(2(z-\mathrm{i}z_0))} \,.$$

The properties of this function are analysed as follows. In a first step, one decomposes the function $(z - \mathrm{i}z_0)^{-1}$ in the first factor and in the exponent in terms of real and imaginary parts:

$$\frac{1}{z - \mathrm{i}z_0} \equiv \frac{1}{R(z)} + \mathrm{i}\frac{2}{kW^2(z)} \,. \tag{4.108a}$$

The real functions $R(z)$ and $W(z)$, which are defined by this ansatz, are

$$R(z) = z\left(1 + \frac{z_0^2}{z^2}\right) \,, \tag{4.108b}$$

$$W(z) = \sqrt{\frac{2z_0}{k}}\sqrt{1 + \frac{z^2}{z_0^2}} \equiv W_0\sqrt{1 + \frac{z^2}{z_0^2}} \,. \tag{4.108c}$$

In a second step, one rewrites the first factor in $a^{(1)}(\boldsymbol{x})$ in a way which is easier to interpret:

$$\frac{a}{z - \mathrm{i}z_0} = \frac{a}{(-\mathrm{i}z_0)} \frac{1}{1 + \mathrm{i}z/z_0} \equiv a^{(1)} w(z) \,, \tag{4.108d}$$

where the following abbreviations were introduced:

$$a^{(1)} = \frac{a}{(-\mathrm{i}z_0)} \,, \qquad w(z) = \frac{1}{1 + \mathrm{i}z/z_0} \,.$$

The function $w(z)$ is complex as well and may be decomposed in terms of modulus and phase:

$$w(z) = \left(1 + \frac{z^2}{z_0^2}\right)^{-1/2} e^{-i\phi(z)} , \quad \phi(z) = \arctan\left(\frac{z}{z_0}\right) . \tag{4.108e}$$

Inserting these intermediate steps and the definitions expressed by (4.108a)–(4.108e), the solution constructed in this way takes a form which is amenable to direct interpretation:

$$u^{(1)}(\boldsymbol{x}) = a^{(1)} \frac{W_0}{W(z)} e^{-\varrho^2/W^2(z)} e^{i[kz-\phi(z)+k\varrho^2/(2R(z))]} . \tag{4.109}$$

In this expression, $a^{(1)}$ is a constant, but in general complex, amplitude, and the constant W_0 is given by

$$W_0 = \sqrt{\frac{2z_0}{k}} = \sqrt{\frac{\lambda z_0}{\pi}} ; \tag{4.110}$$

the functions R and W are given by (4.108b) and (4.108c), respectively, and the phase $\phi(z)$ is given by (4.108e).

4.6.3 Analysis of the Gaussian Solution

The solution given by (4.109) is invariant under rotations about the z-axis. Thus, in cylindrical coordinates, it depends on the variables ϱ and z but not on the azimuth. Its intensity as a function of ϱ and z is given by

$$I(\varrho, z) = I_0 \left(\frac{W_0}{W(z)}\right)^2 e^{-2\varrho^2/W^2(z)} , \quad I_0 = \left|a^{(1)}\right|^2 . \tag{4.111}$$

For fixed values of z, this is a Gaussian curve in the variable ϱ, which is the narrowest for $z = 0$ and whose width increases with increasing z. Figure 4.13 shows the radial distributions $I(\varrho, z)/I_0$ for $z = 0$, $z = z_0$ and $z = 2z_0$ as functions of the variable $\zeta = \varrho/W_0$. It is the specific Gaussian shape of the solution given by (4.109) which gave rise to its name.

Taking ϱ to be zero and fixed, the ratio

$$\frac{I(\varrho = 0, z)}{I_0} = \frac{1}{1 + (z/z_0)^2}$$

is a function of $u = z/z_0$, as shown in Fig. 4.14.

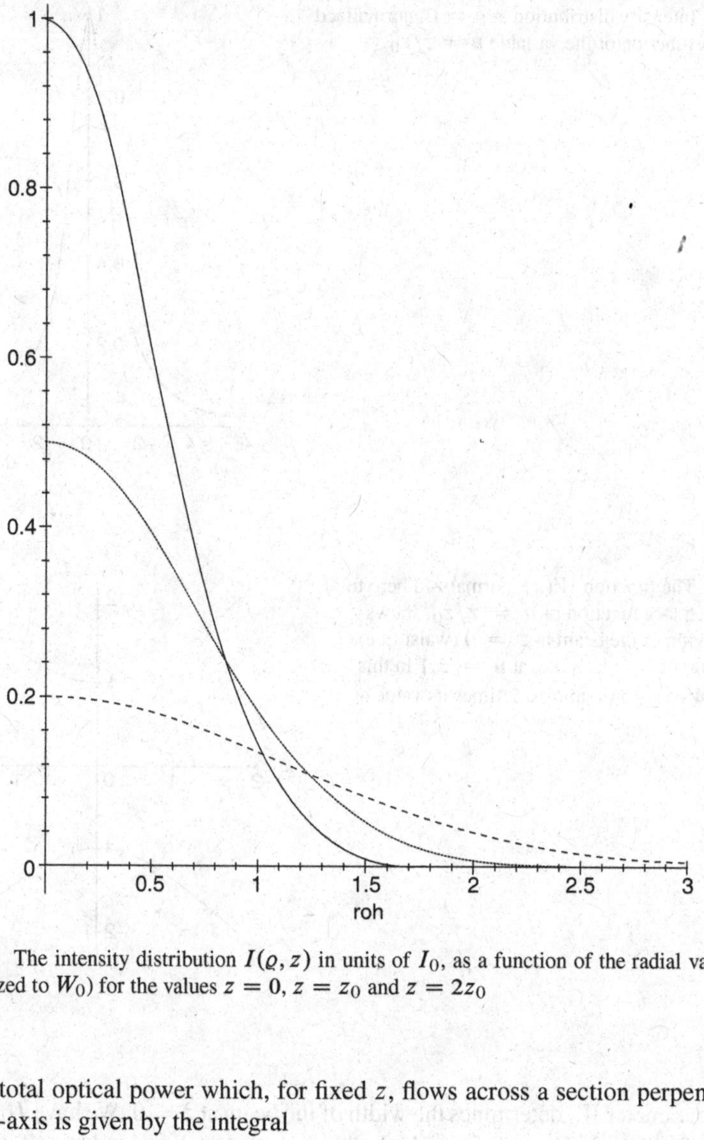

Fig. 4.13 The intensity distribution $I(\varrho, z)$ in units of I_0, as a function of the radial variable ϱ (normalized to W_0) for the values $z = 0$, $z = z_0$ and $z = 2z_0$

The total optical power which, for fixed z, flows across a section perpendicular to the z-axis is given by the integral

$$P = 2\pi \int_0^\infty \varrho \, \mathrm{d}\varrho \; I(\varrho, z)$$

$$= 2\pi I_0 \frac{W_0^2}{W^2(z)} \int_0^\infty \varrho \, \mathrm{d}\varrho \; \mathrm{e}^{-2\varrho^2/W^2(z)} = \frac{1}{2} I_0 \left(\pi W_0^2\right) . \qquad (4.112)$$

As expected, this expression is independent of z.

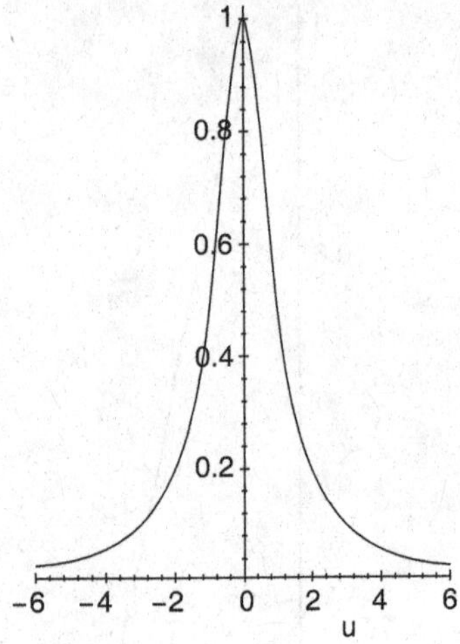

Fig. 4.14 Intensity distribution at $\varrho = 0$, normalized to I_0, as a function of the variable $u = z/z_0$

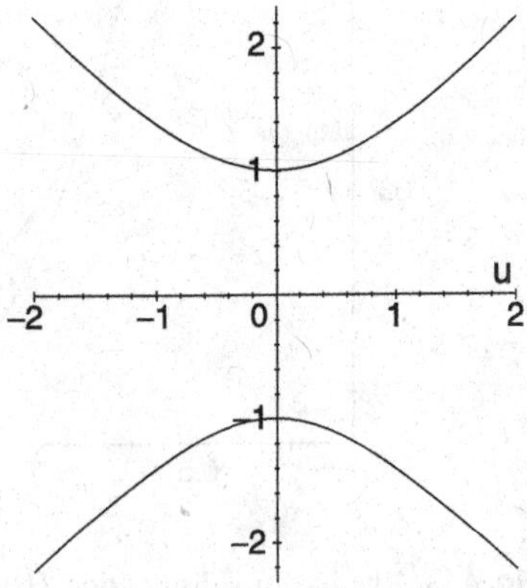

Fig. 4.15 The function $W(z)$, normalized here to W_0, plotted as a function of $u = z/z_0$, shows the narrowing of the beam at $z = 0$ (waist line). At the points $z = \pm z_0$, i.e. at $u = \pm 1$ in this diagram, $W(z)$ grows up to $\sqrt{2}$ times its value at $z = 0$

The parameter W_0 determines the width of the beam at $z = 0$: We have $I(\varrho, 0) = I_0 \exp\{-2\varrho^2/W_0^2\}$. The radius at which the intensity has decreased to half its value at $\varrho = 0$ has the value $\varrho_H = (\sqrt{\ln 2}/\sqrt{2})W_0$.

Figure 4.15 shows the function $W(z)/W_0$ in terms of the variable z/z_0. This figure illustrates two properties: The quantity W_0 characterizes the radius of the waist line of the beam at $z = 0$, and therefore it seems justified to interpret the product (πW_0^2) as the *beam size*. At $z = \pm z_0$ (in Fig. 4.15 these are the points ± 1), the radius of the beam has increased to $W_0\sqrt{2}$. The distance $(2z_0)$ is called the *confocal parameter*. Finally, one may calculate the angle between the asymptotes of the curve $W(z)$ as a function of z and thereby estimate the divergence of such a beam. For $z \gg z_0$ and using (4.108c) and (4.110), one has

$$W(z) \simeq (W_0/z_0)z = z\tan\theta \simeq z\,\theta\,, \quad (z \gg z_0)\,,$$
$$\tan\theta = \frac{W_0}{z_0} = \frac{\lambda}{\pi W_0} \simeq \theta\,. \tag{4.113}$$

Thus, the smaller the wave length λ and the larger the diameter $(2W_0)$ of the waist, the smaller will be the *divergence* (2θ). To take an example, consider a He–Ne laser with $\lambda = 6.33\cdot 10^{-7}$ m and with $W_0 = 5\cdot 10^{-5}$ m. Then one has $\theta \simeq 0.23$ deg. If this laser beam is directed towards the Moon, i.e. if we choose $z = 3.5\cdot 10^8$ m, it will have a diameter of $W(z) = 1.41\cdot 10^6$ m on the Moon's surface.

In studying the wave fronts of the solution given by (4.109), i.e. the surfaces of equal phase $\Phi(\varrho, z) =$ const, the following observation is helpful. The total phase of this solution

$$\Phi(\varrho, z) \equiv kz - \phi(z) + \frac{k\varrho^2}{R(z)}\,, \tag{4.114a}$$

evaluated at $\varrho = 0$, is

$$\Phi(\varrho = 0, z) = kz - \phi(z)\,. \tag{4.114b}$$

Besides the term kz from the plane wave, it contains a z-dependent shift which takes the values

$$\phi(-\infty) = -\frac{\pi}{2}\,, \quad \phi(0) = 0\,, \quad \phi(+\infty) = +\frac{\pi}{2}$$

at $\pm\infty$ and at 0, respectively. This shape can be read from the perspective representation in Fig. 4.16. Letting ϱ increase and moving to a small distance from the z-axis, one identifies immediately two limiting cases:

For $z \gg z_0$ one has $R(z) \simeq z$, and one recovers the approximate solution given by (4.106b). The wave fronts $\Phi(\varrho, z) =$ const. are approximately the same as those of the spherical wave given by (4.106a).

In turn, at $z = 0$, the function R becomes infinite, and the phase $\Phi(\varrho, 0) = 0$ is independent of ϱ, i.e. the wave front is a section of a vertical straight line.

As also shown in Fig. 4.16, the functions $R(z)$, (4.108b), and $\phi(z)$, (4.108e), change little when $\Phi(\varrho, z)$ is kept fixed. Thus, taking the function $R(z)$ to be constant, one sees that the surfaces $\Phi(\varrho, z) = 2\pi c$ with constant c are paraboloids

$$z + \frac{\varrho^2}{2R} \simeq c\lambda + \phi\frac{\lambda}{2\pi}$$

whose curvature is determined by R:

$$\frac{\mathrm{d}^2 z}{\mathrm{d}\varrho^2} \simeq -\frac{1}{R}\,.$$

These paraboloids are axially symmetric about the z-axis. For positive z their curvature is negative; for negative values of z it is positive. At $z = \pm z_0$, the function R

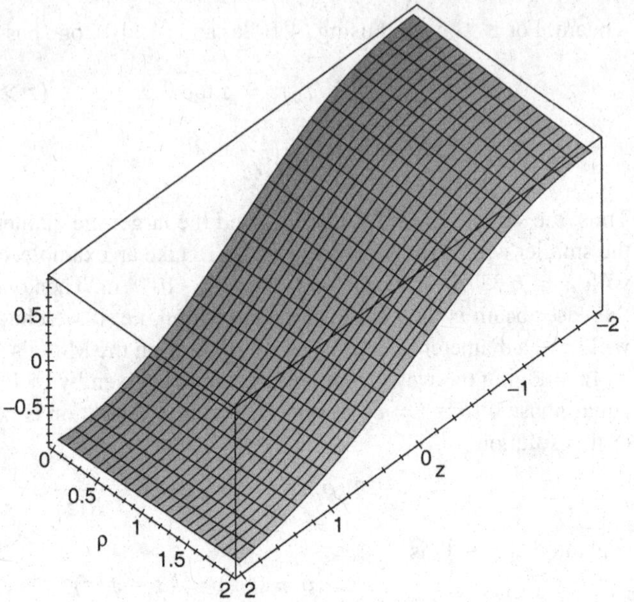

Fig. 4.16 Three-dimensional representation of the phase $\Phi(\varrho, z)$. For $\varrho = 0$ it runs from $-\pi/2$ to $+\pi/2$. If one uses only constant values of $\Phi(\varrho, z) = \text{const.}$, then z varies little

has the smallest modulus, whereas the modulus of the curvature is the largest. This clarifies the physical interpretation of the function $R(z)$: It determines the curvature radius of the wave fronts of the Gaussian beam.

4.6.4 Further Properties of the Gaussian Beam

Gaussian beams are of special interest for the optics of laser beams because in suitably constructed optical instruments, a Gaussian beam goes over into a Gaussian beam. We show this for the example of a thin double-convex lens (see Saleh and Teich (1991) for more details).

Consider a beam of the type expressed by (4.109) which is concentrated about $z = 0$. As sketched in Fig. 4.17, this beam is incident on a thin double-convex lens which is located at $z_L \neq 0$ and is held perpendicular to the optical axis. The double-convex lens may be thought of as a combination of two plano-convex lenses with equal and opposite curvature radii, one of which is oriented towards the right, the other towards the left of the figure. Regarding formula (4.86) for the phase shift as a function of $\varrho = \sqrt{x^2 + y^2}$, there is no modification. Only expression (4.84) for the focal distance is replaced by

$$\frac{1}{f} = \frac{2(n-1)}{r} . \tag{4.115}$$

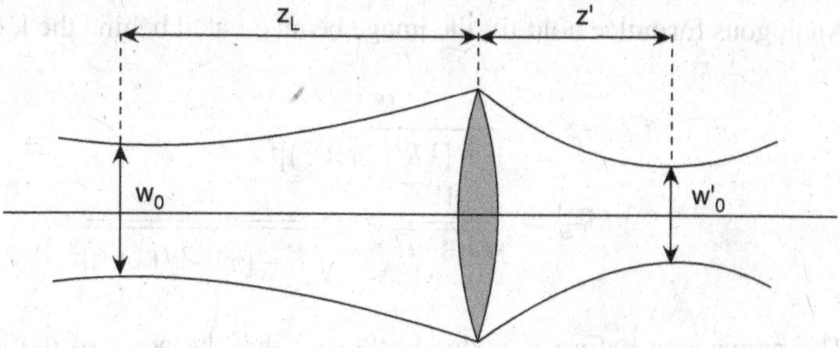

Fig. 4.17 A Gauss beam remains a Gauss beam after traversing a thin bi-convex lens

The phase, (4.114a), of the incident Gaussian beam receives an additional term from the factor $\exp\{-ik\varrho^2/(2f)\}$ of equation (4.86), so that the total phase of the transmitted wave at $z = z_L$ is equal to

$$\Phi'(\varrho, z_L) = kz_L - \phi(z_L) + \frac{k\varrho^2}{R'(z_L)}$$
$$= kz_L - \phi(z_L) + \frac{k\varrho^2}{R(z_L)} - \frac{k\varrho^2}{2f} . \tag{4.116}$$

One concludes from this that $R(z)$ and $R'(z)$ at the location z_L of the lens are related by the well-known equation

$$\frac{1}{f} = \frac{1}{R(z)} - \frac{1}{R'(z)} \qquad \text{at } z = z_L$$

of ray optics. At the same point, the functions $W(z)$ and $W'(z)$ are equal: $W'(z_L) = W(z_L)$.

If functions R and W are given at an arbitrary location z, it is not difficult to calculate the distance to the waist and the beam radius W_0 from these data. Let $u := z/z_0$, and use definitions (4.108b), (4.108c) and (4.110) to obtain

$$W^2 = W_0^2(1+u^2) = \frac{\lambda}{\pi} z_0(1+u^2) , \quad R = z_0 \frac{1}{u}(1+u^2) .$$

This shows that the ratio $\pi W^2/(\lambda R)$ equals u. From this follow an equation for the distance z and, by (4.108c), an equation which determines W_0. These are, respectively,

$$z = \frac{z_0}{u} \frac{(1+u^2)}{(1+1/u^2)} = \frac{R}{1+[\lambda R/(\pi W^2)]^2} , \tag{4.117a}$$

$$W_0 = \frac{W}{\sqrt{1+u^2}} = \frac{W}{\sqrt{1+[\pi W^2/(\lambda R)]^2}} . \tag{4.117b}$$

Analogous formulae hold for the image beam created behind the lens, viz.

$$-z' = \frac{R'}{1 + [\lambda R'/(\pi W^2)]^2} , \tag{4.118a}$$

$$W'_0 = \frac{W}{\sqrt{1+u^2}} = \frac{W}{\sqrt{1 + [\pi W^2/(\lambda R')]^2}} . \tag{4.118b}$$

The minus sign before z' is due to the fact that the waist of the image beam lies beyond the lens, i.e. to the right of it. Using these formulae it is now easy to relate the parameters of the original beam and the transmitted beam. Using the abbreviations

$$r := \frac{z_0}{z-f} , \qquad A_r := \frac{f}{|z-f|} , \tag{4.119a}$$

one defines the *magnification*

$$A := \frac{A_r}{\sqrt{1+r^2}} . \tag{4.119b}$$

The radius and the position of the waist of the transmitted beam are given by, respectively,

$$W'_0 = A\, W_0 , \qquad z' - f = A^2(z-f) . \tag{4.120}$$

Points z'_0 and z_0, where the curvature radii are minimal, are related by $z'_0 = A^2 z_0$, and the divergence of the transmitted beam is related to the divergence of the original beam by $(2\theta') = (2\theta)/A$, see (4.113).

It is instructive to study the limit of geometric optics in these formulae. If

$$z - f \gg z_0 ,$$

then the lens is located far from the confocal parameter. Parameter r is small compared to 1, and we have $A \simeq A_r$. The beam is again approximately a spherical wave. As for the parameters of the transmitted beam, the formulae of (4.120) simplify as follows:

$$W'_0 \simeq A_r W_0 , \tag{4.121a}$$

$$\frac{1}{f} \simeq \frac{1}{z'} + \frac{1}{z} , \tag{4.121b}$$

$$A \simeq A_r = \frac{f}{|z-f|} . \tag{4.121c}$$

We note once more this important result:

A Gaussian beam traversing one or several lenses is mapped to a Gaussian beam.

As an example, consider a set-up in which the lens is located at $z = 0$, i.e. exactly at the waist line of the incoming beam. The equations of (4.120) show that the image beam is focused to a waist radius of

$$W_0' = \frac{W_0}{\sqrt{1 + (z_0/f)^2}}$$

and that $z' = f/(1 + (f/z_0)^2)$. In the limit of geometric optics, these equations reduce to $W_0 \simeq \theta f$ and $z' \simeq f$. In this limit, the focussing effect becomes even more pronounced.

Remarks

1. By combining two lenses, one behind the other, an incoming Gaussian beam can be made broader or smaller, depending on the choice of focal lengths. By arranging a sequence of parallel identical lenses, one obtains an optical instrument which allows for the transport of a Gaussian beam over large distances without modification to its shape. Thus, such an arrangement serves the purpose of beam transport in laser physics.
2. The Helmholtz equation in paraxial approximation has further, more general solutions which are useful for the description of laser beams. One can even make use of *complete* systems of functions such as, for instance, the Hermite polynomials known from quantum mechanics, for expanding solutions in terms of these polynomials. For more information and details, see Saleh and Teich (1991).

Local Gauge Theories

5

5.1 Introduction

Despite being a principle of *classical* field theory, gauge invariance of electrodynamics revealed its deep significance and found its far-reaching interpretation only in relation to quantum mechanics of electrons and the Schrödinger equation. In this chapter, we study the generalization of the concept of a locally invariant gauge theory to non-Abelian gauge groups constructed by following the model of Maxwell theory. This generalization may seem a little academic at first glance because, besides the Maxwell field, it contains further massless gauge fields which are unknown to macroscopic physics. However, it becomes physically realistic if it is combined with the phenomenon of spontaneous symmetry breaking. Both concepts, non-Abelian gauge theory and spontaneous symmetry breaking, initially are purely classical concepts. At the same time, one lays the (classical) foundations for the gauge theories of the fundamental interactions which nowadays are generally accepted and whose validity has been confirmed by numerous experiments. This chapter describes the foundations for the construction of such a theory, within a classical (i.e. nonquantum) framework. Only when introducing fermionic particles (such as quarks and leptons) does the quantization of gauge theories become mandatory.

5.2 Klein–Gordon Equation and Massive Photons

A particularly simple example of a Lorentz covariant field theory which lives on Minkowski space $M = \mathbb{R}^4$ is given by the Lagrange density (3.17a)

$$\mathcal{L}(\phi(x), \partial_\mu \phi(x)) = \frac{1}{2}\left[\partial_\mu \phi(x)\partial^\mu \phi(x) - \kappa^2 \phi^2(x)\right] - \varrho(x)\phi(x) \,. \tag{5.1}$$

F. Scheck, *Classical Field Theory*, Graduate Texts in Physics,
DOI 10.1007/978-3-642-27985-0_5, © Springer-Verlag Berlin Heidelberg 2012

The constants $\hbar$ and c that we introduced there had the purpose of assigning the correct physical dimension to the Lagrange density, viz. (energy/volume). However, as the Euler–Lagrange equations are linear and homogeneous in $\mathcal{L}$, the factor $1/(\hbar c)$ cancels out and may be dropped without loss of generality. The points of the spacetime continuum are denoted by $x \in \mathbb{R}^4$. As before, the derivatives are decomposed according to

$$\partial_\mu = (\partial_0, \nabla)^T, \qquad \partial^\mu = g^{\mu\nu}\partial_\nu = (\partial_0, -\nabla)^T ;$$

$\phi(x)$ is a Lorentz scalar field, i.e. a field which does not change under the action of a Lorentz transformation $\Lambda \in L_+^\uparrow$,

$$x \mapsto x' = \Lambda x : \qquad \phi(x) \mapsto \phi'(x') = \phi(x) .$$

The quantity $\varrho(x)$, which is also scalar, represents an external source – in the same sense as charge and current densities are external sources in Maxwell's equations for the radiation field. The constant κ is a parameter with the dimension of an inverse length. In quantum theory, κ is the reciprocal of the Compton wavelength

$$\frac{1}{\kappa} = \frac{\lambda}{2\pi} = \frac{\hbar}{mc} \tag{5.2}$$

of a particle with mass m. In natural units where both the speed of light and Planck's constant take the value 1, $c = 1$ and $\hbar = 1$, the parameter $\kappa = m$ is simply the mass without any other factor.

The Euler–Lagrange equations (3.16), of which there is only one in this example, yield the equation of motion (3.17b)

$$\left(\Box + \kappa^2\right)\phi(x) = -\varrho(x) . \tag{5.3a}$$

We repeat it here for the following reasons:

(i) By inserting a plane wave as in (4.4a) into the homogeneous version of equation (5.3a),

$$\left(\Box + \kappa^2\right)\phi(x) = 0 , \tag{5.3b}$$

i.e. $\phi(t, \boldsymbol{x}) = \mathrm{e}^{-\mathrm{i}\omega t}\,\mathrm{e}^{\mathrm{i}\boldsymbol{k}\cdot\boldsymbol{x}}$, one obtains the dispersion relation

$$\omega^2 - \boldsymbol{k}^2 - \kappa^2 = 0 . \tag{5.4a}$$

Upon multiplication by $\hbar^2$ and using (5.2), this becomes a relation between the energy $E = \hbar\omega$, the momentum $\boldsymbol{p} = \hbar\boldsymbol{k}$ and the mass m

$$E^2 = \boldsymbol{p}^2 + \left(mc^2\right)^2 , \tag{5.4b}$$

which is simply the relation between energy and momentum of a free particle with mass m, well known from special relativity.

(ii) It is not mandatory to postulate equation (5.3b) only for scalar fields. It could be stipulated equally well for a vector field

$$V(t, \boldsymbol{x}) = \boldsymbol{v}\, \mathrm{e}^{-\mathrm{i}\omega t}\, \mathrm{e}^{\mathrm{i}\boldsymbol{k}\cdot\boldsymbol{x}} ,$$

where $\boldsymbol{v}$ is a constant vector or, more generally, a four-vector field. The Klein–Gordon equation then applies to each of the four components $V^{\mu}(t, \boldsymbol{x})$ of the vector field individually. A similar statement applies to tensor fields of any degree and, in fact, also to spinor fields. Every component satisfies the Klein–Gordon equation on its own.

(iii) A static solution of the inhomogeneous equation (5.3a) with the static source

$$\varrho(t, \boldsymbol{x}) \equiv \varrho(\boldsymbol{x}) = g\delta(\boldsymbol{x})$$

is easily found. In a static situation, using $\Box = (1/c^2)\partial^2/\partial t^2 - \boldsymbol{\Delta}$, (5.3a) reduces to the time-independent differential equation

$$(\boldsymbol{\Delta} - \kappa^2)\phi(\boldsymbol{x}) = \varrho(\boldsymbol{x}) = g\delta(\boldsymbol{x}) .$$

Without specific boundary conditions the solution reads

$$\phi^{\text{stat}}(\boldsymbol{x}) = -\frac{g}{4\pi}\frac{\mathrm{e}^{-\kappa r}}{r} . \tag{5.5}$$

This solution can be derived more directly by solving the corresponding algebraic equation obtained by Fourier transformation and by calculating the inverse Fourier transform of its solution (Exercise 5.1). Alternatively, it may be obtained from the corresponding Green function of the Helmholtz equation (4.8) or (4.32) by analytic continuation in the (initially real) wave number k:

$$k \longrightarrow \mathrm{i}\kappa : \; -\frac{\mathrm{e}^{\mathrm{i}kr}}{4\pi r} \longrightarrow -\frac{\mathrm{e}^{-\kappa r}}{r} .$$

A pointlike source whose "strength" is g and which is located in the origin creates a field (5.5) which decreases exponentially with increasing distance. The rate of decay is determined by the Compton wavelength corresponding to the mass m. The larger the mass, the faster the fall-off of the solution. In turn, if the mass vanishes, then the solution given by (5.5) takes the form of the Coulomb potential, which falls off as $1/r$. At the same time, the Klein–Gordon equation goes over into wave equation (1.45).

Let us return to Maxwell theory and try to introduce a mass term of the kind considered in (5.3a). The simplest way to do this is by modifying the Lagrange density (3.28) as follows:

$$\mathcal{L}_{\text{Proca}}(A_\tau, \partial_\sigma A_\tau) = -\frac{1}{16\pi} F_{\mu\nu}(x) F^{\mu\nu}(x) + \frac{\lambda^2}{8\pi} A_\mu(x) A^\mu(x) - \frac{1}{c} j^\mu(x) A_\mu(x) . \tag{5.6}$$

On the way to the Euler–Lagrange equations one calculates

$$\frac{\partial \mathcal{L}_{\text{Proca}}}{\partial A_\tau} = -\frac{1}{c} j^\tau + \frac{\lambda^2}{4\pi} A^\tau(x) ,$$

$$\frac{\partial \mathcal{L}_{\text{Proca}}}{\partial(\partial_\sigma A_\tau)} = -\frac{1}{4\pi} F^{\sigma\tau}(x) .$$

For two of the three terms on the right-hand side these are the same calculations as in Sect. 3.4. In the new term, which is proportional to λ^2, one must take into account a factor of 2 because

$$A_\tau(x) A^\tau(x) = A_\tau(x) g^{\tau\lambda} A_\lambda(x) ,$$

where the sum must be taken over repeated contragredient indices. Thus, the equations of motion which replace the inhomogeneous Maxwell equations are

$$\partial_\sigma F^{\sigma\tau}(x) + \lambda^2 A^\tau(x) = \frac{4\pi}{c} j^\tau . \tag{5.7}$$

If one represents the field strength tensor by potentials,

$$F^{\sigma\tau}(x) = \partial^\sigma A^\tau(x) - \partial^\tau A^\sigma(x) ,$$

and assumes that the potential satisfies the Lorenz condition $\partial_\mu A^\mu(x) = 0$, there remains the differential equation

$$(\Box + \lambda^2) A^\tau(x) = \frac{4\pi}{c} j^\tau . \tag{5.8}$$

The Lagrange density, (5.6), is called the *Proca Lagrange density*, after A. Proca, who discussed this equation in the 1930s.

▶ **Remarks**

1. In comparing the Lagrange densities given by (5.1) for the scalar field and (5.6) for the Proca vector field, there is a certain analogy: Both contain a mass term which reads $-(\kappa^2/2)\phi^2(x)$ in one case, and $(\lambda^2/8\pi) A_\mu(x) A^\mu(x)$ in the other. Both Lagrange densities contain an interaction term with an external

source given by the scalar density $\varrho(x)$ in (5.1) and the current density $j^\mu(x)$ in (5.6). It is suggestive to interpret the first term in (5.6),

$$-\frac{1}{16\pi} F_{\mu\nu}(x) F^{\mu\nu}(x) ,$$

as the kinetic term of the vector field, which in the case of Maxwell fields is given by $(1/8\pi)(\boldsymbol{E}^2 - \boldsymbol{B}^2)$ (see (3.25a)). This term in the Lagrange density, (3.28), of Maxwell theory, including its characteristic sign, yields the postive-definite energy density, expressed by (3.49a), of the free Maxwell field.

2. If the photon had a nonvanishing mass, then equation (5.8) would be the correct equation of motion for photons, replacing the wave equation. An instructive discussion of this equation of motion and its physical consequences, including a list of original publications on this topic, is found in Jackson (1999).

What matters for the subjects treated in this chapter is the loss of gauge invariance of the Lagrange density $\mathcal{L}_{\text{Proca}}$. Whereas the first and last terms in (5.6) are gauge invariant, as shown in Sect. 3.5.2, this is not true for the mass term: Under the action of a gauge transformation

$$A_\tau(x) \longmapsto A'_\tau(x) = A_\tau(x) - \partial_\tau \Lambda(x) ;$$

the quadratic term in the potentials becomes

$$\begin{aligned} A_\tau(x) A^\tau(x) &\mapsto A'_\tau(x) A'^\tau(x) \\ &= A_\tau(x) A^\tau(x) - 2 A_\tau(x) \partial^\tau \Lambda(x) + \partial_\tau \Lambda(x) \partial^\tau \Lambda(x) , \end{aligned}$$

which cannot be reduced to its original form. On the one hand, the physical significance of this result is that the potential $A_\mu(x)$ acquires physical reality – in contrast to Maxwell theory. On the other hand, it establishes a direct relation between the masslessness of the photon, the infinite range of the Coulomb potential and the gauge invariance of Maxwell's equations. We note the important result:
A genuine mass term in the Lagrange density of Maxwell theory is incompatible with gauge invariance.

5.3 The Building Blocks of Maxwell Theory

Maxwell's theory of electromagnetic phenomena is the prototype of a gauge theory and serves as a guide for constructing all other gauge theories which are relevant for the description of fundamental interactions. Therefore, before turning to more general gauge theories, it would be helpful to recall the building blocks that are used in constructing Maxwell theory. We do this according to the following outline.

The Underlying Spacetime

As far as is has been developed until this point, Maxwell's theory presupposes a flat Minkowski space $M = \mathbb{R}^4$ as the model of spacetime. This space has dimension 4 and is equipped with the metric

$$\mathrm{g} = \mathrm{diag}(1, -1, -1, -1) . \tag{5.9}$$

More precisely, Minkowski space is a flat semi-Riemannian manifold with *index* $\nu = 1$. In fact, a better notation would be $\mathbb{R}^{(1,3)}$, in order to emphasize the special role of the time variable. The definition of the notion of index is as follows:

▶ **Definition 5.1 Index of a bilinear form** On a vector space V with finite dimension $n = \dim V$, let there be a nondegenerate symmetric bilinear form. The index of the bilinear form is the codimension of the largest subspace W of V,

$$\nu = \dim V - \dim W , \tag{5.10}$$

on which the metric is *definite*, positive-definite or negative-definite.[1]

The metric of Minkowski space is a bilinear form of this kind. Given the signs as chosen by (5.9), the metric is negative-definite on three-dimensional space. It would be positive-definite had we chosen the convention $\mathrm{g} = \mathrm{diag}(-1, 1, 1, 1)$. In either case, this subspace has dimension 3, and its codimension is $\nu = \dim M - 3 = 4 - 3 = 1$, independent of the choice of convention for the metric.

The causal structure on M is determined by the Poincaré or the Lorentz group of transformations,

$$x \longmapsto x' = \Lambda x + a , \quad \text{with} \quad \Lambda^T \mathrm{g} \Lambda = \mathrm{g} .$$

Among others, this structure manifests itself in the retardation effects in the propagation of electromagnetic signals.

The Variables

In the vacuum, the tensor fields $F_{\mu\nu}(x)$ of electromagnetic field strengths are the essential observables of the theory. When matter is present, they are replaced by the tensor fields $\mathcal{F}_{\mu\nu}(x)$ and are supplemented by the charge and current densities $j^\mu(x) = (c\varrho(x), \boldsymbol{j}(x))^T$ of matter, which appear as the driving source terms in Maxwell's equations. The vector potential $A_\mu(x)$, though somewhat problematic, is an important variable, too. It is contained in the Lagrangian describing the coupling of the radiation field to matter, but it is not directly observable. It *may* be a four-vector field, depending on the class of gauges one has chosen, but it may have a more complicated transformation behaviour without disturbing the Lorentz covariance of the theory.

[1] The codimension of a subspace W of the finite-dimensional vector space V is $\mathrm{codim}\, W := \dim V - \dim W$.

Once the partition of Minkowski space $\mathbb{R}^{(1,3)}$ into a space part $\mathbb{R}^3$ and time axis $\mathbb{R}_t$ is fixed, through the choice of a class of reference frames, the tensor field $F_{\mu\nu}(x)$ or $\mathcal{F}_{\mu\nu}(x)$ is decomposed into the electric field quantities $\boldsymbol{E}(t,\boldsymbol{x})$ or $\boldsymbol{D}(t,\boldsymbol{x})$, respectively, and the magnetic field quantities $\boldsymbol{B}(t,\boldsymbol{x})$ or $\boldsymbol{H}(t,\boldsymbol{x})$, respectively. These are indispensable for testing the theory against experiment, but they have a complicated transformation behaviour if the frame of reference is changed by a special Lorentz transformation.

Gauge Transformations, Structure Group and Gauge Group

As noted in Sect. 3.5.2 and in a remark in Sect. 2.3.5, Maxwell theory is invariant under *global* gauge transformations

$$G = \mathrm{U}(1) = \left\{ \mathrm{e}^{\mathrm{i}\alpha} \middle| \alpha \in \mathbb{R} \quad (\mathrm{mod}\ 2\pi) \right\} , \tag{5.11a}$$

as well as under *local* gauge transformations

$$\mathfrak{G} = \left\{ \mathrm{e}^{\mathrm{i}\alpha(x)} \middle| \alpha \in \mathfrak{F}(\mathbb{R}^{(1,3)}) \quad (\mathrm{mod}\ 2\pi) \right\} , \tag{5.11b}$$

where $\mathfrak{F}(\mathbb{R}^{(1,3)})$ denotes the set of smooth functions on Minkowski space. The group proper, i.e. the group given by (5.11a) in the case being discussed here, will be called the *structure group* in the sequel. The infinite-dimensional group, (5.11b), which is constructed from it is called the *gauge group*. It determines the explicit form of the gauge transformations.

As was shown in Sect. 3.5.2, the structure group of Maxwell theory is a U(1), i.e. an Abelian, group. This group has one generator only, which we denote by $\mathbb{1}$.

Geometric Structure of Maxwell Theory

From a geometrical point of view, the potential is a one-form $A_\mu(x)\,\mathrm{d}x^\mu$ over Minkowski space. The action of a gauge transformation, (2.59), generated by the function $\chi(x)$ is an affine transformation of the potential, viz.

$$A'_\tau(x) = A_\tau(x) - \partial_\tau \chi(x) .$$

Written in terms of exterior forms, this reads

$$\omega_{A'} = \omega_A - \mathrm{d}\chi ,$$

where $\omega_A = A_\mu(x)\,\mathrm{d}x^\mu$ is defined as in (2.82). Obviously, this is an infinitesimal transformation and shows that the potential is not only a one-form but also an element of the Lie algebra of the gauge group.[2]

[2] Of course, the sign of the gauge function $\chi(x)$ is not relevant because the gauge function is an arbitrary smooth function. I have chosen the minus sign here in order to be in accord with standard conventions as in Sect. 3.5.2 and with the signs in (3.39a).

For the sequel it is useful to define the one-form for the potential as in (2.88a), i.e. by inserting a reference charge q – e. g. the elementary charge e – and a factor i:[3]

$$A := \mathrm{i}q\,\omega_A = \mathrm{i}q\,A_\mu(x)\,\mathrm{d}x^\mu\ . \tag{5.12}$$

By multiplying also the gauge function by these factors, the gauge transformation has the same form as it did previously:

$$A \longmapsto A' = A + \mathrm{d}\Lambda\ , \qquad \text{with} \quad \Lambda(x) = -\mathrm{i}q\,\chi(x)\ . \tag{5.13}$$

Equations (2.88b) and (2.88c) for the covariant derivative and the curvature form (two-form of field strengths), respectively, apply as previously. They are repeated here:

$$D = d + A\ , \tag{5.14a}$$

$$D^2 = (\mathrm{d}A) + A \wedge A = (\mathrm{d}A) = F\ ,$$

$$\Bigl(F = \mathrm{i}q \sum_{\mu<\nu} F_{\mu\nu}(x)\,\mathrm{d}x^\mu \wedge \mathrm{d}x^\nu\Bigr)\ . \tag{5.14b}$$

Example 3.4 of the Schrödinger equation coupled to the radiation field showed that the term

$$-\frac{1}{2m}(D\psi^*)(D\psi) \tag{5.15}$$

remains invariant under local gauge transformations if, concurrently with transformation (5.13) of the potential, the wave function is transformed as follows:

$$\psi(x) \longmapsto \psi'(x) = g(x)\psi(x)\ , \quad \text{with} \quad g(x) = \mathrm{e}^{\mathrm{i}q\chi(x)} = \mathrm{e}^{-\Lambda(x)}\ . \tag{5.16}$$

This is a remarkable observation: The full theory, which yields the Maxwell equations and the Schrödinger equation for a charged particle, is gauge invariant if in the kinetic energy of that particle the ordinary partial derivative is replaced by the covariant derivative.

Remarks

1. The preceding conclusion is not restricted to the Schrödinger equation because, evidently, it is only the kinetic energy, given by (5.15), which matters. This form of the kinetic energy of a matter particle, however, is fairly general (see e. g. (5.1)).
2. The gauge transformation $g(x)$, equation (5.16), which acts on the field $\psi(x)$ is an element of the gauge group $\mathfrak{G}$. Its infinitesimal form is obtained by expanding $g(x)$ in terms of $\chi(x)$, viz.

$$g(x) \simeq 1 + \mathrm{i}q\chi(x)\ , \quad |\chi(x)| \ll 1\ ,$$

[3] The factors $\hbar$ and c which appear there are not essential. In a quantized theory, they can be reinserted at any stage. Alternatively, they may be replaced by 1 by the choice of natural units.

so that the corresponding variation of ψ reads

$$\delta\psi = iq\chi(x)\psi(x) . \qquad (5.16a)$$

At this level one returns to the Lie algebra Lie $(\mathfrak{G})$ of the gauge group. As the group U(1) is Abelian, its Lie algebra has only one generator and the variation $\delta\psi$, equation (5.16a), contains only one single term.

5.4 Non-Abelian Gauge Theories

The schematic summary of Maxwell theory with special emphasis on its gauge invariance and its geometric structure given above suggests one should try the same construction with more general, non-Abelian groups replacing U(1). It is reported that Wolfgang Pauli tried this idea by "gauging" the isospin introduced earlier by Werner Heisenberg, i.e. the Lie group SU(2), but dismissed it for physical reasons (described subsequently). In published form, this construction was proposed by C.N. Yang and R.L. Mills in 1954.[4] Therefore, one also talks about *Yang–Mills theory* as a synonym for a *local gauge theory*.

In this section, we work out this concept following the model of Maxwell's theory. We construct a Lagrange density for a gauge theory and study its coupling to matter fields in its simplest form.

5.4.1 The Structure Group and Its Lie Algebra

For the structure group G which is to replace the U(1) of Maxwell theory, only a *compact* Lie group should be considered because compactness guarantees a (generalized) kinetic term in a Lagrange density which has the correct, physical, sign.

Without striving for too much mathematical rigour, one may characterize compact Lie groups as follows. A (finite-dimensional) Lie group is a smooth manifold G of transformations which satisfy the group axioms and for which the product $g_2 \cdot g_1$ of two elements g_1 and g_2 and the inverse g^{-1} of every element g are smooth functions (i.e. are differentiable functions of the group parameters).

The example of the special orthogonal group SO(3) in three real dimensions, well known from the theory of spinning tops and from nonrelativistic quantum mechanics, illustrates the concept of structure group:

$$\mathrm{SO}(3) = \left\{ \mathrm{R} \in M_3(\mathbb{R}) \,|\, \mathrm{R}^T\mathrm{R} = \mathbb{1}, \det \mathrm{R} = 1 \right\} . \qquad (5.17)$$

The symbol $M_3(\mathbb{R})$ denotes the set of real 3×3 matrices; the term "orthogonal" denotes the property that the inverse of R equals the transposed matrix, and "special"

[4] C.N. Yang and R.L. Mills, Phys. Rev. **96** (1954) 191.

means the restriction to $\det \mathsf{R} = +1$. As it is defined over the reals, a more explicit notation would be SO(3,$\mathbb{R}$).

As a concrete example, every rotation may be characterized by three Eulerian angles, ϕ, θ and ψ, as follows:

$$\mathsf{R} = \mathsf{R}(\phi,\theta,\psi)$$
$$= \begin{pmatrix} [\cos\phi\cos\theta\cos\psi - \sin\phi\sin\psi] & [\sin\phi\cos\theta\cos\psi + \cos\phi\sin\psi] & -\sin\theta\cos\psi \\ [-\cos\phi\cos\theta\sin\psi - \sin\phi\cos\psi] & [-\sin\phi\cos\theta\sin\psi + \cos\phi\cos\psi] & \sin\theta\sin\psi \\ \cos\phi\sin\theta & \sin\phi\sin\theta & \cos\theta \end{pmatrix}, \quad (5.17a)$$

their domain of variation being

$$\phi \in [0, 2\pi], \quad \theta \in [0, \pi], \quad \psi \in [0, 2\pi]. \quad (5.17b)$$

The elements of the group SO(3) depend in a differentiable way on three parameters, each of which is confined to a compact interval. In this case, the group itself is called *compact*.[5] In this connection, it is also important to recall the fact that a function which is defined on a compact set is bounded.

Remark
The Lorentz group provides a counterexample of a noncompact group which is of paramount importance for physics: The special Lorentz transformations depend on a parameter λ which is defined in the infinite interval $[0, \infty)$. This group is not compact.

Let G be a simple or semisimple compact Lie group. Examples which will be studied in detail include the unitary group

$$\mathsf{U}(2) = \left\{\mathsf{U} \in M_2(\mathbb{C}) \,|\, \mathsf{U}^\dagger \mathsf{U} = \mathbb{1}\right\}, \quad (5.18a)$$

as well as its restriction, the unitary, unimodular group

$$\mathsf{SU}(2) = \left\{\mathsf{U} \in M_2(\mathbb{C}) \,|\, \mathsf{U}^\dagger \mathsf{U} = \mathbb{1}, \det \mathsf{U} = 1\right\}. \quad (5.18b)$$

(Again, strictly speaking, these should be denoted U(2, $\mathbb{C}$) and SU(2, $\mathbb{C}$), respectively.) Historically, it was the second of these which was used as a first example for the construction of a gauge theory.

The structure group being the group of global gauge transformations, the identity must of course be contained in the domain of the parameters on which the

[5] This is in agreement with the notion of compactness in set theory: Every infinite-dimensional subset of a compact set M contains a series whose limit is an element of the set.

group elements depend. In the example of rotation group SO(3), this is the element $\mathsf{R}(\phi = 0, \theta = 0, \psi = 0) = \mathbb{1}$. Had one chosen the full rotation group O(3), which contains also orthogonal 3×3 matrices with $\det M = -1$, one would have had to restrict it to its subgroup SO(3), the so-called connection component of unity. Generally, the connection component of unity is the subgroup whose elements can be deformed to the identity $\mathbb{1}$ by continuous variation of the parameters.

The elements of such groups can be written as exponential series in the generators and the parameters:

$$g = \exp\{\mathrm{i} \sum \alpha_k \mathsf{T}_k\} . \tag{5.19}$$

They are smooth functions of the real variables α_k. The generators T_k span the Lie algebra $\mathfrak{g} = \text{Lie}\,(G)$, i.e. they form a basis of the Lie algebra. This algebra is characterized by the commutators

$$[\mathsf{T}_i, \mathsf{T}_j] = \mathrm{i} C_{ij}^k \mathsf{T}_k , \qquad i, j, k = 1, 2, \ldots, \dim \mathfrak{g} \tag{5.20}$$

(where the sum over the repeated indices k is to be taken). The *structure constants* C_{ij}^k are real. They are not fixed once and forever to some canonical form. Indeed, by nonsingular linear transformations of the generators the structure constants can take different forms. However, they have some general properties which are independent of the special choice of basis of the Lie algebra: (i) They are antisymmetric in indices i and j. (ii) From the Jacobi identity for the generators

$$\left[[\mathsf{T}_i, \mathsf{T}_j], \mathsf{T}_k\right] + \text{(two cyclic permutations of } i, j, k) = 0$$

follows an identity for the structure constants, viz.

$$C_{ij}^m C_{mk}^l + C_{jk}^m C_{mi}^l + C_{ki}^m C_{mj}^l = 0 , \tag{5.21}$$

where m is the summation index, all indices taking the values 1 to $\dim \mathfrak{g}$. This relation, too, is called a Jacobi identity.

The freedom in choosing the basis of the Lie algebra can be utilized so as to obtain structure constants which are antisymmetric in all *three* indices. This construction will be given below, together with an example.

In many applications in physics, the *representation theory* of compact Lie groups and of their Lie algebras is of central importance. A representation is a map of a group into a (in general) complex vector space of dimension n,

$$\varrho : G \longrightarrow V : g \longmapsto \mathsf{U}(g) ,$$

which is compatible with the group axioms. At the same time, its Lie algebra is mapped by

$$\varrho : \mathfrak{g} \longrightarrow V : \mathsf{T}_k \longmapsto \mathsf{U}(\mathsf{T}_k) ,$$

such that the $\mathsf{U}(\mathsf{T}_k)$ obey the same commutators as the generators. Thus, the generators, which initially are given in the defining representation only, are represented by finite-dimensional matrices $U_{pq}(\mathsf{T}_k)$, $p,q = 1,2,\ldots,n$. (More precisely, a representation is a homomorphism of $\mathfrak{g}$ into the linear algebra $gl(V)$ of a vector space which preserves the commutators of the generators.) As the hermitian-conjugate generators $\mathsf{T}_k^\dagger$ obey the same Lie algebra, given by (5.20), as the generators T_k,

$$[\mathsf{T}_i, \mathsf{T}_j]^\dagger = [\mathsf{T}_j^\dagger, \mathsf{T}_i^\dagger] = -\mathrm{i}C_{ij}^k \mathsf{T}_k^\dagger = \mathrm{i}C_{ji}^k \mathsf{T}_k^\dagger ,$$

one can represent the T_k by finite-dimensional, *hermitian* matrices.

Two classes of representations are especially important: the *fundamental* representation, also called the *defining* representation, and the *adjoint* representation. The defining representation is the faithful representation (i.e. different from the trivial representation) whose dimension is the lowest. In the older literature, it was also called *spinor representation.* The adjoint representation is defined by the structure constants and, hence, has the dimension of the Lie algebra. Indeed, taking the (l,m)-entry of the matrix $\mathsf{U}(\mathsf{T}_k)$ to be

$$U_{lm}^{(\mathrm{ad})}(\mathsf{T}_k) = -\mathrm{i}C_{kl}^m ,$$

one verifies by means of the Jacobi identity given by (5.21) that these matrices satisfy the commutators given by (5.20) (cf. Exercise 5.6).

Before moving on in this summary, an example may help to illustrate these matters.

Example 5.1 Group SU(2) and its Lie algebra

Every element of SU(2) can be represented by means of two complex numbers u and v whose absolute squares add up to 1:

$$\mathsf{U} = \begin{pmatrix} u & v \\ -v^* & u^* \end{pmatrix} , \qquad |u|^2 + |v|^2 = 1 .$$

This is easily verified. One finds

$$\begin{aligned} \mathsf{U}^\dagger \mathsf{U} &= \begin{pmatrix} u^* & -v \\ v^* & u \end{pmatrix} \begin{pmatrix} u & v \\ -v^* & u^* \end{pmatrix} \\ &= (|u|^2 + |v|^2) \begin{pmatrix} 1 & 0 \\ 0 & 1 \end{pmatrix} = \mathsf{U}\mathsf{U}^\dagger = \mathbb{1}_2 , \\ \det \mathsf{U} &= |u|^2 + |v|^2 = 1 . \end{aligned}$$

By representing the goup element U as an exponential series $\mathsf{U} = \exp\{\mathrm{i}\mathsf{A}\}$, with A an element of the Lie algebra, the following properties are seen to be equivalent:

$$\mathsf{U}^\dagger \mathsf{U} = \mathbb{1}_2 \Longleftrightarrow \mathsf{A}^\dagger = \mathsf{A} \quad \text{and} \quad \det \mathsf{U} = 1 \Longleftrightarrow \operatorname{tr} \mathsf{A} = 0 .$$

The first of these follows from the Baker–Hausdorff–Campbell formula for the product of two exponential series. Indeed, using this formula one has

$$\mathsf{U}^\dagger \mathsf{U} = e^{-i\mathsf{A}^\dagger} e^{i\mathsf{A}} = \exp\{i(\mathsf{A}-\mathsf{A}^\dagger) + \frac{1}{2}[\mathsf{A}^\dagger, \mathsf{A}] + \ldots\} .$$

This is equal to the unit matrix $\mathbb{1}$ if and only if $\mathsf{A}^\dagger = \mathsf{A}$. This result, i.e. A being hermitian, implies the second equivalence. Every hermitian matrix can be diagonalized by a unitary transformation. But when A is diagonal, U is also diagonal. In other terms, U is transformed into

$$\mathsf{U} \longrightarrow \mathring{\mathsf{U}} = \begin{pmatrix} e^{i\lambda} & 0 \\ 0 & e^{-i\lambda} \end{pmatrix}$$

when A becomes $\mathrm{diag}(\lambda, -\lambda)$.

Within the framework of the Lie algebra one recalls that every traceless hermitian 2×2 matrix can be written as a linear combination of the Pauli matrices, given by (4.24), with real coefficients, viz.

$$\mathsf{A} = \begin{pmatrix} c & a - ib \\ a + ib & -c \end{pmatrix} = a\sigma_1 + b\sigma_2 + c\sigma_3 , \qquad a, b, c \in \mathbb{R} . \tag{5.22}$$

Thus the Pauli matrices, repeated here for the sake of convenience,

$$\sigma_1 = \begin{pmatrix} 0 & 1 \\ 1 & 0 \end{pmatrix} , \quad \sigma_2 = \begin{pmatrix} 0 & -i \\ i & 0 \end{pmatrix} , \quad \sigma_3 = \begin{pmatrix} 1 & 0 \\ 0 & -1 \end{pmatrix} , \tag{5.23}$$

are a possible choice for the generators of SU(2) – up to a factor $1/2$. Their commutators are easily worked out to be

$$\left[\left(\frac{\sigma_i}{2}\right), \left(\frac{\sigma_j}{2}\right)\right] = i\varepsilon_{ijk}\left(\frac{\sigma_k}{2}\right) , \tag{5.24}$$

where $\varepsilon_{ijk} = +1\ (-1)$ for even (odd) permutations of $(1, 2, 3)$, and $\varepsilon_{ijk} = 0$ in all other cases. The structure constants are $C_{ij}^{k} = \varepsilon_{ijk}$.

The trivial representation is one-dimensional, and all three generators are zero and trivially satisfy the commutators given by (5.20).

The fundamental or defining representation is the genuine spinor representation. It is two-dimensional. The vector space on which it is realized is spanned by the eigenvectors

$$\begin{pmatrix} 1 \\ 0 \end{pmatrix} \quad \text{and} \quad \begin{pmatrix} 0 \\ -1 \end{pmatrix}$$

of σ_3. The generators are given by (half) the Pauli matrices, (5.23).

The adjoint representation for this example is three-dimensional. Constructing this representation following the rule given above, one finds

$$\mathsf{U}^{(\mathrm{ad})}(\mathsf{T}_1) = -\mathrm{i}\{\varepsilon_{1lm}\} = -\mathrm{i}\begin{pmatrix} 0 & 0 & 0 \\ 0 & 0 & 1 \\ 0 & -1 & 0 \end{pmatrix} = \begin{pmatrix} 0 & 0 & 0 \\ 0 & 0 & -\mathrm{i} \\ 0 & \mathrm{i} & 0 \end{pmatrix}, \tag{5.25a}$$

$$\mathsf{U}^{(\mathrm{ad})}(\mathsf{T}_2) = -\mathrm{i}\{\varepsilon_{2lm}\} = \begin{pmatrix} 0 & 0 & \mathrm{i} \\ 0 & 0 & 0 \\ -\mathrm{i} & 0 & 0 \end{pmatrix}, \tag{5.25b}$$

$$\mathsf{U}^{(\mathrm{ad})}(\mathsf{T}_3) = -\mathrm{i}\{\varepsilon_{3lm}\} = \begin{pmatrix} 0 & -\mathrm{i} & 0 \\ \mathrm{i} & 0 & 0 \\ 0 & 0 & 0 \end{pmatrix}. \tag{5.25c}$$

One verifies that indeed $[\mathsf{U}^{(\mathrm{ad})}(\mathsf{T}_1), \mathsf{U}^{(\mathrm{ad})}(\mathsf{T}_2)] = \mathrm{i}\mathsf{U}^{(\mathrm{ad})}(\mathsf{T}_3)$, as well as the cyclic permutations of this equation, hold true.

The group SU(2) is a *simple* group, that is, it is non-Abelian and has no invariant subgroup. At the level of the Lie algebra, this means that this algebra has no two-sided ideal and that every generator can be related to any other by means of commutators. In other terms, there is no way of splitting the Lie algebra into two subsets such that the structure constants C_{ij}^k vanish whenever index i denotes an element of one of these subsets while index k refers to an element of the other subset.

Remarks

1. We use consistently representations in which the generators are represented by *hermitian* matrices. In the realm of quantum mechanics, there are good reasons for this choice because hermitian matrices may represent physical observables. In the mathematical literature, one uses almost always a *nonhermitian* definition of the generators. This means that in the exponential series given by (5.19) and in the commutators given by (5.20), there is no factor i in the exponent and on the right-hand side, respectively.
2. In classical mechanics, the generators of rotation group SO(3) whose Lie algebra is the same as that of SU(2) are usually represented by real, antisymmetric matrices, see e. g. Eq. (2.68) in [ME]. The relation to the generators as defined here is as follows:

$$\mathsf{J}_i = -\mathrm{i}\mathsf{U}^{(\mathrm{ad})}(\mathsf{T}_i)\,. \quad i = 1, 2, 3\,.$$

The commutators then read

$$\begin{aligned}[\mathsf{J}_1, \mathsf{J}_2] &= -[\mathsf{U}^{(\mathrm{ad})}(\mathsf{T}_1), \mathsf{U}^{(\mathrm{ad})}(\mathsf{T}_2)] = -\mathrm{i}\mathsf{U}^{(\mathrm{ad})}(\mathsf{T}_3) \\ &= \mathsf{J}_3 \quad \text{(cyclically continued)}\,.\end{aligned}$$

The vector space of the adjoint representation of a simple Lie group possesses a natural metric. This is seen by defining the following tensor over the Lie algebra:

$$g_{ij} := \mathrm{tr}\big(\mathsf{U}^{(\mathrm{ad})}(\mathsf{T}_i)\mathsf{U}^{(\mathrm{ad})}(\mathsf{T}_j)\big) = -C_{iq}^{p} C_{jp}^{q} \; . \tag{5.26}$$

This tensor g is symmetric, nondegenerate and positive-definite. Thus it has the properties of a metric. Using the positivity one can find a transformation of the generators such that g becomes diagonal and is equal to the unit matrix in dimension dim $\mathfrak{g}$:

$$\mathrm{g} = \mathrm{diag}(1, 1, \ldots, 1) \; .$$

This metric is called the *Killing metric.* Defining new structure constants by

$$C_{ijk} := C_{ij}^{p} g_{pk} \quad \text{(summing over } p) \; , \tag{5.27}$$

one shows by means of the Jacobi identity that the new structure constants are antisymmetric in all three indices. This is left as an exercise.

An important property of the Lie algebras of simple Lie groups is as follows: By linear transformation, the generators can always be chosen such that one has

$$\mathrm{tr}\big(\mathsf{U}(\mathsf{T}_i)\mathsf{U}(\mathsf{T}_j)\big) = \kappa\, \delta_{ij} \; , \tag{5.28}$$

where the positive constant κ depends on the representation but does not depend on the generator T_i. The proof of this assertion is the topic of Exercise 5.2.

As an example, one verifies that in the fundamental representation of SU(2) one has

$$\mathrm{tr}\big(\mathsf{U}(\mathsf{T}_i)\mathsf{U}(\mathsf{T}_j)\big) = \frac{1}{4}\,\mathrm{tr}\big(\sigma_i\sigma_j\big) = \frac{1}{2}\delta_{ij} \; .$$

This follows, for example, from the formula

$$\sigma_i\sigma_j = \mathbb{1}_2\delta_{ij} + \mathrm{i}\varepsilon_{ijk}\sigma_k$$

and from the fact that the Pauli matrices are traceless, viz.

$$\mathrm{tr}\big(\sigma_i\sigma_j\big) = 2\delta_{ij} + \mathrm{i}\varepsilon_{ijk}\,\mathrm{tr}\big(\sigma_k\big) = 2\delta_{ij} \; .$$

In the adjoint representation of SU(2), one has

$$\mathrm{tr}\big(\mathsf{U}^{(\mathrm{ad})}(\mathsf{T}_i)\mathsf{U}^{(\mathrm{ad})}(\mathsf{T}_j)\big) = 2\delta_{ij} \; .$$

This is verified by means of the explicit expressions (5.25a)–(5.25c).

5.4.2 Globally Invariant Lagrange Densities

It takes only a small step to generalize a Lagrange density of the kind studied in the example given by (5.1) to a set of scalar fields $\Phi = \{\phi^{(1)}, \phi^{(2)}, \ldots, \phi^{(m)}\}$, which span a unitary, reducible or irreducible representation of the structure group in such a way that the new Lagrange density is invariant under all (global) gauge transformations $g \in G$. Of course, it is also possible to combine different species of fields in a globally invariant manner and thereby to obtain a globally gauge-invariant theory. The U(1) Maxwell theory – though not yet interpreted as a gauge group – and its coupling to the Schrödinger field provide an example.

The fields in examples of this kind may also be complex fields. However, these must be combined such that the Lagrange density stays real. For example, in (5.1), one must replace

$$\partial_\mu \phi \partial^\mu \phi - \kappa^2 \phi^2 \quad \text{by} \quad \partial_\mu \phi^* \partial^\mu \phi - \kappa^2 \phi^* \phi \,.$$

An example may illustrate this construction.

Example 5.2

Let the structure group be $G = \mathrm{SO}(3)$. The fields $\Phi = \{\phi^{(\alpha)}\}$, $\alpha = 1, 0, -1$, are assumed to span the triplet representation of the rotation group. Likewise, the external source is replaced by a triplet of sources $\{\varrho^{(\alpha)}\}$ which form an irreducible representation of SO(3) as well. This means that under rotations of the frame of reference, these triplets transform by the D-matrices $\mathsf{D}^{(1)}(\phi, \theta, \psi)$,

$$\phi'_\alpha(x') = \sum_{\beta=-1}^{+1} D^{(1)}_{\alpha\beta}(\theta_i) \phi_\beta(x) \,,$$

where θ_i is an abbreviation for the three Eulerian angles ϕ, θ and ψ. In this example, bilinear forms of the following kind:

$$\sum_\alpha (-)^{1-\alpha} \phi_\alpha \phi_{-\alpha} \,, \quad \sum_\alpha (-)^{1-\alpha} \partial_\mu \phi_\alpha \partial^\mu \phi_{-\alpha} \,, \quad \sum_\alpha (-)^{1-\alpha} \varrho_\alpha \phi_{-\alpha}$$

are invariant under rotations $g \in \mathrm{SO}(3)$. If the Lagrange density is to be globally gauge invariant, it can only contain terms of this kind.

Let us adopt the convention of denoting any globally invariant form by a pair of symbolic round parentheses such as

$$(\Phi, \Phi) \,, \quad (\partial_\mu \Phi, \partial^\mu \Phi) \quad \text{etc.,} \tag{5.29}$$

their explicit expressions depending on the choice of structure group G. Thus, in what follows the round parentheses in (5.29) stand for the generalized scalar product, i.e. for the coupling of the two entries to an invariant term under all transformations $g \in G$. The practical realization of this scalar product depends on the structure group and on the representation one considers. In the case of the rotation group, or the case of the group SU(2), and their irreducible representations of dimension $(2j+1)$, these are the Clebsch–Gordan coefficients for the coupling of the two entries to $J=0$.

To the list of requirements which a Lagrange density must satisfy, besides Lorentz invariance, one now adds invariance under all transformations $g \in G$. In Example 3.4 (atoms in external fields), for instance, the free Lagrange densities

$$\mathcal{L}_{\mathrm{M}} = -\frac{1}{16\pi} F_{\mu\nu}(x) F^{\mu\nu}(x) ,$$

$$\mathcal{L}_{\mathrm{E}} = \frac{1}{2} i\hbar [\psi^* \partial_t \psi - (\partial_t \psi^*)\psi] - U(t,x)\psi^*\psi - \frac{\hbar^2}{2m}(\nabla\psi^*)(\nabla\psi)$$

are invariant[6] under the global U(1) transformation $\psi \mapsto \psi' = e^{i\alpha}\psi$, but they are not yet coupled.

5.4.3 The Gauge Group

Given a structure group G and a Lagrange density which is invariant under this group, the aim is to construct a gauge group $\mathfrak{G}$ of *local* gauge transformations. For that purpose the parameters α_k in (5.19) must be replaced by smooth functions $\alpha_k(x)$ of space and time. From a physical point of view, this entails two new features. On the one hand, it becomes possible to restrict gauge transformations to a finite domain of space or time, i.e. to "rotate" the system under consideration locally and within a finite time interval, without transforming simultaneously other physical systems located far from it. On the other hand, one endows every point $x \in M$ of spacetime with a copy $G(x)$ of the structure group such that this copy is attached to point x like an inner symmetry space. This construction is sketched symbolically in Fig. 5.1. The copies $G(x)$ and $G(y)$ of the structure group with $x \neq y$ are disjoint spaces. Expressed in geometric terms, this construction is a *principal fibre bundle* with spacetime M as base manifold and G, the structure group, as typical fibre:[7]

$$P(M,G) . \tag{5.30}$$

It would take us too long to describe here the wonderful differential geometric description of principal fibre bundles and the zoo of geometric objects which can be

[6] There is no Lorentz invariance in this example because ψ obeys a nonrelativistic equation of motion. Lorentz transformations are replaced by rotations in $\mathbb{R}^3$ with respect to which both Lagrange densities are invariant.

[7] A precise definition of a fibre bundle in differential geometry may be found, e. g., in [ME], Sect. 4.7.

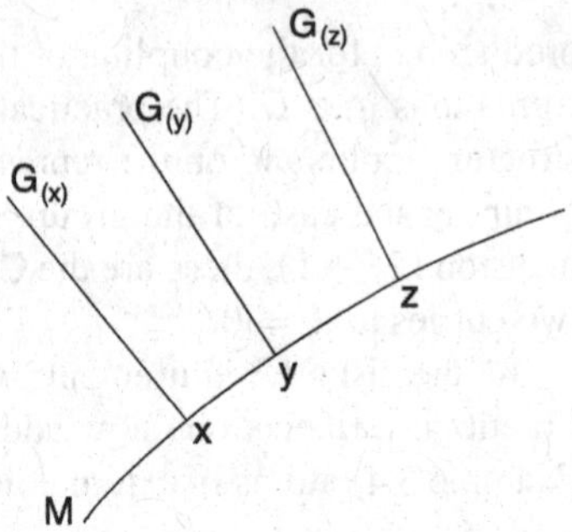

Fig. 5.1 In each of the points $x, y, z, \ldots$ of the spacetime manifold one finds a local copy of the structure group. The finite-dimensional structure group is replaced by the infinite-dimensional gauge group, and the whole setting becomes a principal fibre bundle

defined on them, so the interested reader is referred to the mathematical literature which develops this geometry in the light of Yang–Mills theory. We note, however, that once one becomes familiar with this framework, matters become considerably more transparent than by defining the gauge group and gauge transformations purely formally.

5.4.4 Potential and Covariant Derivative

Because the copies $G(x)$ and $G(x + \mathrm{d}x)$ of the structure group in the neighbouring points x and $x + \mathrm{d}x$, respectively, differ from each other, the representations $\Phi(x)$ and $\Phi(x + \mathrm{d}x)$ live in disjoint vector spaces. Therefore, a given component $\phi_i^{(x)}$ in x cannot be compared directly with the same component $\phi_i^{(x+\mathrm{d}x)}$ in the neighbourhood. Asking the question which transformation takes ϕ_i in the representation defined over the point $x \in M$ to the *same* component in the representation over $x + \mathrm{d}x$ poses a geometric problem which pertains to the same category as the question of parallel transport of vectors from the tangent space over x to the tangent space over the neighbouring $x + \mathrm{d}x$. Thus, it seems justified to talk about *parallel transport* also in the problem posed here. Parallel transport is not given a priori. Rather, at least for nonflat base manifolds, it must be defined appropriately.

Let $N = \dim \mathfrak{g}$ be the dimension of the Lie algebra of the structure group G, and let T_k denote its generators. One defines the generalized potential as follows:

$$A := \mathrm{i}q \sum_{k=1}^{N} A^{(k)} \mathsf{T}_k \,, \quad (N = \dim \mathfrak{g}) \,. \tag{5.31}$$

This definition contains a generalized "charge", i.e. the coupling constant q whose physical interpretation will have to be identified later, as well as a factor i. As the objects $A^{(k)}$ are assumed to be real, factor i renders the generalized potential A hermitian. The quantities $A^{(k)}$, of which there are as many as there are generators of the Lie algebra, are one-forms over $\mathbb{R}^{(1,3)}$,

$$A^{(k)} = A^{(k)}_\mu(x)\,\mathrm{d}x^\mu\,, \quad k = 1, 2, \ldots, \dim \mathfrak{g}\,, \tag{5.32}$$

where the four coefficients $A^{(k)}_\mu(x)$, $\mu = 0, \ldots, 3$ are smooth real functions of spacetime. The way potential A is defined in (5.31) is seen to have a double nature: On the one hand, by (5.32), it is a one-form on spacetime; on the other hand, because of its linear dependence on the generators, it is an element of the Lie algebra $\mathfrak{g}$. One says that the potential given by (5.31) is a Lie algebra valued one-form.

It does not matter for this definition in which representation of the structure group G the generators T_k are assumed to be given. It may be the adjoint representation, but it might as well be, for example, the reducible or irreducible representation spanned by the multiplet Φ of scalar fields. In a specific physical context, this representation is chosen, and when writing the latter in a matrix representation U, equation (5.31) takes the form

$$\mathsf{U}(A) := \mathrm{i}q \sum_{k=1}^{N} A^{(k)}\mathsf{U}(\mathsf{T}_k)\,. \tag{5.33a}$$

This is a matrix whose entries are ordinary one-forms. The same definition, freed of the specific matrix representation $\mathsf{U}(\mathsf{T}_k)$ and written in terms of the "genuine" fields $A^{(k)}_\mu(x)$, then reads

$$A := \mathrm{i}q \sum_{k=1}^{N} \mathsf{T}_k \sum_{\mu=0}^{3} A^{(k)}_\mu(x)\,\mathrm{d}x^\mu\,. \tag{5.33b}$$

The geometric role of this potential emerges when showing that it acts as carrier of parallel transport. Within our example of scalar fields, for example, the difference between a given component in two neighbouring base points on M is given by

$$\phi_i^{(x+\mathrm{d}x)} = \phi_i^{(x)} - \sum_{j=1}^{m} U_{ij}(A)\phi_j^{(x)} = \sum_{j=1}^{m} \{\delta_{ij} - U_{ij}(A)\}\phi_j^{(x)}\,, \tag{5.34}$$

with m the dimension of the representation. At first, this is no more than an ansatz which must be shown to yield what one expects.

It seems meaningful to require that a parallel transport commute with any local transformation $g(x) \in \mathfrak{G}$. We now show that this is true provided the transformation behaviour of A is as follows:

$$A \longmapsto A' = gAg^{-1} + g\,\mathrm{d}(g^{-1})\,, \tag{5.35a}$$

or, writing the same transformation rule in terms of the components of the one-form,

$$A_\mu(x) \longmapsto A'_\mu = g(x)A_\mu(x)g^{-1}(x) + g(x)\partial_\mu g^{-1}(x)\,. \tag{5.35b}$$

This affine transformation might be surprising at first because it mixes a conjugation of the type

$$\mathcal{O} \longmapsto \mathsf{R}\mathcal{O}\mathsf{R}^{-1} ,$$

well known from transformations of operators (e. g. of quantum mechanics), and a gauge transformation proper in the sense of Maxwell theory. Indeed, for the case $g(x) \in \mathrm{U}(1)$, i.e. $g(x) = \exp\{\mathrm{i}\alpha(x)\}$, one has

$$g A_\mu g^{-1} = A_\mu \,, \quad \text{but} \quad g(x)\partial_\mu g^{-1}(x) = -\mathrm{i}\partial_\mu \alpha(x) \,;$$

$A'_\mu(x) = A_\mu(x) - \mathrm{i}\partial_\mu \alpha(x)$ is a local gauge transformation of Maxwell theory.

Given ansatz (5.34), one applies an arbitrary local gauge transformation $g(x)$ to $\phi^{(x)}$ and $g(x + \mathrm{d}x)$ to $\phi^{(x+\mathrm{d}x)}$. Equation (5.34) describes a parallel transport if it commutes with $g \in G$, that is to say, if

$$\mathsf{U}\big(g(x + \mathrm{d}x)\big)\big(\mathbb{1} - \mathsf{U}(A)\big) = \big(\mathbb{1} - \mathsf{U}(A')\big)\mathsf{U}\big(g(x)\big)$$

holds true. Obviously, this relation must be independent of the kind and the dimension of the representation. Therefore, it can be formulated quite generally for g and A in abstract form:

$$g(x + \mathrm{d}x)\big(\mathbb{1} - A\big) = \big(\mathbb{1} - A'\big)g(x) \,. \tag{5.36}$$

Expanding $g(x + \mathrm{d}x)$ around point x to the first order,

$$g(x + \mathrm{d}x) \simeq g(x) + \partial_\alpha g(x)\,\mathrm{d}x^\alpha \equiv g + \mathrm{d}g \,,$$

equation (5.36) yields the condition $\mathrm{d}g - gA = -A'g$ or, after multiplication by g^{-1} from the right,

$$\begin{aligned} A' &= gA\,g^{-1} - (\mathrm{d}g)g^{-1} \\ &= gA\,g^{-1} + g\,\mathrm{d}g^{-1} \,. \end{aligned}$$

In the second step, the derivative of $gg^{-1} = \mathbb{1}$

$$\mathrm{d}\big(gg^{-1}\big) = \mathrm{d}\mathbb{1} = 0 = \big(\mathrm{d}g\big)g^{-1} + g\,\mathrm{d}g^{-1}$$

was used to replace $-(\mathrm{d}g)g^{-1}$ by $g\,\mathrm{d}g^{-1}$. This is the transformation behaviour anticipated in (5.35a).

Starting from the potential given by (5.31) one constructs the *covariant derivative,* following the example expressed by (2.88b) of Maxwell theory:

$$D_A := \mathrm{d} + A \,. \tag{5.37}$$

When specialized to the m-dimensional representation space spanned by the scalar fields of the example above, this means, in concrete terms, the replacement

$$\partial_\mu \Phi(x) \longmapsto \Big\{\partial_\mu \mathbb{1} + iq \sum_{k=1}^{N} A_\mu^{(k)}(x)\mathsf{U}(\mathsf{T}_k)\Big\}\Phi(x) . \tag{5.37a}$$

The behaviour of the covariant derivative under a local gauge transformation is straight conjugation and, hence, is simpler than the behaviour of the potential itself. One has

$$D_{A'} = g D_A\, g^{-1} \quad \text{while} \quad A' = g A g^{-1} + g\,\mathrm{d}g^{-1} . \tag{5.38}$$

This is shown as follows. For the sake of simlicity, return to our model field Φ subject to a local transformation $\Phi' = \mathsf{U}(g)\Phi$. Then calculate

$$\begin{aligned} D_{A'}\Phi' &= \big(\mathbb{1}\mathrm{d} + \mathsf{U}(A')\big)\mathsf{U}(g)\Phi \\ &= \big(\mathrm{d}\mathsf{U}(g)\big)\Phi + \mathsf{U}(g)\,\mathrm{d}\Phi \\ &\quad + \mathsf{U}(g)\Big\{\mathsf{U}(A)\mathsf{U}^{-1}(g) + \big(\mathrm{d}\mathsf{U}^{-1}(g)\big)\Big\}\mathsf{U}(g)\Phi \\ &= \mathsf{U}(g)\big[\mathrm{d} + \mathsf{U}(A)\big]\Phi \\ &\quad + \Big\{\big(\mathrm{d}\mathsf{U}(g)\big) + \mathsf{U}(g)\big[\mathrm{d}\big(\mathsf{U}^{-1}(g)\mathsf{U}(g)\big) - \mathsf{U}^{-1}(g)\big(\mathrm{d}\mathsf{U}(g)\big)\big]\Big\}\Phi , \end{aligned}$$

where in the last step one uses once more the identity

$$\mathrm{d}\big(\mathsf{U}^{-1}(g)\mathsf{U}(g)\big) = 0 = \big(\mathrm{d}\mathsf{U}^{-1}(g)\big)\mathsf{U}(g) + \mathsf{U}^{-1}(g)\,\mathrm{d}\mathsf{U}(g) .$$

The whole term in the curly brackets of the bottom line is equal to zero, so that there remains the relation

$$D_{A'}\Phi' = \mathsf{U}(g) D_A \Phi = \mathsf{U}(g) D_A \mathsf{U}^{-1}(g)\Phi' .$$

Both sides now contain the same field Φ', and one concludes

$$D_{A'} = \mathsf{U}(g) D_A \mathsf{U}^{-1}(g) . \tag{5.39}$$

Of course, the field Φ of the model is simply an auxiliary tool which may be chosen arbitrarily. This means that (5.39) holds in *all* representations, thereby proving (5.38). The covariant derivative transforms under gauge transformations like an operator. In particular, constructing a term $(D_A\Phi, D_A\Phi)$ in the manner desribed above, this term is invariant not only under global but also under *local* gauge transformations.

5.4.5 Field Strength Tensor and Curvature

In a next step, and still following the example of electrodynamics, the covariant derivative D_A is used to construct the curvature form given by (2.88c):

$$F := D_A^2 = (\mathrm{d}A) + A \wedge A\,. \tag{5.40}$$

At variance with Maxwell theory is a new property here: In the case of non-Abelian groups, the second term, the exterior product of A with itself, does not vanish. Writing all sums explicitly, as an exception, one obtains from (5.33b)

$$\begin{aligned} A \wedge A &= -q^2 \sum_{k,l=1}^{N} \mathsf{T}_k \mathsf{T}_l \sum_{\sigma,\tau=0}^{3} A_\sigma^{(k)}(x) A_\tau^{(l)}(x)\, \mathrm{d}x^\sigma \wedge \mathrm{d}x^\tau\,, \\ &= -q^2 \sum_{k,l=1}^{N} [\mathsf{T}_k, \mathsf{T}_l] \sum_{\mu<\nu=0}^{3} A_\mu^{(k)}(x) A_\nu^{(l)}(x)\, \mathrm{d}x^\mu \wedge \mathrm{d}x^\nu\,, \\ &= -\mathrm{i}q^2 \sum_{k,l,m=1}^{N} C_{klm} \mathsf{T}_m \sum_{\mu<\nu=0}^{3} A_\mu^{(k)}(x) A_\nu^{(l)}(x)\, \mathrm{d}x^\mu \wedge \mathrm{d}x^\nu\,. \end{aligned} \tag{5.41}$$

The crucial aspect is the fact that A is an element of the Lie algebra. Ordering the base two-forms $\mathrm{d}x^\sigma \wedge \mathrm{d}x^\tau$ as usual, i.e. by increasing index, by exchanging the indices σ and τ in the term with $\sigma > \tau$, one sees that $\mathrm{d}x^\mu \wedge \mathrm{d}x^\nu$ is multiplied by the factor

$$\sum_{k,l} A_\mu^{(k)}(x) A_\nu^{(l)}(x) (\mathsf{T}_k \mathsf{T}_l - \mathsf{T}_l \mathsf{T}_k)\,,$$

which does not vanish in a non-Abelian theory.

In view of the definition given by (5.33b), it is suggestive to decompose the two-form of (5.40) in terms of tensor fields of the kind $F_{\mu\nu}(x)$ on spacetime and of base two-forms, including in the definition the same factors as in (5.33b). Thus, we define

$$F = \mathrm{i}q \sum_{k=1}^{N} \mathsf{T}_k \sum_{\mu<\nu} F_{\mu\nu}^{(k)}(x)\, \mathrm{d}x^\mu \wedge \mathrm{d}x^\nu\,. \tag{5.42}$$

This definition, too, reveals a double nature: F is a two-form on spacetime but at the same time takes values in the Lie algebra of the gauge group. The tensor fields $F_{\mu\nu}^{(k)}(x)$, of which there are precisely $N = \dim \mathfrak{g}$ species, are direct generalizations of the field strength tensor of electrodynamics. From the decomposition, (5.42), of F and the decomposition, (5.33b), of A, using the result given by (5.41), these tensor fields are given by

$$F_{\mu\nu}^{(k)}(x) = \partial_\mu A_\nu^{(k)}(x) - \partial_\nu A_\mu^{(k)}(x) - q \sum_{m,n=1}^{N} C_{kmn} A_\mu^{(m)}(x) A_\nu^{(n)}(x)\,. \tag{5.43}$$

The first term on the right-hand side is familiar from Maxwell theory, cf. (2.58); the second term is new. It contains the coupling constant ("charge"), the structure constants of the group and the *product* of two vector potentials pertaining to different generators, T_m and T_n. Already at this point one realizes that non-Abelian gauge theories contain nonlinearities, in contrast to Maxwell's equations.

Indeed, the field strength F describes a kind of curvature in the principal fibre bundle $P(M, G)$ with base manifold $\mathbb{R}^{(1,3)}$ (spacetime) and typical fibre G (structure group). To see this, recall the setting of ordinary Riemannian manifolds. Whether or not a Riemannian manifold is curved is tested by parallel transport of tangent vectors from $p \in M$ to $q \in M$ via different geodesics and by comparison of the results. For a flat manifold such as, for example, $\mathbb{R}^n$, the result is independent of the path from p to q. This space has no curvature. On a curved manifold such as, for example, the sphere S^{n-1} in $\mathbb{R}^n$, the result of parallel transport depends on the choice of great circles and meridians along which one performs parallel transport of tangent vectors. The difference between the results along different paths is a measure of the nonvanishing curvature. Another alternative is to study the behaviour of tangent vectors on round trips by parallel transport from p to q along one path and returning from q back to p along another path.

As in Riemannian geometry, a "round trip" can be organized by means of the parallel transport given by (5.34). For the sake of simplicity, we do this locally, i.e. in the immediate neighbourhood of the point $x \in \mathbb{R}^4$, and in the way sketched in Fig. 5.2. We perform parallel transport from x to $z = x + \mathrm{d}x + \mathrm{d}y$ and back to the start in x but choose the path to z different from the path back to x.

We express A by its components $A_\mu(x)$, i.e. write (5.33b) as

$$A = A_\mu(x)\,\mathrm{d}x^\mu\,, \quad \text{with} \quad A_\mu(x) = \mathrm{i}q \sum \mathrm{T}_k A_\mu^{(k)}(x)\,. \tag{5.44}$$

Equation (5.35b) is utilized without reference to the special representation spanned by the fields ϕ_i. One calculates the difference of the parallel transport along one or the other of the two paths drawn in Fig. 5.2 as follows:

$$\begin{aligned}&\Big(\mathbb{1} - A_\nu(x + \mathrm{d}x)\,\mathrm{d}y^\nu\Big)\Big(\mathbb{1} - A_\mu(x)\,\mathrm{d}x^\mu\Big)\\ &-\Big(\mathbb{1} - A_\mu(x + \mathrm{d}y)\,\mathrm{d}x^\mu\Big)\Big(\mathbb{1} - A_\nu(x)\,\mathrm{d}y^\nu\Big)\,.\end{aligned}$$

Expanding $A_\nu(x + \mathrm{d}x)$ and $A_\mu(x + \mathrm{d}y)$ around point x to the first order in $\mathrm{d}x$ and in $\mathrm{d}y$, this difference is found to be

$$\begin{aligned}&= -\mathrm{d}x^\mu\,\mathrm{d}y^\nu\Big(\partial_\mu A_\nu(x) - \partial_\nu A_\mu(x) + A_\mu(x)A_\nu(x) - A_\nu(x)A_\mu(x)\Big)\\ &= -\mathrm{d}x^\mu\,\mathrm{d}y^\nu\Big(\partial_\mu A_\nu(x) - \partial_\nu A_\mu(x) + \big[A_\mu(x), A_\nu(x)\big]\Big)\\ &= -F_{\mu\nu}(x)\,\mathrm{d}x^\mu\,\mathrm{d}y^\nu\,.\end{aligned}$$

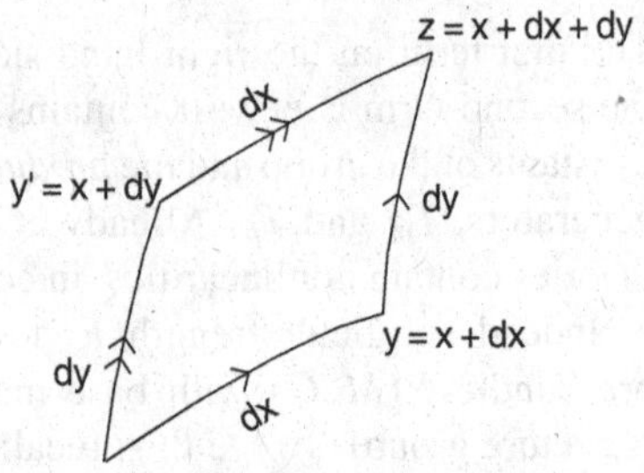

Fig. 5.2 One may move from x to $z = x + \mathrm{d}x + \mathrm{d}y$ via either $y = x + \mathrm{d}x$ or $y' = x + \mathrm{d}y$. The two paths are combined into a nontrivial round trip which starts at x and ends at the same point

In close analogy to (5.44), in the case of the potential, the quantities

$$F_{\mu\nu}(x) = iq \sum_{k=1}^{N} \mathrm{T}_k F_{\mu\nu}^{(k)}(x) \tag{5.45}$$

are (Lie algebra valued) coefficients of F. This result shows very clearly that F describes a curvature indeed. Thus, in a geometrical picture, the tensor fields are curvatures.

We note, furthermore, that the behaviour of F with respect to gauge transformations is again by conjugation. This behaviour was derived for the covariant derivative in (5.38). Obviously, it also applies to the square of D_A and, via the definition given by (5.40), to F,

$$F \longmapsto F' = gFg^{-1} . \tag{5.46}$$

These results and the covariant derivative of matter fields provide the basic tools for the construction of gauge-invariant Lagrange functions.

5.4.6 Gauge-Invariant Lagrange Densities

On the basis laid down in the two preceding sections, it is not difficult to construct Lagrange densities which are invariant under Lorentz transformations as well as under a given gauge group $\mathfrak{G}$. Already at this level one realizes how far Maxwell theory can take us as a model whose physical interpretation is well understood.

Suppose the following tools of the theory are given:

- A compact Lie group G and the infinite-dimensional gauge group $\mathfrak{G}$ over space-time $\mathbb{R}^{(1,3)}$ which is built on G;
- A potential A, equation (5.33b), which is both an element of the Lie algebra $\mathfrak{g}$ and a one-form over Minkowski space $\mathbb{R}^{(1,3)}$;
- A set of (matter) fields $\Phi(x) = \{\phi_1(x), \ldots, \phi_m(x)\}$ which span a reducible or irreducible representation of G. Here, for the sake of simplicity, we assume these to be scalar fields. However, a largely identical construction is possible also for fields whose transformation behaviour under Lorentz transformations is more

complicated. (We have in mind matter fields which describe particles in quantum mechanics having nonvanishing spin.)

Starting from A, equation (5.33b), one constructs the covariant derivative D_A, equation (5.37), and from the latter the curvature form F, equation (5.40). As can be seen from (5.42), this object is a two-form taking values in the Lie algebra. Neglecting the matter field Φ for the moment and following the example of Maxwell theory, the only possibility is to couple F with itself in a way that guarantees Lorentz invariance and gauge invariance. In fact, there are several ways of implementing this prescription, all of which lead to the same result.

(i) The exterior product of the two-form F with its Hodge dual $*F$ can be integrated over the manifold $M = \mathbb{R}^{(1,3)}$. This is a typical geometric approach which we did not describe in the case of Maxwell theory, however, and which we do not wish to work out further.[8] Furthermore, as will be explained below, the trace over the adjoint representation of the Lie algebra must be taken.

(ii) Guided by the example of Maxwell theory, use the two-form

$$F_{\mu\nu} = \mathrm{i}q \sum_{k=1}^{N} \mathsf{T}_k F^{(k)}_{\mu\nu}(x) \tag{5.47a}$$

with $F = F_{\mu\nu}\, \mathrm{d}x^\mu \wedge \mathrm{d}x^\nu$ by (5.42), to construct the Lorentz invariant $F_{\mu\nu}F^{\mu\nu}$. To obtain an invariant under all gauge transformations $g \in \mathfrak{G}$, one must take the trace over the adjoint representation of G. Written more explicitly, the Lie algebra-valued two-form F reads

$$F = F_{\mu\nu}\, \mathrm{d}x^\mu \wedge \mathrm{d}x^\nu = \mathrm{i}q \sum_{k=1}^{N} \mathsf{U}^{(\mathrm{ad})}(\mathsf{T}_k) \sum_{\mu<\nu} F^{(k)}_{\mu\nu}(x)\, \mathrm{d}x^\mu \wedge \mathrm{d}x^\nu \; . \tag{5.47b}$$

Therefore, a trace must be taken of the product of the two $N \times N$ matrices $\mathsf{U}^{(\mathrm{ad})}(\mathsf{T}_k)$ and $\mathsf{U}^{(\mathrm{ad})}(\mathsf{T}_l)$. As both the potential A and the quantity F are elements of the Lie algebra, one often writes $\mathrm{tr}(F_{\mu\nu}F^{\mu\nu})$ for short.

(iii) A third possibility consists in defining scalar products of exterior forms on Minkowski space. A natural choice well adapted to Lorentz covariance is

$$\langle \mathrm{d}x^\mu | \mathrm{d}x^\nu \rangle = \kappa_1 g^{\mu\nu} \,,$$
$$\langle \mathrm{d}x^\mu \wedge \mathrm{d}x^\nu | \mathrm{d}x^\sigma \wedge \mathrm{d}x^\tau \rangle = \kappa_2 \left(g^{\mu\sigma} g^{\nu\tau} - g^{\mu\tau} g^{\nu\sigma} \right) .$$

[8] This geometric approach makes use, in an essential way, of the fact that $*F \wedge F$ is a 4-form and, hence, can be integrated over a manifold with dimension 4. The dimension of the base manifold is essential in this context. Integration on arbitrary smooth manifolds is not developed in this book.

Note, however, that scalar products are not really fixed by the geometry because the constants κ_1 and κ_2 may be chosen at will. Furthermore, as the dx^μ one-forms are lengths, these constants must have physical dimensions, viz.

$$[\kappa_1] = \text{length}^{-2}\,, \quad [\kappa_2] = \text{length}^{-4}\,.$$

Finally, as in the two alternatives described previously, one must take the trace over the product of two generators in the adjoint representation.

By either of the three foregoing methods one arrives at the term $\mathrm{tr}(F_{\mu\nu}F^{\mu\nu})$, which is Lorentz and $\mathfrak{G}$-invariant. This term yields a generalized "kinetic" term in the Lagrange density provided it obeys an important normalization condition. We explain this normalization condition by means of (5.43), which we write as

$$F^{(k)}_{\mu\nu}(x) = f^{(k)}_{\mu\nu}(x) - q \sum_{m,n=1}^{N} C_{kmn} A^{(m)}_\mu(x) A^{(n)}_\nu(x)\,, \tag{5.48a}$$

$$f^{(k)}_{\mu\nu}(x) := \partial_\mu A^{(k)}_\nu(x) - \partial_\nu A^{(k)}_\mu(x)\,. \tag{5.48b}$$

The invariant $I := \mathrm{tr}(F_{\mu\nu}F^{\mu\nu})$ contains in a first term the contraction of the tensor field $f_{\mu\nu}$, defined in (5.48b), with itself,

$$\sum_{k=1}^{N}\sum_{l=1}^{N} \mathrm{tr}(\mathsf{U}^{(\mathrm{ad})}(\mathsf{T}_k)\mathsf{U}^{(\mathrm{ad})}(\mathsf{T}_l))\, f^{(k)}_{\mu\nu}(x) f^{(l)\mu\nu}(x)\,.$$

According to (5.28), the trace in this expression depends not on the generator but on the group representation. This means that we must know the normalization of the generators in the adjoint representation, i.e. we need to know the numerical value of the constant $\kappa^{(\mathrm{ad})}$. The terms $f^{(k)}_{\mu\nu}(x) f^{(k)\mu\nu}(x)$ take the same role as the Lagrange density, (3.36a), of free Maxwell theory whose physical interpretation was worked out in Sect. 3.4. It is this comparison which fixes the normalization of the invariant $\mathrm{tr}(F_{\mu\nu}F^{\mu\nu})$ in the Lagrange density: As in the Maxwell case, this term must appear multiplied by $-1/16\pi$ (using Gaussian units, or $-1/4$ using natural units). This fixes the first part of the Lagrange density. It reads

$$\mathcal{L}_{\mathrm{YM}} = -\frac{1}{16\pi q^2 \kappa^{(\mathrm{ad})}} \mathrm{tr}\big(F_{\mu\nu}F^{\mu\nu}\big)\,. \tag{5.49}$$

The subscript "YM" stands for "Yang–Mills", the Lie algebra-valued field $F_{\mu\nu}$ is defined as in (5.47a) or (5.47b), division by q^2 being due to the choice of having inserted a factor q in the definition of $F_{\mu\nu}$.

Finally, the coupling of the N Yang–Mills fields to matter is introduced by following the model of Maxwell theory. For simplicity, we discuss this aspect within the example of the multiplet Φ whose components span a representation of G and all of which are Lorentz scalar fields. How such a multiplet is to be coupled to a bilinear G-invariant term depends on the group and on the representation. As was stated earlier, we abbreviate this scalar product by big round parentheses.

Regarding the matter terms in the Lagrange density, we assume that all the $\phi_k(x)$ components pertain to the same mass parameter, (5.2). In this framework, the Lagrange density describing matter fields will initially be of the form of (5.1), viz.

$$\mathcal{L}_{\Phi}^{(0)} = \frac{1}{2}\Big[\big(\partial_\mu \Phi(x), \partial^\mu \Phi(x)\big) - \mu^2 (\Phi, \Phi)\Big] - W\big(\Phi(x)\big) , \qquad (5.50)$$

where $W(\Phi)$ is a potential energy density, i.e. a kind of self-coupling among the scalar fields ϕ_k which is globally gauge invariant.

The Lagrange density, (5.50), is globally gauge invariant but not locally gauge invariant because the kinetic energy density $(\partial_\mu \Phi, \partial^\mu \Phi)$ is not gauge invariant. To construct a gauge-invariant theory, two modifications must be applied to it. Firstly, the ordinary derivative must be replaced by the covariant derivative (5.37),

$$\mathrm{d}\Phi \to D_A \Phi \quad \text{or}$$

$$\partial_\mu \Phi(x) \to \Big\{\mathbb{1}\partial_\mu + \mathrm{i}q \sum_{k=1}^{N} \mathsf{U}^{(\Phi)}(\mathsf{T}_k) A_\mu^{(k)}(x)\Big\}\Phi(x) .$$

Secondly, the gauge fields which are contained in A must be endowed on their own with a "kinetic" term of the kind expressed by (5.49). Thus, the completed, locally invariant theory is defined by a Lagrange density of the form

$$\begin{aligned}\mathcal{L} = &-\frac{1}{16\pi q^2 \kappa^{(\mathrm{ad})}} \operatorname{tr}\big(F_{\mu\nu} F^{\mu\nu}\big) \qquad (5.51)\\ &+\frac{1}{2}\Big[\big(D_\mu \Phi(x), D^\mu \Phi(x)\big) - \mu^2 (\Phi, \Phi)\Big] - W\big(\Phi(x)\big) .\end{aligned}$$

It contains the physics of the gauge fields, of the original scalar fields, and their coupling to the gauge fields in a gauge-invaraint manner. The Lagrange density, (5.51), possesses a high degree of symmetry with respect to the gauge group $\mathfrak{G}$.

Remarks

1. As has been emphasized repeatedly, the structure group, as the basis of the theory, must be compact. Only in this case, the Killing metric (5.26) is positive-definite. Recalling the calculation of the field energy (3.30a) of Maxwell fields, one sees that the signs in (3.36a) and (5.49) are responsible for the fact that the energy density, and thus the total energy content of free Maxwell or Yang–Mills fields, is positive. If one had allowed for a noncompact structure group, the Killing metric in its diagonalized form would have contained positive as well as negative entries, i.e.

$$\mathbf{g} = \operatorname{diag}(\varepsilon_1, \varepsilon_2, \dots, \varepsilon_N) \quad \text{with} \quad \varepsilon_i = \pm 1 .$$

As a consequence, the kinetic energy of at least some of the vector fields in the Yang–Mills theory would have the wrong sign, that is, they would be assigned negative field energies.
2. By construction, the Lagrange density, (5.51), is invariant also under *global* gauge transformations, i.e. under any element $g \in G$ of the structure group. In some sense, this is a weaker form of gauge invariance.

5.4.7 Physical Interpretation

Starting from the Lagrange density as given by (5.51) and a decomposition of the tensor field $F_{\mu\nu}$, in a given frame of reference, into G-electric and G-magnetic fields, one derives the generalized Maxwell equations for non-Abelian gauge theories. The equations of motion which one obtains have a similar structure as those of electrodynamics. They are composed of a "radiation" part and a "matter" part, the latter appearing as inhomogeneous or source terms. Non-Abelian gauge theories describe fundamental interactions whose ranges are microscopically small and, hence, which play no role in classical, macroscopic physics. This means that, although these theories are defined at the level of classical, nonquantum physics and, hence, are closely related to the theory of Maxwell fields, they obtain their *physical* realization in the framework of quantum field theory. The gauge potentials $A_\mu^{(k)}(x)$, which, strictly speaking, are not observables, are nevertheless helpful auxiliary quantities in the quantization of the theory and in the interpretation of the gauge fields in terms of vector bosons as carriers of fundamental interactions. This is in close analogy to the photon, which is understood as the carrier of electric and magnetic interactions. Therefore, it is sufficient to discuss the Lagrange density proper, including the couplings it contains, instead of working out the Euler Lagrangê equations.

Gauge Invariance and Massless Gauge Bosons

It is a striking property of the Lagrange density, (5.51), that none of the N fields $A_\mu^{(k)}$ possesses a mass term. As in Maxwell theory, cf. Sect. 5.2, any primordial mass term

$$\frac{1}{8\pi}\lambda^{(k)\,2} A_\mu^{(k)}(x) A^{(k)\,\mu}(x) \tag{5.52}$$

would ruin gauge invariance. This means that, after quantization, locally gauge-invariant, non-Abelian theories describe massless gauge bosons. This was why Pauli rejected this construction. At that time, no massless gauge bosons other than the photon were known.

This general statement and Pauli's criticism are justified as long as one considers the physics of gauge fields in isolation. In this framework, there cannot be mass terms of the kind given by (5.52) without losing local gauge invariance. Surprisingly, the rescue from this apparent uselessness, from the perspective of fundamental interactions of nature, comes from the interplay of pure gauge theory with matter fields and their self-interaction $W(\Phi)$, in the framework of *spontaneous symmetry breaking*. If the pure Yang–Mills theory is supplemented by genuine physical scalar fields, by arranging their potential energy such that it has an absolute minimum at (infinitely many) $\Phi = \Phi^{(0)} \neq 0$, the Lagrange density remains gauge invariant and describes vector fields some of which become massive. This phenomenon of spontaneous symmetry breaking means that the physical ground state has less symmetry than the Lagrange density defining the theory. Thus, the gauge invariance is not lost; it is "hidden" in the physically realized state. It is for this reason that one often calls this phenomenon *hidden symmetry*.

The mechanism of spontaneous symmetry breaking was discovered in the mid-1960s by Higgs, Kibble, Brout, and Engler.[9] It was of paramount importance in the development of the gauge theory describing electroweak interactions. We return to this in somewhat more detail below.

Interactions of the Gauge Bosons

The generic Lagrange density, (5.51), differs from the Lagrange density of electrodynamics by an aspect which is of central importance for physics. Compare the Lagrange density $\mathcal{L}_{\text{YM}}$, equation (5.49), of pure Yang–Mills theory and $\mathcal{L}_{\text{M}}$, equation (3.36a), of electrodynamics without external sources, recalling the definitions of the field strength tensors, (5.43) and (2.58), respectively. The Yang–Mills theory is constructed from a non-Abelian group G, whereas Maxwell theory is a U(1) gauge theory. The Euler–Lagrange equations of motion contain the derivatives of the Lagrange density by the potentials and by the spacetime derivatives thereof:

$$\frac{\partial \mathcal{L}_{\text{YM}}}{\partial A_\tau^{(k)}} \quad \text{and} \quad \frac{\partial \mathcal{L}_{\text{YM}}}{\partial(\partial_\mu A_\tau^{(k)})} .$$

Formally, the structure is the same as in Maxwell theory. However, in Yang–Mills theory the equations of motion obtained are nonlinear. Indeed, writing the Lagrange density, (5.49), somewhat more explicitly, taking the trace yields a Kronecker delta δ_{kl} for the generators so that there remain diagonal terms of the following kind:

$$\begin{aligned} F_{\mu\nu}^{(k)} F^{(k)\,\mu\nu} &= \left(\partial_\mu A_\nu^{(k)} - \partial_\nu A_\mu^{(k)}\right)\left(\partial^\mu A^{(k)\,\nu} - \partial^\nu A^{(k)\,\mu}\right) \\ &\quad - 2q\left(\partial_\mu A_\nu^{(k)} - \partial_\nu A_\mu^{(k)}\right) \sum_{m,n=1}^{N} C_{kmn} A^{(m)\,\mu} A^{(n)\,\nu} \\ &\quad + q^2 \sum_{m,n=1}^{N} \sum_{p,q=1}^{N} A_\mu^{(m)} A_\nu^{(n)} A^{(p)\,\mu} A^{(q)\,\nu} . \end{aligned}$$

[9] P.W. Higgs, Phys. Lett. **12** (1964) 132 and Phys. Rev. **145** (1966) 1156; F. Engler and R. Brout, Phys. Rev. Lett. **13** (1964) 321; G.S. Guralnik, C.R. Hagen and T.W.B. Kibble, Phys. Rev. Lett. **13** (1964) 585; T.W.B. Kibble, Phys. Rev. **155** (1967) 1554.

The first term on the right-hand side is quadratic in the gauge potentials; it has the same form as the kinetic term of Maxwell theory. The second term, which is proportional to charge q, is cubic in the potentials. Upon insertion into the equations of motion it leads to an interaction of the gauge fields among themselves, which is new. The third term is proportional to q^2 and contains the product of four gauge potentials. It describes another interaction of the gauge fields – even without yet having introduced any matter fields. This is an essential difference between the Abelian U(1) theory of Maxwell's equations and a non-Abelian gauge theory. The latter contains cubic and quartic couplings of the gauge fields, leads to nonlinearities in the equations of motion, and describes physically significant interactions among gauge fields.

5.4.8 *More on the Gauge Group

In this book, we develop gauge-invariant classical field theories exclusively on flat manifolds such as the well-known Minkowski space $\mathbb{R}^{(1,3)}$. However, there is nothing to prevent our extending these constructions to more general differentiable manifolds, i.e. studying electrodynamics or non-Abelian gauge theory on curved spacetimes. For instance, one is confronted with this situation as soon as one includes the gravitational interaction, besides electromagnetic and other interactions acting on microscopic scales (the so-called weak and strong interactions). The presence of masses and energy densities in the universe has the effect that spacetime can no longer be the simple Minkowski space but must be a semi-Riemannian manifold with index 1 which, loosely speaking, is more curved the higher the mass and energy density is. On a local scale, the physics of Maxwell's equations and of non-Abelian interactions remains the same as on $\mathbb{R}^{(1,3)}$. However, on larger scales and in the presence of gravitational fields, it changes. In this situation, one will have to formulate classical gauge theory on semi-Riemannian manifolds which are not flat, i.e. do not look like Minkowski space.

This perspective reaches far into differential geometry. In fact, historically, these concepts were developed in mathematics and in theoretical physics, initially largely independently, and it took some time before it was realized that the objects which one studied had different names but were, in fact, by their very nature the same. It would take us too far from the physical framework of classical field theory and go beyond the scope of this book to work out in a mathematically rigorous manner the differential geometric aspects of gauge theories. Therefore, I restrict this ad libitum section to a few remarks which serve merely to pique the curiosity of mathematically oriented readers and to invite them to read more about these matters.

Spacetime is assumed to be an orientable, connected Riemannian or semi-Riemannian manifold. A Riemannian manifold is a pair $(M, \mathbf{g})$ consisting of a smooth manifold M (in physics, as a rule, its dimension is $\dim M = 4$) and a metric $\mathbf{g}$ which is a nondegenerate, positive-definite or negative-definite bilinear form on every tangent space. Its index, as defined earlier, is zero. The manifold is

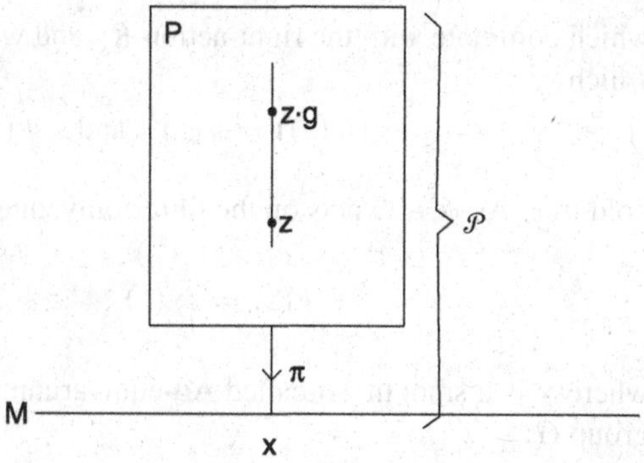

Fig. 5.3 The principal fibre bundle $(P \underset{\pi}{\rightarrow} M, G)$ contains a fibre above every point z of the base manifold, i.e. an inner space where the structure group acts freely. The base point x, onto which the whole fibre is projected, does not feel this group action

called semi-Riemannian if the metric is no longer definite or, more precisely, if the index ν is different from zero and is the same on all tangent spaces. In physics, the case of interest is $\nu = 1$. Such manifolds are also called Lorentz manifolds. Differential geometers like to assume, furthermore, that M is compact. This assumption, for which there are good mathematical grounds, is not very convenient for physics because physical spacetimes, in general, are not compact.

The arena for the construction of a Yang–Mills theory is the principal fibre bundle

$$\mathcal{P} = \left(P \underset{\pi}{\rightarrow} M, G\right), \tag{5.53a}$$

whose base is spacetime M and whose typical fibre is structure group G. As sketched in Fig. 5.3, map π is the projection from P to M, which maps every point z of the fibre to the base point $\pi(z)$. This is a surjective mapping from $\mathcal{P}$ to M. Locally, the principal fibre bundle $\mathcal{P}$ is isomorphic to the direct product $M \times G$,

$$\mathcal{P} \cong M \times G \,. \tag{5.53b}$$

Globally, matters can be more complicated whenever more than one chart is needed for the description of M.

The structure group acts within the fibres by free action from the right:[10]

$$R_g z = z \cdot g \,, \; z \in P \,, g \in G \,. \tag{5.54}$$

When defined in the framework of differential geometry, the gauge group $\mathfrak{G}$ is the group of those automorphisms of the principal bundle $\mathcal{P}$ which induce the identity on basis M. In other terms, these are automorphisms,

$$\psi \in \mathfrak{G} \,, \quad \psi : \mathcal{P} \longrightarrow \mathcal{P} \,, \tag{5.55}$$

[10] The action from the *right* is the conventional choice in differential geometry. Unfortunately, it is not in agreement with the practice in physics where one prefers to have symmetries act from the left. Right action, then, occurs only with contragredient transformation behaviour.

which commute with the right-action R_g and which map each fibre to itself, i.e. for which

$$\pi\big(\psi(z)\big) = \pi(z) \quad \text{and} \quad \psi(z \cdot g) = \psi(z) \cdot g \tag{5.55a}$$

hold true. As $\psi \in \mathfrak{G}$ acts on the fibres only, one has

$$\psi(z) = z\gamma(z)\,, \quad z \in \mathcal{P}\,, \psi \in \mathfrak{G}\,, \tag{5.55b}$$

where γ is a smooth, so-called Ad-equivariant map which maps $\mathcal{P}$ to the structure group G:

$$\gamma : P \to G : z \mapsto \gamma(z)\,, \quad \text{with}$$
$$\gamma(z \cdot g) = \text{Ad}\, g^{-1}\big(\gamma(z)\big) = g^{-1}\gamma(z)g\,. \tag{5.55c}$$

Relation (5.55b) defines a bijection $\gamma : \mathcal{P} \to G$, so that the gauge group $\mathfrak{G}$ can be identified with the set of these mappings. Thus, one has, for example, $(\gamma\gamma')(z) = \gamma(z)\gamma'(z)$. Obviously, the gauge group is infinite-dimensional. It is represented in every fibre by a copy of the structure group.

In a local perspective, the principal bundle $\mathcal{P}$ has the structure $M \times G$, with M the basis (spacetime) and G the structure group. Let $x \in M$ be a point of the base manifold, and let π be the projection, as before. The fibre $\pi^{-1}(x)$ over point x is isomorphic to the structure group G. This may alternatively be interpreted and applied as follows.

Let $p \in \mathcal{P}$ be a point of $\mathcal{P}$, $p = (x, \pi^{-1}(x))$, with the two entries x on M, and $\pi^{-1}(x)$ in the fibre. We write (5.54) in a somewhat different form and note that there is an isomorphism between G and the fibre $\pi^{-1}(x)$ over x, that is, for $z \in \pi^{-1}(x)$ and $g \in G$ the function

$$\varrho_p : G \longrightarrow \pi^{-1}(x) \in \mathcal{P} : g \longmapsto \varrho_p(g) = z \cdot g$$

is the announced mapping of the structure group to the given fibre. Consider now the corresponding tangent map in the group element $e = \mathbb{1}$ of G, i.e. the tangent map restricted to $\mathfrak{g}$, the Lie algebra of G. This mapping maps the Lie algebra to the tangent space $T_p\mathcal{P}$,

$$T_e\varrho_p : \mathfrak{g} \hookrightarrow T_p\mathcal{P}\,.$$

In this process, $\mathfrak{g}$ is embedded in the tangent space $T_p\mathcal{P}$. The subspace of $T_p\mathcal{P}$, which is identified in this manner

$$G_p := T_e\varrho_p(\mathfrak{g})\,, \tag{5.56}$$

is called the *vertical subspace* of $T_p\mathcal{P}$. Figure 5.3 illustrates this nomenclature.

The local isomorphism given by (5.53b), in the perspective of the tangent spaces, tells us that $T_p\mathcal{P}$ is isomorphic to the following decomposition:

$$T_p\mathcal{P} \cong T_{\pi(p)}M \oplus \mathfrak{g} . \tag{5.57}$$

The isomorphism of $\mathfrak{g}$ and G_p, equation (5.56), is obviously canonical. This is not true, however, for the rest: An isomorphism between $T_{\pi(p)}M$ and a subspace of $T\mathcal{P}$ must be defined. The specific choice to be made here is equivalent to the definition of a *connection*. The connection, in turn, defines the *covariant derivative*. The choice of a connection on the given principal bundle $\mathcal{P}$ means that for every point $p \in \mathcal{P}$ one defines a subspace $Q_p \subset T_p\mathcal{P}$ which has the following properties:

(i) The tangent space in p is the direct sum of G_p and Q_p:

$$T_p\mathcal{P} = Q_p \oplus G_p . \tag{5.58a}$$

(ii) The action of the structure group on Q_p satisfies the condition

$$Q_{p\cdot g} = TR_g(Q_p) . \tag{5.58b}$$

(iii) The subspace Q_p is differentiable in $p \in \mathcal{P}$.

A space Q_p defined in this way is called a *horizontal subspace* of $T_p\mathcal{P}$. The mapping $p \mapsto Q_p$ is called a *connection* on $\mathcal{P}$.

Condition (ii) guarantees that the decomposition given by (5.58a) is invariant under the right action of the structure group:

$$\begin{aligned} T_{p\cdot g}\mathcal{P} &= TR_g(T_p\mathcal{P}) = TR_g\big(Q_p \oplus G_p\big) \\ &= TR_g(Q_p) \oplus TR_g(G_p) = Q_{p\cdot g} \oplus G_{p\cdot g} . \end{aligned}$$

The condition that the map $p \mapsto Q_p$ must be differentiable is needed for the connection to yield a covariant derivative.

We can do no more than to sketch the remainder of this construction and to make it plausible by means of qualitative arguments. In selecting the horizontal subspaces in the points $p \in \mathcal{P}$ in the way described above, the different tangent spaces become comparable. For instance, one thereby determines where the identity $\mathbb{1}$ of G is located in the respective tangent spaces. In terms of physics, one might say that the inner symmetry spaces over points x of the base manifold become comparable to each other.

The close relationship to a standard notion in differential geometry is obvious: the definition of Riemannian manifolds. It is well known that tangent vectors from two different, disjoint tangent spaces $T_x M$ over x and $T_y M$ over y cannot be compared directly unless the manifold is a Euclidean space, $M = \mathbb{R}^n$. Only in this special case is parallel transport of a tangent vector from $T_x M$ to $T_y M$ defined in a natural way. In all other cases, the connection and, hence, the covariant derivative must be defined within the framework of some general rules.

In the present case, the connection is a one-form on $\mathcal{P}$ and takes values in the Lie algebra $\mathfrak{g}$. In a geometric language, one talks about a connection form like

$$\omega \in \Omega^1(\mathcal{P}, \mathfrak{g}) .$$

This is an element of the space of one-forms $\Omega^1(\mathcal{P})$, which at the same time is an element of the Lie algebra. The relationship to what is called a *gauge field*, or *gauge potential*, is established by local sections. Let

$$\sigma : U \subset M \longrightarrow \mathcal{P} \tag{5.59}$$

be a differentiable section, and let ω be a connection form on $\mathcal{P}$. Then the pullback of ω to U,

$$A^{(\sigma)} := \sigma^* \omega , \tag{5.60}$$

is a one-form on U which is defined by σ, viz. $A^{(\sigma)} \in \Omega^1(U, \mathfrak{g})$. This one-form is the gauge field pertaining to the connection form ω, the section σ, and $U \subset M$. If there are no global sections on M, meaning that the atlas for M consists of more than one chart, then in every local domain U_i of the open covering $\{U_i\}$ of M, one must know the local representative ω_i such that the set $\{\omega_i\}$ yields a representation of ω in a bundle chart. This is done by means of the bundle chart

$$\Phi : \pi^{-1}(U_i) \longrightarrow U_i \times G$$

and the so-called unit section

$$x \in U_i \longmapsto \sigma_i(x) := \Phi^{-1}(x, \mathbb{1})$$

by calculation of $\omega_i := \sigma_i^* \omega$. Finally, making sure that in the transition from chart U_i to chart U_j the one-forms ω_i and ω_j transform by the transition map one obtains what is called a Cartan connection valid for the whole of M. The corresponding gauge field, which obeys the equations of motion, is then given in terms of charts.

The gauge theories described in this chapter are defined on $M = \mathbb{R}^{(1,3)}$. In this case, the analysis reduces to the simplified construction worked out in previous sections. However, already in describing a (hypothetical) magnetic monopole on flat base space one needs a description of the sphere S^2 in terms of charts. The gauge potential is composed of at least two parts, valid on different charts, which are combined by means of the transition mappings.

5.5 The U(2) Theory of Electroweak Interactions

The U(2) gauge theory of electroweak interactions combines Maxwell theory with the theory of weak interactions. Here we venture on a journey to the physics of elementary particles and to quantum field theory. We note, however, that the essential features of this theory can be discussed and understood in the framework of *classical* field theory, without dwelling on the techniques of its quantization. In this section, we describe this theory as a concrete example of a non-Abelian gauge theory, along the principles developed Sect. 5.4. Similarly, the mechanism of spontaneous symmetry breaking mentioned in Sect. 5.4.7 can be understood in this framework and is illustrated by an important example from physics.

5.5.1 A U(2) Gauge Theory with Massless Gauge Fields

Every element of the unitary group U(2) in two complex dimensions can be written in the form

$$\mathrm{U} = \mathrm{e}^{\mathrm{i}\alpha} \begin{pmatrix} u & v \\ -v^* & u^* \end{pmatrix} \quad \text{with} \quad \alpha \in \mathbb{R}\,, \quad u, v \in \mathbb{C} \text{ and } |u|^2 + |v|^2 = 1\,. \tag{5.61}$$

The corresponding Lie algebra is spanned by four generators chosen to be

$$\mathrm{T}_0 = \sigma_0 = \begin{pmatrix} 1 & 0 \\ 0 & 1 \end{pmatrix}, \quad \mathrm{T}_i = \frac{1}{2}\sigma_i\,, \quad (i = 1, 2, 3)\,, \tag{5.62}$$

where σ_i are the Pauli matrices, (5.23). From (5.24) the commutators are

$$\left[\mathrm{T}_0, \mathrm{T}_i\right] = 0\,, \tag{5.63a}$$

$$\left[\mathrm{T}_i, \mathrm{T}_j\right] = \mathrm{i}\varepsilon_{ijk}\mathrm{T}_k\,. \tag{5.63b}$$

Thus, there is a generator which commutes with all others and which generates a U(1) factor of the gauge group. This factor is manifest in (5.61) in the phase factor $\exp\{\mathrm{i}\alpha\}$.

The corresponding gauge group $\mathfrak{G}$, the potentials and the covariant derivative are constructed as was described in full generality in Sects. 5.4.3 and 5.4.4. In the adjoint representation, the numerical factor κ from (5.28) equals 2. The Lagrange density, (5.51), for the field strength tensor, (5.45), and for a multiplet of scalar fields reads

$$\begin{aligned} \mathcal{L} = &-\frac{1}{32\pi q^2}\operatorname{tr}\left(F_{\mu\nu}F^{\mu\nu}\right) \\ &+\frac{1}{2}\left[\left(D_\mu\Phi(x), D^\mu\Phi(x)\right) - \mu^2(\Phi,\Phi)\right] - W\left(\Phi(x)\right)\,, \end{aligned} \tag{5.64}$$

where $F_{\mu\nu}$ and D_μ are given by

$$F_{\mu\nu} = \mathrm{i}q \sum_{k=0}^{3} \mathsf{U}^{(\mathrm{ad})}(\mathsf{T}_k) F_{\mu\nu}^{(k)}(x) \quad \text{and} \tag{5.64a}$$

$$D_\mu \Phi = \{\mathbb{1}\partial_\mu + \mathrm{i}q \sum_{k=0}^{3} \mathsf{U}^{(\Phi)}(\mathsf{T}_k) A_\mu^{(k)}(x)\}\Phi(x) . \tag{5.64b}$$

They contain the generator T_0 of the U(1) factor as well as the generators $(\sigma_k/2)$ of SU(2), both in the adjoint representation (gauge fields) and in another representation decribing the multiplet Φ.

Without the terms containing the scalar fields, the Lagrange density, (5.64), describes four, initially massless, gauge bosons, two of which can also be replaced by the linear combinations

$$W_\mu^{(\pm)}(x) := \frac{1}{\sqrt{2}}\left(A_\mu^{(1)}(x) \pm \mathrm{i}A_\mu^{(2)}(x)\right) . \tag{5.65}$$

This replacement is done in view of a later interpretation which assigns these new gauge fields $W_\mu^{(\pm)}$ to particles carrying electric charges ± 1. By this redefinition the generators T_1 and T_2 are replaced by the linear combinations[11]

$$\mathsf{T}_+ := \mathsf{T}_1 + \mathrm{i}\mathsf{T}_2 , \quad \mathsf{T}_- := \mathsf{T}_1 - \mathrm{i}\mathsf{T}_2 . \tag{5.66a}$$

In the defining representation of SU(2), this replacement means defining

$$\frac{1}{2}(\sigma_1 + \mathrm{i}\sigma_2) =: \sigma_+ = \begin{pmatrix} 0 & 1 \\ 0 & 0 \end{pmatrix} , \; \frac{1}{2}(\sigma_1 - \mathrm{i}\sigma_2) =: \sigma_- = \begin{pmatrix} 0 & 0 \\ 1 & 0 \end{pmatrix} . \tag{5.66b}$$

The sum of the terms with $k = 1$ and $k = 2$ in the covariant derivative (5.64b) becomes

$$\begin{aligned} &\mathsf{U}^{(\Phi)}(\mathsf{T}_1)A_\mu^{(1)}(x) + \mathsf{U}^{(\Phi)}(\mathsf{T}_2)A_\mu^{(2)}(x) \\ &= \frac{1}{\sqrt{2}}\left\{\mathsf{U}^{(\Phi)}(\mathsf{T}_-)W_\mu^{(+)}(x) + \mathsf{U}^{(\Phi)}(\mathsf{T}_+)W_\mu^{(-)}(x)\right\} . \end{aligned} \tag{5.66c}$$

Of course, these new fields $W_\mu^{(\pm)}$ have no mass terms either; otherwise the local gauge invariance of the Lagrange density given by (5.64) would be violated.

At first, it seems more difficult to determine what the role of the two gauge fields $A_\mu^{(3)}(x)$ and $A_\mu^{(0)}(x)$ could be in a theory defined by (5.64) and which is meant to

[11] In quantum mechanics, these operators are called "raising and lowering operators". Note that using these linear combinations does not mean that one has complexified the Lie algebra of the structure group or the structure group itself.

describe the Maxwell fields and the vector fields of weak interactions. Maxwell theory is a "genuine" gauge theory; its gauge group is $\mathfrak{G} = \mathrm{U}(1)$. This U(1), which, for the sake of clarity, we mark by a subscript "e.m." for "electromagnetic", $\mathrm{U}(1)_{\mathrm{e.m.}}$, *may* be identical with the U(1) factor of U(2). However, it might also be constructed from a linear combination of this part and the 3-component of SU(2). This cannot be decided a priori. However, at this level, we may take into account empirical information which says that electromagnetic interactions are mediated by *massless* photons but that weak interactions are due to the exchange of three *massive* vector particles, $W^{(+)}$, $W^{(-)}$ and Z^0, two of which are charged, while the third is electrically neutral (as indicated by the superscripts (+), (−) and 0, respectively). The two charged particles $W^{(+)}$ and $W^{(-)}$ are expected to be partners in a triplet, i.e. in the adjoint representation of SU(2), whereas the neutral Z^0 will be the heavy partner of the photon.

With this remark in mind one, is led to try a general linear combination of the initially massless gauge fields $A_\mu^{(3)}$ and $A_\mu^{(0)}$ and thus to construct new neutral fields, viz.

$$A_\mu^{(\gamma)}(x) = A_\mu^{(0)}(x)\cos\theta_{\mathrm{W}} - A_\mu^{(3)}(x)\sin\theta_{\mathrm{W}}\,, \tag{5.67a}$$

$$A_\mu^{(Z)}(x) = A_\mu^{(0)}(x)\sin\theta_{\mathrm{W}} + A_\mu^{(3)}(x)\cos\theta_{\mathrm{W}}\,. \tag{5.67b}$$

With this ansatz one entertains the hope that the first of these, equation (5.67a), might be the vector potential of Maxwell theory, and that the second, equation (5.67b), might become the vector field describing the Z^0 vector boson. The mixing angle θ_{W}, named after Steven Weinberg, who introduced it in the theory of electroweak interactions, remains a free parameter at this stage and will have to be taken from experiment. We note that in the special case $\theta_{\mathrm{W}} = 0$, the U(1) field is identical with the Maxwell field, whereas the Z^0 field is the third partner of the two W-fields, spanning the triplet representation of SU(2).

This model raises an important question: Can one arrange the theory defined by (5.64) in such a way that it retains its full gauge invariance and yet the fields $W_\mu^{(\pm)}$ and $A_\mu^{(Z)}$ receive mass terms?

5.5.2 Spontaneous Symmetry Breaking

The multiplet $\Phi = \{\phi^{(1)}, \dots \phi^{(m)}\}$ spans a representation of structure group G, and the action $\mathsf{U}^{(\Phi)}(g)\Phi$ of an element $g \in G$ on Φ is well defined. Before returning to the U(2) model, (5.64), proper, we discuss the more general case of a theory which contains an arbitrary non-Abelian part and a multiplet of scalar fields and which is defined by a Lagrange density of the kind given by (5.51). The self-interaction $W(\phi)$ in (5.51) or (5.64) is the crucial input for the phenomenon developed in this section. We assume this part of the Lagrange density to have the following properties:

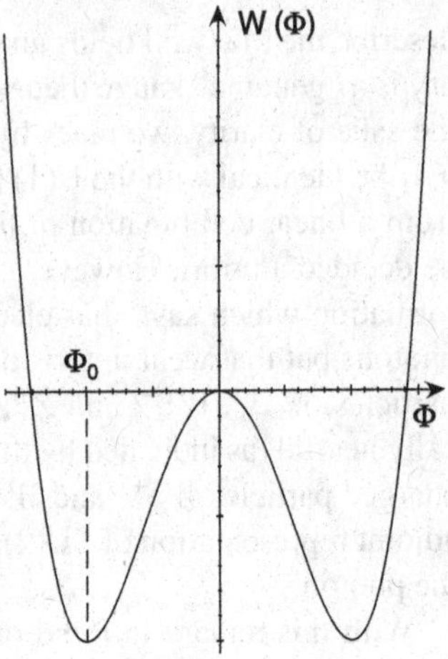

Fig. 5.4 The function $w(x) = -ax^2 + bx^4$ of the real variable x has an absolute minimum at $x = \pm\sqrt{a/(2b)}$. Replacing the real variable x by the complex variable $z = x + iy$, one obtains the graph of the function $w(z)$ by rotating this curve about the ordinate. This generates a 2-surface which looks like a sombrero or the bottom of a wine bottle

(i) The potential $W(\phi)$ is invariant under the whole structure group:

$$W(\mathsf{U}(g)\Phi) = W(\Phi) \quad \text{for all} \quad g \in G\,.$$

(This is still invariance with respect to a global symmetry!).

(ii) The potential $W(\phi)$ has an absolute minimum at $\Phi_0 = \{\phi_0^{(1)}, \ldots \phi_0^{(m)}\}$.

(iii) This minimum is degenerate, i.e. the configuration Φ_0 is not invariant under the entire structure group G.

When these assumptions are met, one talks about *spontaneously broken symmetry*. What follows is an example satisfying all three assumptions. We denote the G-invariant coupling of the multiplet Φ to itself symbolically by round parentheses $(\cdots,\cdots)$. Let

$$W(\Phi) = -\frac{\mu^2}{2}(\Phi, \Phi) + \frac{\lambda}{4}(\Phi, \Phi)^2 + C \quad \text{with } \lambda > 0\,. \tag{5.68}$$

As the factor multiplying the bilinear term has a negative sign whereas the quartic term has a positive sign, we have the analogue of the one-dimensional function

$$w(x) = -a\,x^2 + bx^4 \quad \text{with} \quad a, b > 0\,. \tag{5.69}$$

This function has a relative maximum at $x = 0$ and an absolute minimum at $x = \pm\sqrt{a/(2b)}$ which is degenerate (Fig. 5.4). The absolute minimum of (5.68) is reached when

$$(\Phi_0, \Phi_0) = \frac{\mu^2}{\lambda} \equiv v^2\,. \tag{5.70}$$

Obviously, this minimum is degenerate since only the bilinear form given by (5.70) is fixed, but not Φ_0. The configuration Φ_0, by itself, is not invariant under the action of structure group G. Inserting v^2 from (5.70), the function $W(\Phi)$ can equivalently be written as

$$W(\Phi) = \frac{\lambda}{4}\Big((\Phi,\Phi) - v^2\Big)^2 - \frac{\lambda}{4}v^4 + C\,, \tag{5.71}$$

a form exhibiting more clearly the minimum and its degeneracy. In this classical framework, the additive constant is irrelevant and may be dropped.

There appears a remarkable phenomenon in the framework of a local gauge theory. As $W(\Phi)$ takes its absolute minimum at arguments which all have the same modulus but which do not coincide with the origin $\Phi = \mathbf{0}$, in reality it is not the field Φ which is the dynamical field (with m components), but rather its difference from Φ_0,

$$\Theta(x) := \Phi(x) - \Phi_0\,. \tag{5.72}$$

This phenomenon may be visualized qualitatively as follows. The energetically favoured state of the system described by the Lagrange function (5.51) will be at the minimum of the "potential" $W(\Phi)$ or in its immediate neighbourhood.[12] In point mechanics, this situation is comparable to the potential energy $U(z) = (1/2)m\omega^2 z^2$, with $z = x - x_0$, of a harmonic oscillator where the back-driving force is directed towards the point $x_0 \neq 0$, and not to the origin. Or, if one wishes to take account of the degeneracy, it is comparable to a potential energy of the kind $U(\boldsymbol{x}) = \lambda(\boldsymbol{x}^2 - \boldsymbol{x}_0^2)^2/4$, where the force that follows from it, $\boldsymbol{K} = -\lambda(\boldsymbol{x}^2 - \boldsymbol{x}_0^2)\boldsymbol{x}$, changes sign at $|\boldsymbol{x}| = |\boldsymbol{x}_0|$. In both cases, $z = x - x_0$ and $\boldsymbol{z} = \boldsymbol{x} - \boldsymbol{x}_0$, respectively, are the physically meaningful variables.

Inserting (5.72) into the "kinetic" term $(D_\mu\Phi, D^\mu\Phi)$ in the Lagrange density given by (5.51),[13] one obtains a term

$$\frac{1}{2}\big(D_\mu\Phi_0, D^\mu\Phi_0\big) = \frac{1}{2}\big(\mathsf{U}^{(\phi)}(A_\mu)\Phi_0, \mathsf{U}^{(\phi)}(A^\mu)\Phi_0\big)\,, \quad \text{with} \tag{5.73}$$
$$A_\mu = \mathrm{i}q\sum_k \mathsf{T}_k A_\mu^{(k)}(x)\,,$$

which is seen to have the form of a mass term for at least some of the hitherto massless gauge fields! Note, however, that gauge invariance is not violated. Indeed, gauge invariance means that (i) the fields $F_{\mu\nu}$ come in in the form of the first term on the right-hand side of (5.51); (ii) the field Φ, or the field Θ as defined by (5.72), appears with the covariant derivative; and (iii) the mass term and the potential of field Φ are (globally) invariant under structure group G. All of these conditions are fulfilled. Therefore, the Lagrange density is still gauge invariant. However, it prefers a ground state which has less symmetry than the theory on which it is built.

[12] As one sees, the notion of "potential" is used in two different meanings: on the one hand as a gauge potential in the sense of electrodynamics, on the other hand as a potential or potential energy of the scalar fields in the sense of classical mechanics. This should not be a serious source of misunderstandings.

[13] The quotation marks are meant to emphasize that this term contains more than the kinetic energy of the scalar field.

This kind of symmetry breaking is fundamentally different from an explicit perturbation of the original symmetry where the theory is defined by a Lagrange density

$$\mathcal{L} = \mathcal{L}_0 + \mathcal{L}' ,$$

where the term $\mathcal{L}'$ is small compared to $\mathcal{L}_0$, in a sense to be defined more precisely. The Proca Lagrange density (5.6),

$$\mathcal{L}_{\text{Proca}} = \mathcal{L}_{\text{Maxwell}} + \mathcal{L}' \quad \text{with} \quad \mathcal{L}' = \frac{\lambda^2}{8\pi} A_\mu(x) A^\mu(x) ,$$

provides an example, if the hypothetical photon mass λ is not zero but small. The big primordial symmetry of $\mathcal{L}_{\text{Maxwell}}$ is lost. Nevertheless, the influence of the perturbation $\mathcal{L}'$ might possibly be estimated in the framework of perturbation theory.

With symmetry breaking as developed in this section, the original gauge invariance is not lost. Only the state(s) of lowest energy has (have) less visible symmetry than the Lagrange density. Therefore, one calls this kind of symmetry breaking *spontaneous symmetry breaking*. As the symmetry is not really broken but acts "secretly" in the physical states of the theory, L. O'Raifeartaigh coined the notion of *hidden symmetry* (O'Raifeartaigh 1998).

The phenomenon of spontaneous symmetry breaking in a local gauge theory can be analyzed more concisely on the basis of the theory's group structure. In a sense, it will become clear below how one can *tune* the pattern of spontaneous breaking. In particular, one can fix the number of gauge fields which are to be provided with a mass.

The assumptions listed at the beginning of this section contain the condition that the minimum of $W(\Phi)$ be degenerate: If there is more than one single configuration $\Phi_0 = \{\phi_0^{(1)}, \ldots, \phi_0^{(m)}\}$, then there exists at least one element $g \in G$ of the structure group for which

$$\sum_{a=1}^{m} U_{ab}^{(\Phi)}(g)\phi_0^{(b)} \neq \phi_0^{(a)} . \tag{5.74a}$$

Such a g shifts the point Φ_0 to Φ_0' for which the potential still assumes its minimum. Expressing this element g by generators of the Lie algebra of G,

$$g = \exp\{\mathrm{i} \sum_{k=1}^{N} \alpha_k \mathrm{T}_k\} ,$$

(5.74a) is equivalent to the statement that there is at least one generator T_i whose action on Φ_0 does not yield zero,

$$\mathrm{U}^{(\Phi)}(\mathrm{T}_i)\Phi_0 \neq 0 . \tag{5.74b}$$

One then carries out the following construction. The generators T_j are replaced by linear combinations

$$\mathsf{S}_i = \sum_{j=1}^{N} a_{ij}\mathsf{T}_j \,, \tag{5.75}$$

with $\mathsf{a} = \{a_{ij}\}$ a nonsingular, constant matrix such that the new generators fall into two classes, viz.:

a) A group of generators $\{\mathsf{S}_1, \ldots, \mathsf{S}_F\}$ whose action on Φ_0 gives zero:

$$\mathsf{S}_i \Phi_0 = 0 \,, \quad i \in (1, 2, \ldots, F); \quad \text{and} \tag{5.75a}$$

b) A group of generators $\{\mathsf{S}_{F+1}, \ldots, \mathsf{S}_N\}$ which shift Φ_0 in a nontrivial way:

$$\mathsf{S}_j \Phi_0 \neq 0 \,, \quad j \in (F+1, F+2, \ldots, N) \,. \tag{5.75b}$$

It is not difficult to verify that the elements of the first class (a) generate a subgroup $H \subset G$ of G. This subgroup consists of all elements of the form

$$h = \exp\{\mathrm{i} \sum_{i=1}^{F} \alpha_i \mathsf{S}_i\} \,. \tag{5.76}$$

All such $h \in H$ have in common that they leave invariant an arbitrarily chosen position Φ_0 of the minimum:

$$\mathsf{U}^{(\Phi)}(h)\Phi_0 = (\mathbb{1} + \sum_{i=1}^{F} \alpha_i \mathsf{U}^{(\Phi)}(\mathsf{S}_i) \ldots)\Phi_0 = \Phi_0 \,, \quad h \in H \subset G \,. \tag{5.77}$$

Pictorially speaking, these are transformations which do not lead out of the bottom of the bottle of $W(\Phi)$ at its lowest point. They do not change the energy of these states of lowest energy. This means, however, that the subgroup $H \subset G$ which is spanned by the elements h remains a genuine symmetry. For this reason, one talks of a *residual symmetry*.

In contrast, the remaining generators in class (b) shift a given Φ_0 away from the minimum. They act, in a pictorial description, against the walls of the potential, transversally to the set of points Φ_0. For these generators the G-invariant mass term, (5.73), is different from zero, so that certain linear combinations of the initially massless gauge fields obtain nonvanishing masses.

The structure group G, which defines the basic symmetry of the theory, has the Lie algebra $\mathfrak{g} = \mathrm{Lie}\,(G)$ whose dimension is

$$\dim \mathfrak{g} = N \,.$$

The Lie algebra $\mathfrak{h} = \text{Lie}\,(H)$ of residual group H has dimension

$$\dim \mathfrak{h} = F \; .$$

This analysis reveals a remarkable result:

The number n_γ of those gauge fields which remain massless after spontaneous breaking of the primordial symmetry and the number n_m of those which become massive depend only on the dimensions of the Lie algebras of the structure group G and the residual group. They are

$$n_\gamma = \dim \mathfrak{h} = F \; , \tag{5.78a}$$

$$n_\text{m} = \dim \mathfrak{g} - \dim \mathfrak{h} = N - F \; . \tag{5.78b}$$

These numbers are independent of the nature of the multiplet of scalar fields.

The number of gauge fields remaining massless is denoted by n_γ to remind us that these are photonlike fields.

Remarks

Within this classical framework, one need not fix the precise form of the potential $W(\Phi)$. It is sufficient to know that $W(\Phi)$ has absolute minima and that it satisfies conditions (i) to (iii). Thus, at this stage, no explicit functional form must be chosen for $W(\Phi)$. However, in the framework of quantized field theory, there are further restrictions if one requires the theory to yield well-defined results at all finite orders of perturbation theory. The requirement of what is called *renormalizability* does not allow for powers of Φ higher than four. If this is taken into account, then one is left with the specific form given by equation (5.68) which satisfies all conditions including this additional requirement.

It should now be clear that the amount of spontaneous symmetry breaking can indeed be *tuned.* The dimension of the Lie algebra $\mathfrak{g}$ of the structure group G yields the total number N of vector fields. Without minimal coupling to scalar fields Φ and without the self-interactions $W(\Phi)$, these vector fields remain photonlike, i.e. massless. Keeping track of the conditions mentioned previously, the choice of the potential $W(\Phi)$ determines the pattern of spontaneous symmetry breaking,

$$G \longrightarrow H \subset G \; ,$$

and hence the number of gauge fields remaining massless. The specific multiplet spanned by the scalar fields $\Phi = \{\phi^{(1)}, \dots, \phi^{(m)}\}$ does not enter as long as it does not violate the conditions on the potential.

5.5.3 Application to the U(2) Theory

With these preparations we can now tackle the question raised at the end of Sect. 5.5.2 and provide a conclusive answer. If the U(2) theory, (5.64), is to describe electrodynamics and the weak interactions and if the photon is the only gauge boson which stays massless, then the spontaneous symmetry breaking must be adjusted such that in the end, only the group $U(1)_{e.m.}$ remains a true gauge symmetry. The symmetry is broken spontaneously according to the pattern

$$G = U(2) \cong U(1) \times SU(2) \longrightarrow H = U(1)_{e.m.} \, . \tag{5.79}$$

In other terms, the original symmetry is "hidden" following the scheme of (5.79). Note that the $U(1)_{e.m.}$ of electrodynamics, in general, is not identical with the U(1) factor of the gauge group $\mathfrak{G} = U(2)$.

Using ansatz (5.65) for the W fields and ansatz (5.67a) and (5.67b) for the fields $A^{(\gamma)}_\mu$ and $A^{(Z)}_\mu$, or its inverse,

$$A^{(0)}_\mu(x) = A^{(\gamma)}_\mu(x) \cos\theta_W + A^{(Z)}_\mu(x) \sin\theta_W \, , \tag{5.80a}$$

$$A^{(3)}_\mu(x) = -A^{(\gamma)}_\mu(x) \sin\theta_W + A^{(Z)}_\mu(x) \cos\theta_W \, , \tag{5.80b}$$

and introducing the raising and lowering generators of (5.66a), the action of A on Φ is given by

$$\begin{aligned} U^{(\Phi)}(A_\mu)\Phi &= iq \sum_{k=0}^{3} A^{(k)}_\mu(x) U^{(\Phi)}(T_k)\Phi \\ &= iq \Bigg\{ \frac{1}{\sqrt{2}} \left[W^{(-)}_\mu(x) U^{(\Phi)}(T_+) + W^{(+)}_\mu(x) U^{(\Phi)}(T_-) \right] \\ &\quad + A^{(Z)}_\mu(x) U^{(\Phi)}(T_3 \cos\theta_W + T_0 \sin\theta_W) \\ &\quad + A^{(\gamma)}_\mu(x) U^{(\Phi)}(-T_3 \sin\theta_W + T_0 \cos\theta_W) \Bigg\} \Phi \, . \end{aligned} \tag{5.81}$$

More specifically, let Φ span a unitary irreducible representation of SU(2). One knows that such representations can be characterized by two numbers t and m_t which satisfy the following eigenvalue equations:

$$U^{(\Phi)}(\boldsymbol{T}^2)\Phi = \sum_{k=1}^{3} U^{(\Phi)}(\boldsymbol{T}^2_k)\Phi = t(t+1)\Phi \tag{5.82a}$$

$$U^{(\Phi)}(\boldsymbol{T}_3)\Phi = m_t \Phi \, , \tag{5.82b}$$

where t takes the values $(0, 1/2, 1, 3/2, \ldots)$, whereas m_t, for a given value of t, takes the values $m_t = -t, -t+1, \ldots, t-1, t$. This is the analogue of a representation of the rotation group, with t replacing the eigenvalue j of angular momentum and m_t replacing the projection quantum number m_j. The action of the generator T_0,

$$\mathsf{U}^{(\Phi)}(\mathsf{T}_0)\Phi = t_0\Phi\,, \tag{5.82c}$$

is not fixed. However, the following reasoning is helpful: As T_0 commutes with all other generators and as the action of the raising and lowering generators $\mathsf{U}^{(\phi)}(\mathsf{T}_\pm)$ does not lead out of the multiplet, all components of Φ must have the same value t_0.

The last term on the right-hand side of formula (5.81) is particularly interesting. Suppose the shift operation given by (5.72) is performed in the term of (5.81), $\Phi(x) = \Theta(x) + \Phi_0$. This has both a desirable and an undesirable effect: The term $(D_\mu\Phi, D^\mu\Phi)$ of the Lagrange density, (5.64), contains, among others, (5.81) in the form of a bilinear coupling to a scalar. As noted earlier, in particular, the constant Φ_0 yields quadratic terms with constant factors for $W^{(\pm)}$ and for $Z^{(0)}$, i.e. the expected mass terms for these vector fields. This is the desired property. At the same time, however, the Maxwell field $A_\mu^{(\gamma)}(x)$ appears multiplied by the factor

$$\mathsf{U}^{(\Phi)}\big(-\mathsf{T}_3\sin\theta_{\mathrm{W}} + \mathsf{T}_0\cos\theta_{\mathrm{W}}\big)\Phi_0\,, \tag{5.83}$$

which possibly is not zero. This is the undesirable feature. Setting the constant Φ_0 to zero, $\Phi_0 = 0$, is no solution because then also the first effect is lost. The only way out of this dilemma is to choose the set of constant values $\Phi_0 = \{\phi_0^{(1)}, \ldots, \phi_0^{(m)}\}$ such that for exactly one component, say the ith one, the entry is different from zero, whereas it is zero for all others,

$$\Phi_0 = \big\{0, 0, \ldots, \phi_0^{(i)} = v \neq 0, 0, \ldots\big\}\,, \tag{5.84}$$

and, furthermore, to fix the eigenvalue of $\mathsf{U}^{(\Phi)}(\mathsf{T}_0)$ for this component (and hence for all other components) such that

$$t_0^{(i)} = t_3^{(i)}\tan\theta_{\mathrm{W}}\,. \tag{5.85}$$

With this choice a nonvanishing component $\phi_0^{(i)}$ will have the eigenvalue zero of the linear combination given by (5.83). Denoting the value of this component $\phi_0^{(i)}$ by v and inserting (5.84) into (5.81), the G-invariant scalar constructed from $\mathsf{U}^{(\Phi)}(A_\mu)\Phi_0$ is given by

$$\begin{aligned}
&\Big(\mathsf{U}^{(\Phi)}(A_\mu)\Phi_0, \mathsf{U}^{(\Phi)}(A^\mu)\Phi_0\Big)\\
&= q^2\bigg\{\frac{1}{2}\Big(\Phi_0, \mathsf{U}^{(\Phi)}(\mathsf{T}_+\mathsf{T}_- + \mathsf{T}_-\mathsf{T}_+)\Phi_0\Big)\, W_\mu^{(-)}(x)W^{(+)\mu}(x)\\
&+ \Big(\Phi_0, \big[\mathsf{U}^{(\Phi)}(\mathsf{T}_3\cos\theta_{\mathrm{W}} + t_3^{(i)}\sin\theta_{\mathrm{W}}\tan\theta_{\mathrm{W}})\big]^2\Phi_0\Big)\, A_\mu^{(\mathrm{Z})}(x)A^{(\mathrm{Z})\mu}(x)\bigg\}\,.
\end{aligned}$$

In this construction, we have used the fact that all components of Φ belong to the same eigenvalue given by (5.85). Terms which are quadratic in T_+ or in T_- do not contribute because Φ_0 has only one nonvanishing component, $\phi_0^{(i)}$, and because $\mathsf{U}^{(\Phi)2}(\mathsf{T}_\pm)$ would link the ith component to the $(i \pm 2)$-d components, both of which vanish.

Taking account of the relation

$$\frac{1}{2}(\mathsf{T}_+\mathsf{T}_- + \mathsf{T}_-\mathsf{T}_+) = \sum_{k=1}^{3} \mathsf{T}_k^2 - \mathsf{T}_3^2 = \boldsymbol{T}^2 - \mathsf{T}_3^2\,, \tag{5.86}$$

inserting (5.84) for Φ_0 and noting that $\cos\theta_W + \sin\theta_W \tan\theta_W = 1/\cos\theta_W$, one obtains

$$\begin{aligned}&\left(\mathsf{U}^{(\Phi)}(A_\mu)\Phi_0, \mathsf{U}^{(\Phi)}(A_\mu)\Phi_0\right)\\ &= q^2 v^2 \left\{\left[t(t+1) - (t_3^{(i)})^2\right] W_\mu^{(-)}(x) W^{(+)\mu}(x)\right.\\ &\qquad \left. + \frac{1}{\cos^2\theta_W}(t_3^{(i)})^2 A_\mu^{(Z)}(x) A^{(Z)\mu}(x)\right\}.\end{aligned} \tag{5.87}$$

This is a remarkable result. By construction, the Maxwell field remains massless. In other terms, spontaneous breaking is adjusted in such a way that of the original gauge group $\mathfrak{G} = \mathrm{U}(2)$ there remains only the residual gauge group $\mathfrak{H} = \mathrm{U}_{\text{e.m.}}(1)$ as a gauge symmetry. As a result, the gauge group of Maxwell theory is found to be generated by a linear combination of the U(1) generator T_0 and the 3-component T_3 of the Lie algebra of SU(2), viz.

$$-\mathsf{T}_3 \sin\theta_W + \mathsf{T}_0 \cos\theta_W =: \mathsf{T}_{\text{e.m.}}\,. \tag{5.88}$$

The remaining three gauge fields of the theory defined by (5.64) obtain nonvanishing mass terms. The $W^{(+)}$ and $W^{(-)}$ fields have the same mass, which is proportional to

$$m_W^2 \propto \tfrac{1}{2} q^2 v^2 \left[t(t+1) - (t_3^{(i)})^2\right], \tag{5.89a}$$

whereas the Z field receives a mass proportional to

$$m_Z^2 \propto q^2 v^2 \cos^{-2}\theta_W (t_3^{(i)})^2\,, \tag{5.89b}$$

with the same numerical factors. This yields an important relation:

$$\frac{m_W^2}{m_Z^2 \cos^2\theta_W} = \frac{t(t+1) - (t_3^{(i)})^2}{2(t_3^{(i)})^2}\,. \tag{5.90}$$

Once the parameter θ_W is fixed, the ratio of the two masses depends only on the assignment of the multiplet Φ to a representation of the structure group G.

Remarks and Comments

1. In what may be called the *standard model of electroweak interactions*, one chooses the Φ field to be a *doublet*:

$$t = \frac{1}{2}\,, \quad t_3^{(i)} = \frac{1}{2}\,. \tag{5.91}$$

With this choice the ratio given by(5.90) is equal to

$$\varrho := \frac{t(t+1) - (t_3^{(i)})^2}{2(t_3^{(i)})^2} = 1\,. \tag{5.92}$$

Note, however, that the choice of the multiplet for Φ is not predicted by the model. Experimentally, the quantity $\cos^2\theta_W$, with θ_W being the Weinberg angle, is an empirical quantity, as are the masses of W and Z, which have been determined in various experiments. The measured values m_W, m_Z and $\sin\theta_W$ are in very good agreement with the ratio given by (5.92).

2. For other components $\theta^{(j)} \neq \theta^{(i)}$ of the dynamical scalar field, (5.72), $\Theta = \{\theta^{(1)}, \ldots, \theta^{(m)}\}$, the factor multiplying $A_\mu^{(\gamma)}(x)$ in (5.81) is not zero. These fields describe electrically charged particles. Therefore, this suggests that one should interpret the product of coupling constant q and $\sin\theta_W$ as a negative elementary charge, though this must await a deeper justification in the quantized version of the theory:

$$-q\sin\theta_W \equiv -e\,. \tag{5.93}$$

In the quantum theory of the standard model, the coupling constants responsible for the weak interactions are also fixed (see e. g. Scheck (2012)).

3. It seems odd that the generator given by (5.88) of the $U(1)_{e.m.}$ depends on the parameter θ_W, the ratio of the two terms in (5.88) being fixed by the requirement of spontaneous symmetry breaking. This flaw is easily repaired by means of a simple redefinition: The generators of the original U(1) and of $U(1)_{e.m.}$ may be replaced by, respectively,

$$\mathsf{Y} := -2\frac{\cos\theta_W}{\sin\theta_W}\mathsf{T}_0\,, \tag{5.94a}$$

$$\mathsf{Q} := \mathsf{T}_3 + \frac{1}{2}\mathsf{Y}\,. \tag{5.94b}$$

The operator Q, by (5.83) and (5.93), then becomes the electric charge in units of the elementary charge,

$$q\big(-\mathsf{T}_3\sin\theta_W + \mathsf{T}_0\cos\theta_W\big) = e\big(\mathsf{T}_3 + \frac{1}{2}\mathsf{Y}\big),$$

whereas Y is an alternative for the generator of the U(1) factor in G. Denoting the eigenvalue of Y in the multiplet Φ by the real number y, the requirement given by (5.85) for ϕ_0 simplifies to

$$y^{(i)} = -2t_3^{(i)} . \tag{5.95}$$

In elementary particle physics, Y is called a *weak hypercharge.* The scalar fields which correspond to the classical field Φ are called *Higgs particles.*

4. As in the case of Maxwell's equations, one may study the behaviour of the gauge theory, (5.64), under space reflection, time reversal and charge reflection C. The result is that it is found to be invariant under all three discrete symmetries. As C relates the fields $W_\mu^{(+)}$ and $W_\mu^{(-)}$, it is not surprising that the two must have the same mass term.

5.6 Epilogue and Perspectives

We conclude this chapter with a few more remarks and an outlook which may stimulate the reader to further study and reading.

Local Gauge Theory in the Classical Framework

The concept of local gauge theory penetrates deep into the physics of fundamental interactions and of elementary particles. The property of gauge invariance, which was discovered in the framework of *classical* electrodynamics, has become very important for *all* interactions known to us. The aspects of the standard model of electroweak interactions which we developed in Sect. 5.5 are still classical to a large extent. Non-Abelian gauge field theories need to be quantized when one introduces, in a second step, fields for electrons, nucleons, quarks and other matter particles and if one sets out to interpret the quantized fields in terms of particles with definite properties. But even in this third step, one follows the example of Maxwell theory, albeit in a technically more involved manner. Of course, one must investigate whether the structures developed in a classical setting remain meaningful in the corresponding quantum field theory or, if the need arises, how they are modified by the process of quantization.

Spontaneous Symmetry Breaking in Other Areas

The phenomenon of spontaneous symmetry breaking is encountered also in other areas of physics and in a diversity of manifestations. The case of a continuous symmetry and invariance under a Lie gauge group described in this chapter exhibits strongly geometric aspects. For example, it is of interest to study the space of all gauge potentials, from the point of view of differential geometry, and the action of the (infinite-dimensional) gauge group on them. Pure Yang–Mills theory, in particular with self-dual field strengths (curvatures), is a rich field of reserach in mathematics which gave many results of a purely mathematical nature but also many links to theoretical physics.

A particularly nice example for spontaneous symmetry breaking in a purely classical system is provided by a rotating star under the action of self-gravitation. The star is modelled by an incompressible fluid, and the aim of the analysis is to find its shape as a function of the angular velocity. This problem was first studied by C.G.J. Jacobi, and a complete solution is found in a work of D.H. Constantinescu, L. Michel and L.A. Radicati.[14]

Outlook

According to our present knowledge, all realistic theories describing weak, electromagnetic and strong interactions are local gauge theories of the kind described in this chapter. Also, Einstein's theory of general relativity, which is a purely classical theory, has many features of a geometric theory with a large amount of local symmetry. In this case, the gauge group is the group of diffeomorphisms on a semi-Riemannian manifold with dimension 4. General relativity, in many of its aspects, bears strong resemblance to local gauge theories but differs from them in other characteristic properties. This is one of the reasons why there is still no generally accepted quantized version of general relativity and why it seems difficult to unify gravitation with the other interactions in a geometric way.

In contrast to the local, non-Abelian gauge theories, general relativity is primarily a theory describing macroscopic physics, viz. the physics of large assemblies of masses and energies, and of physically viable universes. As such and in these areas, general relativity has been tested in various applications. It was found to be in overwhelming agreement with observations and passed all tests.

[14] D.H. Constantinescu, L. Michel, L.A. Radicati, Journal de Physique **40** (1979) 147.

Classical Field Theory of Gravitation 6

6.1 Introduction

The classical field theories developed in the preceding chapters all have in common that they are formulated on a *flat* spacetime, i.e. on a four-manifold which is a Euclidean space and which locally is decomposable into a direct product $M^4 = \mathbb{R}^3 \times \mathbb{R}$ of a physical space $\mathbb{R}^3_x$ of motions, and a time axis $\mathbb{R}_t$. The first factor is the three-dimensional space as it is perceived by an observer at rest while the time axis displays the (coordinate) time that he/she measures on his/her clocks. This spacetime is endowed with the Poincaré group as the invariance group of physical laws and inherits the corresponding specific causality structure. In the case of small velocities, $|\boldsymbol{v}| \ll c$, the Poincaré group is replaced by the Galilei group, and the causality structure becomes trivial. Equations of motion that relate physical observables, must be *form invariant* or, in other terms, must transform in a *covariant* way. The light cone in every point $x \in M$ of spacetime classifies all events y into those that are in causal relation with x, and those for which this does not hold. Flat spacetime is singled out by the fact that all light cones are parallel, i.e. they are related by translations. Every observer, by the choice of a frame of reference, defines a globally valid coordinate system which allows the observer to compare physical observables in different world points $x = (ct_x, \boldsymbol{x})^T$ and $y = (ct_y, \boldsymbol{y})^T$.

While these concepts are overwhelmingly successful in the description of classical mechanics, of classical and quantum electrodynamics, but also of electroweak and strong interactions in particle physics, and are brilliantly confirmed by many experimental tests, they fail in the description of the gravitational interaction. Strangely enough, gravitation which marked the beginning of the development of mechanics, and thereby of the whole of theoretical physics, in its full beauty and its vast scope, cannot be described on a globally flat space such as Minkowski space. In this chapter we give plausibility arguments why this is so, and we develop the geometric foundations for Einstein's equations for gravitation. These equations relate the geometry of spacetime with the energy-momentum content of matter and (non-

F. Scheck, *Classical Field Theory*, Graduate Texts in Physics,
DOI 10.1007/978-3-642-27985-0_6, © Springer-Verlag Berlin Heidelberg 2012

gravitational) radiation present in the universe. We study characteristic solutions of Einstein's equations and analyze their properties.

6.2 Phenomenology of Gravitational Interactions

Considering gravitational interaction at the same level as the other fundamental interactions of nature, i.e. like macroscopic electromagnetic interactions as developed in this book, their quantized form, the electroweak and strong interactions visible at the scales of elementary particle physics, some of its properties are conspicuous. In contrast to the other interactions, gravitation is always *attractive*[1], it is *universal* and it obeys an *equivalence principle* which does not apply to other interactions.

6.2.1 Parameters and Orders of Magnitude

Newton's constant carries a physical dimension, viz.
(length^3 × mass^{-1} × time^{-2}). It has the numerical value

$$G = (6.67428 \pm 0.00067) \cdot 10^{-11}\, \mathrm{m^3\, kg^{-1}\, s^{-2}}\,. \tag{6.1}$$

The force is *attractive* for all massive bodies. This is known from daily experience with falling objects, from the motion of the planets of the solar system in finite orbits, and from the choice of the branch of the hyperbola of the Kepler problem that a comet moves along. Regarding antimatter the gravitational interaction with matter is also *attractive,* as demonstrated and tested with antiprotons (the antiparticles of protons).

In order to develop a feeling for the number (6.1) one may compare, for example, the gravitational interaction between a proton and an antiproton with their static electric interaction. The proton and antiproton have the same mass but equal and opposite charges. Thus, the gravitational force and the Coulomb force acting between them, are, respectively, (in SI units)

$$\boldsymbol{F}_{\mathrm{G}} = -G m_{\mathrm{P}}^2 \frac{1}{r^2} \hat{\boldsymbol{r}}\,, \qquad \boldsymbol{F}_{\mathrm{C}} = -\kappa_{\mathrm{C}} e^2 \frac{1}{r^2} \hat{\boldsymbol{r}}\,,$$

where $\boldsymbol{r}$ and $r = |\boldsymbol{r}|$ denote the relative coordinate and its modulus, respectively. The ratio of these forces both of which are attractive, is independent of the direction and the distance. With $m_{\mathrm{P}} = 1.6726 \cdot 10^{-27}$ kg, $e = 1.6022 \cdot 10^{-19}$ C, and $\kappa_{\mathrm{C}} = 1/(4\pi\varepsilon_0) = c^2 \cdot 10^{-7}$ it has the value

$$R_{\mathrm{GC}} := \frac{G m_{\mathrm{P}}^2}{\kappa_{\mathrm{C}} e^2} = 0.81 \cdot 10^{-36}\,. \tag{6.2}$$

[1] This is true in "reasonable" universes and away from singularities. In the neighbourhood of rotating black holes there are spacetime regions where gravitation becomes repulsive.

This number whose smallness might surprise the reader, shows that gravitation is by far the weakest of the fundamental interactions. At the scales of microscopic particle physics it plays no role, in general, and can often be neglected.[2] Nevertheless, letting some Meissen porcelain fall on the kitchen floor, or falling oneself from a cherry tree can be catastrophic, because relatively large masses are involved and, in contrast to Coulomb forces, the like sign of the forces does not allow for any compensation of partial forces.

There are other ways of demonstrating the orders of magnitude. It is known that the square of the elementary charge, Planck's quantum of action, and the velocity of light are combined to the dimensionless fine structure constant α which characterizes the strength of the electromagnetic interactions. With

$$e = 4.8032 \cdot 10^{-10}\,\text{esu} \quad \text{and}$$
$$\hbar c = 197.33\,\text{MeV fm} = 3.16153 \cdot 10^{-17}\,\text{erg cm}$$

(thus using Gaussian units), one has

$$\alpha := \frac{e^2}{\hbar c} = 0.0072973 = \frac{1}{137.036} \,. \tag{6.3a}$$

From the Newton constant G, equation (6.1), from $\hbar c$, and using some reference mass M, one defines by analogy the dimensionless quantity $GM^2/(\hbar c)$. Inserting the mass of the proton, as an example, one finds

$$\alpha_G := \frac{Gm_P^2}{\hbar c} = 5.9 \cdot 10^{-39} \,. \tag{6.3b}$$

This number is smaller than the fine structure constant (6.3a) by the same ratio (6.2).

Noting that G has physical dimension (energy × length/mass^2) and that $\hbar c$ has dimension (energy × length) one can form a quantity with the dimension of a mass. This defines what is called the *Planck mass*

$$M_{\text{Pl}} := \sqrt{\frac{\hbar c}{G}} = 1.221 \cdot 10^{19}\,\text{GeV} = 2.177 \cdot 10^{-8}\,\text{kg} \,. \tag{6.4}$$

When expressed in terms of the mass of the proton and of the ratio (6.2) one has $M_{\text{Pl}} = m_P/\sqrt{\alpha R_{\text{GC}}}$. This is a value that should be measurable with a chemist's balance and which is larger by many orders of magnitude than typical masses in the zoo of elementary particles. Alternatively one might wish to give a Compton wave length corresponding to this mass, thus obtaining

$$\lambda_{\text{Pl}} = \frac{2\pi\hbar c}{M_{\text{Pl}}c^2} = (2\pi)1.62 \cdot 10^{-35}\,\text{m} \,. \tag{6.5}$$

[2] This is not completely true. In horizontally mounted circular accelerators where charged particles are transported over large distances one must take into account their free fall in the gravitational field of earth.

This quantity is called the *Planck length*. Its significance is not really clear. Nevertheless, one concludes that it is smaller by many orders of magnitude than typical ranges of the weak or the strong interactions which lie rather in the range 10^{-18} m. Presumably the Planck length (6.5) indicates the scale at which our model of spacetime in terms of a smooth manifold breaks down and where general relativity as formulated at the classical level must be replaced by a quantized theory of gravitation.

▶ **Remark**

Like the Coulomb interaction the gravitational interaction has infinite range. Both, the Coulomb potential and the gravitational potential are proportional to $1/r$. If in the future one succeeds in quantizing gravitation the carriers of this interaction will be *gravitons,* which are massless like the photons of quantum electrodynamics. In contrast to photons which have spin 1, gravitons have spin 2. Thus, the Planck length (6.5) is not to be understood as the range of this force.

There are qualitative arguments which show that general relativity and quantum theory may be incompatible at small distances of the order of (6.5). The idea is the following: On a smooth manifold M one can localize events $x \in M$ with arbitrary accuracy, at least in principle. Heisenberg's uncertainty relation then implies that this will involve arbitrarily large energy-momentum densities. These, in turn, upon insertion into Einstein's equations, cause locally strong curvatures of spacetime which upset the assumption from which one started. Arguments of this type lead to the conjecture that spacetime loses its character of a smooth manifold at small distances and is replaced by something new such as, perhaps, a space whose elements (points) no longer commute.

6.2.2 Equivalence Principle and Universality

Consider a mechanical system consisting of a sun with mass $m_\odot$ and a light planet whose mass is so small that it practically does not perturb the field created by the sun. If for a while m_I denotes the *inertial mass*, and m_G denotes the *gravitational mass* of the planet, and if $\boldsymbol{x}$ and $\boldsymbol{x}_\odot$ are the positions of the planet and the sun, respectively, nonrelativistic Newtonian theory yields the force

$$\dot{\boldsymbol{p}} = m_\mathrm{I}\ddot{\boldsymbol{x}} = -\frac{G m_\mathrm{G} m_\odot}{|\boldsymbol{x}-\boldsymbol{x}_\odot|^2}\frac{\boldsymbol{x}-\boldsymbol{x}_\odot}{|\boldsymbol{x}-\boldsymbol{x}_\odot|} . \tag{6.6}$$

Experience tells us that inertial and gravitational masses are, in essence, the same so that, by suitable choice of physical units, one can take them to be equal,

$$m_\mathrm{I} = m_\mathrm{G} . \tag{6.7}$$

The mass factors cancel in (6.6), and the resulting motion of the (light) planet is independent of its mass. This empirical equality of heavy and inertial mass is called *weak equivalence principle.* The same property is also an expression of the *universality* of gravitation: The motion of a test body in a given gravitational field does not depend on its mass and on its inner composition.

The following simple example illustrates the universality of gravitation and, at the same time, the equivalence principle:

Example 6.1 Test particle in a static homogeneous field

Consider a certain number of test particles in a homogeneous and static gravitational force field $\boldsymbol{K}^{(i)} = m_i \boldsymbol{g}$, where m_i denotes the mass of the i-th particle, while $\boldsymbol{g}$ is the acceleration field which is independent of time and position. Furthermore, the particles are assumed to be subject to internal forces $\boldsymbol{F}_{ji}$. In an *inertial* system K the equations of motion read

$$m_i \ddot{\boldsymbol{x}}^{(i)} = m_i \boldsymbol{g} + \sum_{j \neq i} \boldsymbol{F}_{ji} \,.$$

We now replace the inertial system with an *accelerated* system K′ that is uniformly accelerated with the acceleration field $\boldsymbol{g}$ so that the transformation formulae for the time and space variables are

$$t_y = t_x \,, \quad \boldsymbol{y} = \boldsymbol{x} - \frac{1}{2} \boldsymbol{g} t_x^2 \,.$$

The same equations of motion then read

$$m_i \ddot{\boldsymbol{y}}^{(i)} = \sum_{j \neq i} \boldsymbol{F}_{ji} \,.$$

What immediately strikes us is the fact that the gravitational field completely disappeared from these equations. For example, if by appropriate measurements one finds that the freely falling particles have vanishing acceleration two interpretations are possible

(a) either there are no forces at all, the observer is located in an inertial system,
(b) or there is a static and homogeneous gravitational field but the frame of reference is not an inertial system. The frame falls freely with the particles.

As a matter of principle, there is no way to distinguish between the two interpretations of the empirical result.

In general, the acceleration field $\boldsymbol{g}$ will depend on the position $\boldsymbol{x}$ where it is measured, and possibly also on time. In this case it will no longer be possible, by transformation to an accelerated frame, to make this field vanish *globally.*

Example 6.2 Star-like field of a point mass
Consider the spherically symmetric acceleration field outside a spherically symmetric mass distribution whose total mass is $m_\odot$,

$$b(x) = -\frac{Gm_\odot}{r^2}\hat{x}\,, \quad (r = |x|)\,.$$

The test particles have very small masses, $m_i \ll m_\odot$, and fall along radial directions in the direction of the centre of the force field, but certainly not on parallel orbits, cf. Fig. 6.1. An observer who falls freely in this field will realize that two test particles (without mutual forces) do not have constant distance but rather approach each other in the course of time. However, depending on how accurately he or she can measure orbits in the immediate neighbourhood, motions will look similar to the ones in Example 6.1, in a strictly local perspective: The orbits of free fall appear as practically parallel. By means of purely local measurements one cannot decide whether the frame of reference is accelerated and falls freely, or whether one is located in an inertial system where a genuine gravitation field is acting.

These examples are helpful in developing our intuition for a more precise formulation of the equivalence principle. Furthermore, they hint at a possible mathematical realization in terms of a geometry describing spacetime. Imagine an arbitrary smooth gravitational field $g(x)$ is given which depends on position and time. The following *local* property shall hold true:

▶ **Definition 6.1 Strong equivalence principle** In every point x of spacetime M one can always find a *local* inertial system such that in a sufficiently small neighbourhood $U \subset M$ of x the physical equations of motion take precisely the form known from special relativity in unaccelerated frames of reference.

▶ **Remarks**

1. In a neighbourhood of any point $x \in U \subset M$ one can always define coordinates which are such that the gravitational field is no longer felt. Locally

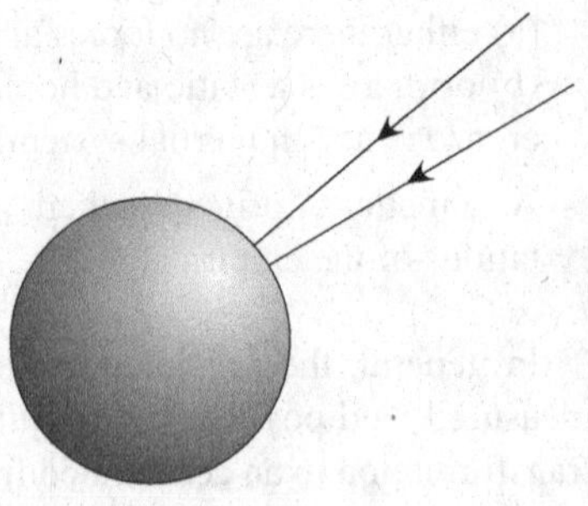

Fig. 6.1 Two test particles falling freely in the gravitational field of a spherically symmetric mass distribution, approach each other

special relativity holds in the way we are used to. What "small" means with regard to the neighbourhood depends on the gravitational field and on the accuracy of local measurements. In our terrestrial neighbourhood the dominant gravitational field is the one of the sun. A dimensionless parameter which characterizes the red shift of sun light caused by gravitation is $z = \Delta\Phi_N/c^2$. Here $\Delta\Phi_N$ is the difference of the Newtonian potential at the position of the earth and the position of the sun. Denoting by $M_\odot$ and by $R_\odot$ the mass and the radius of the sun, respectively, whose numerical values are

$$M_\odot = 1.989 \cdot 10^{30}\,\text{kg} \quad \text{and} \quad R_\odot = 6.960 \cdot 10^5\,\text{km}\,,$$

and inserting the major half-axis of the earth's orbit

$$1\,\text{AU} = 1.496 \cdot 10^8\,\text{km}$$

for the distance earth–sun (AU stands for *astronomical unit*), one obtains

$$\begin{aligned} z = \frac{\Delta\Phi_N}{c^2} &= \frac{GM_\odot}{c^2}\left\{\frac{1}{R_\odot} - \frac{1}{1\,\text{AU}}\right\} \\ &\simeq \frac{GM_\odot}{R_\odot c^2} = 2.12 \cdot 10^{-6}\,. \end{aligned} \tag{6.8}$$

We are dealing here with a very small effect which shows that at the position of the earth there is neither a strong nor a rapidly varying gravitational field. Presumably, the neighbourhood alluded to in the equivalence principle where special relativity remains applicable, is rather large. This estimate makes plausible why (relativistic) mechanics, electrodynamics, and non-Abelian gauge theories on flat Minkowski space are correct and why these theories are confirmed in applications to terrestrial experiments.

2. In *local* domains the metric and causal structure is the one of Minkowski space and is characterized by the so-called *flat metric* $g_{\mu\nu}(x) \simeq \eta_{\mu\nu}$. The latter notation η is introduced for the sake of clarity, and will be distinguished from the x-dependent full metric $\mathrm{g}(x)$,

$$\eta_{\mu\nu} = \begin{pmatrix} 1 & 0 & 0 & 0 \\ 0 & -1 & 0 & 0 \\ 0 & 0 & -1 & 0 \\ 0 & 0 & 0 & -1 \end{pmatrix}. \tag{6.9}$$

In the *large* the metric structure of spacetime varies as a function of x, $\mathrm{g}(x) = \{g_{\mu\nu}(x)\}$, depending on the mass and energy densities which are present in the universe. Neither from a mathematical standpoint nor from a physical point of view is it meaningful to insist on defining a globally valid and globally applicable frame of reference.

3. A more concise mathematical definition of the strong equivalence principle (that we shall develop in Sect. 6.5.6 below) is the following: In every point x_0 of spacetime one can construct a frame of reference such that

$$g_{\mu\nu}(x_0) = \eta_{\mu\nu} \text{ and } \left.\frac{\partial g_{\mu\nu}(x)}{\partial x^\alpha}\right|_{x_0} = 0\,, \quad (\alpha = 0, 1, 2, 3)\,. \tag{6.10}$$

   Coordinates in this class are called *Gaussian coordinates*, or *normal coordinates.*
4. The key to the construction of normal coordinates is provided by curves of free motion across spacetime, i.e. the world lines along which test particles move without disturbing the given gravitational field. Curves of free motion are those for which the acceleration vanishes in every point of the orbit. From the point of view of geometry these are curves whose length is an extremum. These are called *geodesics.*
   Consider then geodesics passing through the point x, and their tangent vectors in x. These tangent vectors can be utilized to construct a frame of reference for which the reduction (6.10) holds.
5. One more remark about the choice of $\eta_{\mu\nu}$: In some texts on general relativity and a large part of the mathematical literature the choice $\tilde{\eta}_{\mu\nu} = \text{diag}(-1, 1, 1, 1)$ is made instead of (6.9). As we know from special relativity this convention is as good as the one chosen in this book. The only essential feature is that the entry for precisely one coordinate has a sign different from the others. There is only one time direction but there are three (or more) space directions.

Riemannian geometry provides the framework for describing n-dimensional spacetimes exhibiting a causal structure, with one time coordinate and $n-1$ space coordinates, and, in particular, allows one to formalize the strong equivalence principle. An adequate model is a *semi-Riemannian manifold* (M, g). It consists of a pair of a smooth manifold M with dimension $\dim M = 4$ and index $\nu = 1$, and a metric g. (Recall the definition of the index in (5.10).) These are manifolds that are equipped with a unique prescription for parallel transport, the requirement (6.10) turns out to be a property that can be proved.

6.2.3 Red Shift and Other Effects of Gravitation

There are many well-understood experimental results showing that the spacetime of our universe deviates from flat Euclidean space. Historically, the first effects of general relativity to be known were the deflection of light rays passing close to the sun and the perihelion precession of the planet Mercury. *Deflection of light* by the sun will be calculated in Sect. 6.7.2 below. It shows that light rays, called *null geodesics* in geometry, are influenced by large concentrations of masses. The *precession of the perihelion* of Mercury concerns a phenomenon in celestial mechanics that has been known since about 1880: The planet Mercury whose Kepler orbit has the second-largest eccentricity in the solar system, $\varepsilon = 0.2056$, exhibits a precession of its perihelion which cannot be explained in its full size by perturbations from the other planets (in fact, mostly Venus, earth, and Jupiter). Celestial mechanics predicts 531″ per century. The observed shift is larger than that by about 43″, i.e. the perihelion is advancing somewhat faster than predicted by classical mechanics.

Among other confirmed effects of general relativity we quote the following:

Example 6.3 Kinematic and gravitational Doppler effect

Like in nonrelativistic mechanics there is a kinematic Doppler effect in special relativity theory (SRT). This effect is calculated in this example. By means of simple plausibility arguments we show that there must be an additional, new Doppler effect due to general relativity.

Consider two unaccelerated observers A and B in flat Minkowski space who move apart with constant velocity $\boldsymbol{v} = v\hat{\boldsymbol{e}}_1$. Observer A sends two successive signals travelling with the velocity of light and which are separated by the time interval T in his/her coordinates. The signals propagate in the (x^1, x^0)-plane to the right in Fig. 6.2, under 45°. Observer B records these signals and notes that in his/her coordinate time they are separated by the interval T'. Setting $T' = \kappa T$ one realizes easily that κ determines the *red shift* of light which is defined, for a spectral line of given wave length, by

$$z := \frac{\lambda_D - \lambda_S}{\lambda_S} = \kappa - 1 . \tag{6.11}$$

Index "S" stands for the "source", and index "D" for the "detector". The signals are reflected at B back to A who measures the time interval T'' between the first and the second signal, as sketched in Fig. 6.2. As the two observers are equivalent only their relative velocity matters. Therefore, one has $T'' = \kappa T' = \kappa^2 T$. The parameter κ is easily calculated by choosing the trajectories of A and B such that they intersect at the point O of Fig. 6.3. Both A and B take this point to be the origin of their coordinate times. With the times T, T', and T'' as shown in Fig. 6.3, observer A concludes that the event a whose time coordinate, in his/her frame, is

$$t_a = \frac{1}{2}(T + T'') = \frac{1}{2}(1 + \kappa^2)T ,$$

must be simultaneous with event b on the world line of B. This allows him/her to calculate the distance from A to B at this time,

$$d\big|_{(A)} = \frac{c}{2}(T'' - T) = \frac{c}{2}(\kappa^2 - 1)T .$$

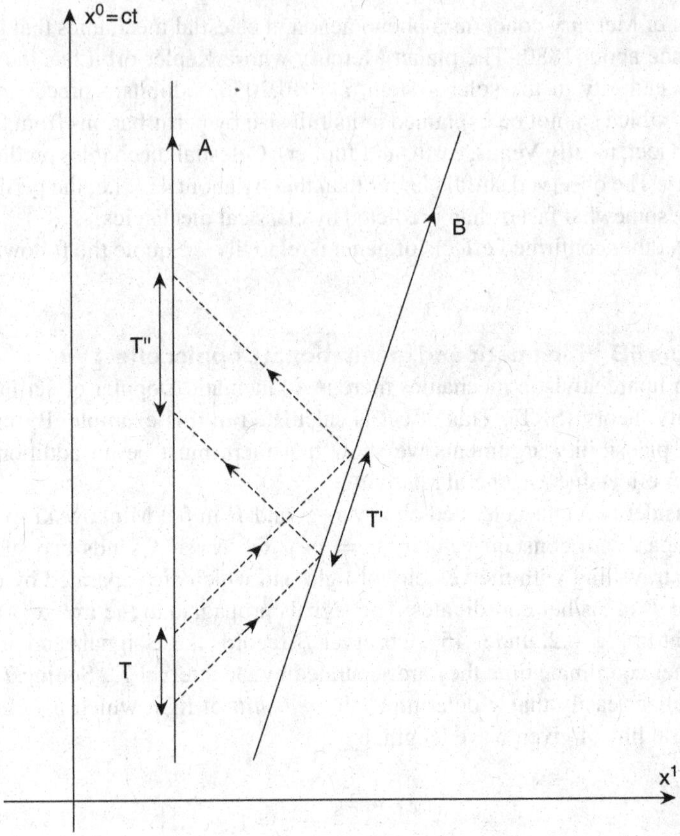

Fig. 6.2 Two unaccelerated observers who move apart with constant relative velocity $\boldsymbol{v} = v\hat{\boldsymbol{e}}_1$, exchange two light signals

He/she can also determine the modulus of the velocity relative to B, about which he/she knows that it has the value βc,

$$\frac{1}{c}v\Big|_{(A)} = \frac{d}{t_a} = \frac{\kappa^2 - 1}{\kappa^2 + 1} = \beta \, .$$

If B is departing from A, as assumed in the example, there follows a *red*shift, (i.e. $\kappa > 1$),

$$\kappa_{\text{red}} = \sqrt{\frac{1+\beta}{1-\beta}} = \gamma + \sqrt{\gamma^2 - 1} \, . \tag{6.12a}$$

Fig. 6.3 The same observers as in Fig. 6.2 have chosen the intersection point O of their world lines for the origin of their coordinate times

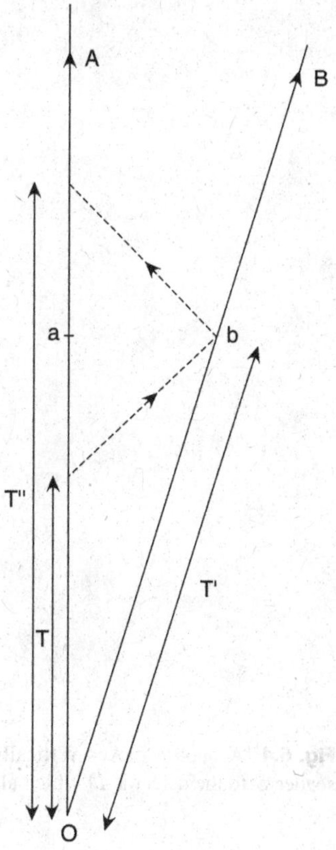

In turn, if B approaches A there is a *blue*shift, $(0 < \kappa < 1)$,

$$\kappa_{\text{blue}} = \sqrt{\frac{1-\beta}{1+\beta}} = \gamma - \sqrt{\gamma^2 - 1}\,. \tag{6.12b}$$

One verifies that in either case $\gamma(\kappa) = (\kappa^2 + 1)/(2\kappa)$ and that $\kappa_{\text{blue}} = 1/\kappa_{\text{red}}$. Note that these red and blue shifts occur already in flat Minkowski space. One deals here with the special relativistic version of the Doppler effect known from nonrelativistic kinematics.

Beyond this purely kinematic effect there is a red shift that is due to gravitation. The following thought experiment may help to understand this phenomenon. As a consequence of the equivalence principle, Definition 6.1, the effects of a homogeneous gravitational field (in an inertial frame of reference) cannot be distinguished from those observed in a uniformly accelerated frame in a space without field. The source S and the detector D, placed vertically one above the other in the field $\boldsymbol{g} = -g\hat{\boldsymbol{e}}_3$ as shown in Fig. 6.4, have the distance

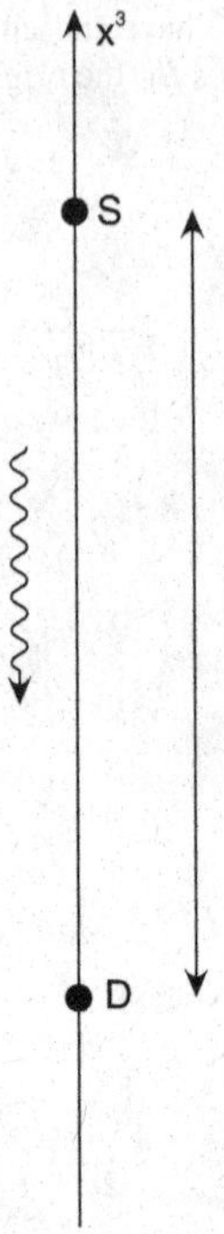

Fig. 6.4 A photon moves vertically and downwards in the earth's gravitational field from the sender S to the detector D which also falls freely

h. At time $t = 0$ when S emits a photon of given wave length λ, this set-up is supposed to be at rest. The photon reaches the detector D after the time of flight $\Delta t \simeq h/c$. At this instant of time D is already falling with velocity $\boldsymbol{v} = -v\hat{\boldsymbol{e}}_3$ where $v = g\Delta t \simeq gh/c$. The photon moves *along* the field and, therefore, is blue-shifted. This shift can be estimated by means of (6.12b) in the weakly relativistic case where one has

$$\kappa = \frac{1-\beta}{\sqrt{1-\beta^2}} \simeq 1 - \beta \,.$$

From the definition (6.11) one concludes

$$z \simeq -\frac{v}{c} = -\frac{gh}{c^2} = -\frac{\Delta\Phi_{\rm N}}{c^2} \,,$$

where $\Delta\Phi_{\rm N}$ is the difference of the Newtonian potential between source and detector. Although this effect is estimated here from a kinematic shift, the equivalence principle tells us that, in reality, it stems from the difference of the gravitational potential between source and detector and, hence, that it is a new effect.

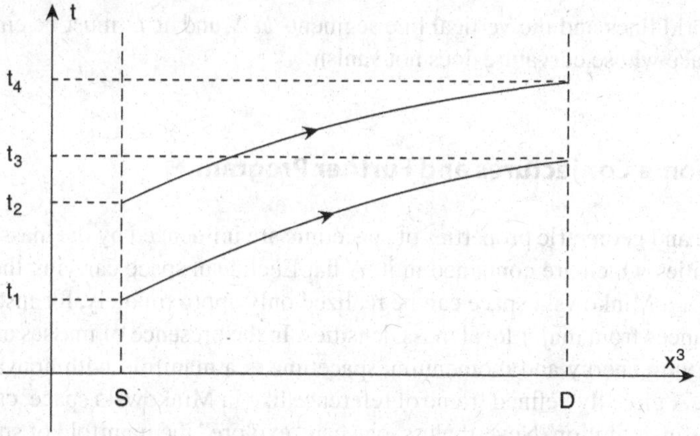

Fig. 6.5 The sender S emits two light signals at times t_1 and t_2, respectively. These signals are detected in detector D at times t_3 and t_4, respectively

Another thought experiment which is closely related to the previous one, shows that null geodesics, in the presence of gravitation, cannot be straight lines. Let $x(\tau)$ be a physical orbit in Minkowski space. In every point of this world line the velocity is timelike or, at most, lightlike. On a comoving clock time intervals are calculated by means of the formula

$$\Delta = \frac{1}{c}\int_{\tau_1}^{\tau_2} \mathrm{d}\tau \sqrt{\eta_{\mu\nu}\frac{\mathrm{d}x^\mu}{\mathrm{d}\tau}\frac{\mathrm{d}x^\nu}{\mathrm{d}\tau}}\,, \quad (\eta_{\mu\nu} = \mathrm{diag}(1,-1,-1,-1))\,. \tag{6.13}$$

This formula can no longer be correct if there is a gravitational field in the background.

Example 6.4 Light rays in a gravitational field

In a static and homogeneous acceleration field $\boldsymbol{g} = -g\hat{\boldsymbol{e}}_3$ a source emits two signals with (dominant) frequency ν_S at times t_1 and t_2, respectively. Figure 6.5 shows the orbits of the two signals in a position vs. time diagram. They are recorded by the detector at times t_3 and t_4, respectively. As the set-up is static the two world lines shown in the figure must be parallel, independently of their detailed functional form. Therefore, the time interval $\Delta t^{(D)} = t_4 - t_3$ is the same as $\Delta t^{(S)} = t_2 - t_1$ for the source. The number of pulses emitted from S, $N = \nu_S \Delta t^{(S)}$, is the same as the number of pulses $N = \nu_D \Delta t^{(D)}$ detected by D. As the time intervals are equal also the frequencies are equal, $\nu_D = \nu_S$. Therefore, no red or blue shift is predicted in the detector – in contrast to the result of the preceding example. One concludes from this that the surface enclosed by the two world lines and the vertical line segments at S and at D must be embedded in a space whose curvature does not vanish.

6.2.4 Some Conjectures and Further Program

The shape and geometric properties of spacetime are influenced by the mass and energy densities which are contained in it. A flat Euclidean space carrying the causal structure of a Minkowski space can be realized only approximately, for instance, at large distances from major local mass densities. In the presence of masses and other fields carrying energy and momentum, spacetime is a manifold with nonvanishing curvature. A globally defined frame of reference like in Minkowski space, cannot be a meaningful definition. Nevertheless, one can "explore" the manifold of spacetime by means of massive and massless test particles which do not modify the gravitational field in an essential way. These particles move along geodesics, i.e. orbits which correspond to free fall in a curved manifold, and, hence, yield information about the shape of spacetime.

Gravitation is not just another interaction, for example like Maxwell theory on flat space. It is contained in the geometry, and the structure of spacetime determines all effects of inertia and gravitation. The energy-momentum tensor field of matter and of nongravitational fields plays the role of the source term in the equations which determine the metric field $\mathrm{g}(x)$. This will turn out to be the essential hypothesis.

The program of the following sections is based on these considerations: We start by investigating models for the tensor field describing some given energy and momentum densities. This tensor field which will be termed "matter", drives the gravitational field and its geometry. Next we collect and discuss the most important geometric objects which live on smooth manifolds and which do not refer to any kind of embedding space. One of the requirements will be that the universe be describable "out of itself" and shall not be part of a fictitious larger embedding space.

Equipped with this knowledge manifolds can be found which fulfill the equivalence principle and, hence, which may serve as models for physical spacetime. Semi-Riemannian geometry provides all the tools which are needed to formulate Einstein's equations.

6.3 Matter and Nongravitational Fields

We begin still within the concept of a flat Minkowski space and assume that matter and radiation fields can be described by a Lagrange density of the type of (3.42). The action of translations in space and in time yield the tensor field (3.44). If the Lagrange density is *invariant* under translations then the energy-momentum tensor field fulfills the four conservation laws (3.45). This is the assertion of Noether's theorem. As we explained in the example of Maxwell theory this tensor field, referring to an arbitrarily chosen inertial frame, describes the energy and momentum densities as well as their flux densities.

Two examples for a tensor field which describes the energy and momentum content in a manifestly covariant way are:
The tensor field (3.23a) for a real scalar field,

$$T_{\phi}^{\mu\nu}(x) = \partial^{\mu}\phi(x)\partial^{\nu}\phi(x) - \frac{1}{2}\eta^{\mu\nu}\left(\partial_{\lambda}\phi(x)\eta^{\lambda\eta}\partial_{\eta}\phi(x) - \kappa^2\phi^2(x) - 2\varrho(x)\phi(x)\right), \tag{6.14}$$

and the Maxwell tensor field (3.47),

$$T_{\mathrm{M}}^{\mu\nu}(x) = \frac{1}{4\pi}\left\{F^{\mu\sigma}(x)\eta_{\sigma\tau}F^{\tau\nu}(x) + \frac{1}{4}\eta^{\mu\nu}F_{\alpha\beta}(x)F^{\alpha\beta}(x)\right\}. \tag{6.15}$$

Both tensor fields are symmetric, $T^{\mu\nu}(x) = T^{\nu\mu}(x)$. While the first of them, (6.14), can be no more than a useful model on microscopic scales and certainly is not suitable for describing macroscopic matter densities, the second, (6.15), is relevant for classical macroscopic situations whenever there are high densities of Maxwell fields. In order to describe the effective action of matter on the geometry of the universe one needs simple models for the energy-momentum tensor field. The following examples describe two such models.

Example 6.5 Energy-momentum tensor for dust
A swarm of noninteracting particles can be modelled by a cloud of dust whose particles move with a common local velocity. It is assumed that there is no pressure and no viscosity. With these assumptions there exists a local rest system at every point x in which the energy density is $\varrho_0(x)c^2$, with $\varrho_0(x)$ the mass density. The dust moving with local velocity $\boldsymbol{v}$, the density becomes $\varrho = \varrho_0\gamma^2$, where the first factor γ is due to the length contraction of the reference volume, while the second one is due to the relativistic increase in mass. For $T^{\mu\nu}$ one makes the ansatz

$$T^{\mu\nu} = \varrho_0 u^{\mu}u^{\nu}, \tag{6.16}$$

where $u = (\gamma c, \gamma\boldsymbol{v})^T$ is the velocity (2.41b). One convinces oneself that this ansatz is well-founded by verifying its properties. One has $T^{00} = \varrho c^2 = \varrho_0\gamma^2 c^2$, as expected. The tensor field is symmetric, $T^{\nu\mu} = T^{\mu\nu}$. The conservation law $\partial_{\mu}T^{\mu\nu} = 0$ is verified as follows. By the continuity equation,

$$\partial_{\mu}T^{\mu 0} = c\left[c\partial_0\varrho + \boldsymbol{\nabla}(\varrho\boldsymbol{v})\right] = 0.$$

Using this equation one calculates

$$\begin{aligned}\partial_\mu T^{\mu i} &= c\partial_0(\varrho v^i) + \nabla(\varrho v^i \boldsymbol{v}) \\ &= \varrho\, c\partial_0 v^i + \varrho\, \boldsymbol{v}\cdot\nabla v^i = \varrho\frac{\mathrm{d}v^i}{\mathrm{d}t} = 0\,.\end{aligned}$$

The last step follows from the assumption that there are no external forces.

Example 6.6 Ideal fluid

Let ϱ_0 denote the mass density, p the pressure density, and u the four-velocity. We assume the tensor field T to have the form

$$T^{\mu\nu} = \left(\frac{p}{c^2} + \varrho_0\right) u^\mu u^\nu - p\,\eta^{\mu\nu}\,, \tag{6.17a}$$

and verify that it has the expected properties. In a local rest system K_0 the energy-momentum tensor field $\mathsf{T}_{(0)}$ has the entries

$$T^{00}_{(0)} = \varrho_0\, c^2\,, \quad T^{0i}_{(0)} = 0 = T^{i0}_{(0)}\,, \quad T^{ik}_{(0)} = p\,\delta^{ik}\,. \tag{6.17b}$$

It is symmetric and it is conserved, i.e. it fulfills $\partial_\mu T^{\mu\nu}_{(0)} = 0$. For $\nu = 0$ this follows from $\partial_0 T^{00}_{(0)} + \partial_i T^{i0}_{(0)} = c^2\partial_0\varrho_0 = 0$ and from the assumption that the mass density is locally static. For $\nu = k$ one has $\partial_0 T^{0k}_{(0)} + \partial_i T^{ik}_{(0)} = \partial_k p$. This is also equal to zero. Indeed, a nonvanishing gradient of the pressure would cause a nonvanishing flow of the fluid – in contradiction to the assumption of a local rest system with respect to which the situation is static. Finally, for $u = (c, 0, 0, 0)^T$ the equation (6.17a) goes over into (6.17b).

It suffices then to verify by explicit calculation that the special Lorentz transformation $\mathsf{L}(\boldsymbol{v})$, when applied to $\mathsf{T}_{(0)}$, yields (6.17a),

$$\mathsf{L}(\boldsymbol{v})\mathsf{T}_{(0)}\mathsf{L}^T(\boldsymbol{v}) = \mathsf{T}\,.$$

Indeed, using (2.34) or (2.44), one has

$$\begin{aligned}T^{00} &= \gamma^2 c^2\left(\varrho_0 + \frac{p}{c^2}\beta^2\right) = \gamma^2 c^2\left(\varrho_0 + \frac{p}{c^2}\right) - p\,, \\ T^{i0} &= \gamma^2 c\left(\varrho_0 + \frac{p}{c^2}\right) v^i\,, \\ T^{ik} &= p\,\delta^{ik} + \gamma^2\varrho_0\, v^i v^k + 2p\frac{\gamma^2}{(1+\gamma)c^2} v^i v^k + p\frac{\gamma^4\beta^2}{(1+\gamma)^2 c^2} v^i v^k \\ &= p\,\delta^{ik} + \left(\frac{p}{c^2} + \varrho_0\right)\gamma^2 v^i v^k\,.\end{aligned}$$

In the first and in the last lines use was made of the relation $\beta^2 = (\gamma - 1)(\gamma + 1)/\gamma^2$. These results are seen to be coordinate expressions of the tensor field (6.17a).

Both models (6.16) and (6.17a) are easily extended to curved manifolds and both of them are used in cosmological models built on Einstein's equations. Two essential changes should be noted in these generalizations. Firstly, the flat Minkowski metric $\eta_{\mu\nu}$ in (6.17a) is replaced by the x-dependent metric tensor field $g_{\mu\nu}(x)$. Furthermore, all derivatives ∂_μ in the conservation equations on flat space are replaced by *covariant* derivatives D_μ which depend on the structure of the curved spacetime. The conservation laws are modified in that there can be changes in the energy-momentum densities. However, the total balance is restored by interaction with the gravitational background. This is worked out in Sect. 6.5.2, in Sect. 6.5.3, and in Sect. 6.6.1.

6.4 Spacetimes as Smooth Manifolds

The spacetime of classical gravitation is described by a four-dimensional smooth manifold equipped with a special metric structure which contains the equivalence principle (in its strong version). In this section we recapitulate the essential tools and notions that one needs for the description of smooth manifolds and of the objects defined on them. For a more extensive presentation we refer to [ME], Chap. 5. These definitions are followed by the definition and analysis of semi-Riemannian manifolds which are used in the theory of gravitation.

6.4.1 Manifolds, Curves, and Vector Fields

A distinctive property of a smooth manifold M with dimension n is that, locally, it looks like the Euclidean space $\mathbb{R}^n$: It can be covered by a countable set of open subsets $U_1, U_2, \ldots \subset M$ such that every point $x \in M$ is contained in at least one U_i. For every subset U_i there is a homeomorphism φ_i (i.e. a mapping which is invertible and continuous in either direction) which maps U_i into an open neighbourhood of the image $y = \varphi_i(x)$ in a copy of $\mathbb{R}^n$, viz. $\varphi_i : U_i \to \varphi(U_i)$, $U_i \subset M$ and $\varphi_i(U_i) \subset \mathbb{R}^n$. The original point on M is denoted by x. Its image in $\mathbb{R}^n$ is denoted by y, or, if the need arises, by $y^{(i)}$ with φ_i being the aforementioned homeomorphism.

This construction yields an *atlas* of *charts*, or local *coordinate systems*. Any two charts overlap smoothly, that is to say, every transition mapping $(\varphi_k \circ \varphi_i^{-1})$ is a diffeomorphism, which, by definition, links the image $\varphi_i(U_i)$ in the i-th copy of $\mathbb{R}^n$ with the image $\varphi_k(U_k)$ in the k-th copy of $\mathbb{R}^n$. (Recall that a diffeomorphism is an invertible mapping which is smooth and whose inverse is also smooth.) As a further assumption, let every chart that has smooth overlap with all other charts, be contained in the atlas. This defines a differentiable structure $(M, \mathcal{A})$ consisting of a manifold and a *complete atlas* or *maximal atlas*[3].

Remarks

1. In order to better visualize a differentiable structure $(M, \mathcal{A})$ it is helpful to illustrate the definitions by drawings. In doing so it is particularly important to keep track of the direction of the mappings φ_i, φ_i^{-1}, and $\varphi_k \circ \varphi_i^{-1}$, i.e. from where to where they point. Concrete examples of differentiable manifolds such as the torus T^2, the sphere S^2, or the group SO(3), also provide good illustrations. (These examples are worked out in [ME], Chap. 5.)
2. Note that here and in what follows we do not assume the manifold M to be embedded like a hypersurface in a space of higher dimension. This corresponds to the idea that the physical universe is not placed in an ambient, preexistent space, but exists by itself and its inner properties. Its geometry, its metric properties must be describable exclusively by intrinsic properties. This is an essential difference between the gravitational interaction and all other fundamental interactions whose theory assumes spacetime to be given with a specific causal structure.
3. The fact that the manifold is described by a "patchwork" of charts in $\mathbb{R}^n$-s does not contradict the previous remark. Charts are no more than imagined artefacts – comparable to the concept of phase space in mechanics – which are meant to facilitate the description of M and whose use reflects our inability to visualize at once a Moebius band, Klein's bottle, or a mug with 27 handles.
4. As a rule, physical theories are formulated in terms of equations of motion and, hence, assume that the manifold on which they are based carry a differentiable structure. One must keep in mind, however, that, a priori, differentiation on M proper is not defined. With no more than real analysis on $\mathbb{R}^n$ at our disposal local charts are an essential tool.

Functions on Manifolds

The notion of a function on a flat space $\mathbb{R}^n$ is familiar from real analysis: The function $f : \mathbb{R}^n \to \mathbb{R}$ is said to be smooth if f is differentiable infinitely many times. The generalization of this notion to curved manifolds is straightforward and follows this special case. A *smooth function* on an n-dimensional differentiable manifold is a mapping

$$f : M \longrightarrow \mathbb{R} \tag{6.18}$$

from M to the real axis for which $(f \circ \varphi_i^{-1})$ is a smooth function on the open neighbourhood $\varphi_i(U_i) \subset \mathbb{R}^n$ for all U_i. Figure 6.6 illustrates this prescription.

[3] As we consider only smooth manifolds in what follows we write manifold, for short, but always have in mind smooth manifolds.

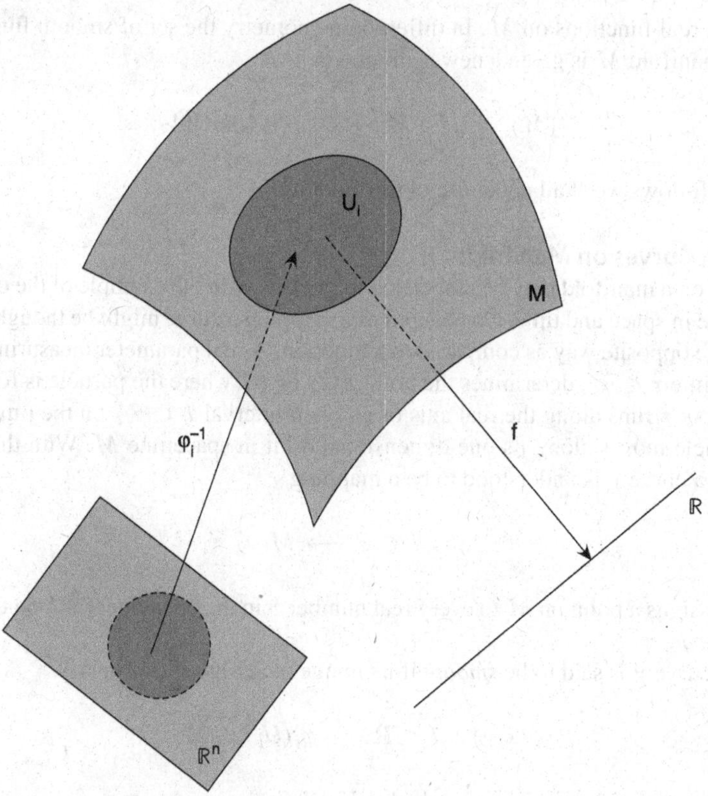

Fig. 6.6 A function f maps an open subset $U_i \subset M$ onto the real axis. Adding the inverse of the chart map φ_i, the composition $(f \circ \phi_i^{-1})$ is seen to be a function on $\mathbb{R}^n$

Simple yet important examples are provided by the *coordinate functions* $(f^\mu \circ \varphi_i)$, $\mu = 0, 1, 2, \ldots, n-1$, by means of which U_i is mapped again onto the subset $\varphi_i(U_i)$ of $\mathbb{R}^n$. More specifically the image $y = \varphi_i(x)$ of x, via the mapping $f^\mu \circ \varphi_i(x)$, is mapped to its μ-th coordinate $y^\mu = f^\mu(\varphi_i(x))$[4] which, indeed, is a real number. Expressed in symbols, one has

$$(f^\mu \circ \varphi_i) : M \to \mathbb{R} : x \mapsto y^\mu = f^\mu(\varphi_i(x)) , \ (\mu \text{ fixed}) . \qquad (6.19a)$$

In coordinates the image $y = \varphi_i(x)$ of $x \in M$ is given by

$$\varphi_i(x) = \left(y^0, y^1, \ldots, y^{n-1}\right) , \qquad (6.19b)$$

[4] The notation and the numbering of the coordinates is chosen here in view of semi-Riemannian spacetimes. Books on differential geometry usually use Latin indices which run from 1 to n.

i.e. by n real functions on M. In differential geometry the set of smooth functions on the manifold M is given a new symbol, viz.

$$\mathfrak{F}(M) := \{f : M \to \mathbb{R} \mid f \text{ is smooth}\} .$$

In what follows we shall make use of this notation.

Smooth Curves on Manifolds

A *curve* on a manifold may be associated in physics with the example of the orbit of a particle in space and time. Considered as a mapping a curve might be thought of as going the opposite way as compared to a function. A real parameter measuring, say, proper time $\tau \in \mathbb{R}_\tau$, determines the point $x(\tau)$ on M where the particle is found at time τ. As τ runs along the real axis or an open interval $I \subset \mathbb{R}_\tau$ on the time axis, the particle moves along its one-dimensional orbit in spacetime M. With this idea in mind a curve γ is understood to be a mapping

$$\gamma : I \subset \mathbb{R} \longrightarrow M , \tag{6.20}$$

which assigns a point on M to every real number τ in the open interval I on the real axis, $\tau \in I$.

The curve γ is said to be *smooth* if its image in $\mathbb{R}^n$ has this property,

$$(\varphi_i \circ \gamma) : I \subset \mathbb{R} \longrightarrow \varphi_i(U_i) \subset \mathbb{R}^n .$$

See the image of the curve as sketched in Fig. 6.7.

Smooth Vector Fields

At every point x of an n-dimensional manifold there are two vector spaces attached to it: The tangent space T_xM of all tangent vectors in x, and its dual T_x^*M, the cotangent space, whose elements are the linear maps of tangent vectors to the real numbers. The directional derivative of a smooth function $f \in \mathfrak{F}(M)$ in the direction of the tangent vector $v \in T_xM$ provides a good guideline for the definition of smooth vector fields. Denote this derivative by $v(f)$. This quantity is a real number. One concludes that tangent vectors act on smooth functions, or, in symbols, $v : \mathfrak{F}(M) \to \mathbb{R}$, and their images are elements of the real axis. Like all derivatives directional derivatives are $\mathbb{R}$-linear and satisfy the Leibniz rule. This example is helpful in formulating a more abstract definition:

▶ **Definition 6.2 Tangent vectors** A tangent vector $v \in T_xM$ is a real-valued function $v : \mathfrak{F}(M) \to \mathbb{R}$ which has the following properties

$$v(c_1 f_1 + c_2 f_2) = c_1 v(f_1) + c_2 v(f_2) , \quad c_1, c_2 \in \mathbb{R} , \tag{6.21a}$$
$$v(f_1 f_2) = v(f_1) f_2(x) + f_1(x) v(f_2) , \quad f_1, f_2 \in \mathfrak{F}(M) . \tag{6.21b}$$

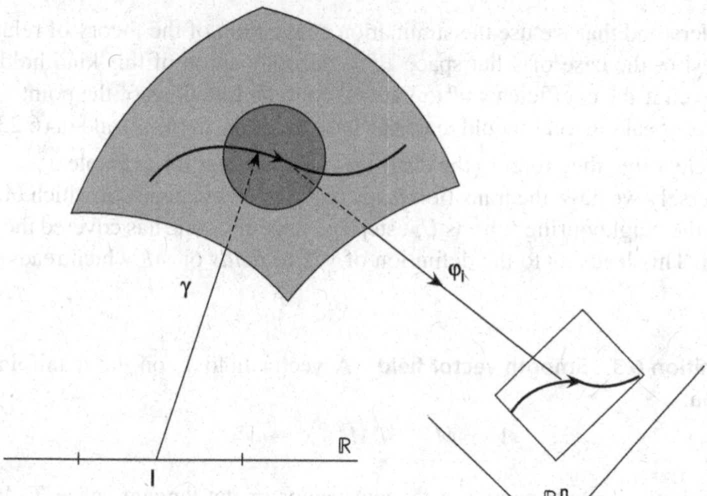

Fig. 6.7 A curve γ on M has as its image the curve $(\varphi_i \circ \gamma)$ in $\mathbb{R}^n$. Using this construction one can tell whether γ is smooth

The first property is called $\mathbb{R}$-linearity, the second property is the product rule or Leibniz rule.

Let $\varphi_i = (\varphi_i^0, \varphi_i^1, \ldots, \varphi_i^{n-1})$ be a chart for the open subset $U_i \subset M$ of $x \in M$ and let $g \in \mathfrak{F}(M)$ be a smooth function. Define then

$$\left.\frac{\partial g}{\partial y^\mu}\right|_x \equiv \partial_\mu\big|_x\, g := \frac{\partial(g \circ \varphi_i^{-1})}{\partial f^\mu}(\varphi_i(x)) \,, \tag{6.22}$$

$\mu = 0, 1, \ldots, n-1$, where f^μ are coordinate functions as introduced above (see also (6.19a)). Then

$$\partial_\mu\big|_x : \mathfrak{F}(M) \to \mathbb{R} : g \mapsto \partial_\mu g(x) \tag{6.23}$$

is a tangent vector of M in the point x. The symbol $\partial_\mu|_x$ is a useful abbreviation for the operation that is defined more explicitly in (6.22). The set of these tangent vectors $(\partial_0, \partial_1, \ldots \partial_{n-1})$ spans a basis of the tangent space $T_x M$. Like in the case of $M = \mathbb{R}^n$ that we dealt with in Sect. 2.2.2, every tangent vector in x can be expanded in terms of the base fields ∂_μ,

$$v = \sum_{\mu=0}^{n-1} v^\mu(x)\partial_\mu \equiv v^\mu(x)\partial_\mu \,.$$

(It is understood that we use the summation convention of the theory of relativity.) In contrast to the case of a flat space $\mathbb{R}^n$ a decomposition of this kind holds only locally, even if the coefficients $v^\mu(x)$ are taken to be functions of the point $x \in U_i$. So, strictly speaking, one should mark the base fields ∂_μ in (6.22) and in (6.23) such that it is clear that they refer to the chart (φ_i, U_i), writing, for example $\partial_\mu^{(i)}$.

Conversely we have the transition maps $(\varphi_k \circ \varphi_i^{-1})$ by means of which one continues to the neighbouring subsets U_k, step-by-step, until one has covered the whole manifold. This leads us to the definition of vector *fields* on M which reads as follows:

Definition 6.3 Smooth vector field A vector field V on the manifold M is a function,

$$V : M \to TM : x \mapsto V_x , \tag{6.24}$$

which assigns to every point x a tangent vector in the tangent space $T_x M$. The vector field is *smooth* if its action on a smooth function $f \in \mathfrak{F}(M)$

$$(Vf)(x) = V_x(f) \tag{6.24a}$$

is again a smooth function for all $f \in \mathfrak{F}(M)$.

Vector fields are smooth mappings $V : \mathfrak{F}(M) \to \mathfrak{F}(M)$ which map a smooth function to another smooth function. As an example from physics one might think of the flow of an ideal fluid in a given vessel. In every point of the interior the vector field identifies a unique, well-defined vector which in the example is the local velocity of the fluid. If one moves smoothly from the point of reference to its neighbourhood the flow changes in a continuous and differentiable way.

Expressed in coordinates of a local chart (φ_i, U_i) one has the decomposition

$$V = \sum_{\mu=0}^{n-1} \left(V\varphi_i^\mu\right) \partial_\mu^{(i)} . \tag{6.24b}$$

This is often written, in a somewhat simplifying notation, $V = \sum V^\mu(x)\partial_\mu$, or, even shorter and using the summation convention, $V = V^\mu(x)\partial_\mu$. The coefficients $V^\mu(x)$ are smooth functions.

Remarks

1. The set of all smooth vector fields on a differentiable manifold M is usually denoted by the symbol $\mathfrak{V}(M)$ or by $\mathfrak{X}(M)$. In this chapter we use a third notation which renders their tensor character more explicit, viz.

$$V \in \mathfrak{T}_0^1(M) ,$$

where $\mathfrak{T}^r_s(M)$ are the smooth tensor fields which are r-fold *contra*variant, and s-fold *co*variant. This notation emphasizes that vector fields are contravariant tensor fields of degree 1.
The objects which are dual to vector fields, i.e. the one-forms, are elements of the space $\mathfrak{X}^*(M) = \mathfrak{T}^0_1(M)$, i.e. they are covariant tensor fields of degree 1.
2. Smooth functions and smooth curves on manifolds are special cases of smooth mappings between differentiable structures $(M, \mathcal{A})$ and $(N, \mathcal{B})$,

$$\Phi : (M, \mathcal{A}) \longrightarrow (N, \mathcal{B}) .$$

In the case of functions the target manifold is the space $\mathbb{R}$ equipped with what is called the canonical differential structure. In the case of curves the starting manifold is $\mathbb{R}$, the target manifold is M. Any such mapping induces a well-defined mapping between the corresponding tangent spaces.

The tangent vector field of a smooth curve $\gamma : \mathbb{R} \to M$ on the manifold M is an important example of a smooth vector field. On the real axis $\mathbb{R}$ there is only one base field $\partial = \mathrm{d}/\mathrm{d}u$ so that

$$\frac{\mathrm{d}}{\mathrm{d}u}(\tau) \in T_\tau(\mathbb{R})$$

is the unit vector in the point τ which points in the positive direction. Its image is the velocity vector $\dot{\gamma}(\tau)$ in the point $\gamma(\tau)$ on M whose action on a smooth function $f \in \mathfrak{F}(M)$ is calculated by means of

$$\dot{\gamma}(\tau) f = \frac{\mathrm{d}(f \circ \gamma)}{\mathrm{d}u}(\tau) . \tag{6.25a}$$

Expressed in local coordinates (φ_i, U_i) the following formula applies

$$\dot{\gamma} = \sum_{\mu=0}^{n-1} \frac{\mathrm{d}(\varphi_i^\mu \circ \gamma)}{\mathrm{d}u}(\tau)\, \partial_\mu \big|_{\gamma(\tau)} . \tag{6.25b}$$

One may ask a closely related question: Given a smooth vector field $V \in \mathfrak{T}^1_0(M)$, is there a curve $\alpha : I \subset \mathbb{R} \to M$ which satisfies the differential equation $\dot{\alpha} = V_\alpha$, i.e. for which at any time $\tau \in I$ the velocity vector $\dot{\alpha}$ coincides with the tangent vector V_α? This leads to the

Definition 6.4 Integral curve of a vector field The curve $\alpha : I \to M$ is called the *integral curve* of the vector field $V \in \mathfrak{T}^1_0(M)$ if $\dot{\alpha} = V_\alpha$, or, more precisely, if

$$\dot{\alpha}(\tau) = V_{\alpha(\tau)} \tag{6.26}$$

holds for all $\tau \in I$.

The equation (6.26) represents a dynamical system and is the general geometric form of a typical equation of motion of physics. To quote an example let α be a physical orbit of mechanics in phase space, $\alpha = (q(t), p(t))^T$, and V the Hamiltonian vector field $V = (\partial H/\partial p, -\partial H/\partial q)^T$ so that the explicit form of (6.26)

$$\begin{pmatrix} \dot{q}(t) \\ \dot{p}(t) \end{pmatrix} = \begin{pmatrix} \frac{\partial H}{\partial p} \\ -\frac{\partial H}{\partial q} \end{pmatrix}\Bigg|_t ,$$

yields the well-known canonical equations of motion. The local velocity at the point (q, p) and at the time t coincides with the tangent vector calculated at this same point and time.

Remark

The existence and uniqueness theorem for ordinary differential equations of first order guarantees that the integral curve which passes through the point $x_0 = \alpha(\tau_0)$ is determined uniquely. Without loss of generality one can set $\tau_0 = 0$. Starting from this point and constructing the maximal extension of α on M yields the *maximal integral curve* of the vector field through $x_0 = \alpha(0)$. If every maximal integral curve is defined on the whole axis $\mathbb{R}_\tau$, that is, if the interval I in Definition 6.4 can be extended to the whole space $\mathbb{R}$, the vector field is said to be *complete*.

6.4.2 One-Forms, Tensors, and Tensor Fields

The definition of exterior one-forms for curved manifolds is very similar to the case of flat space $M = \mathbb{R}^n$, see Sect. 2.2.2a. At the point $x \in M$ the form $\omega : T_x M \to \mathbb{R}$ is a linear map from the tangent space into the real numbers. In every chart there are base one-forms $(\mathrm{d}x^0, \mathrm{d}x^1, \ldots, \mathrm{d}x^{n-1})$, dual to the base vector fields ∂_μ, $\mu = 0, \ldots, n-1$. As before the duality relation reads

$$\mathrm{d}x^\mu(\partial_\nu) = \frac{\partial}{\partial x^\nu} x^\mu = \delta^\mu_\nu . \tag{6.27}$$

As such, ω is an element of the cotangent space $T_x^* M$ in x. An arbitrary one-form can be expanded in terms of base one-forms, like in (2.10),

$$\omega = \sum_{\mu=0}^{n-1} \omega_\mu \, \mathrm{d}x^\mu , \tag{6.28a}$$

where the factors ω_μ are real coefficients. The difference as compared to the case of an $\mathbb{R}^n$ is that although in (6.28a) one can replace these numbers by smooth functions

$\omega_\mu(x)$, thereby obtaining the *smooth* one-form on M

$$\omega = \sum_{\mu=0}^{n-1} \omega_\mu(x)\, dx^\mu \,, \tag{6.28b}$$

this representation initially holds only in a local chart (φ_i, U_i). One obtains a well-defined smooth one-form on M only if this local representation can be continued to the whole manifold by means of a complete atlas. These considerations suggest the

▶ **Definition 6.5 Smooth one-form** A smooth one-form on the manifold M is a function

$$\omega : M \to T_x^* M \; : x \mapsto \omega_x \in T_x^* M \,, \tag{6.29}$$

which assigns to every point x an element ω_x of the cotangent space $T_x^* M$. The one-form is said to be *smooth* if its action $\omega(V)$ on a vector field V is a smooth function for all $V \in \mathfrak{T}_0^1(M)$.

The action of a one-form $\omega = \sum_{\mu=0}^{n-1} \omega_\mu(x)\, dx^\mu$ on a vector field $V = \sum_{\nu=0}^{n-1} V^\nu(x)\partial_\nu$ follows from (6.28b) and (6.27),

$$\omega(V) = \sum_{\mu=0}^{n-1}\sum_{\nu=0}^{n-1} \omega_\mu(x) V^\nu(x)\, dx^\mu(\partial_\nu) = \sum_{\mu=0}^{n-1} \omega_\mu(x) V^\mu(x) \,.$$

One-forms are covariant tensor fields of degree 1, $\omega \in \mathfrak{T}_1^0$.

The rules of the calculus with exterior products that we developed in Sect. 2.2.2, remain unchanged. One constructs exterior forms of grade k, with $k = 1, 2, \ldots, n$ as we did in the case of Euclidean spaces, and expand them in terms of base k-forms $dx^0 \wedge dx^1 \wedge \cdots dx^{n-1}$ in local charts.

The *exterior derivative* is a local operation. Therefore, it has the same properties as for a flat space[5] $\mathbb{R}^n$.

Tensors with more than one index as well as tensors which carry both covariant and contravariant indices appear in various contexts. Before we return to the tensor fields on spacetime which are relevant for physics, we summarize the definitions and mathematical properties of tensor fields. The characteristic property of tensors as well as of tensor fields is their *multilinearity*. A tensor with r contravariant indices and s covariant indices at the point $x \in M$ maps r one-forms and s tangent vectors onto a real number,

$$\begin{aligned}(\mathsf{T}^r_s)_x \,:\, (T_x^* M)^r \times (T_x M)^s &\longrightarrow \mathbb{R} \\ : \omega^1, \ldots, \omega^r, V_1, \ldots V_s &\mapsto (\mathsf{T}^r_s)_x \left(\omega^1, \ldots, \omega^r, V_1, \ldots V_s\right) .\end{aligned}$$

This mapping is linear in all its arguments.

[5] The Poincaré lemma on manifolds which are not singly connected, holds for star-like domains only, see remark 3 in Sect. 2.2.2.

If one lets the base point x move over M the right-hand side becomes a function. The following definition is a straightforward generalization of smooth vector fields (6.24) and of smooth one-forms (6.29):

▶ **Definition 6.6 Smooth tensor fields** A tensor field of the type (r, s), i.e. a tensor field which is r-fold *contra*variant and s-fold *co*variant, is a multilinear map

$$\mathsf{T}^r_s : (T^*M)^r \times (TM)^s \longrightarrow \mathfrak{F}(M), \tag{6.30}$$

which assigns to every set of r one-forms from the cotangent space T^*_xM and to every set of s vector fields from the tangent space T_xM a function

$$\omega^1, \dots, \omega^r, V_1, \dots, V_s \mapsto \left(\mathsf{T}^r_s\right)_x \left(\omega^1, \dots, \omega^r, V_1, \dots V_s\right). \tag{6.30a}$$

The tensor field is said to be *smooth* if this function is a smooth function.

▶ **Remarks**

1. The set of smooth tensor fields of type (r, s) over the manifold M is denoted by $\mathfrak{T}^r_s(M)$. Its elements $\mathsf{T}^r_0 \in \mathfrak{T}^r_0(M)$ are called contravariant tensor fields of degree r, while the elements $\mathsf{T}^0_s \in \mathfrak{T}^0_s(M)$ are called covariant tensor fields of degree s. If both indices are different from zero one talks about mixed tensor fields.
2. Two important examples are $\mathfrak{T}^1_0 \equiv \mathfrak{V}(M) \equiv \mathfrak{X}(M)$, the smooth vector fields, and $\mathfrak{T}^0_1 \equiv \mathfrak{X}^*(M)$, the smooth one-forms. The set of smooth functions can also be written $\mathfrak{F}(M) = \mathfrak{T}^0_0(M)$.
   Addition of two tensor fields is meaningful only if they are of the same type (r, s).
   The tensor product of T^r_s and $\mathsf{T}^{r'}_{s'}$ is again a tensor field and its type is $(r + r', s + s')$. If it is evaluated on $r + r'$ one-forms and on $s + s'$ vector fields one obtains

$$\begin{aligned}\left(\mathsf{T}^r_s \otimes \mathsf{T}^{r'}_{s'}\right)&\left(\omega^1, \dots, \omega^{r+r'}, V_1, \dots, V_{s+s'}\right) \\ &= \mathsf{T}^r_s\left(\omega^1, \dots, \omega^r, V_1, \dots, V_s\right) \\ &\quad \times \mathsf{T}^{r'}_{s'}\left(\omega^{r+1}, \dots, \omega^{r+r'}, V_{s+1}, \dots, V_{s+s'}\right),\end{aligned} \tag{6.31}$$

   i.e. the product of the two tensor fields both of which are evaluated on the appropriate number of one-forms and of vector fields, respectively.
3. If one evaluates the mixed tensor field (r, s) only on s vector fields but on no one-forms,

$$\mathsf{T}^r_s\Big(\underbrace{\cdot, \dots\dots, \cdot}_{r \text{ empty slots}}, V_1, \dots, V_s\Big) =: L$$

the result is a contravariant tensor field of degree r. Thus, L is a map which is linear in each component

$$L : (\mathfrak{X}(M))^s \longrightarrow (\mathfrak{X}(M))^r .$$

In a similar way the object

$$G := \mathrm{T}^r_s\left(\omega_1, \ldots, \omega_r, \underbrace{\cdot, \ldots, \cdot}_{s \text{ empty slots}}\right)$$

is a map from $(\mathfrak{X}^*(M))^r$ to $(\mathfrak{X}^*(M))^s$.

4. Like all tensor fields these mappings are not only linear but also $\mathfrak{F}(M)$-linear, i.e. linearity is fulfilled not only if the arguments are multiplied by real numbers but also when they are multiplied by functions. One says that L, G, etc. are *$\mathfrak{F}(M)$-modules.*
5. The contravariant tensor fields T^r_0 and the covariant tensor fields T^0_s, in general, have no special symmetry character. Symmetric and antisymmetric tensor fields are subsets, and a few examples are discussed below. For instance, the exterior forms $\eta \in \Lambda^k(M)$ of degree k are nothing else than antisymmetric covariant tensor fields $\eta \in \mathfrak{T}^0_k(M)$.

The following definition concerns a symmetric tensor field of central importance for relativity theory and physics in general:

Definition 6.7 Metric field Suppose the manifold is such that it admits a metric. A metric on M is a smooth covariant tensor field $\mathrm{g} \in \mathfrak{T}^0_2(M)$ which is symmetric and nondegenerate. In more detail this means: In every point $x \in M$ one has

(i) $g(v, w)|_x = g(w, v)|_x$ for all $v, w \in T_xM$,
(ii) if $g(v, w)|_x = 0$ for fixed v and for all $w \in T_xM$, then v can only be the null vector, $v = 0$.

6.4.3 Coordinate Expressions and Tensor Calculus

Local coordinate expressions of tensor fields are obtained by evaluating them on base one-forms and base vector fields in charts (φ_i, U_i). Let T^r_s be a tensor field of type (r, s), $\mathrm{T}^r_s \in \mathfrak{T}^r_s(M)$. In the domain of a chart one applies T^r_s to r base one-forms and to s base fields thus obtaining the functions

$$t^{\mu_1 \ldots \mu_r}_{\nu_1 \ldots \nu_s}(x) = \mathrm{T}^r_s\left(\mathrm{d}x^{\mu_1}, \ldots, \mathrm{d}x^{\mu_r}, \partial_{\nu_1}, \ldots, \partial_{\nu_s}\right) . \tag{6.32}$$

This is the form of tensors as one knows them from elementary tensor analysis. Making use of the rule

$$\mathrm{d}x^\mu(\partial_\nu) = \partial_\nu(x^\mu) = \delta^\mu_\nu ,$$

the tensor field, locally, is seen to be a linear combination of tensor products of r base fields and s base one-forms,

$$\mathsf{T}^r_s = \sum_{\mu_1\cdots\mu_r=0}^{n-1} \sum_{\nu_1\cdots\nu_s=0}^{n-1} t^{\mu_1\ldots\mu_r}_{\nu_1\ldots\nu_s}(x) \\ (\partial_{\mu_1} \otimes \ldots \otimes \partial_{\mu_r}) \otimes (\mathrm{d}x^{\nu_1} \otimes \ldots \otimes \mathrm{d}x^{\nu_s}) . \tag{6.33}$$

A few examples seem in order to illustrate these matters.

(i) A covariant tensor field of degree 2 has the representation

$$\mathsf{T}^0_2 = \sum_{\mu,\nu=0}^{n-1} t_{\mu\nu}(x)\,\mathrm{d}x^\mu \otimes \mathrm{d}x^\nu . \tag{6.34a}$$

This is the general case when the tensor field is neither symmetric nor antisymmetric. However, if the coefficients are antisymmetric, $t_{\nu\mu} = -t_{\mu\nu}$, then

$$\mathsf{T}^0_2 = \sum_{\mu<\nu} t_{\mu\nu}(x)\,(\mathrm{d}x^\mu \otimes \mathrm{d}x^\nu - \mathrm{d}x^\nu \otimes \mathrm{d}x^\mu) \\ = \sum_{\mu<\nu} t_{\mu\nu}(x)\,\mathrm{d}x^\mu \wedge \mathrm{d}x^\nu . \tag{6.34b}$$

We rediscover here the coordinate representations of two-forms.

(ii) A coordinate representation of the metric has the same form (6.34a)

$$\mathrm{g}(x) = \sum_{\mu,\nu=0}^{n-1} g_{\mu\nu}(x)\,\mathrm{d}x^\mu \otimes \mathrm{d}x^\nu , \tag{6.35a}$$

with symmetric coefficients, $g_{\nu\mu}(x) = g_{\mu\nu}(x)$. The matrix $\{g_{\mu\nu}(x)\}$ is usually called the *metric tensor.* Except for the case of a flat space such as $M = \mathbb{R}^n$ this expression holds only locally, i.e. in the charts (φ_i, U_i) of a complete atlas. The metric tensor is obtained by evaluating g on base fields, following (6.32).

$$\mathrm{g}(\partial_\sigma, \partial_\tau) = \sum_{\mu,\nu=0}^{n-1} g_{\mu\nu}(x)\,\mathrm{d}x^\mu(\partial_\sigma)\,\mathrm{d}x^\nu(\partial_\tau) \\ = \sum_{\mu,\nu=0}^{n-1} g_{\mu\nu}(x)\delta^\mu_\sigma \delta^\nu_\tau = g_{\sigma\tau}(x) , \tag{6.35b}$$

with $\sigma, \tau = 0, \ldots, n-1$.

The metric being nondegenerate is equivalent with the matrix $\{g_{\mu\nu}(x)\}$ being nowhere singular. Thus, in every point $x \in M$ it possesses an inverse which is denoted by $g^{\mu\nu}(x)$,

$$g^{\mu\nu}(x)g_{\nu\tau}(x) \equiv \sum_{\nu=0}^{n-1} g^{\mu\nu}(x)g_{\nu\tau}(x) = \delta^{\mu}_{\tau} . \qquad (6.36)$$

(On the left-hand side, as before, we have used the summation convention.)

(iii) On a chart a mixed tensor field of type (1, 3) has the representation

$$\mathsf{T}^1_3 = \sum_{\mu=0}^{n-1} \sum_{\lambda,\sigma,\tau=0}^{n-1} t^{\mu}_{\lambda\sigma\tau}(x)\, \partial_\mu \otimes \mathrm{d}x^\lambda \otimes \mathrm{d}x^\sigma \otimes \mathrm{d}x^\tau . \qquad (6.37)$$

Coordinate expressions like this one, or like in (6.33), (6.34a) and (6.35a), are unique on the subset U_i. Using the transition maps $(\varphi_k \circ \varphi_i^{-1})$ they can be continued to other subsets that overlap with U_i and, hence, to the whole (maximal) atlas. Therefore, expressions of the kind of (6.33) are valid and useful representations of tensor fields.

Summations over pairs of covariant and contravariant indices are well known from the theory of special relativity, to witness, expressions of the form $a_\mu b^\mu$, $\eta_{\mu\nu}T^{\mu\nu}$, or others, where Lorentz vectors and tensors are combined to invariants. This operation of *contraction* exists also for tensor fields on curved manifolds, in a coordinate-free form. One proceeds as follows: For a tensor field which is given by the tensor product of a vector field V and a one-form ω, $\mathsf{T}^1_1 = V \otimes \omega$, contraction is defined to be the action of ω on V. The result is a function on M,

$$C(\mathsf{T}^1_1) = C(V \otimes \omega) := \omega(V)(x) .$$

In this case contraction converts the (1, 1)-tensor to a function, i.e. written in symbols, $C : \mathfrak{T}^1_1 \to \mathfrak{T}^0_0$. For a tensor field of type (1, 1) there is only one possibility of contraction. Writing an arbitrary tensor field T^1_1 in local coordinates, viz. $\mathsf{T}^1_1 = \sum t^{\mu}_{\nu}\, \partial_\mu \otimes \mathrm{d}x^\nu$, we have

$$\begin{aligned} C\left(\mathsf{T}^1_1\right) &= \sum_{\mu,\nu=0}^{n-1} t^{\mu}_{\nu}(x) C\left(\partial_\mu \otimes \mathrm{d}x^\nu\right) \\ &= \sum_{\mu,\nu=0}^{n-1} t^{\mu}_{\nu}(x)\delta^{\nu}_{\mu} = \sum_{\mu=0}^{n-1} t^{\mu}_{\mu}(x) . \end{aligned}$$

Thus, when worked out in terms of components, one obtains the prescription familiar from special relativity. This is also true for more general mixed tensor fields $\mathsf{T}^r_s \in \mathfrak{T}^r_s$. However, one needs to know which covariant index is to be contracted with which contravariant index. Suppose, for example, that for a given tensor field T^2_3 the first upper index is to be contracted with the third lower index. This means that in the function

$$\mathsf{T}^2_3\left(\omega^1, \omega^2, V_1, V_2, V_3\right)$$

the one-form ω^1 and the vector field V_3 should be contracted,

$$\omega^1(V_3) = \sum_{\mu=0}^{n-1}\sum_{\nu=0}^{n-1} \omega^1_\mu V_3^\nu \, \mathrm{d}x^\mu(\partial_\nu) = \sum_{\mu=0}^{n-1} \omega^1_\mu V_3^\mu \, .$$

The tensor field T^2_3 thereby becomes a tensor field with $r = 1$ and $s = 2$, $C^1_3(\mathsf{T}^2_3) \in \mathfrak{T}^1_2(M)$. Calculating this new tensor field in coordinates, its expansion coefficients are

$$\begin{aligned}(C^1_3\mathsf{T}^2_3)^\mu_{\ \sigma\tau}(x) &= (C^1_3\mathsf{T}^2_3)\left(\mathrm{d}x^\mu, \partial_\sigma, \partial_\tau\right) \\ &= \sum_{\lambda=0}^{n-1} \mathsf{T}^2_3\left(\mathrm{d}x^\lambda, \mathrm{d}x^\mu, \partial_\sigma, \partial_\tau, \partial_\lambda\right) = \sum_{\lambda=0}^{n-1} t^{\lambda\mu}_{\ \ \sigma\tau\lambda}(x) \, .\end{aligned}$$

In components the rules of calculus are the same as in special relativity: One contracts indices in an invariant way by taking an equal upper and lower index and by summing over all values of this index.

At this point we know already a great deal about calculus of tensor fields: *Addition* – only tensor fields of the same type can be added –, *multiplication* by means of the tensor product (6.31)), *contraction*, and their $\mathfrak{F}(M)$*-multilinearity*. What is missing is a universal rule of how to take derivatives of tensor fields. For example, the Lie derivative is a rule that allows one to differentiate a geometrical object along the flux of a vector field. It is applicable equally well to functions, to vector fields, to one-forms, or more generally, to tensor fields of type (r, s). Here the aim is to formulate rules for any such derivative, both in coordinates and in a coordinate-free formulation. A derivative of this kind is generally denoted by $\mathcal{D}$. However, it is understood that its explicit form depends on the type (r, s) of the tensor field on which it acts.

▶ **Definition 6.8 Tensor derivation** A *tensor derivation* $\mathcal{D}$ is a mapping of tensor fields which does not change the type (r, s),

$$\mathcal{D} \equiv \mathcal{D}^r_s \, : \, \mathfrak{T}^r_s(M) \longrightarrow \mathfrak{T}^r_s(M) \, , \tag{6.38}$$

and which has the following properties:

(i) With S and T, two arbitrary smooth tensor fields, one has

$$\mathcal{D}(\mathsf{S}\otimes\mathsf{T}) = (\mathcal{D}\mathsf{S})\otimes\mathsf{T} + \mathsf{S}\otimes(\mathcal{D}\mathsf{T}) \,, \tag{6.39a}$$

(Leibniz rule);

(ii) The derivation commutes with any possible contraction,

$$\mathcal{D}(C(\mathsf{T})) = C(\mathcal{D}\mathsf{T}) \,. \tag{6.39b}$$

Remarks

1. The space of derivations and the space of vector fields are isomorphic. Therefore, for a given derivation $\mathcal{D}$ there is a unique vector field $V \in \mathfrak{T}^1_0$ such that $\mathcal{D}g = Vg$ holds for all $g \in \mathfrak{F}(M)$.
2. The tensor product of a function g with a tensor field T^r_s with $(r,s) \neq (0,0)$ is the ordinary product, $g \otimes \mathsf{T}^r_s = g\mathsf{T}^r_s$.
3. With g a function and T^r_s an (r,s)-tensor field the Leibniz rule takes the form $\mathcal{D}(g\mathsf{T}^r_s) = (\mathcal{D}g)\mathsf{T}^r_s + g\mathcal{D}\mathsf{T}^r_s$. Note that the symbol $\mathcal{D}$ on the left-hand side and in the second term on the right-hand side means $\mathcal{D}^r_s$, while in the first term of the right-hand side it stands for $\mathcal{D}^0_0$.

Return once more to Definition 6.6 which concerns the evaluation of T^r_s with r arbitrary one-forms and s vector fields. Inserting coordinate representations in a chart,

$$\omega^i = \sum \omega^i(x)_\mu \,\mathrm{d}x^\mu \,, \quad V_k = \sum V^\nu_k(x)\partial_\nu$$

one finds

$$\begin{aligned}&\mathsf{T}^r_s\left(\omega^1,\ldots,\omega^r,V_1,\ldots,V_s\right)\\ &\quad= \sum t^{\mu_1\cdots\mu_r}_{\nu_1\cdots,\nu_s}(x)\,\omega^1_{\mu_1}\ldots\omega^r_{\mu_r}V^{\nu_1}_1\ldots V^{\nu_s}_s \,.\end{aligned}$$

On the right-hand side the sum over all indices is taken, every covariant index is contracted with a contravariant one. Thus, the right-hand side, in fact, is the combined contraction of the tensor product

$$\mathsf{T}^r_s \otimes \omega^1 \otimes \cdots \otimes \omega^r \otimes V_1 \otimes \cdots \otimes V_s \,.$$

Taking then the derivation $\mathcal{D}$ of this tensor product and using the Leibniz rule (6.39a), one obtains a sum of terms in which $\mathcal{D}$ moves successively "to the right" and acts on each of the factors of the tensor product once. By the assumption (6.39b) $\mathcal{D}$ commutes with all contractions. This yields a general formula which is of great importance in practice.

How to calculate a tensor derivation

$$\begin{aligned}\mathcal{D}\left[\mathsf{T}^r_s\left(\omega^1,\ldots,\omega^r,V_1\ldots,V_s\right)\right] \\ &= \left(\mathcal{D}\mathsf{T}^r_s\right)\left(\omega^1,\ldots,\omega^r,V_1\ldots,V_s\right) \\ &+\sum_{i=1}^{r}\mathsf{T}^r_s\left(\omega^1,\ldots,(\mathcal{D}\omega^i),\ldots,\omega^r,V_1,\ldots,V_s\right) \\ &+\sum_{k=1}^{s}\mathsf{T}^r_s\left(\omega^1,\ldots,\omega^r,V_1,\ldots,(\mathcal{D}V_k),\ldots,V_s\right) . \end{aligned} \tag{6.40}$$

On the left-hand side of this equation $\mathcal{D}$ is applied to a function, that is to say, it is realized as $\mathcal{D}^0_0$. In the first term of the right-hand side it is the derivation $\mathcal{D}^r_s$ that acts, in the second group of terms it is $\mathcal{D}^0_1$, while in the third group it is $\mathcal{D}^1_0$.

The importance of the universal formula (6.40) becomes evident if one realizes that it allows one to deduce the explicit form of the derivation $\mathcal{D}^r_s$ for $r > 0$ and $s > 0$ if one knows its action on *functions* and on *vector fields*. Indeed, knowing $\mathcal{D}^0_0$ and $\mathcal{D}^1_0$, one obtains $\mathcal{D}^0_1$ from (6.40): For an arbitrary one-form ω and for every vector field V this equation yields

$$(\mathcal{D}\omega)(V) = \mathcal{D}(\omega(V)) - \omega(\mathcal{D}V) . \tag{6.40a}$$

The left-hand side contains the derivation $\mathcal{D}^0_1$ one wishes to obtain. In the first term on the right-hand side the derivative of a function is taken, that is, this term contains $\mathcal{D}^0_0$, while the second term contains the derivation of V, that is, it contains $\mathcal{D}^1_0$.

An important special case of (6.40) is the following: Apply the derivation to a tensor field S^1_s, with an empty slot at the place of the first argument. Then $\mathsf{S}^1_s(\cdot, V_1,\ldots,V_s)$ is a vector field so that (6.40) yields

$$\begin{aligned}&\mathcal{D}\left(\mathsf{S}^1_s(\cdot, V_1,\ldots,V_s)\right) \\ &\quad = (\mathcal{D}\mathsf{S})(\cdot, V_1,\ldots,V_s) + \sum_{i=1}^{s}\mathsf{S}(\cdot, V_1,\ldots,(\mathcal{D}V_i),\ldots,V_s) . \end{aligned} \tag{6.40b}$$

Before working out an example for a tensor derivation I quote here an important lemma whose proof may be found, for example in [O'Neill 1983].

Construction theorem for tensor derivations
Given a vector field $V \in \mathfrak{T}^1_0(M)$ and an $\mathbb{R}$-linear function $\delta : \mathfrak{T}^1_0(M) \to \mathfrak{T}^1_0(M)$ which has the property

$$\delta(fX) = (Vf)\,X + f\delta(X) \qquad (6.41a)$$

for all functions $f \in \mathfrak{F}(M)$ and for all vector fields $X \in \mathfrak{T}^1_0(M)$. Then there exists a unique tensor derivation $\mathcal{D}$ with the properties

$$\mathcal{D}^0_0 = V : \mathfrak{F}(M) \to \mathfrak{F}(M) \quad \text{and} \qquad (6.41b)$$

$$\mathcal{D}^1_0 = \delta : \mathfrak{T}^1_0(M) \to \mathfrak{T}^1_0(M)\,. \qquad (6.41c)$$

Example 6.7 Derivative along a flow
In this example we review a tensor derivation which is of special importance for physics and for geometry: The derivative along the flow of a given vector field V. (See also, e. g. [ME], Sects. 5.5.5 and 5.5.6.)

▶ **Definition 6.9 Lie derivative** For every fixed vector field $V \in \mathfrak{T}^1_0(M)$ let the action of the Lie derivative L_V on functions $f \in \mathfrak{F}(M)$ and on vector fields $X \in \mathfrak{T}^1_0(M)$ be defined by the equations, respectively,

$$L_V(f) = Vf \quad \text{for all } f \in \mathfrak{F}(M)\,, \qquad (6.42a)$$

$$L_V(X) = [V, X] \quad \text{for all } X \in \mathfrak{T}^1_0(M)\,. \qquad (6.42b)$$

Expressed in words, the Lie derivative L_V applied to a function is equal to the action of the fixed vector field V on this function. When applied to a vector field X it gives the commutator of V with X, $[V, X] = VX - XV$. In the first case (6.42a) Vf is again a function, in the second case (6.42b) the commutator $[V, X]$ is again a vector field.

By calculating the action of L_V on the vector field fX, with f a function, and X an arbitrary vector field, one sees that the Lie derivative fulfills the assumption (6.41a) of the construction theorem:

$$\begin{aligned} L_V(fX) &= [V, fX] = VfX - fXV \\ &= (Vf)\,X + fVX - fXV = (Vf)\,X + f\,[V, X] \\ &= (Vf)\,X + fL_VX = L_V(f)X + fL_VX\,. \end{aligned}$$

In passing from the first to the second line the product rule for the action of V on (fX) was used.

Equation (6.40a) yields the action of the Lie derivative on an arbitrary smooth one-form as follows

$$\begin{aligned}(L_V\omega)(X) &= L_V(\omega(X)) - \omega(L_V X) \\ &= V(\omega(X)) - \omega([V,X]) \ . \end{aligned} \tag{6.43}$$

It is an easy exercise to express the coordinate-free formulae obtained above in terms of local charts. In a given chart (φ_i, U_i) and using the summation convention, let $V = v^\mu \partial_\mu$, $W = w^\nu \partial_\nu$ be two vector fields, and $\omega = \omega_\sigma \, \mathrm{d}x^\sigma$ a one-form. The defining equations (6.42a) and (6.42b) yield

$$L_V f = v^\mu \partial_\mu f \ , \tag{6.44a}$$

$$L_V W = \left[v^\mu \partial_\mu, w^\nu \partial_\nu\right] = \left\{v^\mu (\partial_\mu w^\nu) - w^\mu (\partial_\mu v^\nu)\right\} \partial_\nu \ , \tag{6.44b}$$

$$L_V \omega = \left\{v^\mu \left(\partial_\mu \omega_\sigma\right) + \omega_\mu \left(\partial_\sigma v^\mu\right)\right\} \mathrm{d}x^\sigma \ . \tag{6.44c}$$

As expected, (6.44a) gives the directional derivative of the function f in the direction of V. In deriving (6.44b) use was made of the fact that the base fields commute, $[\partial_\mu, \partial_\nu] = 0$. Finally equation (6.44c) is obtained by first calculating the action of $(L_V\omega)$ on an arbitrary vector field W,

$$\begin{aligned}(L_V\omega)(W) &= V(\omega(W)) - \omega([V,W]) \\ &= v^\mu \partial_\mu (\omega_\sigma w^\sigma) - \omega_\mu \left(v^\sigma \partial_\sigma w^\mu - w^\sigma \partial_\sigma v^\mu\right) \\ &= v^\mu w^\sigma (\partial_\mu \omega_\sigma) + w^\sigma \omega_\mu (\partial_\sigma v^\mu) \ . \end{aligned}$$

As this vector field is arbitrary, the last equation must hold for every component. This gives (6.44c).

Remarks

1. These formulae simplify even further for base vector fields and for base one-forms in a chart

$$L_V \partial_\mu = \left[V, \partial_\mu\right] = -\left(\partial_\mu v^\nu\right) \partial_\nu \ , \tag{6.45a}$$

$$L_V \, \mathrm{d}x^\mu = (\partial_\nu v^\mu) \, \mathrm{d}x^\nu \ . \tag{6.45b}$$

2. In a coordinate-free notation (6.45b) is contained in the formula

$$L_V \, \mathrm{d}f = \mathrm{d}(L_V f) \ . \tag{6.46}$$

This formula is proved starting from the equation $Xf = \mathrm{d}f(X)$, with X a vector field and f a function.

It remains to write down the Lie derivative of arbitrary tensor fields in the coordinates of a chart. Starting from the general formula (6.40) for a tensor field $\mathsf{T}^r_s \in \mathfrak{T}^r_s(M)$ and inserting the representation (6.32) or (6.33), one obtains

$$\begin{aligned}
\left(L_V \mathsf{T}^r_s\right)^{\mu_1\cdots\mu_r}_{\nu_1\cdots\nu_s} &= v^\lambda \left(\partial_\lambda t^{\mu_1\cdots\mu_r}_{\nu_1\cdots\nu_s}\right) \\
&\quad - t^{\lambda,\mu_2\cdots\mu_r}_{\nu_1\cdots\nu_s} \left(\partial_\lambda v^{\mu_1}\right) - \ldots - t^{\mu_1\cdots\mu_{r-1}\lambda}_{\nu_1\cdots\nu_s} \left(\partial_\lambda v^{\mu_r}\right) \\
&\quad + t^{\mu_1\cdots\mu_r}_{\lambda\nu_2\cdots\nu_s} \left(\partial_{\nu_1} v^\lambda\right) + \ldots + t^{\mu_1\cdots\mu_r}_{\nu_1\cdots\nu_{s-1}\lambda} \left(\partial_{\nu_s} v^\lambda\right) . \qquad (6.47)
\end{aligned}$$

Note that (6.40) is solved for the first term on its right-hand side and that the formulae (6.44a), (6.45a), and (6.45b) were inserted. As a test, one verifies that equations (6.44b) and (6.44c) are obtained as special cases of (6.47).

The construction theorem has an important corollary:

Comparison of two tensor derivations
If the tensor derivations $\mathcal{D}_1$ and $\mathcal{D}_2$ coincide on the set of smooth functions $\mathfrak{F}(M)$ as well as on the set of smooth vector fields $\mathfrak{T}^1_0(M)$ then they are equal, $\mathcal{D}_1 = \mathcal{D}_2$.

6.5 Parallel Transport and Connection

In this section one learns how to move vectors by parallel transport and how to take covariant derivatives of geometric objects in case the spacetime is curved. The essential tool for this is a *connection*, for which semi-Riemannian geometry offers a special and distinguished choice. We start with a summary of the properties of a metric field, cf. Definition 6.7.

6.5.1 Metric, Scalar Product, and Index

The metric field g, cf. Definition 6.7, at every point $x \in M$, is a mapping

$$\mathrm{g} : T_x M \times T_x M \to \mathbb{R} : v, w \mapsto \mathrm{g}_x(v, w) ,$$

with the property $\mathrm{g}_x(v, w) = \mathrm{g}_x(w, v)$. As g and hence also $\mathrm{g}(v, w)|_x$ is nondegenerate for all x, the metric defines a scalar product on the vector space $T_x M$. This scalar product is denoted by $g_x(v, w)$, or, alternatively, in "bra"- and "ket"-notation, by

$$\langle v|w\rangle \equiv g_x(v, w)\,, \quad v, w \in T_x M\,. \tag{6.48}$$

The scalar product of two vector fields V and W exists also globally, and is denoted by the symbol $\langle V|W\rangle$. This function defines the scalar product (6.48) at every point $x \in M$. The following equivalence is often useful:

Metric equivalence
On a metric manifold the spaces $\mathfrak{X}(M) \equiv \mathfrak{T}^1_0(M)$ and $\mathfrak{X}^*(M) \equiv \mathfrak{T}^0_1(M)$ are isomorphic, or, as one also says, the two spaces are *metrically equivalent.* The isomorphism associates to the vector field $V \in \mathfrak{T}^1_0(M)$ the one-form $\omega \in \mathfrak{T}^0_1(M)$ such that

$$\omega(W) = \langle V|W\rangle \quad \text{for all } W \in \mathfrak{T}^1_0(M) \tag{6.49}$$

holds true. The correspondence $V \longleftrightarrow \omega$ is the announced isomorphism.

It is easy to indicate the isomorphism (6.49) in charts. Using the summation convention one has

$$\omega = \omega_\mu(x)\mathrm{d}x^\mu\,, \quad V = g^{\mu\nu}\omega_\mu(x)\partial_\nu\,, \tag{6.50}$$

where $g^{\mu\nu}(x)$ denotes the inverse of $g_{\mu\nu}(x)$. This is verified by calculation as follows:

$$\begin{aligned}\langle V|\partial_\sigma\rangle &= g^{\mu\nu}\omega_\mu(x)\,\langle\partial_\nu|\partial_\sigma\rangle = g^{\mu\nu}g_{\nu\sigma}\omega_\mu(x)\\ &= \delta^\mu_\sigma\omega_\mu(x) = \omega_\sigma(x) = \omega(\partial_\sigma)(x)\,.\end{aligned}$$

The existence of a metric allows one to decide whether two tangent vectors are orthogonal. One says that $v \in T_x M$ and $w \in T_x M$ are orthogonal if $g_x(v, w) = 0$.

The metric of a *semi*-Riemannian manifold is characterized by the fact that it has a nonvanishing *index*. The index is defined in analogy to the index of a bilinear form, see (5.10). Indeed, if the metric is restricted to a point $x \in M$ it acts like a real bilinear form on the vector space $T_x M$. Thus, the index is the codimension of the largest subspace of $T_x M$ on which $g_{\mu\nu}(x)$ is *definite,* either positive-definite or negative-definite.

A metric manifold (M, g) with a constant nonvanishing index is generally called a *semi-Riemannian manifold.* For the description of physical spacetime with dimension n one wishes to allow for only one time direction but $n - 1$ space directions, i.e. written in symbols, $T_x M \sim \mathbb{R} \times \mathbb{R}^{n-1}$. The largest subspace $\mathbb{R}^{n-1}$ on which the metric is definite for this choice has the dimension $(n-1)$. Thus, the index is

$$\nu = n - (n-1) = 1 \,.$$

Under these special circumstances, i.e. one time axis and $(n-1)$ space axes, one talks about the *signature* imprinted on the space, or, shorter but more specifically, about the *Minkowski signature.* A manifold (M, g) with $\dim M \geqslant 2$ for which $\nu = 1$, is called a *Lorentz manifold.*

We note that general relativity is studied not only with regard to the physical $1+3$ dimensions but also to $1+2$, $1+4$, or even higher spatial dimensions.

6.5.2 Connection and Covariant Derivative

If one wishes to shift a smooth vector field $X = X^\mu(x)\partial_\mu$ defined on $M = \mathbb{R}^n$, from the point $x \in \mathbb{R}^n$ by parallel transport in the direction of the tangent vector $V \in T_x M$, a simple and natural prescription comes to mind: At the point x let V act on the component functions $X^\mu(x)$ and then use the so-obtained functions $V(X^\mu)|_x$ to construct the transported vector $V(X^\mu)\partial_\mu$, viz.

$$V(X^\mu)\partial_\mu \equiv \sum_{\mu=0}^{n-1} V(X^\mu)\partial_\mu = V^\nu(x)\frac{\partial X^\mu}{\partial x^\nu}\partial_\mu =: D_V(X) \,. \tag{6.51}$$

This prescription generates a new vector field which is denoted by $D_V(X)$, or, for short, by $D_V X$. The operation $D_V X$ is called the *natural covariant derivative* of X with respect to the vector field V. In this case, that is to say still over Euclidean or semi-Euclidean space, the covariant derivative has the following properties:

When D_V acts on a linear combination of vector fields then one has, obviously,

$$D_V(c_1 X_1 + c_2 X_2) = c_1 D_V(X_1) + c_2 D_V(X_2) \,. \tag{6.52a}$$

If one replaces V by a sum $V_1 + V_2$, or multiplies V by a function f one has, respectively,

$$D_{(V_1+V_2)}X = D_{V_1}X + D_{V_2}X \,, \quad D_{(fV)}(X) = f D_V(X) \,. \tag{6.52b}$$

However, if one multiplies the vector field X on which D_V is acting, by a function g one must apply the product rule in (6.51), i.e. one has

$$D_V(gX) = (Vg)\,X + g D_V(X) \,. \tag{6.52c}$$

In formula (6.52c) the product rule

$$\frac{\partial (gX^\mu)}{\partial x^\nu} = \frac{\partial g}{\partial x^\nu} X^\mu + g \frac{\partial X^\mu}{\partial x^\nu}$$

or rule (6.21b) is applied.

On a manifold with nonvanishing curvature the simple formula (6.51) no longer holds, there is no longer a natural definition of a covariant derivative. On the contrary, the tangent spaces $T_x M$ and $T_{x'} M$ at two different points x and $x' \neq x$ are disjoint vector spaces whose elements cannot be compared without further tools at our disposal. In other words, in the general case *parallel transport* is not defined in an obvious way but requires an additional rule of how to implement it. Expressed differently again, there are many possibilities to shift a tangent vector in x by parallel transport! It is a question relating to the physics of gravitation whether it singles out a specific connection compatible with the equivalence principle.

Example 6.8 Parallel transport on a sphere

Consider the three great circles on the sphere S^2 which are drawn in Fig. 6.8, and a tangent vector v_A in the point A. Great circles are geodesics on S^2. Now transport the vector once along the path $(a) + (b)$, and once along the path (c) such that it includes always the same angle with the tangent of the geodesic along which it moves. This is a prescription which fixes the connection.

Upon arriving at point N by parallel transport, the original vector v_A has differing directions, $v_N^{(a)+(b)}$ for path $(a)+(b)$, and $v_N^{(c)}$ for path (c), as shown in the figure. This difference is an indication for and a measure of the curvature of the base manifold. (In order to follow up how parallel transport works on v_A it is useful to decompose v_A in a component tangent to the great circle (a) and a component tangent to the great circle (c). These components are easy to follow.)

The symbol D, without the subscript V, denotes a linear *connection*, which maps two vector fields, say V and X, onto a new vector field, say $D_V(X)$. The example of the flat manifold from which we started, shows that a connection must obey a number of rules as follows,

Definition 6.10 Connection A connection D is a mapping

$$D : \mathfrak{X}(M) \times \mathfrak{X}(M) \to \mathfrak{X}(M) , \tag{6.53}$$

which has the following properties: It is $\mathfrak{F}(M)$-linear in its first argument, i.e.

$$D_{V_1+V_2}(X) = D_{V_1}(X) + D_{V_2}(X) , \tag{6.54a}$$

$$D_{(gV)}(X) = g\,(D_V(X)) , \quad g \in \mathfrak{F}(M) . \tag{6.54b}$$

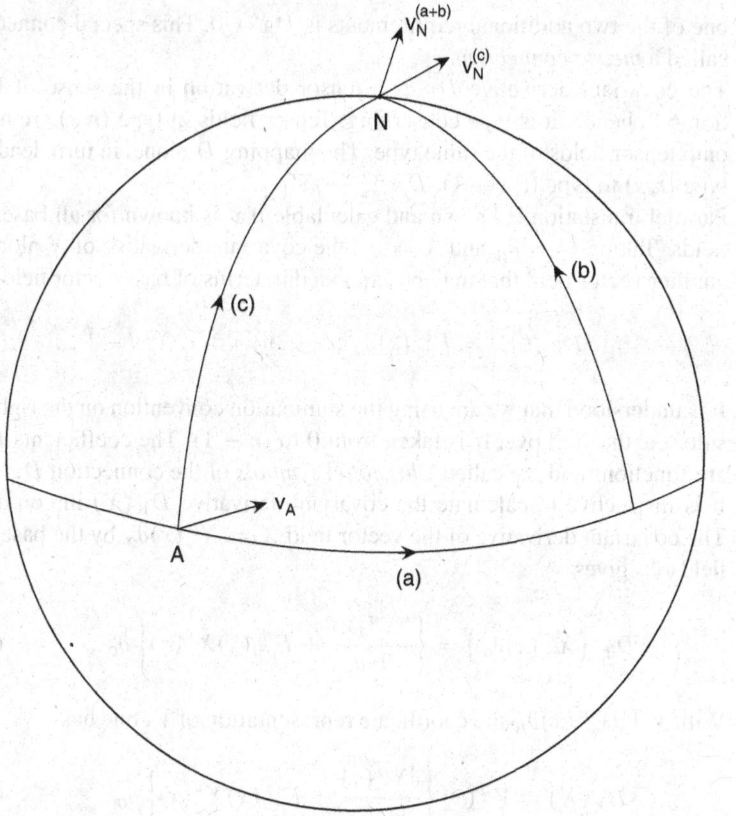

Fig. 6.8 A tangent vector v_A is shifted by parallel transport along two different paths from point A to point N, but always following great circles. As a result one finds two different images – a hint to the curvature of the manifold $M = S^2$

It is only $\mathbb{R}$-linear in its second argument but obeys the Leibniz rule

$$D_V(c_1 X_1 + c_2 X_2) = c_1 D_V(X_1) + c_2 D_V(X_2), \tag{6.55a}$$

$$D_V(fX) = (Vf)X + f D_V(X), \tag{6.55b}$$

with $V, X, X_1, X_2 \in \mathfrak{X}(M)$ and $g, f \in \mathfrak{F}(M)$.

Remarks

1. It is important to realize that a connection, in general, is not fixed by the postulates (6.54a)–(6.55b). Further requirements are needed to fix it uniquely. It is the "miracle" of Riemannian and of semi-Riemannian geometry that there exists a unique connection which, when applied as a tensor derivation to the metric g, gives zero and which has vanishing torsion (see (6.59) below). Thus,

one of the two additional requirements is $Dg = 0$. This special connection is called a *metric connection*.

2. The covariant derivative D_V is a tensor derivation in the sense of Definition 6.8, hence, it is type conserving. Tensor fields of type (r, s) are mapped onto tensor fields of the same type. The mapping D alone, in turn, leads from type (r, s) to type $(r, s+1)$, $D : \mathfrak{T}^r_s \to \mathfrak{T}^r_{s+1}$.
3. Parallel translation is known and calculable if it is known for all base vector fields. Taking $V = \partial_\mu$ and $X = \partial_\nu$, the covariant derivative of X along V is another vector field that may be expanded in terms of base vector fields,

$$D_{\partial_\mu}(\partial_\nu) = \Gamma^\sigma_{\mu\nu}(x)\partial_\sigma\,, \quad \mu, \nu = 0, 1, \ldots n-1\,. \tag{6.56}$$

It is understood that we are using the summation convention on the right-hand side, i.e. the sum over σ is taken from 0 to $(n-1)$. The coefficients $\Gamma^\sigma_{\mu\nu}(x)$ are functions and are called *Christoffel symbols* of the connection D.

4. It is instructive to calculate the covariant derivative $D_V(X)$ in coordinates. The covariant derivative of the vector field $X = X^\sigma(x)\partial_\sigma$ by the base vector fields ∂_μ gives

$$D_{\partial_\mu}\left(X^\sigma(x)\partial_\sigma\right) = \left\{\frac{\partial X^\sigma(x)}{\partial x^\mu} + \Gamma^\sigma_{\mu\nu}(x)X^\nu(x)\right\}\partial_\sigma\,. \tag{6.57a}$$

With $V = V^\mu(x)\partial_\mu$ the coordinate representation of V, one has

$$D_V(X) = V^\mu(x)\left\{\frac{\partial X^\sigma(x)}{\partial x^\mu} + \Gamma^\sigma_{\mu\nu}(x)X^\nu(x)\right\}\partial_\sigma\,. \tag{6.57b}$$

Comparison with (6.51) shows that the Christoffel symbols on a flat manifold $M = \mathbb{R}^n$ are zero.

5. There is a useful notation for derivatives which is often used: Ordinary derivatives that describe parallel transport on flat manifolds are abbreviated by a comma,

$$X^\sigma{}_{,\mu} \equiv \frac{\partial X^\sigma(x)}{\partial x^\mu}\,, \tag{6.57c}$$

while covariant derivatives in coordinate expressions are abbreviated by a semicolon,

$$X^\sigma{}_{;\mu} \equiv \frac{\partial X^\sigma(x)}{\partial x^\mu} + \Gamma^\sigma_{\mu\nu}X^\nu(x)\,. \tag{6.57d}$$

Note that this kind of derivative also appears in $\mathbb{R}^n$ when one uses curvilinear coordinates.

6. These formulae, together with the general coordinate expression (6.33) for tensor fields, yield useful formulae for the covariant derivative of tensor fields. From (6.57b) follows

$$D_V(\partial_\sigma) = V^\mu \Gamma^\nu_{\mu\sigma}\partial_\nu\,.$$

The action on a base one-form is calculated from (6.40),

$$D_V(\mathrm{d}x^\tau)(\partial_\sigma) = V\,\mathrm{d}x^\tau(\partial_\sigma) - \mathrm{d}x^\tau\,(D_V(\partial_\sigma)) = -V^\mu \Gamma^\tau_{\mu\sigma}\,.$$

The first term on the right-hand side vanishes because $\mathrm{d}x^\tau(\partial_\sigma)$ is a constant which gives zero when V is applied to it. In the second term the previous formula is inserted and the duality relation $\mathrm{d}x^\tau(\partial_\nu) = \delta^\tau_\nu$ is used. From the coordinate representation (6.33) and with (6.40) solved for the first term on the right-hand side, one obtains the following useful formula,

$$t^{\mu_1\ldots\mu_r}_{\nu_1\ldots\nu_s\,;\,\rho} = t^{\mu_1\ldots\mu_r}_{\nu_1\ldots\nu_s\,,\,\rho} + \Gamma^{\mu_1}_{\rho\sigma}\, t^{\sigma\mu_2\ldots\mu_r}_{\nu_1\ldots\nu_s} + \Gamma^{\mu_2}_{\rho\sigma}\, t^{\mu_1\sigma\mu_3\ldots\mu_r}_{\nu_1\ldots\nu_s} + \ldots$$
$$- \Gamma^\tau_{\rho\nu_1}\, t^{\mu_1\ldots\mu_r}_{\tau\nu_2\ldots\nu_s} - \Gamma^\tau_{\rho\nu_2}\, t^{\mu_1\ldots\mu_r}_{\nu_1\tau\nu_3\ldots\nu_s} - \ldots\,. \tag{6.57e}$$

On the left-hand side the semicolon stands for the covariant derivative – as defined in the preceding remark. On the right-hand side the comma stands for an ordinary partial derivative (repeated indices are to be summed over).
Two examples may be useful to illustrate this somewhat involved master formula. For a tensor field A of type $(2,0)$ whose components are denoted by $A^{\mu\nu}(x)$ we have

$$A^{\mu\nu}{}_{;\rho} = A^{\mu\nu}{}_{,\rho} + \Gamma^\mu_{\rho\sigma} A^{\sigma\nu} + \Gamma^\nu_{\rho\tau} A^{\mu\tau}\,. \tag{6.57f}$$

For a $(0,2)$-tensor field B whose components are $B_{\mu\nu}$ one has

$$B_{\mu\nu\,;\,\rho} = B_{\mu\nu\,,\,\rho} - \Gamma^\sigma_{\rho\mu} B_{\sigma\nu} - \Gamma^\sigma_{\rho\nu} B_{\mu\sigma}\,. \tag{6.57g}$$

7. The action of the covariant derivative along the vector field V on a function is the same as the directional derivative (see also the first remark following Definition 6.8),

$$D_V(f) \equiv (D_V)^0_0(f) = V(f)\,, \quad f \in \mathfrak{F}(M)\,. \tag{6.58}$$

The action on functions and on vector fields determines the action on one-forms, see (6.40a). By the construction theorem the covariant derivative is a tensor derivation and, hence, can be applied to arbitrary smooth tensor fields of type (r,s). In particular, it can be applied to the metric $\mathbf{g} \in \mathfrak{T}^0_2(M)$. We make use of this below.

6.5.3 Torsion and Curvature Tensor Fields

For every linear connection (6.53) one defines a *tensor field of torsion,* also called *torsion* for short, $\mathsf{T} \in \mathfrak{T}^1_2(M)$ by its action on two vector fields. One should keep in mind that a general tensor field S_2^1 of this type, when evaluated on a one-form and two vector fields, yields a function, viz. $\mathsf{S}_2^1(\omega, X_1, X_2) \in \mathfrak{F}(M)$. However, if the one-form is left out and is replaced by an empty slot then $\mathsf{S}_2^1(\cdot, X_1, X_2)$ is again a vector field.

▶ **Definition 6.11 Torsion** To every linear connection D belongs a tensor field of type $(1, 2)$ which is defined by the mapping

$$\begin{aligned} \mathsf{T} &: \mathfrak{X}(M) \times \mathfrak{X}(M) \to \mathfrak{X}(M) \\ &: X, Y \longmapsto \mathsf{T}(X, Y) = D_X(Y) - D_Y(X) - [X, Y] \end{aligned} \tag{6.59}$$

for all smooth vector fields X and Y. Here $D_X Y$ and $D_Y X$ are the covariant derivatives of Y along X, and of X along Y, respectively, while $[X, Y]$ is the commutator.

Furthermore, to the chosen linear connection D one associates a tensor field of type $(1, 3)$ which yields a vector field when it is applied to three vector fields but to no one-form. It is defined as follows.

▶ **Definition 6.12 Riemann curvature field** For every linear connection there exists a tensor field of type $(1, 3)$, called the *Riemann curvature field*, which is defined by its action on three arbitrary smooth vector fields $X, Y, Z \in \mathfrak{X}(M)$:

$$\begin{aligned} \mathsf{R} &: \mathfrak{X}(M) \times \mathfrak{X}(M) \times \mathfrak{X}(M) \to \mathfrak{X}(M) \\ &: X, Y, Z \longmapsto \mathsf{R}(X, Y, Z) = [D_X, D_Y]\, Z - D_{[X,Y]} Z\ . \end{aligned} \tag{6.60}$$

The resulting vector field is often written as $\mathsf{R}(X, Y, Z)$, in case one wishes to exploit its symmetry properties, but also as $\mathsf{R}(X, Y)Z$ or $\mathsf{R}_{XY} Z$.

▶ **Remarks**

1. For a geometric interpretation of the torsion tensor we refer to [Nakahara 2003, Sect. 7.3]. The Levi-Civita connection that will be defined below, has vanishing torsion. This is one of its defining properties. If this is so the Christoffel symbols are symmetric in their lower two indices. This becomes obvious if one inserts two base fields ∂_μ and ∂_ν in (6.59) and recalls that ∂_μ and ∂_ν commute. One has

   $$\mathsf{T}(\partial_\mu, \partial_\nu) = D_{\partial_\mu}(\partial_\nu) - D_{\partial_\nu}(\partial_\mu) = \left(\Gamma^\sigma_{\mu\nu}(x) - \Gamma^\sigma_{\nu\mu}(x)\right) \partial_\sigma\ .$$

   If the torsion is equal to zero everywhere one concludes

   $$\Gamma^\sigma_{\mu\nu}(x) = \Gamma^\sigma_{\nu\mu}(x) \qquad \text{(holds when } \mathsf{T} \equiv 0\text{)}\ . \tag{6.61}$$

2. The tensor field (6.60), Riemann's curvature field, can be interpreted more directly by studying its restriction to planes in the tangent spaces $T_x M$, i.e. to two-dimensional subspaces. This yields the so-called *sectional curvature* which is akin to the classical Gaussian curvature of a curve, see for example [O'Neill 1983]. From a physical point of view this tensor field is relevant when one studies neighbouring geodesics and calculates the forces which cause particles on such geodesics to approach each other or to move apart. Forces of this kind which are "transverse" to orbits of free fall, are also called *tidal forces*.

In Sect. 6.5.7 we return to the Riemann curvature tensor field, in its form needed for general relativity, and analyze its properties.

6.5.4 The Levi-Civita Connection

On Riemannian and semi-Riemannian manifolds there exists a special connection whose torsion field vanishes and which gives zero when applied to the metric. These are its distinguishing features. Although we will continue to denote it by D, one should keep in mind that one is talking about this special connection named after Tullio Levi-Civita (1873–1941). It is defined as follows:

Levi-Civita connection

The Levi-Civita connection is defined on a Riemannian or semi-Riemannian manifold (M, g),

$$D : \mathfrak{X} \times \mathfrak{X} \longrightarrow \mathfrak{X} : V, W \longmapsto D_V(W) .$$

It is $\mathfrak{F}(M)$-linear in its first argument (V), cf. (6.54a) and (6.54b). In the second argument (W) it is $\mathbb{R}$-linear, (6.55a), and obeys the Leibniz rule (6.55b). Beyond these postulates which apply to any connection, it has two more properties which fix it uniquely: The torsion (6.59) calculated from it vanishes identically, i.e. one has

$$\mathrm{T} \equiv 0 , \quad \text{hence } [V, W] = D_V(W) - D_W(V) . \tag{6.62}$$

Furthermore, it obeys the *Ricci condition:* For any triple of vector fields X, V, and W one has

$$X \langle V|W \rangle = \langle D_X(V)|W \rangle + \langle V|D_X(W) \rangle . \tag{6.63}$$

▶ Remarks

1. What we have summarized here is the content of a central theorem of Riemannian geometry: The requirements (6.54a)–(6.55b), supplemented by the additional conditions (6.62) and (6.63), fix the connection uniquely.
2. The Levi-Civita connection defines a tensor derivation as studied in Sect. 6.4 and, more specifically, in Definition 6.8. Thus, the Ricci condition (6.63) may be interpreted as a condition on the covariant derivative of the metric as follows. Evaluating the metric on two vector fields V and W yields the function $\mathrm{g}(V, W)$ that can also be written as $\langle V|W\rangle$, cf. (6.48). Apply then the general formula (6.40) to the covariant derivative of this function along X,

$$D_X\left(\mathrm{g}(V, W)\right) \equiv D_X \langle V|W\rangle \; .$$

This gives

$$\begin{aligned} D_X \langle V|W\rangle = &\left[D_X(\mathrm{g})\right](V, W) \\ &+ \mathrm{g}\left(D_X(V), W)\right) + \mathrm{g}\left(V, D_X(W)\right) \; . \end{aligned}$$

The left-hand side shows the covariant derivative of a function. The first term on the right-hand side contains the derivative of the $(0, 2)$ tensor field $\mathrm{g} \in \mathfrak{T}^0_2(M)$, while the second and third terms contain the action of D_X on vector fields. When applied to *functions* the action of D_X is the same as the action of the vector field X on this function, $D_X \langle V|W\rangle = X \langle V|W\rangle$. Furthermore,

$$\mathrm{g}\left(D_X(V), W)\right) = \langle D_X(V)|W\rangle \quad \text{and}$$
$$\mathrm{g}\left(V, D_X(W)\right) = \langle V|D_X(W)\rangle$$

are just different ways of writing these functions. Thus, the requirement (6.63) is equivalent to the requirement that the covariant derivative of the metric vanish,

$$\begin{aligned} &(D_X \mathrm{g}) = 0 \quad \Longleftrightarrow \\ &X \langle V|W\rangle = \langle D_X(V)|W\rangle + \langle V|D_X(W)\rangle \; . \end{aligned} \tag{6.64}$$

Thus, another way of stating the Ricci condition is to require the Levi-Civita connection to be *metric*. The covariant derivative of the metric shall be identically zero.

6.5.5 Properties of the Levi-Civita Connection

Let us summarize once more the properties of the Levi-Civita connection:

The Levi-Civita connection is defined uniquely by the following properties:

(i) $D_V(W)$ is $\mathfrak{F}(M)$-linear in the vector field V, i.e. the relations (6.54a) and (6.54b) apply;
(ii) In its argument W it is $\mathbb{R}$-linear only, i.e. fulfills (6.55a);
(iii) It satisfies the Leibniz rule (6.55b);
(iv) The corresponding torsion is identically zero, see (6.62);
(v) The Levi-Civita connection is a metric connection, $D_X(\mathrm{g}) = 0$. Equivalently, it fulfills the Ricci condition (6.64),

$$X\,\langle V|W\rangle = \langle D_X(V)|W\rangle + \langle V|D_X(W)\rangle$$

for all vector fields X, V, and W.

From the five properties above one derives a formula which is useful for many calculations. It reads as follows:

Formula of Koszul
For any three vector fields V, W, and X, using the bracket notation of (6.49), one has

$$\begin{aligned} 2\,\langle D_V(W)|X\rangle &= V\,\langle W|X\rangle + W\,\langle X|V\rangle - X\,\langle V|W\rangle \\ &\quad - \langle V|[W,X]\rangle + \langle W|[X,V]\rangle + \langle X|[V,W]\rangle\ . \end{aligned} \tag{6.65}$$

Here $[W,X]$ is the commutator of the vector fields W and X which is again a vector field.

Proof of the Koszul formula: Collect the first three terms on the right-hand side of (6.65) in one group, the last three in another group, and calculate their sum. For the first three one has, using (v)

$$\begin{aligned} &V\,\langle W|X\rangle + W\,\langle X|V\rangle - X\,\langle V|W\rangle \\ &\quad = \langle D_V(W)|X\rangle + \langle W|D_V(X)\rangle + \langle D_W(X)|V\rangle \\ &\qquad + \langle X|D_W(V)\rangle - \langle D_X(V)|W\rangle - \langle V|D_X(W)\rangle\ . \end{aligned}$$

The remaining three terms are worked on making use of (iv), by replacing the commutators with differences of covariant derivatives by means of (6.62),

$$\begin{aligned} &-\langle V|[W,X]\rangle + \langle W|[X,V]\rangle + \langle X|[V,W]\rangle \\ &\quad = -\langle V|D_W(X)\rangle + \langle V|D_X(W)\rangle + \langle W|D_X(V)\rangle \\ &\qquad - \langle W|D_V(X)\rangle + \langle X|D_V(W)\rangle - \langle X|D_W(V)\rangle \, . \end{aligned}$$

Adding these intermediate results and taking account of the symmetry of the metric, all terms but two cancel pairwise. The two terms that do not cancel are

$$\langle D_V(W)|X\rangle + \langle X|D_V(W)\rangle = 2\,\langle D_V(W)|X\rangle \, .$$

This, indeed, is the left-hand side of the relation (6.65).

Remarks

1. The Koszul formula (6.65) is useful in proving the existence and uniqueness of the Levi-Civita connection. Uniqueness is easily shown: Suppose there were two connections $D' \neq D$ which fulfill the formula (6.65) for all vector fields V, W, and X. As the metric is nondegenerate equation (6.65) implies that $D'_V(W) = D_V(W)$ for all V and W.
   The proof of its existence is a bit more involved. Denote the right-hand side of the Koszul formula by $\omega(V, W, X)$. Keeping V and W fixed, $X \to \omega(V, W, X)$ is a one-form. As the spaces $\mathfrak{X}(M)$ and $\mathfrak{X}^*(M)$ are isomorphic there is a unique vector field Y such that for all $X \in \mathfrak{X}(M)$ the equation $2\,\langle Y|X\rangle = \omega(V, W, X)$ holds. There is no objection against denoting the vector field Y that is constructed this way, by $Y = D_V(W)$. If one then confirms the five axions (i)–(v) the existence of the Levi-Civita connection is proved. (The reader is invited to complete this proof!)
2. The Koszul formula may be used to derive in a few steps an important formula for the Christoffel symbols. Replacing the three arbitrary vector fields by the base fields $V = \partial_\mu$, $W = \partial_\nu$, and $X = \partial_\rho$ one obtains

   $$2\big\langle D_{\partial_\mu}(\partial_\nu)\big|\partial_\rho\big\rangle = \partial_\mu g_{\nu\rho} + \partial_\nu g_{\mu\rho} - \partial_\rho g_{\mu\nu} \, .$$

   Here the chart representation (6.35a) of the metric

   $$\langle \partial_\sigma|\partial_\tau\rangle = \mathrm{g}(\partial_\sigma, \partial_\tau) = g_{\sigma\tau}$$

   was inserted and use was made of the fact that the base fields commute. With the definition (6.56) of the Christoffel symbols the left-hand side is equal to

   $$2\big\langle D_{\partial_\mu}(\partial_\nu)\big|\partial_\rho\big\rangle = 2\Gamma^{\tau}_{\mu\nu} g_{\tau\rho} \quad \text{(summed over } \tau) \, .$$

Multiply then this equation by the inverse $g^{\rho\sigma}$ of the metric tensor and take the sum over ρ. This yields the formula

$$\Gamma^{\sigma}_{\mu\nu} = \frac{1}{2} g^{\sigma\rho} \left\{ \partial_\mu g_{\nu\rho} + \partial_\nu g_{\mu\rho} - \partial_\rho g_{\mu\nu} \right\} . \tag{6.66}$$

Thus, the Christoffel symbols are functions of the *first* derivatives of the metric and they contain its inverse. On a flat spacetime, i.e. on a manifold with constant metric they are identically zero.

3. The symmetry of the Christoffel symbols in the two lower indices μ and ν is obvious from the coordinate expression (6.66), $\Gamma^{\sigma}_{\mu\nu} = \Gamma^{\sigma}_{\nu\mu}$. One should note, however, that the coordinate expression (6.66) refers to the base fields ∂_μ which commute among each other, $[\partial_\mu, \partial_\nu] = 0$. There are other possibilities to choose a local basis, for instance a vielbein, i.e. a comoving coordinate system on geodesics, also called *repère mobile*, whose elements do not commute.
4. A vector field V is called *parallel* if its covariant derivative $D_X(V)$ is equal to zero for all $X \in \mathfrak{X}(M)$. When applied to base fields ∂_μ this shows that the Christoffel symbols measure the extent to which these are not parallel.

Example 6.9 Cylinder coordinates in $\mathbb{R}^3$

Nonvanishing Christoffel symbols appear even on flat space $\mathbb{R}^3$ in case one uses curvilinear coordinates. This example discusses cylinder coordinates

$$y^1 \equiv r , \quad y^2 \equiv \phi ; \quad y^3 \equiv z ,$$

which are related to cartesian coordinates by

$$x^1 = r \cos\phi , \quad x^2 = r \sin\phi , \quad x^3 = z .$$

For the base fields one has

$$\begin{aligned}
\partial_{y^1} &= \cos\phi \, \partial_{x^1} + \sin\phi \, \partial_{x^2} , \\
\partial_{y^2} &= r \left(-\sin\phi \, \partial_{x^1} + \cos\phi \, \partial_{x^2} \right) , \\
\partial_{y^3} &= \partial_{x^3} .
\end{aligned}$$

In this new basis the metric tensor is

$$\begin{aligned}
g_{11} &= g(\partial_{y^1}, \partial_{y^1}) = \cos^2\phi \, g(\partial_{x^1}, \partial_{x^1}) + \sin^2\phi \, g(\partial_{x^2}, \partial_{x^2}) \\
&\quad + 2 \sin\phi \cos\phi \, g(\partial_{x^1}, \partial_{x^2}) = \cos^2\phi + \sin^2\phi = 1 , \\
g_{22} &= r^2 , \\
g_{33} &= 1 ,
\end{aligned}$$

and $g_{ij} = 0$ for $i \neq j$. Thus, the metric tensor and its inverse are, respectively,

$$g_{ij} = \text{diag}(1, r^2, 1) , \quad g^{ij} = \text{diag}(1, 1/r^2, 1) .$$

Inserting these formulae in the expression (6.66) the nonvanishing Christoffel symbols are found to be

$$\Gamma^2_{12} = \frac{1}{r} = \Gamma^2_{21} , \quad \Gamma^1_{22} = -r .$$

Among the covariant derivatives of base fields which are not equal to zero there remain

$$D_{\partial_\phi}(\partial_\phi) = -r\,\partial_r , \quad D_{\partial_\phi}(\partial_r) = -\sin\phi\,\partial_{x^1} + \cos\phi\,\partial_{x^2} = D_{\partial_r}(\partial_\phi) .$$

Of course, the base field $\partial_z = \partial_{x^3}$ is parallel. Furthermore, one expects the covariant derivatives $D_{\partial_z}(\partial_r)$ and $D_{\partial_z}(\partial_\phi)$ to be zero because the base fields ∂_r and ∂_ϕ do not change if one shifts them along the direction of ∂_z.

6.5.6 Geodesics on Semi-Riemannian Spacetimes

We learnt in Sect. 6.5.2 and Sect. 6.5.4 that linear connections, among other useful features, allow one to compare vectors from different tangent spaces. A case of special interest is the test whether such vectors are parallel. This leads back to a remark that we made in the preceding section: A vector field V is said to be *parallel* if its covariant derivative $D_X(V)$ vanishes for all $X \in \mathfrak{X}(M)$.

Let $\alpha : I \to M$ be a smooth curve on the spacetime $(M, \mathbf{g})$, and let $\tau \in I \subset \mathbb{R}_\tau$ be the "time" that parametrizes the curve. Then $\alpha(\tau)$ is a one-dimensional manifold and a submanifold of M, as sketched in Fig. 6.9. The Levi-Civita connection which is defined on M, induces the covariant derivative of an arbitrary vector field $Z \in \mathfrak{X}(\alpha)$ on α by the tangent vector field $\dot{\alpha}$, $\dot{Z} = D_{\dot{\alpha}}(Z)$.

In a local chart both Z and $\dot{\alpha}$ may be decomposed as usual,

$$Z = Z^\mu \partial_\mu , \qquad \dot{\alpha} = a^\sigma \partial_\sigma .$$

The coefficients a^σ, $\sigma = 0, 1, \ldots, n-1$, are functions to be calculated from the composition of the curve α (which maps $I \subset \mathbb{R}_\tau$ to M) and the coordinate function φ^σ (which maps M to $\mathbb{R}^n$), $a^\sigma = \mathrm{d}(\varphi^\sigma \circ \alpha)/\mathrm{d}\tau$. In order to calculate $\dot{Z} = D_{\dot{\alpha}}(Z)$ one utilizes, in essence, a variant of the formula (6.57b). One has

$$\dot{Z} = \frac{\mathrm{d}Z^\mu}{\mathrm{d}\tau}\partial_\mu + Z^\mu D_{\dot{\alpha}}(\partial_\mu) .$$

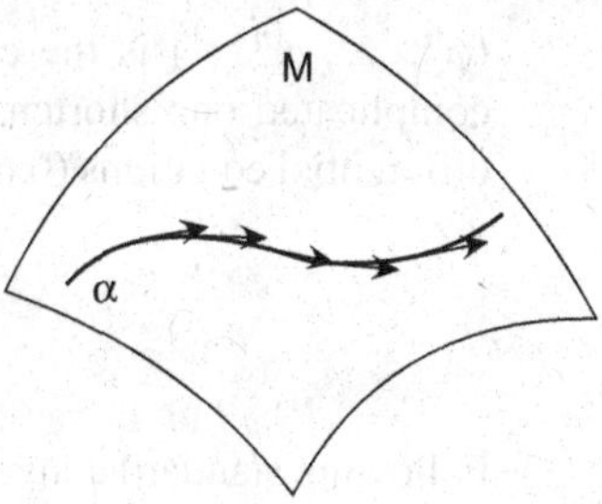

Fig. 6.9 A smooth curve α on M and elements of its (one-dimensional) tangent spaces. The curve itself is a one-dimensional manifold

The second term is transformed by making use of the $\mathfrak{F}(M)$-linearity of D and by inserting the defining equation (6.56) of the Christoffel symbols:

$$D_{\dot{\alpha}}(\partial_\mu) = D_{(a^\sigma \partial_\sigma)}(\partial_\mu) = a^\sigma D_{\partial_\sigma}(\partial_\mu) = a^\sigma \Gamma^\tau_{\sigma\mu} \partial_\tau \, .$$

Renaming the summation indices in the second term and inserting $a^\sigma = \mathrm{d}(\varphi^\sigma \circ \alpha)/\mathrm{d}\tau$, one obtains

$$\dot{Z} = \left\{ \frac{\mathrm{d}Z^\mu}{\mathrm{d}\tau} + \Gamma^\mu_{\rho\sigma} \frac{\mathrm{d}(\varphi^\rho \circ \alpha)}{\mathrm{d}\tau} Z^\sigma \right\} \partial_\mu \, . \tag{6.67}$$

The vector field Z is said to be *parallel* along the curve α if $\dot{Z} = 0$ holds everywhere on the curve.

Consider now, as a special case, the tangent vector field of the given curve, i.e. $Z = \dot{\alpha}$. Its derivative $\dot{Z} = \ddot{\alpha}$ is the *acceleration field.* This consideration shows clearly how to define geodesics in this framework. Speaking in terms of physics in general and mechanics in particular, geodesics are curves of free motion, that is to say, orbits without acceleration. From the point of view of geometry geodesics are curves whose length is extremal. That there is a relation between these two notions is well known from mechanics.

► **Definition 6.13 Geodesics** A geodesic is a curve $\gamma : I \to M$ whose tangent vector field $\dot{\gamma}$ is *parallel*. When written in terms of local charts, a geodesic satisfies the differential equation of second order

$$\frac{\mathrm{d}^2}{\mathrm{d}\tau^2}(\varphi^\mu \circ \gamma) + \Gamma^\mu_{\rho\sigma} \frac{\mathrm{d}(\varphi^\rho \circ \gamma)}{\mathrm{d}\tau} \frac{\mathrm{d}(\varphi^\sigma \circ \gamma)}{\mathrm{d}\tau} = 0 \, , \tag{6.68}$$
$$\mu = 0, 1, \ldots, n-1 \, .$$

► **Remarks**

1. Strictly speaking, the second term of (6.68) contains the derivatives of the functions $(\varphi^\rho \circ \gamma)(\tau) = \varphi^\rho(\gamma(\tau))$ which are defined on $\mathbb{R}^n$ and where $\varphi =$

$(\varphi^0, \dots, \varphi^{(n-1)})$ is the chart mapping on $U \subset M$. As this notation looks complicated one shortens it by using the notation $\varphi^\rho(\gamma(\tau)) \equiv y^\rho(\tau)$. The differential equations (6.68) for geodesics then take a form easy to remember, viz.

$$\ddot{y}^\mu + \Gamma^\mu_{\rho\sigma}\, \dot{y}^\rho\, \dot{y}^\sigma = 0 \,. \tag{6.68a}$$

Following standard conventions of mechanics the dot stands for the derivative by τ.

2. Consider a *massive* test particle whose mass is so small that it does not modify, for all practical purposes, the geometry of the given spacetime $(M, \mathbf{g})$. The equivalence principle tells us that in every point $x \in M$ the equation of free motion is the same as in flat space, i.e.

$$\frac{\mathrm{d}^2 y^\mu}{\mathrm{d}\tau^2} = 0 \,. \tag{6.69}$$

Except for a factor c that we choose to be 1 anyhow, the variable τ is the arc length for which $\mathrm{d}\tau^2 = g_{\mu\nu}\,\mathrm{d}y^\mu\,\mathrm{d}y^\nu$, or

$$g_{\mu\nu}\frac{\mathrm{d}y^\mu}{\mathrm{d}\tau}\frac{\mathrm{d}y^\nu}{\mathrm{d}\tau} = 1 \,. \tag{6.70a}$$

One convinces oneself by means of a calculation (that we skip here) that the condition (6.70a) is compatible with the differential equations (6.68a).

3. For a *massless* particle the equation of motion must be the same as (6.69). However, as such a particle never finds a rest system and, hence, cannot be assigned a proper time, the condition (6.70a) can no longer hold. Instead, one replaces τ by a parameter λ which is chosen such that with $\mathrm{d}^2 y^\mu / \mathrm{d}\lambda^2 = 0$ the condition

$$g_{\mu\nu}\frac{\mathrm{d}y^\mu}{\mathrm{d}\lambda}\frac{\mathrm{d}y^\nu}{\mathrm{d}\lambda} = 0 \tag{6.70b}$$

is fulfilled. In other terms, the equations of motion for a massive and for a massless particle are the same. All that changes is the initial condition which is given by (6.70a) in the first, massive case, by (6.70b) in the second, massless case.

4. In semi-Riemannian geometry the equivalence principle is realized in the following way. In a given point $x \in U \subset M$ one associates to every tangent vector $v \in T_x M$ the geodesic which goes through x and which has initial velocity v,

$$\exp_x \;:\; T_x M \to M \;:\; v \mapsto \exp_x(v) = \gamma_v(1) \,. \tag{6.71}$$

This mapping is called *exponential mapping*. Regarding the geodesic pertaining to the initial velocity τv, one shows that

$$\exp_x(\tau v) = \gamma_{\tau v}(1) = \gamma_v(\tau) \,.$$

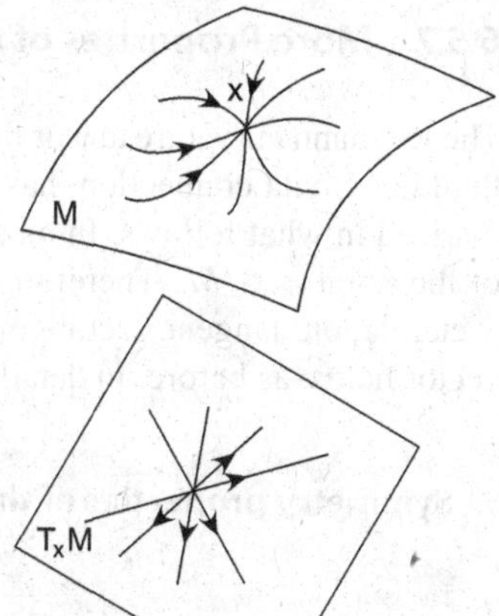

Fig. 6.10 Geodesics passing through the point $x \in M$ form a star domain. Also shown are the initial velocities in the tangent space $T_x M$ at x. By choosing a base system in $T_x M$, the corresponding geodesics then yield a normal system

This means that the exponential mapping maps straight lines through the origin of $T_x M$ onto geodesics on M[6], (cf. Fig. 6.10). With $\{e_\mu\}$ an orthonormal basis on $T_x M$, i.e. if $\langle e_\mu | e_\nu \rangle = \eta_{\mu\nu}$ holds and if $\exp^{-1}(x) = \varphi^\mu(x) e_\mu$ one shows: The functions $(\varphi^0, \ldots, \varphi^{n-1})$ yield a chart for the subset U containing x such that one has

$$g_{\mu\nu}(x) = \eta_{\mu\nu} \, , \qquad \Gamma^\sigma_{\mu\nu}(x) = 0 \, . \tag{6.72}$$

Special coordinates of this kind are called *Gaussian coordinates* or *normal coordinates.* (For more details and a proof of the construction see, e. g. [O'Neill 1983].)

The result is rather intuitive because the force-free equation of motion has the simple form (6.69) along any geodesic. Comparison with (6.68a) shows that the Christoffel symbols at the point x are zero in Gaussian coordinates. Thus, with this construction the equivalence principle becomes obvious and, hence, finds its most accurate expression.

At every point of the semi-Riemannian manifold (M, g) there exists a coordinate system in which the metric tensor has the form of the flat metric, $g_{\mu\nu}(x) = \eta_{\mu\nu}$, and in which the Christoffel symbols vanish, $\Gamma^\sigma_{\mu\nu}(x) = 0$.

[6] The existence and uniqueness theorem for ordinary differential equations guarantees that this mapping is a diffeomorphism for star-like domains of the kind sketched in Fig. 6.10. Thus, the inverse of the exponential mapping exists and maps geodesics through the point x to straight lines through the origin of the vector space $T_x M$.

6.5.7 More Properties of the Curvature Tensor

The Riemann curvature tensor field, constructed following definition (6.60), with D the Levi-Civita connection, has a number of symmetry properties which are summarized in what follows. In most cases it is sufficient to consider a neighbourhood of the point $x \in M$. Therefore, in the following relations lower case letters x, y, v etc. denote tangent vectors of the space $T_x M$, while capital letters refer to full vector fields, as before. In detail one has

Symmetry properties of the Riemann tensor field

$$\mathsf{R}_{XY} = -\mathsf{R}_{YX} , \tag{6.73a}$$

$$\left(\mathsf{R}_{xy} v \middle| w\right) = -\left(\mathsf{R}_{xy} w \middle| v\right) , \tag{6.73b}$$

$$\mathsf{R}_{xy} z + \mathsf{R}_{yz} x + \mathsf{R}_{zx} y = 0 , \tag{6.73c}$$

$$\left(\mathsf{R}_{xy} v \middle| w\right) = \left(\mathsf{R}_{vw} x \middle| y\right) , \tag{6.73d}$$

$$(D_Z \mathsf{R})(X, Y) + (D_X \mathsf{R})(Y, Z) + (D_Y \mathsf{R})(Z, X) = 0 . \tag{6.73e}$$

The first of these is obvious from the general definition (6.60). The remaining relations are specific for the Levi-Civita connection because they make use of the Ricci condition (6.63) or (6.64) and of the requirement that the torsion vanish, cf. (6.62). The proofs for the relations (6.73a)–(6.73e) in this beautiful coordinate-free form can be found, for example, in [O'Neill 1983].

The relations (6.73c) and (6.73e) are called *first* and *second Bianchi identity*, respectively.

The same symmetry relations can also be derived from a coordinate expression of the curvature tensor field. For this purpose we evaluate definition (6.60) (with D the Levi-Civita connection!) with three base vector fields, using the following definition of the coefficients[7]

$$R_{\partial_\mu \partial_\nu}(\partial_\rho) = R^{\tau}_{\ \rho\mu\nu} \partial_\tau . \tag{6.74}$$

The right-hand side defines the coefficients of R. The left-hand side is calculated by means of (6.56): Noting that $[\partial_\mu, \partial_\nu] = 0$ and using (6.57a) (with $X^\alpha = \delta^\alpha_\rho$) one

[7] It is important to keep in mind the order in which the indices are written. Some texts on general relativity use a different definition of the coefficients.

has

$$R_{\partial_\mu \partial_\nu}(\partial_\rho) = D_{\partial_\mu}\left(D_{\partial_\nu}(\partial_\rho)\right) - D_{\partial_\nu}\left(D_{\partial_\mu}(\partial_\rho)\right)$$
$$= \partial_\mu \Gamma^\sigma_{\nu\rho}\partial_\sigma + \Gamma^\sigma_{\nu\rho}\Gamma^\tau_{\mu\sigma}\partial_\tau - (\mu \leftrightarrow \nu)$$
$$= \left\{\partial_\mu \Gamma^\tau_{\nu\rho} + \Gamma^\tau_{\mu\sigma}\Gamma^\sigma_{\nu\rho} - (\mu \leftrightarrow \nu)\right\}\partial_\tau \,. \tag{6.75}$$

Comparison with definition (6.74) then yields the

Riemann curvature tensor in coordinates:

$$R^\tau_{\rho\mu\nu} = \partial_\mu \Gamma^\tau_{\nu\rho} - \partial_\nu \Gamma^\tau_{\mu\rho} + \Gamma^\tau_{\mu\sigma}\Gamma^\sigma_{\nu\rho} - \Gamma^\tau_{\nu\sigma}\Gamma^\sigma_{\mu\rho} \,. \tag{6.76}$$

► **Remark**

While the Christoffel symbols depend only on the inverse metric and the *first* derivatives of the metric, formula (6.76) shows that the Riemann curvature tensor is determined by *first* and *second* derivatives of the metric. The equivalence principle tells us that in Gaussian coordinates the first derivatives of the metric vanish at the point x but it does not say that every curvature vanishes! In terms of physics this means that the motion of a test particle along a geodesic looks locally like free motion in flat space, but neighbouring geodesics, in general, are not parallel. They may attract or repel each other.

We close this subsection by counting the number of independent entries of the (1, 3)-tensor $R^\tau_{\rho\mu\nu}$ by making use of its symmetry properties. One way of doing this is to express the formula (6.76) obtained above, in the form $g_{\lambda\tau}R^\tau_{\rho\mu\nu} \equiv R_{\lambda\rho\mu\nu}$, in terms of first and second derivatives of the metric only, inserting the coordinate formula (6.66). This yields an expression for $R_{\lambda\rho\mu\nu}$ whose symmetry properties can be read off. (This is a useful exercise!) Here, as an alternative, we utilize the symmetry properties in the form of (6.73a)–(6.73e).

Starting from the coordinate formula (6.74) one calculates expressions such as, for example, $\mathsf{R}_{xy}(z)$ and expands them in terms of the components of the three tangent vectors. For instance, one has

$$\left(\mathsf{R}_{xy}z\middle|w\right) = R^\sigma_{\rho\mu\nu}x^\mu y^\nu z^\rho w_\sigma = R_{\sigma\rho\mu\nu}x^\mu y^\nu z^\rho w^\sigma \,,$$

where the first index of R was lowered by means of the metric tensor. The first relation (6.73a) shows that the (0, 4)-tensor $R_{\sigma\rho\mu\nu}$ is antisymmetric under the exchange $\mu \leftrightarrow \nu$. The relation (6.73d) tells us, in turn, that nothing changes if the pair (μ, ν) is interchanged with the pair (σ, ρ). Therefore, $R_{\sigma\rho\mu\nu}$ is also antisymmetric under the exchange $\sigma \leftrightarrow \rho$. This is the relation (6.73b). These antisymmetries are usually denoted by square brackets,

$$R_{[\sigma\rho][\mu\nu]} \, .$$

The first Bianchi identity (6.73c), when expressed in coordinates, yields

$$\begin{aligned} &R^{\tau}_{\ \rho\mu\nu} \left(z^{\rho} x^{\mu} y^{\nu} + x^{\rho} y^{\mu} z^{\nu} + y^{\rho} z^{\mu} x^{\nu}\right) \\ &\qquad = z^{\rho} x^{\mu} y^{\nu} \left(R^{\tau}_{\ \rho\mu\nu} + R^{\tau}_{\ \mu\nu\rho} + R^{\tau}_{\ \nu\rho\mu}\right) = 0 \, . \end{aligned}$$

As this holds for all tangent vectors, or vector fields, the sum over the cyclic permutations of the indices (ρ, μ, ν) of $R^{\tau}_{\ \rho\mu\nu}$ and likewise of $R_{\tau\rho\mu\nu}$ is equal to zero,

$$\begin{aligned} &R^{\tau}_{\ \rho\mu\nu} + R^{\tau}_{\ \mu\nu\rho} + R^{\tau}_{\ \nu\rho\mu} = 0 \quad \text{or} \\ &R_{\tau\rho\mu\nu} + R_{\tau\mu\nu\rho} + R_{\tau\nu\rho\mu} = 0 \, . \end{aligned} \tag{6.77}$$

Now, one counts components as follows: In dimension n antisymmetry $[\mu, \nu]$ in the indices μ and ν both of which are in the set $(0, 1, \ldots, n-1)$, implies that there are

$$\binom{n}{2} \quad \text{independent components.}$$

The same conclusion follows from the antisymmetry $[\tau, \rho]$. Thus, a tensor $R_{[\sigma\rho][\mu\nu]}$ with four indices and antisymmetries as indicated initially has $\binom{n}{2}^2$ independent components. However, one still must subtract the conditions (6.77) from this number. For fixed τ there are $\binom{n}{3}$ possible choices for the triple (ρ, μ, ν). In total, i.e. for $\tau = 0, \ldots, n-1$, equation (6.77) contains

$$n \binom{n}{3}$$

conditions. Subtracting the number of constraints from the number of independent components one obtains the number N_{R} of independent entries of the Riemann curvature tensor, viz.

$$N_{\mathrm{R}} = \binom{n}{2}^2 - n\binom{n}{3} = \frac{1}{12} n^2 (n^2 - 1) \, . \tag{6.78}$$

This is an interesting result:

- In dimension $n = 1$ one has $\mathrm{R} = 0$, there is no Riemann curvature tensor. This might seem surprising at first. To a curve, for example, in the plane $\mathbb{R}^2$ one does ascribe a curvature. However, for the Riemann curvature tensor one finds

$R_{1111} = 0$ because it does not refer to an embedding space (such as, for example, $\mathbb{R}^2$) but describes an intrinsic property of the one-dimensional manifold M. As long as there is no embedding space, or as long as one does not know about an embedding space, one does not perceive any curvature.

- In dimension $n = 2$ a $(0,4)$-tensor without symmetries and without constraints would have 16 components. The Riemann tensor, from (6.78), only has one single component R_{1212}.
- In dimension $n = 3$, that is, with two space dimensions and one time axis, an arbitrary $(0,4)$-tensor has 81 components, the Riemann tensor has only $9 \cdot 8/12 = 6$ components.
- In the case of a Lorentz manifold (s. Sect. 6.5.1) an unrestricted tensor of type $(0,4)$ has already 256 components while the Riemann curvature tensor has $16 \cdot 15/12 = 20$ components.

The Riemann curvature tensor "blossoms out" with increasing dimension and one will not be surprised to learn that general relativity in $3 = 2 + 1$ dimensions and in $4 = 3 + 1$ dimensions exhibit rather different properties.

6.6 The Einstein Equations

After this long journey through semi-Riemannian geometry for models of spacetimes and with our knowledge of the energy-momentum tensor field for matter and nongravitational fields from Sect. 6.3 we are well prepared to render more concrete the conjectures made in Sect. 6.2.4 and to cast the program formulated there into a theory of gravitation. As announced there we choose the source of gravitation to be the tensor field which describes the energy and momentum content of matter and of nongravitational fields. This tensor field as constructed over flat space in Sect. 6.3, is of type $(2,0)$, or, by metric equivalence, of type $(0,2)$. It is symmetric and fulfills a condition on its divergence. The geometric tools of the theory must be tuned to these properties.

6.6.1 Energy-Momentum Tensor Field in Curved Spacetime

The tensor field T for energy and momentum is taken over from the expressions derived in Sect. 6.3, with the following modifications. The flat metric $\eta_{\mu\nu}$ must be replaced by the metric tensor $g_{\mu\nu}(x)$ everywhere, all ordinary partial derivatives must be replaced by covariant derivatives. The tensor field $\mathsf{T} \in \mathfrak{T}^2_0(M)$ is still symmetric, i.e. its coefficients fulfill $T^{\mu\nu} = T^{\nu\mu}$. The condition that it be divergence-free, in components, has the form

$$T^{\mu\nu}(x)_{;\mu} = 0 = T^{\mu\nu}(x)_{;\nu} \,. \tag{6.79a}$$

If one prefers a coordinate-free notation one may write this condition by means of contractions (see Sect. 6.4.3) as follows. Take first $D(\mathsf{T}) \in \mathfrak{T}_1^2$ (note that this is of type $(2, 1)$ as it is D, not D_V that acts!), then contract with C_1^1 or C_1^2 (this is the same because of the symmetry of the tensor),

$$\mathbf{div}\,\mathsf{T} = C_1^1\,(D\mathsf{T}) = C_1^2\,(D\mathsf{T})\,. \tag{6.79b}$$

It is easy to see that the result is a vector field, $\mathbf{div}\,\mathsf{T} \in \mathfrak{X}(M)$. The tensor field T has the components $T^{\mu\nu}$, $D\mathsf{T}$ has components $T^{\mu\nu}{}_{;\rho}$, hence, is of type $(2, 1)$. After contraction of one of the upper indices with the lower index this becomes a tensor field of type $(1, 0)$, i.e. a vector field.

Analogous formulae apply to tensor fields B of type $(0, 2)$. Here one has $D\mathsf{B} \in \mathfrak{T}_3^0$, and $\mathbf{div}\,\mathsf{B} = C_{13}(D\mathsf{B})$ is of type $(0, 1)$, i.e. a one-form whose components read

$$(\mathbf{div}\,\mathsf{B})_\lambda = g^{\mu\nu}\,B_{\mu\lambda\,;\nu} = B^\nu{}_{\lambda;\nu}\,. \tag{6.79c}$$

The metric $\mathsf{g} \in \mathfrak{T}_2^0$ provides an example. Its covariant derivative $D_V(\mathsf{g})$ vanishes for all V,

$$(\mathbf{div}\,\mathsf{g})_\lambda = g^{\mu\nu}\,g_{\mu\lambda\,;\nu} = 0\,.$$

As another example one may study Maxwell's equations on a curved spacetime, see for example [Straumann 2009].

6.6.2 Ricci Tensor, Scalar Curvature, and Einstein Tensor

By contraction of the upper index of the Riemann curvature tensor $\mathsf{R} \in \mathfrak{T}_3^1(M)$ with the second lower index one obtains a symmetric tensor field $\mathsf{R}^{(\mathrm{Ricci})} \in \mathfrak{T}_2^0(M)$, called the

Definition 6.14 Ricci tensor field

$$\mathsf{R}^{(\mathrm{Ricci})} := C_2^1\,(\mathsf{R})\,, \tag{6.80a}$$

whose components are given by

$$R^{(\mathrm{Ricci})}_{\mu\nu} = R^\lambda_{\mu\lambda\nu}\,. \tag{6.80b}$$

Expressed in local coordinates it reads

$$R^{(\mathrm{Ricci})}_{\mu\nu} = \partial_\sigma \Gamma^\sigma_{\mu\nu} - \partial_\nu \Gamma^\sigma_{\sigma\mu} + \Gamma^\tau_{\nu\mu}\Gamma^\sigma_{\sigma\tau} - \Gamma^\tau_{\sigma\mu}\Gamma^\sigma_{\nu\tau}\,. \tag{6.80c}$$

(Note that we used the symmetry $\Gamma^\sigma_{\nu\mu} = \Gamma^\sigma_{\mu\nu}$.) The chart representation (6.80c) follows from the coordinate representation (6.76) of the Riemann curvature tensor.

By contracting the Ricci tensor field once more one obtains a function called the

Definition 6.15 Curvature scalar

$$S := C\left(\mathsf{R}^{(\mathrm{Ricci})}\right) \in \mathfrak{F}(M)\,, \tag{6.81a}$$

which can be expressed in terms of components of the metric and of the Ricci tensor field, viz.

$$S = g^{\mu\nu}(x) R^{(\mathrm{Ricci})}_{\mu\nu}(x)\,. \tag{6.81b}$$

Remarks

1. The Ricci tensor is symmetric. This is seen by means of the following calculation
$$R^{(\mathrm{Ricci})}_{\mu\nu} = g^{\sigma\tau} R_{\tau\mu\sigma\nu} = g^{\sigma\tau} R_{\sigma\nu\tau\mu} = R^{(\mathrm{Ricci})}_{\nu\mu}\,.$$
Here we have interchanged the pairs (τ, μ) and (σ, ν), the property (6.73d) guaranteeing the invariance of the Riemann tensor.
2. There is only one independent contraction of the upper index with one of the lower indices of the curvature tensor. Any other contraction than the one chosen in Definition 6.14 yields the same tensor field, modulo a sign.

We proceed to prove a simple but important relation between the total derivative of the curvature scalar and the divergence of the Ricci tensor. Evaluating the second Bianchi identity (6.73e) with base vector fields $Z = \partial_\rho$, $X = \partial_\mu$, $Y = \partial_\nu$, one has

$$R^{\sigma}_{\tau\mu\nu\,;\rho} + R^{\sigma}_{\tau\nu\rho\,;\mu} + R^{\sigma}_{\tau\rho\mu\,;\nu} = 0\,.$$

Now, choose $\sigma = \rho$, sum over this index from 0 to $n-1$,

$$R^{\sigma}_{\tau\mu\nu\,;\sigma} + R^{\sigma}_{\tau\nu\sigma\,;\mu} + R^{\sigma}_{\tau\sigma\mu\,;\nu} = 0\,,$$

and contract with $g^{\tau\mu}$,

$$g^{\tau\mu}\left\{R^{\sigma}_{\tau\mu\nu\,;\sigma} + R^{\sigma}_{\tau\nu\sigma\,;\mu} + R^{\sigma}_{\tau\sigma\mu\,;\nu}\right\} = 0\,. \tag{6.82}$$

The first of the three terms is a contraction of the Riemann tensor,

$$\begin{aligned} g^{\tau\mu} R^{\sigma}_{\tau\mu\nu} &= g^{\sigma\lambda} g^{\tau\mu} R_{\lambda\tau\mu\nu} = -g^{\sigma\lambda} g^{\tau\mu} R_{\tau\lambda\mu\nu} \\ &= -g^{\sigma\lambda} (R^{(\mathrm{Ricci})})_{\lambda\nu} = -(R^{(\mathrm{Ricci})})^{\sigma}_{\nu}\,. \end{aligned}$$

Applying the covariant derivative by σ the first term yields

$$-(R^{(\mathrm{Ricci})})^{\sigma}_{\nu\,;\sigma} = -\left(\mathbf{div}\,\mathsf{R}^{(\mathrm{Ricci})}\right)_{\nu}\,.$$

The second term gives the same result

$$g^{\tau\mu} R^{\sigma}_{\tau\nu\sigma\,;\mu} = -\left(\mathbf{div}\,\mathsf{R}^{(\mathrm{Ricci})}\right)_{\nu} ,$$

while the third term

$$g^{\tau\mu} R^{\sigma}_{\tau\sigma\mu\,;\nu} = S_{;\nu} ,$$

yields the covariant derivative of the curvature scalar. As S is a function, the covariant derivative is the same as the ordinary partial derivative, $S_{;\nu} = S_{,\nu}$. This calculation gives the proof of two important results. Multiplying the above relation between the partial derivatives of S and the divergence of the Ricci tensor with $\mathrm{d}x^{\nu}$, it tells us that the exterior derivative of the curvature scalar is equal to twice the divergence of the Ricci tensor:

$$\mathrm{d}S = 2\,\mathbf{div}\,\mathsf{R}^{(\mathrm{Ricci})} . \tag{6.83}$$

An alternative, but equivalent result is that the Einstein tensor field, defined by

Definition 6.16 Einstein tensor field

$$\mathsf{G} := \mathsf{R}^{(\mathrm{Ricci})} - \frac{1}{2}\mathbf{g}S , \tag{6.84}$$

is symmetric and has divergence zero. Indeed, one has

$$(\mathbf{div}(\mathbf{g}S))_{\mu} = g^{\sigma\tau}\left(g_{\sigma\mu}S\right)_{;\tau} = g^{\sigma\tau} g_{\sigma\mu} S_{;\tau} .$$

In the last step of these equations we have made use of the fact that the connection is a metric connection, i.e. that $g_{\sigma\mu\,;\tau} = 0$.

The Einstein tensor field (6.84), by construction, has the same properties as the energy-momentum tensor field of matter and nongravitational fields: It is symmetric and its divergence is zero. Conversely, the Einstein tensor field G contains the same geometric information as the Ricci tensor field. This suggests one should try to relate these two tensor fields, the Einstein tensor field and the energy-momentum tensor field.

6.6.3 The Basic Equations

The basic equations to be constructed should relate, on the one hand, the Einstein tensor field describing the geometry of spacetime, and, on the other hand, the energy-momentum tensor field serving as a source term for the former. Clearly, in the limit of an almost flat metric and in nearly static situations the sought-after equations must contain the Newtonian mechanics, for instance in the form of Poisson's equation,

$$\Delta \Phi_{\rm N} = 4\pi G \varrho\,, \tag{6.85}$$

where $\Phi_{\rm N}$ is the Newtonian potential, ϱ is the mass density of matter, and G denotes Newton's constant. Let us assume that the metric tensor deviates but little from the flat Minkowski metric,

$$g_{\mu\nu}(x) = \eta_{\mu\nu} + h_{\mu\nu}(x) \quad \text{with} \quad \left|h_{\mu\nu}\right| \ll 1\,, \tag{6.86}$$

and let us assume, for the sake of simplicity, that the metric is static.

The motion of a massive particle is described by the geodesic equation (6.68a), with $\mu = 0, 1, 2, 3$. For a particle that moves slowly one has (setting the speed of light equal to 1, $c = 1$)

$$\frac{\mathrm{d}x^0}{\mathrm{d}\tau} \simeq 1 \quad \text{and} \quad \frac{\mathrm{d}x^i}{\mathrm{d}\tau} \ll \frac{\mathrm{d}x^0}{\mathrm{d}\tau}\,, \qquad i = 1, 2, 3\,.$$

Thus, the first derivatives $\mathrm{d}x^i/\mathrm{d}\tau$ can be neglected as compared to $\mathrm{d}x^0/\mathrm{d}\tau$. The second derivatives, a priori, are not negligible and can be calculated approximately from the geodesic equation (6.68a). One has (with $c = 1$)

$$\frac{\mathrm{d}^2x^i}{\mathrm{d}t^2} \simeq \frac{\mathrm{d}^2x^i}{\mathrm{d}\tau^2} = -\Gamma^i_{\mu\nu}\frac{\mathrm{d}x^\mu}{\mathrm{d}\tau}\frac{\mathrm{d}x^\nu}{\mathrm{d}\tau} \simeq -\Gamma^i_{00}\,.$$

The only Christoffel symbol that enters here is calculated in the approximation (6.86) to first order in the variables $h_{\mu\nu}$ and with $\eta_{\mu\nu} = \mathrm{diag}(1, -1, -1, -1)$ as follows:

$$\begin{aligned}\Gamma^i_{00} &= \frac{1}{2}g^{i\rho}\left\{2\partial_0 g_{0\rho} - \partial_\rho g_{00}\right\}\\ &\simeq \frac{1}{2}\eta^{i\rho}\left\{2\partial_0 h_{0\rho}(x) - \partial_\rho h_{00}\right\}\\ &= \frac{1}{2}\partial_i h_{00}(x) - \partial_0 h_{0i}(x)\,.\end{aligned}$$

The second term on the right-hand side is zero because we started from a static situation. Therefore, the acceleration is seen to be

$$\frac{\mathrm{d}^2\boldsymbol{x}^i}{\mathrm{d}t^2} \simeq -\Gamma^i_{00} = -\frac{1}{2}\nabla_i h_{00} \mathrel{\widehat{=}} -\nabla_i \Phi_{\rm N}\,.$$

This establishes the relation between the Newtonian potential $\Phi_{\rm N}$ of mechanics in flat spacetime and the metric tensor in the Newtonian limit, viz.

$$g_{00}(\boldsymbol{x}) \simeq 1 + 2\Phi_{\rm N} \equiv 1 + \frac{2}{c^2}\Phi_{\rm N}\,. \tag{6.87a}$$

In the same static limit of weak fields the Ricci tensor (6.80c) is given approximately by (still with $c = 1$)

$$R_{00}^{(\text{Ricci})} \simeq \sum_{i=1}^{3} \partial_i \Gamma_{00}^{i} = \frac{1}{2}\mathbf{\Delta} h_{00} = \frac{1}{2}\mathbf{\Delta} g_{00} = \mathbf{\Delta}\Phi_{\text{N}} .$$

All other components of $\mathsf{R}^{(\text{Ricci})}$ are negligibly small. For $\mathbf{\Delta}\Phi_{\text{N}}$, finally, one may insert the Poisson equation (6.85).

In the case of weak, static fields, and for matter in a nonrelativistic state of motion, the energy-momentum tensor simplifies such that only T_{00} is appreciably different from zero. Using the results of Sect. 6.3 (taking $c = 1$ also there) one has

$$T_{00} \simeq \varrho \,, \quad |T_{\mu\nu}| \ll T_{00} \quad \text{for } (\mu, \nu) \neq (0,0) .$$

Therefore, the Ricci tensor and the energy-momentum tensor must be related by

$$R_{00}^{(\text{Ricci})} \simeq \mathbf{\Delta}\Phi_{\text{N}} \simeq 4\pi G T_{00} . \tag{6.87b}$$

The essential postulate on which general relativity rests is the following: One takes the Einstein tensor field (6.84) to be proportional to the energy-momentum tensor field T,

$$\mathsf{G} = \alpha \mathsf{T} , \tag{6.88a}$$

and makes use of the estimates of the Newtonian limit to determine the constant α. Both tensor fields are symmetric and have divergence zero,

$$\mathbf{div}\, \mathsf{G} = 0 \,, \quad \mathbf{div}\, \mathsf{T} = 0 \,,$$

or, expressed in local coordinates as in (6.79c),

$$G^{\nu}{}_{\lambda;\nu} = R^{(\text{Ricci})\nu}{}_{\lambda;\nu} - \frac{1}{2} S_{;\lambda} = 0 \,, \quad T^{\nu}{}_{\lambda;\nu} = 0 .$$

The postulated equation (6.88a) is generally covariant. Calculating the contraction of both sides of the ansatz (6.88a), noting that the contraction of the metric equals 4,

$$C\,(\mathsf{g})\ (\text{locally } = g^{\mu\nu}(x) g_{\mu\nu}(x)) = 4 \,,$$

one obtains

$$C\,(\mathsf{G}) = C\left(\mathsf{R}^{(\text{Ricci})}\right) - \frac{1}{2} S\, C(\mathsf{g}) = S - 2S = \alpha C\,(\mathsf{T}) .$$

As the curvature scalar is $S = -\alpha C(\mathsf{T}) \equiv -\alpha\, \text{tr}(\mathsf{T})$, the Ricci tensor is seen to be, using the notation $g^{\mu\nu} T_{\mu\nu} = \text{tr}(\mathsf{T})$,

$$R_{\mu\nu}^{(\text{Ricci})} = \alpha \left(T_{\mu\nu} - \frac{1}{2}\, \text{tr}(\mathsf{T}) g_{\mu\nu} \right) . \tag{6.88b}$$

In the limit discussed above only the 00-component of the equations (6.88b) differs from zero in an essential way and one has

$$T_{00} - \frac{1}{2} g_{00}\,(\mathrm{tr}(\mathsf{T})) \simeq \frac{1}{2} T_{00}\,. \tag{6.88c}$$

Combining equations (6.88c), (6.88b), and (6.87b), one concludes that the unknown α must be given by $\alpha = 8\pi G$. The postulate (6.88a) and the above arguments then lead to

Einstein's equations
in coordinate-free form

$$\mathsf{G} \equiv \mathsf{R}^{(\mathrm{Ricci})} - \frac{1}{2}\,\mathsf{g}\,S = 8\pi G\,\mathsf{T}\,, \tag{6.89}$$

or, expressed in local coordinates,

$$G_{\mu\nu} \equiv R^{(\mathrm{Ricci})}_{\mu\nu} - \frac{1}{2} g_{\mu\nu} S = 8\pi G T_{\mu\nu}\,. \tag{6.90}$$

► **Remarks**

1. Reviewing once more the formulae (6.66) for the Christoffel symbols and (6.76) for the Riemann tensor, as well as (6.80c) for the Ricci tensor in local coordinates, one sees that Einstein's equations (6.90) depend on the metric and its first and second derivatives. If, indeed, the metric is the physically relevant field then this observation fits well with the familiar framework of physics according to which equations of motion usually are differential equations of second order.
2. As we emphasized previously the energy-momentum tensor field which acts as a source term in Einstein's equations, contains not only the ordinary matter of the universe but also all nongravitational fields. The contribution of a Maxwell field to energy and momentum is part of it, very much like the contribution of massive objects moving in space and time.
3. If one asks whether Einstein's equations are unique one finds a remarkable answer. With an ansatz of the form $\mathsf{A} = \alpha' \mathsf{T}$, where $\mathsf{A} \in \mathfrak{T}^0_2(M)$ is supposed to depend on g and the first and second derivatives thereof, and, of course, to be divergenceless, one can show that A must have the form

$$\mathsf{A} = \mathsf{G} + \Lambda \mathsf{g}\,, \quad \Lambda \text{ a real constant}, \tag{6.91}$$

where G is the Einstein tensor field (6.84). The equations (6.89) are slightly modified,

$$\mathsf{G} + \Lambda \mathsf{g} = \alpha' \mathsf{T} . \tag{6.92}$$

The constant Λ which possibly enters here, is called the *cosmological constant*. Taking the constant of proportionality α' to be the same as before (the relation $\alpha' = 8\pi G$ applies to the case $\Lambda = 0$) and analyzing the equations (6.92) in the same static Newtonian limit as before, the Newtonian potential is found to satisfy

$$\mathbf{\Delta} \Phi_{\mathrm{N}} = 4\pi G \varrho + \frac{1}{2} \Lambda . \tag{6.93a}$$

A cosmological constant Λ which is not equal to zero, corresponds to a static homogeneous mass density in the universe

$$\varrho_{\mathrm{eff}} = \frac{\Lambda}{4\pi G} . \tag{6.93b}$$

Such a "background density" should manifest itself in experiments and the reader will not be surprised to realize that the cosmological constant continues to be a subject of present-day research. As a remark within a remark let us note that the flat metric $g_{\mu\nu} = \eta_{\mu\nu}$ is a solution of Einstein's equations only if $\Lambda = 0$.

4. One might ask the question whether by using the Ricci tensor field, i.e. a contraction of the Riemann curvature field, one has given away some aspect of the geometry of spacetime and whether there are stronger equations than (6.89) relating geometry and matter. To this question I can offer only an incomplete and somewhat indirect answer by the following remark.
5. The theory of gravitation as described by Einstein's equations (6.89), (6.90), fits into the framework of a classical field theory. We assume that genuine (massive) matter and the Maxwell field (and, possibly, other radiation fields) are described by a Lagrange density

$$\mathcal{L}(\psi, D\psi, \mathsf{g})$$

where ψ stands symbolically for all nongravitational fields, D denotes the covariant derivative constructed from the Levi-Civita connection, and g is the metric. For the Maxwell field, for example, we have

$$\mathcal{L}_{\mathrm{M}} = -\frac{1}{16\pi} F_{\mu\nu} F_{\sigma\tau} g^{\mu\sigma} g^{\nu\tau} .$$

Regarding gravitation, the curvature scalar seems a natural candidate for constructing a Lagrange density from which to derive equations of motion for the metric. With these ideas in mind one writes down the action

$$\iiiint_U \omega \left\{ -\frac{1}{16\pi G} S + \mathcal{L}(\psi, D\psi, g) \right\} , \qquad (6.94)$$

where U is a subset of M while ω denotes the volume element on spacetime M. In the spirit of Hamilton's principle discussed in Sect. 3.3, one varies the fields ψ and the metric g. As one can show, the variation of the action yields both the equations of motion of the fields ψ on curved spacetime and Einstein's equations for gravity, cf., for example, [Straumann 2009]. The integral (6.94) is called the *Hilbert action.*

6. A direct comparison of Maxwell theory and general relativity reveals the following interesting analogy: Gauge invariance of Maxwell theory in interaction with charged matter is guaranteed only if the electromagnetic current density is conserved, see Sect. 3.5.2. One shows that invariance of the Hilbert action under general diffeomorphisms $\phi \in \text{Diff}(M)$ holds only if the energy-momentum tensor field is covariantly conserved, i.e. if locally it fulfills the condition $T_\mu{}^\nu{}_{;\nu} = 0$. Very much like in electrodynamics where both the Maxwell fields A_μ and the particle fields ψ change under gauge transformations, a diffeomorphism

$$\{g, \psi\} \mapsto \{{}^\phi g = \phi^*(g), {}^\phi\psi = \phi^*(\psi)\} \qquad (6.95)$$

acts on the metric as well as on all other fields.
In general relativity the group of diffeomorphisms Diff (M) *on spacetime* M *takes over the role of the gauge group of electrodynamics.*

7. Equation (6.88b) with $\alpha = 8\pi G$ leads to the following observation: If there is no matter and no nongravitational field at all then $T \equiv 0$. Thus, in a vacuum Einstein's equations take the simple form

$$R_{\mu\nu}^{(\text{Ricci})} = 0 . \qquad (6.96)$$

These are the equations that have to be solved if, for instance, one sets out to determine the gravitational field created by a point-like mass distribution.

6.7 Gravitational Field of a Spherically Symmetric Mass Distribution

As a most important example we discuss a static solution of Einstein's equations (6.96) in a vacuum which describes the gravitational field outside a spherically symmetric mass distribution[8]. Spherical symmetry means that the metric is invari-

[8] An instructive, nontechnical discussion can be found in [Rindler 1977]. In section 7.6 of this reference the general form of the metric for static fields is determined.

ant under $(t, \boldsymbol{x}) \mapsto (t, \mathrm{R}\boldsymbol{x})$ with $\mathrm{R} \in \mathrm{SO}(3)$. The manifold which is to describe the spacetime must have the structure

$$\mathbb{R}_t \times \mathbb{R}^+ \times S^2 , \tag{6.97}$$

with S^2 the unit sphere in $\mathbb{R}^3$ with its standard metric

$$\mathrm{d}\Omega^2 := \mathrm{d}\theta \otimes \mathrm{d}\theta + \sin^2\theta \, \mathrm{d}\phi \otimes \mathrm{d}\phi . \tag{6.98}$$

The axis $\mathbb{R}_t$ describes the space of the *Schwarzschild time*, while $\mathbb{R}^+$ describes the space of the *Schwarzschild radius.*

6.7.1 The Schwarzschild Metric

A basis of one-forms is chosen as follows:

$$\omega^0 = e^{a(r)} \, \mathrm{d}t , \; \omega^1 = e^{b(r)} \, \mathrm{d}r , \; \omega^2 = r \, \mathrm{d}\theta , \; \omega^3 = r \sin\theta \, \mathrm{d}\phi , \tag{6.99}$$

with $a(r)$ and $b(r)$ two functions of the Schwarzschild radius r which must be determined from (6.96). (They cannot depend on t because the solution is assumed to be static. Also, a dependence on θ and ϕ would destroy the spherical symmetry!)

When expressed by these one-forms the metric takes the form

$$\mathbf{g} = \eta_{\mu\nu} \, \omega^\mu \otimes \omega^\nu , \quad \left(\eta_{\mu\nu} = \mathrm{diag}(1, -1, -1, -1)\right) . \tag{6.100}$$

A remarkable property of the basis (6.99) is that it defines a repère mobile, i.e. a comoving frame of reference which is orthonormal and is optimally adapted to the Schwarzschild spacetime. The rest of the calculation, though somewhat tedious in its details, is straightforward, see, i.e. [Rindler 1977], or [Straumann 2009]. One calculates the components of the Riemann curvature tensor in this basis, then, from this, the components of the Ricci tensor, and, eventually, the curvature scalar. From the Ricci tensor and the curvature scalar follows the Einstein tensor (6.84). The result is

$$\begin{aligned} G_{00} &= \mathrm{e}^{-2b}\left(\frac{2b'}{r} - \frac{1}{r^2}\right) + \frac{1}{r^2} , \\ G_{11} &= \mathrm{e}^{-2b}\left(\frac{2a'}{r} + \frac{1}{r^2}\right) - \frac{1}{r^2} , \\ G_{22} = G_{33} &= \mathrm{e}^{-2b}\left(a'^2 - a'b' + a'' + \frac{a' - b'}{r}\right) . \end{aligned} \tag{6.101}$$

All components not listed here are equal to zero.

Einstein's equations (6.90) in vacuum yield differential equations for the functions $a(r)$ and $b(r)$. For example, the equation $G_{00} + G_{11} = 0$ yields $a'(r) + b'(r) = 0$ which is immediately integrated to

$$b(r) = -a(r) . \tag{6.102a}$$

There is no integration constant to be added because the metric, for $r \to \infty$, must tend to the Minkowski metric η, so that

$$\lim_{r\to\infty} a(r) = \lim_{r\to\infty} b(r) = 0 \tag{6.102b}$$

must hold. The equation $G_{00} = 0$, taken in isolation, allows one to determine $b(r)$. One has

$$G_{00} = 0 = \mathrm{e}^{-2b}\left(\frac{2b'}{r} - \frac{1}{r^2}\right) + \frac{1}{r^2} , \tag{6.102c}$$

or, multiplying by r^2,

$$1 = \left(1 - 2b'r\right)\mathrm{e}^{-2b} = \frac{\mathrm{d}}{\mathrm{d}r}\left(r\,\mathrm{e}^{-2b}\right) . \tag{6.102d}$$

From this one concludes $r\,\mathrm{e}^{-2b} = r - 2m$, with $2m$ an integration constant which so far remains undetermined. Thus, the result is

$$\mathrm{e}^{-2b(r)} = \mathrm{e}^{+2a(r)} = 1 - \frac{2m}{r} , \tag{6.102e}$$

and the metric (6.100) becomes the

Schwarzschild metric

$$\mathbf{g} = \left(1 - \frac{2m}{r}\right)\mathrm{d}t \otimes \mathrm{d}t - \frac{\mathrm{d}r \otimes \mathrm{d}r}{1 - 2m/r} - r^2\left(\mathrm{d}\theta \otimes \mathrm{d}\theta + \sin^2\theta\,\mathrm{d}\phi \otimes \mathrm{d}\phi\right) . \tag{6.103}$$

▶ **Remarks**

1. In the Newtonian limit the constant of integration $2m$ can be expressed in terms of the total mass M of the localized, spherically symmetric mass distribution. Re-inserting for a moment the velocity of light one has

$$g_{00} \simeq 1 + \frac{2}{c^2}\Phi_{\mathrm{N}} = 1 - \frac{2}{c^2}\frac{GM}{r} .$$

Thus, the constant $2m$ which has the dimension of a length, is found to be

$$r_S := 2m = 2\frac{GM}{c^2} . \tag{6.104}$$

This radius r_S is called the *Schwarzschild radius.*

2. The example of our sun illustrates the relative order of magnitude of the Schwarzschild radius in comparison with the matter radius of the sun. With $M_{\odot} = 1.989 \cdot 10^{30}$ kg and with G from (6.1) one obtains

$$r_S^{(\odot)} = 2.952 \text{ km} \simeq 3 \text{ km} ,$$

i.e. a number which is very small when compared to $R_{\odot} \simeq 7 \cdot 10^5$ km. In the case of the earth whose mass is $M_{\oplus} = 5.9742 \cdot 10^{24}$ kg one finds a Schwarzschild radius of $r_S^{(\oplus)} = 0.887$ cm!
3. In the metric (6.103) only the (t, r)-submanifold deviates from the flat metric, the component S^2 remains unaffected. Therefore, the specific properties of this spacetime will become most transparent by studying radial geodesics.
4. In the Schwarzschild spacetime the radius $r = r_S$ which is approached from the outside plays a special role. The coordinate representation (6.103) of the metric is no longer valid even though examination of the Einstein tensor (6.101) (and likewise of the entries of the curvature tensor) shows that there is no singularity at $r = r_S$. These geometric quantities remain perfectly finite. The correct mathematical interpretation shows that the Schwarzschild radius is no more than a coordinate singularity. However, the interpretation from the point of view of physics shows that the radius $r_S = 2m$ plays an important role as an *event horizon.* One may approach the Schwarzschild radius from exterior space but the local representation (6.103) is no longer valid when one continues r to values which are smaller than r_S.

6.7.2 Two Observable Effects

The solution (6.103) describes the gravitational field outside of a localized, spherically symmetric mass distribution. If the radius R of this distribution is larger than the Schwarzschild radius r_S we have a model for the field created by our sun. If, however, the radius is smaller than r_S then the mass distribution is no longer visible and its boundary cannot be reached by any physical means. This second case provides a model for what is called a *black hole.* In the first case there appear general relativistic effects already in the mildly curved spacetime of our solar system. These are the subject of this section.

Perihelion Precession of a Planet

We study the geodesic equation (6.68a) of a massive test body in the field of a localized spherically symmetric mass distribution, i.e. in the outer space defined by the

Schwarzschild metric (6.103). Using orthogonal coordinates

$$y^0 = t\,, \quad y^1 = r\,, \quad y^2 = \theta\,, \quad y^3 = \phi \tag{6.105a}$$

the geodesic equation reads

$$\ddot{y}^\mu + \Gamma^\mu_{\nu\rho}\dot{y}^\nu\dot{y}^\rho = 0\,.$$

One multiplies this equation by $g_{\lambda\mu}$, takes the sum over μ, but keeps the index λ fixed. Using formula (6.66) for the Christoffel symbols and with

$$\mathbf{g} = \operatorname{diag}\left((1-2m/r), -(1-2m/r)^{-1}, -r^2, -r^2\sin^2\theta\right) \tag{6.105b}$$

one obtains the differential equation

$$g_{\lambda\lambda}\ddot{y}^\lambda + \frac{1}{2}\left[\sum_\mu \partial_\mu g_{\lambda\lambda}\dot{y}^\mu\dot{y}^\lambda + \sum_\nu \partial_\nu g_{\lambda\lambda}\dot{y}^\lambda\dot{y}^\nu\right]$$
$$= \frac{1}{2}\sum_\rho \left(\partial_\lambda g_{\rho\rho}\right)\dot{y}^\rho\dot{y}^\rho\,, \quad (\lambda \text{ fixed!})\,. \tag{6.106a}$$

The dot stands for the derivative by the arc length, or eigentime, s. The two terms in square brackets are equal and may be combined using the chain rule for differentiation by this variable,

$$\frac{1}{2}\left[\cdots\right] = \left(\frac{\mathrm{d}}{\mathrm{d}s}g_{\lambda\lambda}\right)\dot{y}^\lambda\,,$$

so that the left-hand side of (6.106a) is seen to be the derivative of $g_{\lambda\lambda}\,\mathrm{d}y^\lambda/\mathrm{d}s$ by s. Thus, equation (6.106a) becomes

$$\frac{\mathrm{d}}{\mathrm{d}s}\left(g_{\lambda\lambda}\frac{\mathrm{d}y^\lambda}{\mathrm{d}s}\right) = \frac{1}{2}\sum_{\rho=0}^{3}\left(\frac{\partial g_{\rho\rho}}{\partial y^\lambda}\right)\left(\frac{\mathrm{d}y^\rho}{\mathrm{d}s}\right)^2\,, \quad \lambda = 0,1,2,3\,. \tag{6.106b}$$

This equation describes orbits of free motion of a test body (e. g. a light comet) in the field of the given mass distribution (the sun) – assuming, of course, that its own mass does not disturb this field appreciably.

Does this analysis yield the well-known Kepler motion, or something closely related to it?

For $\lambda = 0$, where $y^0 = t$, and for $\lambda = 3$, where $y^3 = \phi$, equation (6.106b) yields two conservation laws because its right-hand side vanishes (none of the components of the metric tensor depends on t or on ϕ!),

$$\frac{\mathrm{d}}{\mathrm{d}s}\left(g_{00}\frac{\mathrm{d}t}{\mathrm{d}s}\right) = 0 \Longrightarrow \left(1-\frac{2m}{r}\right)\dot{t} \equiv E = \text{const.} \tag{6.107a}$$

$$\frac{\mathrm{d}}{\mathrm{d}s}\left(g_{33}\frac{\mathrm{d}\phi}{\mathrm{d}s}\right) = 0 \Longrightarrow r^2\sin^2\theta\,\dot{\phi} \equiv \sin^2\theta L = \text{const.} \tag{6.107b}$$

Equation (6.106b) with $\lambda = 2$ yields the differential equation

$$\frac{\mathrm{d}}{\mathrm{d}s}\left(r^2 \frac{\mathrm{d}\theta}{\mathrm{d}s}\right) = r^2 \sin\theta \cos\theta \, \dot{\phi}^2 \, . \tag{6.107c}$$

Indeed, choosing spatial polar coordinates such that the planet starts with $\theta = \pi/2$ and with initial velocity $\dot{\theta} = 0$, it will remain in this equatorial plane for all times. Therefore, without loss of generality one can choose $\theta = \pi/2$. The motion of the planet is characterized by the constants E and L which are defined by (6.107a) and (6.107b), respectively. The constant E is interpreted as the energy per unit mass at infinity, the constant L as the angular momentum per unit of mass.

It remains to derive the differential equation (6.106b) for $\lambda = 1$, i.e. for the variable r. The left-hand side of this equation reads

$$\frac{\mathrm{d}}{\mathrm{d}s}\left(\frac{-1}{1-2m/r}\dot{r}\right) = -\frac{1}{1-2m/r}\ddot{r} + \frac{2m}{r^2}\frac{1}{(1-2m/r)^2}\dot{r}^2 \, ,$$

its right-hand side gives

$$\begin{aligned}\frac{1}{2}\sum_{\rho}\left(\frac{\partial g_{\rho\rho}}{\partial r}\right)(\dot{y}^{\rho})^2 &= \frac{m}{r^2}\dot{t}^2 + \frac{m/r^2}{(1-2m/r)^2}\dot{r}^2 - r\dot{\phi}^2 \\ &= \frac{m}{r^2}\frac{E^2}{(1-2m/r)^2} + \frac{m/r^2}{(1-2m/r)^2}\dot{r}^2 - \frac{L^2}{r^3} \, .\end{aligned}$$

Subtracting these two expressions, putting the difference equal to zero, and multiplying the whole equation by $2\dot{r}$, one obtains

$$-\frac{2m}{r^2}\frac{E^2}{(1-2m/r)^2}\dot{r} - \frac{2\dot{r}\ddot{r}}{1-2m/r} + \frac{2m}{r^2}\frac{\dot{r}^3}{(1-2m/r)^2} + \frac{2L^2}{r^3}\dot{r} = 0 \, .$$

This is seen to be the time derivative of a constant function,

$$\frac{E^2}{1-2m/r} - \frac{\dot{r}^2}{1-2m/r} - \frac{L^2}{r^2} = C \, . \tag{6.107d}$$

The constant C is obtained by noting that the expression on the left-hand side of (6.107d) equals

$$\sum_{\mu} g_{\mu\mu}\dot{y}^{\mu}\dot{y}^{\mu} = \left(1 - \frac{2m}{r}\right)\dot{t}^2 - \frac{1}{1-2m/r}\dot{r}^2 - r^2\dot{\phi}^2 \, ,$$

and also equals the invariant square of the four-velocity. We always normalized this to c^2 so that, in natural units ($c = 1$), the constant must be $C = 1$.

Thus, one obtains an equation of motion of the form

$$\dot{r}^2 + U(r) = E^2 \quad \text{with } U(r) = \left(1 - \frac{2m}{r}\right)\left(1 + \frac{L^2}{r^2}\right) \tag{6.108}$$

which is already very close to the Kepler problem. In order to understand this we return to the original Kepler problem of nonrelativistic classical mechanics, cf. [ME], Sect. 1.7.2. Remember that there one did not solve the equations of motion for $r(t)$ and $\phi(t)$ (the polar coordinates in the plane) directly but rather for the radius as a function of the azimuth, $r(\phi)$. Furthermore, it was useful to consider the reciprocal $\sigma(\phi) = 1/r(\phi)$ for a while before returning to $r(\phi)$. In detail, one has

$$r = r(\phi)\,, \quad r' \equiv \frac{\mathrm{d}r}{\mathrm{d}\phi} = \frac{\dot{r}}{\dot{\phi}}\,, \quad \dot{r} = r'\dot{\phi} = r'\frac{\ell}{\mu r^2}\,, \tag{6.109a}$$

$$\sigma(\phi) := \frac{1}{r(\phi)}\,, \quad \sigma' \equiv \frac{\mathrm{d}\sigma}{\mathrm{d}\phi} = -\frac{1}{r^2}r'\,. \tag{6.109b}$$

From the laws of energy and angular momentum conservation one derives the differential equation

$$\sigma'^2 + \left(\sigma - \frac{1}{p}\right)^2 = \frac{\varepsilon^2}{p^2} \quad \text{with} \tag{6.110a}$$

$$p = \frac{\ell^2}{A\mu}\,, \quad \varepsilon^2 = 1 + \frac{2E\ell^2}{\mu A^2}\,, \quad A = Gm_0M\,, \tag{6.110b}$$

and where $\mu = m_0M/(m_0+M)$ is the reduced mass. The mass m_0 of the test body (the planet) is assumed to be negligibly small as compared to the mass M. Then

$$p = \frac{(\ell/\mu)^2}{G(M+m_0)} \simeq \frac{(\ell/\mu)^2}{GM} = \frac{L^2}{m}\,, \tag{6.110c}$$

where we took $(\ell/\mu)^2 = L^2$ and inserted the formula (6.104) for half the Schwarzschild radius. Differentiating the differential equation (6.110a) once more by ϕ one obtains two differential equations

$$\sigma'(\phi) = 0\,, \tag{6.111a}$$

$$\sigma''(\phi) + \sigma(\phi) = \frac{1}{p} \simeq \frac{GM}{(\ell/\mu)^2} = \frac{m}{L^2}\,, \tag{6.111b}$$

which are to be understood as alternatives. The first of them describes circular orbits, the second has the known solution

$$\sigma^{(0)}(\phi) = \frac{1}{p}(1 + \varepsilon\cos\phi)\,, \tag{6.111c}$$

in which the polar coordinates are chosen such that the perihelion is attained at $\phi = 0$. The eccentricity is denoted as usual by the symbol ε.

We leave the classical Kepler problem at this point and return to the calculation of the geodesic motion (6.108). Inserting the formula $\dot{r} = r' L/r^2$ and making use of the definitions and formulae (6.109a) and (6.109b) there follows the differential equation

$$\sigma'^2(\phi) + \sigma^2(\phi) = \frac{E^2 - 1}{L^2} + \frac{2m}{L^2}\sigma(\phi) + 2m\sigma^3(\phi) \, . \tag{6.112}$$

Once more, one takes the derivative with respect to ϕ and thereby obtains the alternative differential equations

$$\sigma'(\phi) = 0 \, , \tag{6.113a}$$

$$\sigma''(\phi) + \sigma(\phi) = \frac{m}{L^2} + 3m\sigma^2(\phi) \, . \tag{6.113b}$$

The first of these pertains to circular orbits. Regarding the second equation we note that without the second term on the right-hand side it is identical with the Keplerian equation (6.111b) provided one takes $m = GM$ from (6.104) and identifies $L = \ell/m_0$ as the angular momentum per unit of mass.

In our solar system the additional term $3m\sigma^2(\phi)$ of (6.113b), is a very small perturbation. In the case of the planet Mercury one estimates this term by inserting the unperturbed solution (6.111c) into the function $3m\sigma^2$, expressing p in (6.110c) by the major half-axis a and the eccentricity ε. One has $p = a(1 - \varepsilon^2)$, hence

$$\frac{3m\sigma^2}{(m/L^2)} \simeq \frac{3m}{p} = \frac{3m}{a(1 - \varepsilon^2)} =: \eta \, . \tag{6.114}$$

Now take $m = r_S^{(\odot)}/2 = 1.476\,\text{km}$ and use the orbital data of Mercury, $a = 5.79 \cdot 10^7$ km and $\varepsilon = 0.2056$, to estimate this ratio to be $\eta \simeq 8 \cdot 10^{-8}$. The smallness of this ratio makes it possible to calculate the modification of the Kepler orbits by first-order perturbation theory. The idea is the following: One writes the solution of (6.113b) that we wish to determine as the sum of the original Kepler solution (6.111c) and an additional term $\delta(\phi)$,

$$\sigma(\phi) = \sigma^{(0)}(\phi) + \delta(\phi) \, , \quad \text{with } \sigma^{(0)}(\phi) = \frac{m}{L^2}(1 + \varepsilon\cos\phi) \, .$$

One then inserts this in (6.113b), but neglects the terms $6m\sigma^{(0)}\delta$ and $3m\delta^2$. The function δ then obeys the approximate differential equation

$$\begin{aligned} \delta''(\phi) + \delta(\phi) &= \frac{3m^3}{L^4}\left[1 + 2\varepsilon\cos\phi + \varepsilon^2\cos^2\phi\right] \\ &= \frac{3m}{p^2}\left[1 + 2\varepsilon\cos\phi + \varepsilon^2\cos^2\phi\right] \, . \end{aligned} \tag{6.115a}$$

One would like to arrange matters such that also the corrected solution of (6.115a) has its initial perihelion at $\phi = 0$, i.e. that the first derivative $\delta'(\phi)$ vanishes at $\phi = 0$. The solution of (6.115a) which fulfills this condition reads

$$\delta(\phi) = \frac{3m}{p^2}\left[1 + \varepsilon\phi\sin\phi + \frac{1}{2}\varepsilon^2 - \frac{1}{6}\varepsilon^2\cos(2\phi)\right] . \tag{6.115b}$$

Its first and second derivatives are, respectively,

$$\delta'(\phi) = \frac{3m}{p^2}\left[\varepsilon\sin\phi + \varepsilon\phi\cos\phi + \frac{1}{3}\varepsilon^2\sin(2\phi)\right] ,$$

$$\delta''(\phi) = \frac{3m}{p^2}\left[2\varepsilon\cos\phi - \varepsilon\phi\sin\phi + \frac{2}{3}\varepsilon^2\cos(2\phi)\right] .$$

In calculating the perihelion which follows the first, initial one, one must determine the zero of the first derivative at $\phi = 2\pi + \Delta\phi$,

$$\frac{\mathrm{d}}{\mathrm{d}\phi}\left(\sigma^{(0)}(\phi) + \delta(\phi)\right)$$

$$= \frac{\varepsilon}{p}\left\{-\sin\phi + \frac{3m}{p}\left(\sin\phi + \phi\cos\phi + \frac{\varepsilon}{3}\sin(2\phi)\right)\right\}_{\phi=2\pi+\Delta\phi} \stackrel{!}{=} 0 .$$

As the ratio $3m/p$ is very small one concludes $\Delta\phi \ll 2\pi$. The equation obtained just above yields in very good approximation

$$\Delta\phi \simeq \frac{3m}{p}2\pi = \frac{6\pi m}{a(1-\varepsilon^2)} \equiv 2\pi\eta . \tag{6.116}$$

Inserting the data of Mercury the precession of its perihelion after one complete revolution is found to be

$$\Delta\phi^{(\mathrm{M})} = 5.0265 \cdot 10^{-7}\ \text{radian} . \tag{6.117a}$$

The period of Mercury's orbit is 87.969 d (days). Thus, in the course of one century its perihelion advances by the amount

$$\Delta\phi^{(\mathrm{M})} \cdot \frac{100\,\mathrm{y} \cdot 365\,\mathrm{d}}{87.969\,\mathrm{d}} = 2.08 \cdot 10^{-4}\ \text{rad} \cdot \text{century}^{-1}$$
$$\widehat{=}\ 42.9'' \cdot \text{century}^{-1} . \tag{6.117b}$$

The observed advance of the perihelion is about 574 arcsec per century, see for example [Boccaletti-Pucacco 2001].

Astronomical calculations which take into account the influence of the other planets on the original Kepler ellipse of Mercury yield an amount of about 531″. The difference of 43″ per century is in perfect agreement with the results (6.117b) of general relativity[9].

[9] A detailed discussion of this important effect and of its experimental determination is found, for example, in [Weinberg 1972].

Remarks

1. As perhaps the most important lesson one has learnt that according to general relativity planetary motion is motion along geodesics, i.e. free motion in the gravitational field of a large central mass. In the approximation described above, the equations closely resemble those of the Kepler problem in flat space but are not identical with them.
2. Instead of the geodesic differential equation (6.106a) or (6.106b) one can solve a genuine problem of classical mechanics that leads to the same result. This problem emerges if one recalls that the geodesic equations (6.68a) may be read as the Euler–Lagrange equations pertaining to the Lagrange function

$$\mathcal{L} = \frac{1}{2} g_{\mu\nu}(y) \dot{y}^\mu \dot{y}^\nu = \frac{1}{2} c^2 , \tag{6.118}$$

(see [ME] Section 5.7). Using the coordinates (6.105a) of the Schwarzschild metric (6.103) and taking $c = 1$, one has

$$2\mathcal{L} = \left(1 - \frac{2m}{r}\right) \dot{t}^2 - \frac{\dot{r}^2}{1 - 2m/r} - r^2 \left(\dot{\theta}^2 + \sin^2\theta \dot{\phi}^2\right) . \tag{6.119a}$$

The Euler–Lagrange equation in the variables θ and $\dot{\theta}$ is identical with (6.107c). As we argued before, without loss of generality the coordinates may be chosen such that $\theta = \pi/2$ and $\dot{\theta} = 0$ for all times. The Lagrange function then simplifies to

$$2\mathcal{L} = \left(1 - \frac{2m}{r}\right) \dot{t}^2 - \frac{\dot{r}^2}{1 - 2m/r} - r^2 \dot{\phi}^2 . \tag{6.119b}$$

In this Lagrange function the variables ϕ and t are cyclic. Therefore, there are two conserved quantities, viz.

$$-\frac{\partial \mathcal{L}}{\partial \dot{\phi}} = r^2 \dot{\phi} = L , \quad \frac{\partial \mathcal{L}}{\partial \dot{t}} = \dot{t} \left(1 - \frac{2m}{r}\right) = E . \tag{6.120}$$

These are the constants of the motion that we already encountered in (6.107b) and (6.107a), respectively.

Null Geodesics and Deflection of Light

Starting from the second remark above one can reduce the calculation of null geodesics to a problem of mechanics. The only difference to the previous case is that now one must take $\mathcal{L} = 0$. The polar coordinates in the plane are again chosen such that $\theta = \pi/2$ and $\dot{\theta} = 0$. One inserts the conserved quantities (6.120) into the analogue of (6.119b), thus obtaining

$$\frac{E^2}{(1-2m/r)} - \frac{\dot{r}^2}{1-2m/r} - \frac{L^2}{r^2} = 0\,, \quad \text{or}$$

$$\dot{r}^2 + \frac{L^2}{r^2}\left(1 - \frac{2m}{r}\right) = E^2\,. \tag{6.121}$$

In this case, too, it is useful to study the representation $r(\phi)$ of the radius as a function of the azimuth,

$$\dot{r} = r' \frac{L}{r^2}\,,$$

whereby (6.121) goes over into the analogue of (6.112),

$$\sigma'^2(\phi) + \sigma^2(\phi) = \frac{1}{b^2} + 2m\sigma^3(\phi)\,, \quad \left(b := \frac{L}{E}\right)\,. \tag{6.122a}$$

Differentiation by ϕ yields two alternative equations, viz.

$$\sigma'(\phi) = 0 \quad \text{or} \tag{6.122b}$$

$$\sigma''(\phi) + \sigma(\phi) = 3m\sigma^2(\phi)\,. \tag{6.122c}$$

In the second of these the term on the right-hand side is very small for scattering of light by the sun. Comparing the quantity $3m\sigma^2(\phi)$ with $\sigma(\phi)$ for a light ray grazing the edge of the sun one finds

$$\frac{3m\sigma^2(\phi)}{\sigma} = \frac{3m}{r} \leqslant \frac{3r_S^{(\odot)}}{2R_\odot} \simeq 1 \cdot 10^{-5}\,.$$

Without this term the solution of (6.122a) and (6.122c) which starts with $\phi = 0$ and whose impact parameter is b, reads

$$\sigma^{(0)}(\phi) = \frac{1}{b}\sin\phi\,.$$

This solution is represented by the dashed straight line in Fig. 6.11. The deviation $\delta(\phi)$ from this unperturbed solution satisfies the differential equation

$$\delta''(\phi) + \delta(\phi) \simeq \frac{3m}{b^2}\left(1 - \cos^2\phi\right)\,.$$

The solution of this differential equation which satisfies the same initial condition, reads

$$\delta(\phi) \simeq \frac{3m}{2b^2}\left(1 + \frac{1}{3}\cos(2\phi)\right)\,.$$

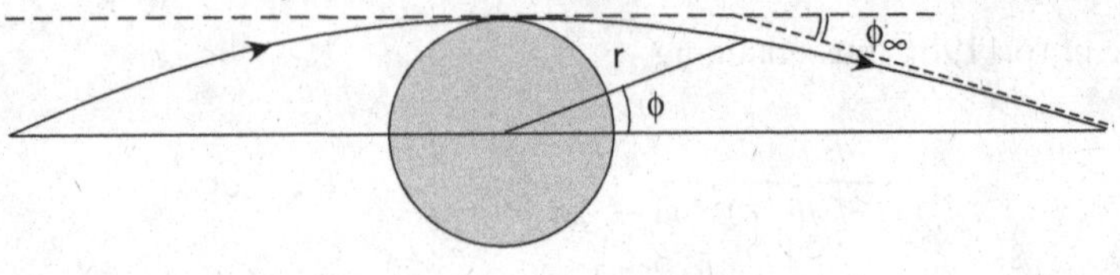

Fig. 6.11 A light beam is deflected by a massive ball. The figure shows a beam grazing the surface of the ball

Thus, the full solution is given approximately by

$$\sigma(\phi) = \sigma^{(0)}(\phi) + \delta(\phi)$$
$$\simeq \frac{1}{b}\sin\phi + \frac{3m}{2b^2}\left(1 + \frac{1}{3}\cos(2\phi)\right) . \tag{6.123}$$

One can now calculate the deflection of a light ray incident from infinity with impact parameter b that would follow a straight line if there were no perturbation. With $r \to \infty$, i.e. $\sigma \to 0$, one has approximately $\sin\phi \simeq \phi$ and $\cos(2\phi) \simeq 1$, and therefore

$$\frac{1}{b}\phi + \frac{3m}{2b^2}\frac{4}{3} \simeq 0 , \quad \text{or} \quad \phi_\infty \simeq -\frac{2m}{b} . \tag{6.124a}$$

The total deflection of a light ray between $-\infty$ and $+\infty$ is found to be

$$\Delta = 2\phi_\infty \simeq \frac{4m}{b} . \tag{6.124b}$$

If one inserts here $2m = r_S^{(\odot)}$ and takes the impact parameter to be the radius of the sun, $b = R_\odot$, one finds $\Delta \simeq 1.7''$. This prediction is in good agreement with measurements of light deflection by the sun (see e. g. [Will 2005]).

6.7.3 The Schwarzschild Radius is an Event Horizon

Upon approaching the Schwarzschild radius (6.104) from the outside one confirms that all components of the Riemann curvature tensor remain *finite*. We did not calculate this tensor explicitly here but one may infer this regular behaviour from the components of the Einstein tensor given in (6.101). The physical interpretation is that the tidal forces remain perfectly finite at $r = r_S$. An observer traversing that radius from outside-in will not notice anything special occurring to him or her. Expressed in mathematical terms, this radius can "only" be a coordinate singularity. However, what happens there in reality? In order to answer this question one calculates the geodetic orbit of a massive particle which moves in the radial direction towards the origin, both in a comoving frame of reference, and from the perspective of an observer who is at rest relative to the origin, (i.e. the centre of the mass distribution).

For radial motion one has $L = 0$, i.e. one must solve the differential equation (6.108) with $U(r) = (1 - 2m/r)$ for the initial condition $(r = R, \dot{r} = 0)$. The energy parameter E is determined from (6.108), using the condition $\dot{r} = 0$, giving $E^2 = 1 - 2m/R$. One thereby obtains a modified form of the differential equation (recall that $\dot{r}$ is the derivative of r with respect to proper time s),

$$\dot{r}^2 = \frac{2m}{r} - 1 + E^2 \quad \text{or} \quad \dot{r}^2 = 2m\left(\frac{1}{r} - \frac{1}{R}\right) . \tag{6.125}$$

With s the proper time of the freely falling observer, the second equation yields the differential equation

$$\mathrm{d}s = \left(\frac{2m}{r} - \frac{2m}{R}\right)^{-1/2} \mathrm{d}r . \tag{6.125a}$$

This is a differential equation with separable variables whose solution can be written formally as a quadrature,

$$s = \int \mathrm{d}r \, \frac{1}{\sqrt{(2m/r) - (2m/R)}} . \tag{6.125b}$$

The solution which satisfies the initial condition given above, is a cycloid. We parametrize this orbit by means of a parameter η which takes its values in the interval $[0, \pi]$,

$$r = \frac{1}{2} R\,(1 + \cos\eta) , \tag{6.126a}$$

$$s = \sqrt{\frac{R^3}{8m}}\,(\eta + \sin\eta) , \quad \text{with} \quad 0 \leqslant \eta \leqslant \pi . \tag{6.126b}$$

This is verified as follows: Calculate $\mathrm{d}r$ and $\mathrm{d}s$ from (6.126a) and (6.126b), respectively, and from there

$$\dot{r} \equiv \frac{\mathrm{d}r}{\mathrm{d}s} = -\frac{\sin\eta}{\sqrt{R/(2m)}\,(1 + \cos\eta)} .$$

The initial condition is obviously satisfied. Furthermore, by calculating

$$\dot{r}^2 - \frac{2m}{r} = \frac{2m}{R} \frac{\sin^2\eta - 2 - 2\cos\eta}{1 + 2\cos\eta + \cos^2\eta} = -\frac{2m}{R} ,$$

one sees that the differential equation (6.125) is fulfilled.

The orbit of free motion

$$\text{begins at } \eta = 0 : \ (r = R,\ s = 0) ,$$

$$\text{it ends at } \eta = \pi : \ \left(r = 0,\ s = \frac{\pi}{2}\sqrt{\frac{R^3}{2m}}\right) .$$

The freely moving observer reaches the origin $r = 0$ after some *finite* proper time. At the moment of passage through the Schwarzschild radius $r = r_S = 2m$ his/her proper time is obtained from (6.126b) with $\eta \equiv \eta_S$ as obtained from (6.126a), viz.

$$\eta_S = \arccos\left(\frac{4m - R}{R}\right) .$$

The same motion looks completely different from the perspective of an observer B_0 at rest, that is, an observer who stays behind at the starting point $r = R$, so to speak "at home", while the voyager on the geodesic plunges towards the singularity at $r = 0$. Let his/her coordinate time be denoted by t. Using the chain rule one concludes

$$\dot{r} = \frac{\mathrm{d}r}{\mathrm{d}t}\dot{t} = \frac{\mathrm{d}r}{\mathrm{d}t}\frac{E}{1 - 2m/r} . \tag{6.127}$$

It turns out that the following auxiliary variable is a useful definition:

$$r^* := r + 2m \ln\left(\frac{r - 2m}{2m}\right) . \tag{6.128}$$

One calculates its derivative by the coordinate time t,

$$\frac{\mathrm{d}r^*}{\mathrm{d}t} = \frac{\mathrm{d}r}{\mathrm{d}t}\frac{1}{1 - 2m/r} \quad \text{from which} \quad \dot{r} = E\frac{\mathrm{d}r^*}{\mathrm{d}t}$$

follows. Inserting this in (6.125) one obtains a relation which is sufficient for our discussion,

$$\left(E\frac{\mathrm{d}r^*}{\mathrm{d}t}\right)^2 = \frac{2m}{r} + E^2 - 1 .$$

Letting r tend to $2m$ from above the right-hand side of this relation is approximated by E^2. Equation (6.128) shows that in this limit the function r^* tends to minus infinity, $r^* \to -\infty$, while its derivative by the coordinate time tends to -1, $\mathrm{d}r^*/\mathrm{d}t \simeq -1$. From this one has approximately $r^* \simeq -t + c$, with c a constant. Using again definition (6.127) one concludes

$$2m\left\{1 + \ln\left(\frac{r - 2m}{2m}\right)\right\} \simeq -t + c$$

and from this, eventually, obtains the approximate solution

$$r(t) \simeq 2m + \text{const.}\ \mathrm{e}^{-t/(2m)} . \tag{6.129}$$

The result (6.129) tells us that, as seen from the perspective of the observer who stayed at home, the Schwarzschild radius is reached only after infinite time.

It is also instructive to study the equation of motion for radially falling *massless* particles. Equation (6.121) with $L = 0$ taken together with (6.127) yields the relation

$$\frac{dr}{dt} = \pm\left(1 - \frac{2m}{r}\right) . \tag{6.130}$$

This shows that the local light cones narrow more and more, the closer one approaches the Schwarzschild radius. As a consequence, the observer at rest will eventually receive no more signals from the freely falling observer. Furthermore, he or she has no means of seeing what happens beyond the horizon, at $r < r_S$.

6.8 Some Concluding Remarks

1. With the perihelion precession and the deflection of light that we calculated from the static Schwarzschild metric, we came to know two classic tests of general relativity which historically belong to the first important successes of this theory of gravitation. Both effects refer to a situation in which the gravitational field is static and is weak in the sense that real spacetime deviates but little from a flat Minkowski space. Nowadays there are a number of further tests which also include systems with *strong* gravitational fields as well as *time-dependent, oscillatory* fields. A complete up-to-date summary is found in [Will 2005].
2. The discussion of the Schwarzschild metric of the preceding section is incomplete because in the form (6.103) given there it does not apply for values of the radius $r \leqslant r_S$.
   Continuation to values below the Schwarzschild radius is perfectly possible by, geometrically speaking, a change of charts. This continuation shows, however, that the roles of the time and the radial variables are interchanged, that is to say that a static solution is converted into a nonstatic one. By the same token, this continuation yields a model which can be used to describe a *black hole* i.e. a spherically symmetric mass distribution which is so dense that its Schwarzschild radius is larger than its geometrical radius.
3. There are a few more analytic solutions of Einstein's equations. Among them there are solutions that describe rotating black holes. Solutions of this class are relevant for models that provide a basis for (classical) cosmologies. Of special interest is the study of singularities of solutions and their possible relation to quantum properties of the gravitational field.
4. Another important branch of general relativity concerns the analysis and experimental verification of *gravitational waves.* The direct comparison with electromagnetic waves is particularly interesting, both with regard to their similarities and to their differences.

With these comments I leave the fascinating field of the theory of gravitation as a classical field theory and refer to the many excellent monographs on this subject. With the experience gained in this chapter and, in particular, the acquired knowledge of the geometric foundations of general relativity, I sincerely hope, the reader is well prepared for studying some of these monographs in greater detail.

Bibliography

Abramowitz, M., Stegun, I. A.: *Handbook of Mathematical Functions* (Dover, New York 1968)

Arnol'd, V. I.: *Mathematical Methods of Classical Mechanics*, (Springer, New York, Berlin, Heidelberg, 2nd ed. 1989)

Boccaletti, D., Pucacco, G.: *Theory of Orbits 1 + 2* (Springer-Verlag, Berlin 2001, 1999)

Bronstein, I. N., Semendjajew, K. A., Musiol, G. Muelig, H.: *Handbook of Mathematics* (Springer, New York, Berlin, Heidelberg, 5th ed. 2011)

Flanders, H.: *Differential forms with applications to the physical sciences* (Academic Press, New York 1963)

Hehl, F. W., Obukhov, Yu. N.: *Foundations of Electrodynamics*, (Birkhäuser, Basel 2003),

Honerkamp, J., Römer, H.: *Theoretical Physics* (Springer, Berlin, Heidelberg 1993)

Jackson, J. D.: *Classical Electrodynamics*, (John Wiley, New York, 3d ed. 1999)

Landau, L., Lifshitz, E. M.: *The Classical Theory of Fields*, Vol. 2 (Butterworth-Heinemann, Oxford 1975)

Electrodynamics of Continuous Media, Vol. 8 (Butterworth-Heinemann, Oxford 1984)

Misner, Ch. W., Thorne, K. S., Wheeler, J. A.: *Gravitation* (W. H. Freeman & Co, San Francisco 1973)

O'Neill, B.: *Semi-Riemannian Geometry* (Academic Press, New York 1983)

O'Neill, B.: *The Geometry of Black Holes* (AK Peters, Massachusetts 1995)

O'Raifeartaigh, L.: *Group Structure of Gauge Theories* (Cambridge Monographs on Mathematical Physics, Cambridge 1986)

Pais, A.: *"Subtle is the Lord ...", The Science and Life of Albert Einstein* (Oxford University Press, New York 1982)

Rindler, W.: *Essential Relativity* (Springer, New York 1977)

Scheck, F.: *Mechanics – From Newton's Laws to Deterministic Chaos*, quoted as **ME** in the text (Springer, Berlin, New York, Heidelberg 5th ed. 2010)

Scheck, F.: *Quantum Physics*, quoted as **QP** in the text (Springer, Berlin, New York, Heidelberg 2007)

Scheck, F.: *Electroweak and Strong Interactions – Phenomenology, Concepts, Models* (Springer-Verlag, Berlin, Heidelberg 3d ed. 2012)

Sommerfeld, A.: *Vorlesungen über Theoretische Physik*, Bände 3 und 6, (DVB Wiesbaden 1947)

Straumann, N.: *General Relativity – With Applications to Astrophysics* (Springer, Berlin, Heidelberg 2004)

Thirring, W.: *Classical Mathematical Physics* (Springer, Berlin, New York, Heidelberg 1997)

F. Scheck, *Classical Field Theory*, Graduate Texts in Physics,
DOI 10.1007/978-3-642-27985-0,

Weinberg, S.: *Gravitation and Cosmology* (John Wiley & Sons, New York 1972)
Will, C. M.: *The Confrontation between General Relativity and Experiment* (Los Alamos archive, gr-qc 0510072, 2005), published in *Living Rev. Rel.* 9:3 (2005)

Some Historical Remarks

The history of classical and quantum field theory is a fascinating story of step by step unification – unification of time and space in Special Relativity, unification of electric and magnetic interactions in Maxwell's theory which initially appeared unrelated, unification of dynamics and geometry in General Relativity, and, more recently, the unification of electromagnetic and weak interactions in the framework of the Standard Model, to name but the most important.

Towards the end of the eighteenth century, the Age of Enlightenment, certain electric phenomena and a few facts about magnetism were known but the two types of phenomena seemed unrelated. Simple electrostatic experiments could be and were performed by laymen and the action of stationary currents was known (from Volta's batteries). These phenomena appeared as curiosities which had no relevance for daily life. Perhaps the only exception were the lightning rods invented by Benjamin Franklin (1706–1790) but these were often rejected by ignorance or mistrust because people thought they would attract lightnings rather than divert them. Magnetism, in turn, was well known in the form of natural magnetic material and was ascribed healing power in medicine. Franz Anton Mesmer (1734–1815), the famous doctor Mesmer, founded the lore of so-called animal magnetism, an early precursor of hypnotherapy. Note, however, that even in his time this treatment was not taken too seriously. No one less than Wolfgang Amadeus Mozart immortalized Mesmer and the "Mesmer stones" in his opera *Cosi fan tutte* (1789/1790) in which Despina, chamber maid of the ladies Fiordiligi and Dorabella, disguised as a medical doctor, tries to cure the enamoured gentlemen Guglielmo and Ferrando by magnetism, viz

"Questo è quel pezzo di calamita, pietra Mesmerica,
ch'ebbe l'origine n'ell Allemagna
che poi si celebre in Francia fù."

(W.A. Mozart, *Cosi fan tutte*, Act I, Scene 16.)[1]

1 "This is that piece of magnet, the Mesmer stone, which had its origin in Germany, and then became so famous in France".

In the preface to his book "The present state of music in France and Italy 1771" Charles Burney, the British musicologist (1726–1824) makes an interesting comparison between electricity and music,

> *"Electricity is universally allowed to be a very entertaining and surprising phenomenon, but it has frequently been lamented that it has never yet, with much certainty, been applied to any very useful purpose.*
> *The same reflexion has often been made, no doubt, as to music. It is a charming resource, in an idle hour, to the rich and luxurious part of the world. But say the sour and the wordly, what is its use to the rest of mankind?...*
> *Music has indeed ever been the delight of accomplished princes, and the most elegant amusements of polite courts: but at present it is so combined with things sacred and important, as well as with our pleasures, that mankind seems wholly unable to subsist without it."*

And, indeed, in this diary of this and subsequent journeys he points to the great importance of music in 18th century life, from the noble people at European courts to the farmers on their fields.
Coulomb's law, i.e. the $1/r^2$-dependence of the force between two charges e_1 and e_2, was discovered in 1785 (**Charles Augustin Coulomb, 1736–1806**), but it took another 35 years before it became known in around 1820 that in reality electric and magnetic phenomena are closely related. The Danish physicist **Hans Christian Ørsted (1777–1851)** reported that electric currents circulating in conducting loops align magnetic needles in their neighbourhood. This first step towards unification of interactions attracted a great deal of attention and stimulated the subsequent quantitative investigations by Biot and Savart (**Jean-Baptiste Biot, 1774–1862; Félix Savart, 1791–1841**) culminating in the law that bears their names. These investigations were followed by the series of famous experiments of **André Marie Ampère (1775–1836)** who showed, among other effects, that small solenoids supporting electric current, behave like linear magnets in the magnetic field of the earth, and who first formulated the forces between current-carrying wires.

The great figures of the classical period of electrodynamics were **Michael Faraday (1791–1867)** and **James Clerk Maxwell (1831–1879)**, the first of them primarily an eminent experimenter who discovered the key experiment of induction, the second of them the architect of the basic equations of electrodynamics in their universal local form. The induction law of 1831 established the first direct and explicit relationship between electric and magnetic fields but it needed Maxwell's concept of the displacement current (33 years later!), obtained from a nonstationary application of the Biot–Savart law, before he could formulate a closed and consistent theory of all electric and magnetic phenomena[2].

[2] I recommend, in particular, the essay by Res Jost *Michael Faraday – 150 years after the discovery of electromagnetic induction,* in R. Jost, *Das Märchen vom elfenbeinernen Turm,* (Springer 1995)

Maxwell's equations, formulated in 1864, found their most exciting and influential application in the experiments carried out in 1888 by **Heinrich Rudolph Hertz (1857–1894)** which proved the existence of electromagnetic waves. The tremendous development from then until the present time, from early wireless telegraphy to modern techniques of global positioning in ships, planes, and cars, and to modern telecommunications of all sorts, presumably is well known to the reader. So one might be tempted to repeat the last sentence of the quotation from Burney, but with modified subjects: ... *at present it is so combined with things sacred and important, as well as with our pleasures, that mankind seems wholly unable to subsist without it.*

The history of the notion of *vector potential* which was essential for the discovery of gauge invariance is less straightforward and transparent. In their historical essay Jackson and Okun[3] showed that first hints are found in the work of Franz E. Neumann and Wilhelm Weber around the middle of the 19th century but it was Gustav Kirchhoff (around 1857) and, about a decade or more later, Hermann von Helmholtz who formulated equations which relate scalar and vector potentials and, from a modern point of view, correspond to special choices of the gauge.

The Danish physicist **Ludvig Valentin Lorenz** (1829–1891) was the first who wrote down retarded potentials of the kind of (4.30), i.e. in a notation in use today,

$$\Phi(t,\boldsymbol{x}) = \iiint \mathrm{d}^3x' \int \mathrm{d}t' \, \frac{\varrho(t',\boldsymbol{x}')}{|\boldsymbol{x}-\boldsymbol{x}'|} \delta\left(t'-t+\frac{1}{c}\left|\boldsymbol{x}-\boldsymbol{x}'\right|\right) , \tag{1a}$$

$$\boldsymbol{A}(t,\boldsymbol{x}) = \iiint \mathrm{d}^3x' \int \mathrm{d}t' \, \frac{\boldsymbol{j}(t',\boldsymbol{x}')}{|\boldsymbol{x}-\boldsymbol{x}'|} \delta\left(t'-t+\frac{1}{c}\left|\boldsymbol{x}-\boldsymbol{x}'\right|\right) , \tag{1b}$$

and who noticed that they fulfill the condition

$$\nabla \cdot \boldsymbol{A}(t,\boldsymbol{x}) + \frac{1}{c}\frac{\partial \Phi(t,\boldsymbol{x})}{\partial t} = 0 . \tag{2}$$

It seems as though the use of gauge transformations was familiar to him because he noted the equivalence of these potentials to those of the class $\nabla \cdot \boldsymbol{A} = 0$. It is a long-standing habit to ascribe the condition (2) to the Dutch physicist H. A. Lorentz (**Hendrik Antoon Lorentz, 1853–1928**). However, as L. V. Lorenz discovered and made use of it about a quarter of a century before H. A. Lorentz, it seems appropriate to correct this misassignment of many textbooks, without belittling the importance and the great achievements of the latter.[4]

The unity of the spectra of all kinds of electromagnetic radiation started to emerge with the proof of the interference of X-rays by W. Friedrich and P. Knipping – following up an idea of Max von Laue. Nowadays we know that, though very

[3] J. D. Jackson, L. B. Okun: *Historical roots of gauge invariance*, Rev. Mod. Phys. 73 (2001) 663.

[4] Both names are found in what is called the Lorenz–Lorentz effect in optics. This effect concerns the density dependence of the index of refraction. There is an analogue of this effect in the interaction of negative pions with nuclear matter, called the Ericson–Ericson effect.

different in appearance, X-rays, visible light, infrared radiation, all belong to the same spectrum of electromagnetic waves.

Of utmost importance, from a modern point of view, is the covariance of Maxwell's equations under the Lorentz group that is based on the principle of the constancy of the speed of light in vacuum, and the discovery of special relativity by Albert Einstein. The qualified symmetry between three-dimensional space and time in special relativity, and the progress from Galilean space with its Newtonian absolute time to Minkowski space, brought about a very specific unification of space and time.

The notion of *gauge invariance* was coined in 1919 by Hermann Weyl (**Hermann Weyl, 1885–1955,** German mathematician and physicist). In his attempt to combine electrodynamics and gravitation he originally had in mind scale transformations of the metric

$$g_{\mu\nu} \longmapsto e^{\lambda(x)} g_{\mu\nu}$$

with *real* functions $\lambda(x)$, i.e. a transformation by which coordinates were really "gauged" in the old sense of the word. Vladimir Fock[5] made an important discovery which often is not fully appreciated but which represents another important step of unification: The combination of U(1) transformations generated by real functions $\chi(t, \boldsymbol{x})$ on the one side, and the action of phases

$$e^{i\alpha(t,x)} \quad \text{with} \quad \alpha(t, \boldsymbol{x}) = \frac{e}{\hbar c}\chi(t, \boldsymbol{x})$$

on wave functions of quantum theory, on the other, as worked out in Sect. 3.5.2, equations (3.38) and (3.39b), and in Sect. 5.3, equation (5.16). Here, the fundamental gauge principle of classical electrodynamics, and the characteristic freedom of phases of quantum theory, are combined into something new: A locally gauge invariant theory of radiation and matter in which the covariant derivative plays a special role in fixing the coupling between the two types of fields.

The next major step of unification is the combination of electrodynamics with the other fundamental interactions in the framework of the so-called minimal standard model of elementary particle physics. In Chapt. 5 I collected the relevant steps of the development for the example of electrodynamics and the weak interactions on a *classical* basis. They lead to the widely ramified field of modern quantum field theory and to present-day research in elementary particle physics, for whose historical development I refer to the appendix of [QP].

General relativity which in a precise sense brings together the geometry of spacetime with the nature of gravitation, is a very special kind of unification of formerly unrelated notions. The other fundamental interactions are thought to be well described when formulated on an inert, given Euclidean spacetime. This, as we have seen, is not true for gravitation for which no consistent theory can be found on

[5] V. Fock, Z. Physik **38** (1926) 242 and **39** (1926) 226.

a preexisting, flat spacetime. In its essential aspects, general relativity is the work of a single person, Albert Einstein. The genesis of this theory, the life of Albert Einstein, and much more, can be found, e. g. in the excellent biography by Abraham Pais [Pais 1982].

Exercises

Exercises Chapter 1

1.1 If in $\mathbb{R}^3$ the cartesian basis $\hat{e}_1, \hat{e}_2, \hat{e}_3$ is replaced by a spherical basis

$$\hat{e}_{\pm} := \mp \frac{1}{\sqrt{2}}\left(\hat{e}_1 \pm \mathrm{i}\hat{e}_2\right), \; \hat{e}_0 := \hat{e}_3 , \tag{A.1}$$

the expansion of a vector reads $V = \sum_{m=-1}^{+1} v^m \hat{e}_m$. Write down the orthogonality relations for the base vectors $\hat{e}_m$, the scalar product $V \cdot W$, and work out the difference between contravariant components v^m and the corresponding covariant components.

1.2 Show that the four-component current density (1.25) satisfies the continuity equation.

1.3 Estimate the mass of a Uranium nucleus in micrograms, knowing that it contains 92 protons and 143 neutrons.
Hint: $m_p c^2 \simeq m_n c^2 \simeq 939$ MeV.

1.4 Calculate the electric field in volt per metre that a muon feels in the 1*s*-state of muonic lead.
Hints: Bohr radius $a_\mathrm{B} = \hbar c/(Z\alpha m_\mu c^2)$, $m_\mu c^2 = 105.6\,\mathrm{MeV}$.

1.5 Prove the formula (1.48a), i.e.

$$\sum_{k=1}^{3} \varepsilon_{ijk}\varepsilon_{klm} = \delta_{il}\delta_{jm} - \delta_{im}\delta_{jl} .$$

1.6 Prove the formula (1.52a) by means of the following construction: Consider two concentric spheres with radii R_i and R_a, respectively, and $R_\mathrm{i} < R_\mathrm{a}$ whose centre is $\boldsymbol{x}$. Choose the reference point $\boldsymbol{x}'$ in the domain between the two spheres, and apply Green's second theorem to the volume enclosed by the spheres, for the functions ψ or ϕ equal to $1/r$ ($r = |\boldsymbol{x} - \boldsymbol{x}'|$). Let then R_{i} go to zero, R_{a} to infinity.

1.7 Determine the normalization factor N of the distribution

$$\varrho_{\text{Fermi}}(r) = \frac{N}{1 + \exp\{(r - c)/z\}} \tag{A.2}$$

such that ϱ integrated over the whole space gives 1.

1.8 Let η be the surface charge density on a given smooth surface F. Prove the relation (1.87a).
Hint: Choose a small "box" across the surface such that its base and its lid are parallel to the surface F and have the size $d\sigma$ while its height is small of third order. Apply Gauss' theorem.

1.9 Prove: On a surface which carries the surface charge density η the tangential component of the electric field is continuous, equation (1.87b).
Hint: Choose a closed rectangular path which cuts the surface such that the edges perpendicular to the surface are much smaller than the edges parallel to the surface. Calculate the electromotive force along that path.

1.10 Prove the properties (1.97g) and (1.97h) using the explicit expressions (1.97a) for the spherical harmonics.

1.11 Derive the relation between the cartesian components Q^{ik} of the quadrupole, equation (1.111c), $i, k = 1, 2, 3$, and its spherical components $q_{2\mu}$.

1.12 Show that the space integral of the electric field strength of a dipole is proportional to the dipole moment,

$$\iiint_V \mathrm{d}^3x \, \boldsymbol{E}_{\text{Dipol}}(x) = -\frac{4\pi}{3}\boldsymbol{d} \, . \tag{A.3}$$

1.13 Given a capacitor consisting of two metallic plates and an electrically polarizable, insulating medium between them, consider the process of discharge after short-circuiting the plates and calculate the displacement current in the medium.

1.14 Construct the additional term $F(\boldsymbol{x}, \boldsymbol{x}')$ in (1.90) which is needed for the Dirichlet Green function to vanish on the sphere.

1.15 A point-like electric dipole $\boldsymbol{d} = d\,\hat{\boldsymbol{e}}_3$ is placed in the centre of a conducting sphere whose radius is R. Calculate the potential and the electric field inside the sphere. What is the field outside of the sphere? What is the charge density on (the surface of) the sphere?

1.16 A point-like electric dipole is located in the point $\boldsymbol{x}^{(0)}$. Show: The potential that it creates as well as its energy in an external potential Φ_a can be described by the effective charge density

$$\varrho_{\text{eff}}(\boldsymbol{x}) = -\boldsymbol{d} \cdot \nabla\delta(\boldsymbol{x} - \boldsymbol{x}^{(0)}) .$$

1.17 A conducting sphere with total charge Q is placed in a homogeneous electric field $\boldsymbol{E}^{(0)} = E_0 \hat{\boldsymbol{e}}_3$. How does the electric field change by the presence of the sphere? What is the distribution of the charge on (the surface of) the sphere?
Hint: Assume the potential to have the form

$$\Phi = f_0(r) + f_1(r)\cos\theta$$

and solve the Poisson equation. Can you give plausibility arguments for this ansatz?

1.18 Calculate the energy contained in the electric field of a spherically symmetric, homogeneous charge distribution (radius R, charge Q). Then calculate the self-energy

$$W = \frac{1}{2}\iiint \mathrm{d}^3x\ \varrho(\boldsymbol{x})\Phi(\boldsymbol{x})$$

of this charge distribution.

1.19 An electron located in the origin is assigned the charge distribution $\varrho = (-e)\delta(\boldsymbol{x})$. One considers a sphere of radius R, its centre taken as the origin, and calculates the energy of the electric field outside the sphere. How must the radius be chosen if this energy is to be equal to the rest energy $m_e c^2$ of the electron? This radius is called the *classical electron radius*.
Answer: $R = e^2/(2m_e c^2) = \alpha\hbar c/(2m_e c^2)$.

1.20 A sphere with radius R is made up of a homogeneous dielectric material with dielectric constant ε_1. The sphere is embedded in a medium which is homogeneous, too, and whose dielectric constant is ε_0. Furthermore, an external electric field $\boldsymbol{E} = E_0 \hat{\boldsymbol{e}}_3$ is applied to it. Calculate the potential inside and outside the sphere. Sketch the equipotential surfaces for the special cases ($\varepsilon_0 \equiv \varepsilon$, $\varepsilon_1 = 1$) and ($\varepsilon_0 = 1$, $\varepsilon_1 \equiv \varepsilon$). In the second case let ε become very large and compare with the potential in Exercise 1.17.

1.21 Two positive charges $q = (e/2)$ and two negative charges $-q$ are placed in four points whose cartesian coordinates (x, y, z) are as given here

$$q_1 = \ q : \ (a, 0, 0) , \qquad q_2 = q : \ (-a, 0, 0) ,$$
$$q_3 = -q : \ (0, b, 0) , \qquad q_4 = -q : \ (0, -b, 0) .$$

Write down the charge distribution by means of δ-distributions. What is the dipole moment of this distribution? Determine the quadrupole tensor $Q_{ij} = \iiint d^3x\, [3x_i x_j - \boldsymbol{x}^2\delta_{ij}]\varrho(\boldsymbol{x})$ and the spectroscopic quadrupole moment

$$Q_0 := \sqrt{\frac{16\pi}{5}} \iiint d^3x\, \varrho(\boldsymbol{x}) r^2 \,.$$

List also the moments $q_{\ell,m}$ for $\ell = 2$ (spherical basis).

1.22 A spherical shell with radius R which carries a constant surface charge density η, rotates with angular velocity ω about an axis through its centre. What is the magnetic field it creates?
Hint: The surface current is given by the expression

$$\boldsymbol{K}(\theta) = \eta\boldsymbol{\omega} \times \boldsymbol{x} = \omega\eta R \sin\theta\, \hat{\boldsymbol{e}}_\phi \,.$$

1.23 A hollow ball with inner radius r and outer radius R, consists of a material with high magnetic permeability μ. This sphere is placed in an external induction field $\boldsymbol{B} = B_0\hat{\boldsymbol{e}}_3$. Calculate the field lines in the presence of this ball. In particular, study the special case $\mu \to \infty$.
Hint: As there is no current density one may derive the fields $\boldsymbol{H}$ and $\boldsymbol{B}$ from a magnetic potential Φ_{magn}. Use the multipole expansion for this potential.

Exercises Chapter 2

2.1 By counting the base k-forms determine the dimension of the space $\Lambda^k(M)$ of k-forms over the manifold M.

2.2 Show: A symmetric tensor $S_{\mu\nu}$ of degree two contracted with another, antisymmetric tensor $A^{\mu\nu}$ of degree two, gives zero.

2.3 With $\varepsilon_{\alpha\beta\gamma\delta}$ the Levi-Civita symbol in dimension four, find summation formulae which correspond to the formulae (1.48a) and (1.48b).

2.4 Let $\boldsymbol{A}(t, \boldsymbol{x})$ be a given vector potential which is not subject to any special boundary condition. If one chooses the gauge function

$$\chi(t, \boldsymbol{x}) = \frac{1}{4\pi} \iiint d^3y\, \frac{1}{|\boldsymbol{x} - \boldsymbol{y}|} \boldsymbol{\nabla}_x \cdot \boldsymbol{A}(t, \boldsymbol{y})$$

in order to replace $\boldsymbol{A}$, what can be said about the divergence of the transformed vector potential $\boldsymbol{A}'$? In case there are no external sources what is the gauge function by means of which one obtains $A_0(t, \boldsymbol{x}) = 0$ without leaving the class of Coulomb gauges?

2.5 Since energy and momentum are conserved, a free electron cannot radiate a light quantum, $e \to e + \gamma$. Prove this by using relativistic kinematics.

Exercises Chapter 3

3.1 Determine the physical dimensions of the quantities $u(t, \boldsymbol{x})$, (3.54a), $\boldsymbol{P}(t, \boldsymbol{x})$, (3.54b), $\boldsymbol{S}(t, \boldsymbol{x})$, (3.54c), and $T_j^k(t, \boldsymbol{x})$, (3.54d).

3.2 Show: In $\mathbb{R}^3$ both δ_{ij} and ε_{ijk} are tensors which are invariant under rotations $\mathrm{R} \in \mathrm{SO}(3)$. In Minkowski space what can you say about $\delta_{\mu\nu}$ and about $\varepsilon_{\mu\nu\sigma\tau}$ with regard to Lorentz transformations?

Exercises Chapter 4

4.1 Which boundary conditions hold for electric fields and for induction fields at boundary surfaces? (See also Exercises 1.8 and 1.9).

4.2 A harmonically oscillating dipole source is described by the current density

$$\boldsymbol{j}(t, \boldsymbol{x}) = -\mathrm{i}\omega\, \boldsymbol{d}\, \delta(\boldsymbol{x})\mathrm{e}^{-\mathrm{i}\omega t} .$$

Determine the corresponding charge density and the *physical* expressions for $\boldsymbol{j}$ and ϱ. Calculate the corresponding vector potential $\boldsymbol{A}_{\mathrm{E1}}$, including its harmonic time dependence. Calculate the electric field and the induction field.

4.3 Given two concentric rings made of a conducting material. The inner ring whose radius is a, carries the homogeneously distributed charge q, the outer ring with radius b carries the charge $-q$. Write down the charge density of this setup, expressed in cylinder coordinates where the z-axis goes through the centre of the rings and is perpendicular to the plane spanned by them.

4.4 The setup of Exercise 4.3 is assumed to rotate about the z-axis with angular velocity ω. Derive the current density and calculate the magnetic dipole moment.

Exercises Chapter 5

5.1 Solve the differential equation

$$(\boldsymbol{\Delta} - \kappa^2)\phi(\boldsymbol{x}) = g\, \delta(\boldsymbol{x}) \tag{A.4}$$

in momentum space, i.e. by means of the ansatz

$$\phi(\boldsymbol{x}) = \frac{1}{(2\pi)^{3/2}} \iiint \mathrm{d}^3k \; \mathrm{e}^{\mathrm{i}\boldsymbol{k}\cdot\boldsymbol{x}} \widetilde{\phi}(\boldsymbol{k}) \,. \tag{A.5}$$

5.2 In a given representation the generators of a compact, simple Lie group have the property (omitting, for simplicity, the symbol U in U(T))

$$\mathrm{tr}\{\mathrm{T}_i, \mathrm{T}_j\} = \kappa \delta_{ij} \,. \tag{A.6}$$

Show that the constant κ, though depending on the representation, does not depend on i and j.

5.3 Construct a Lagrange density for the local gauge theory which is built on the structure group $G = \mathrm{SO}(3)$ and which contains a triplet of scalar fields.

5.4 A local gauge theory built on the structure group

$$G = \mathrm{SU}(p) \times \mathrm{SU}(q) \quad \text{with} \quad p, q > 1 \,,$$

allows for two independent "charges" (coupling constants). Show this by constructing the gauge potential and the covariant derivative.

5.5 A major study project might be this: Study the group theoretical aspects of the publication on the self-gravitation of a rotating star quoted in Sect. 5.6. Investigate analytically the bifurcations reported in this work and illustrate by means of numerical examples.

5.6 Show that the matrices

$$M^{(k)}_{lm} = -\mathrm{i}C^{(m)}_{kl}$$

fulfill the Lie algebra (5.20).

Exercises Chapter 6

6.1 The aim is to show that the $(n-1)$-sphere S^{n-1}_R which has radius R and is embedded in $\mathbb{R}^n$, is a smooth manifold.
Let $N = (0, \ldots, 0, R)$ and $S = (0, \ldots, 0, -R)$ be the north and south poles of the sphere, respectively. Let two charts be defined by the projection from N and from S, respectively, of the points $x \in S^{n-1}_R$ onto the equatorial hypersurface $x^n = 0$. The first chart applies to the subset $U_1 = S^{n-1}_R \backslash \{N\}$, and the second applies to $U_2 = S^{n-1}_R \backslash \{S\}$. Specify the maps φ_i as well as their inverses for

$i = 1, 2$. Derive the transition mapping from U_1 to U_2, show that this is a diffeomorphism on the intersection $U_1 \cap U_2$.

6.2 Gravitational redshift: Calculate the relative change of frequency $\Delta\omega/\omega$ of a photon which moves from the tip of a tower with height H to the ground. Compare the shift $\Delta\omega$ for the example $H = 22.5$ m with the natural line width Γ of a spectral line in ^{57}Fe for which $\omega/\Gamma = 3 \cdot 10^{12}$.

6.3 Let $X, Y, Z \in \mathfrak{X}(M)$ be smooth vector fields on the manifold M, $[X, Y]$ etc. their Lie brackets (commutators). Prove the Jacobi identity

$$[X, [Y, Z]] + [Y, [Z, X]] + [Z, [X, Y]] = 0 .$$

Consider the example of $M = \mathbb{R}^2$ with $X = y\partial_x$ and $Y = x\partial_y$. What is their Lie bracket?

6.4 In general, tensor products do not commute. In order to illustrate this fact consider the examples $T^{(i)} \in \mathfrak{T}^0_2$, $i = 1, 2$, with

$$T^{(1)} = \mathrm{d}x^1 \otimes \mathrm{d}x^2 , \quad T^{(2)} = \mathrm{d}x^2 \otimes \mathrm{d}x^1 .$$

Calculate the functions $T^{(i)}(X, Y)$ for

$$X = a^1\partial_1 + a^2\partial_2 , \quad Y = b^1\partial_1 + b^2\partial_2$$

with constant coefficients $a^1, \ldots, b^2$.

6.5 Another way to calculate the covariant derivative of tensor fields of type (0, 1) is the following. For the choice $V = \partial_\mu$ the covariant derivative D_V of a vector field X by V is known to be

$$X^\rho{}_{;\mu} = \partial_\mu X^\rho + \Gamma^\rho_{\mu\nu} X^\nu .$$

Use this formula to calculate $X_{\tau\,;\mu} = g_{\tau\rho} X^\rho{}_{;\mu}$ and make use of the coordinate expression of the Christoffel symbols.

6.6 The Christoffel symbols are not the components of a tensor field: The equations of motion of a free particle in Gaussian (or normal) coordinates at the point $x \in M$ read

$$\frac{\mathrm{d}^2 z^\mu}{\mathrm{d}\tau^2} = 0 \quad \text{with} \quad \mathrm{d}\tau^2 = \eta_{\mu\nu}\,\mathrm{d}z^\mu\,\mathrm{d}z^\nu ,$$

while in any other coordinates they are

$$\frac{\mathrm{d}^2 y^\mu}{\mathrm{d}\tau^2} + \Gamma^\mu_{\rho\sigma} \frac{\mathrm{d}y^\rho}{\mathrm{d}\tau} \frac{\mathrm{d}y^\sigma}{\mathrm{d}\tau} = 0 .$$

Prove the following formulae

$$g_{\mu\nu} = \eta_{\alpha\beta} \frac{\partial z^\alpha}{\partial y^\mu} \frac{\partial z^\beta}{\partial y^\nu} , \quad \Gamma^\mu_{\rho\sigma} = \frac{\partial y^\mu}{\partial z^\alpha} \frac{\partial^2 z^\alpha}{\partial y^\rho \partial y^\sigma} .$$

Derive the transformation formulae for Christoffel symbols under a diffeomorphism $\{y^\mu\} \mapsto \{y'^\mu\}$. The above conclusion follows from the result.

6.7 The semi-Riemannian manifold (M, g) is assumed to have dimension n. Show that the contraction of the metric gives n and that for smooth functions $f \in \mathfrak{F}(M)$ the divergence of $f\mathrm{g}$ is equal to the exterior derivative of f, $\mathbf{div}(f\mathrm{g}) = \mathrm{d}f$.

6.8 A tensor field which is closely related to the Riemann tensor field R is the *Weyl tensor field*. It is a function of the Riemann tensor field R, of the Ricci tensor field $\mathrm{R}^{(\mathrm{Ricci})}$, and of the curvature scalar S. When written in components it is defined as follows

$$\begin{aligned} C_{\mu\nu\sigma\tau} &:= R_{\mu\nu\sigma\tau} + \frac{1}{6} S \left(g_{\mu\sigma} g_{\nu\tau} - g_{\mu\tau} g_{\nu\sigma} \right) \\ &- \frac{1}{2} \left(g_{\mu\sigma} R^{(\mathrm{Ricci})}_{\nu\tau} - g_{\mu\tau} R^{(\mathrm{Ricci})}_{\nu\sigma} - g_{\nu\sigma} R^{(\mathrm{Ricci})}_{\mu\tau} + g_{\nu\tau} R^{(\mathrm{Ricci})}_{\mu\sigma} \right) . \end{aligned}$$

The tensor $C_{\mu\nu\sigma\tau}$ has the same symmetry properties as $R_{\mu\nu\sigma\tau}$. Show: All its contractions give zero. In dimension $n = 4$ it has ten independent components, in dimension $n = 3$ it is identically zero.

If M has dimension 4 and is endowed with a conformally flat metric, i.e. if $g_{\mu\nu} = \phi^2(x)\eta_{\mu\nu}$ holds with $\phi(x)$ a smooth function, the tensor field C is identically zero.

Selected Solutions of the Exercises

Chapter 1 (selected solutions)

1.1 The spherical base vectors $\hat{e}_m$, $m = -1, 0, +1$, have the following properties (please verify!):

$$\hat{e}^*_m = (-)^m \hat{e}_{-m} \qquad \hat{e}^*_m \cdot \hat{e}_{m'} = \delta_{mm'} \ . \tag{1}$$

Expanding vectors in terms of these, $V = \sum_{m=-1}^{+1} v^m \hat{e}_m$, and recalling that V is real, there follows

$$V = \sum v^{m\,*} \hat{e}^*_m = \sum (-)^m v^{-m\,*} \hat{e}_m \ . \tag{2}$$

Equations (1) and (2) show that the basis which is dual to the basis $\{\hat{e}_m\}$ is given by $\hat{e}^m = (-)^m \hat{e}_{-m}$ and that one has $V = \sum v_m \hat{e}^m$ with $v_m = (-)^m v^{-m}$. The scalar product of two vectors reads

$$V \cdot W = \sum_{m=-1}^{+m} v^{m\,*} w^m = \sum_{m=-1}^{+1} v_m w^m \ . \tag{3}$$

Indeed, one verifies the scalar product by calculating

$$\begin{aligned}\sum v_m w^m &= \frac{1}{2}(v^1 - iv^2)(w^1 + iw^2) + v^3 w^3 + \frac{1}{2}(v^1 + iv^2)(w^1 - iw^2) \\ &= v^1 w^1 + v^2 w^2 + v^3 w^3 \ .\end{aligned}$$

This exercise shows that even in a Euclidean space one must distinguish covariant from contravariant transformation behaviour whenever one uses a complex basis instead of a real cartesian basis.

1.2 With a specific partition into space and time components one has

$$\begin{aligned}\partial_0 j^0 &= e\frac{\partial}{\partial t}\delta^{(3)}(\boldsymbol{y}-\boldsymbol{x}(t)) = e\dot{\boldsymbol{x}}\cdot\nabla_x\delta^{(3)}(\boldsymbol{y}-\boldsymbol{x}(t))\\ &= -e\dot{\boldsymbol{x}}\cdot\nabla_y\delta^{(3)}(\boldsymbol{y}-\boldsymbol{x}(t))\,,\\ \partial_i j^i &= e\dot{\boldsymbol{x}}\cdot\nabla_y\delta^{(3)}(\boldsymbol{y}-\boldsymbol{x}(t))\,.\end{aligned}$$

The sum of these expressions gives zero.

1.3 Except for binding effects one has

$$Mc^2 = 235\cdot 939\,\text{MeV} = 3.535\cdot 10^{-8}\,\text{J}\,.$$

Using the conversion formula (see, e. g. [QP] Appendix A.7) $1\,\text{eV c}^{-2} = 1.78266\cdot 10^{-27}\,\mu\text{g}$, one finds the approximate value $M = 3.9\cdot 10^{-16}\,\mu\text{g}$.

Remark
This value is still very small as compared to the Planck mass

$$M_{\text{Planck}} := \sqrt{\frac{\hbar c}{G_{\text{Newton}}}} = 22.2\,\mu\text{g}\,,$$

which could be measured with a chemist's balance.

1.4 The muonic Bohr radius is smaller by the factor m_e/m_μ than the one of the electron. For the example of lead, i.e. with $Z = 82$, it is

$$a_{\text{B}}^{(\mu)}(Z=82) = \frac{\hbar c}{Z\alpha m_\mu c^2} = 3.12\cdot 10^{-15}\,\text{m}\,.$$

This value is smaller than the nuclear radius which is about $7\cdot 10^{-15}$ m. If instead the entire charge of the nucleus were localized in its centre the magnitude of the electric field at the position of the muon would be

$$|\boldsymbol{E}| = \frac{Ze}{r^2} = 1.35\cdot 10^{12}\,\text{V m}^{-1}\,.$$

The realistic value which is smaller than this, can be estimated by a model of the lead nucleus in which the charge distribution is homogeneous and has the radius given above.

1.5 There are different ways to check the relation (1.48a).
(a) For fixed values of i and j also k is fixed. It cannot be equal to i or j. This holds also for l and m which cannot be equal to k. As they must be different from each other there remain the possibilities ($i = l, j = m$) and ($i = m, j = l$) only. The first of these appears with the positive sign, the second appears with the negative sign.

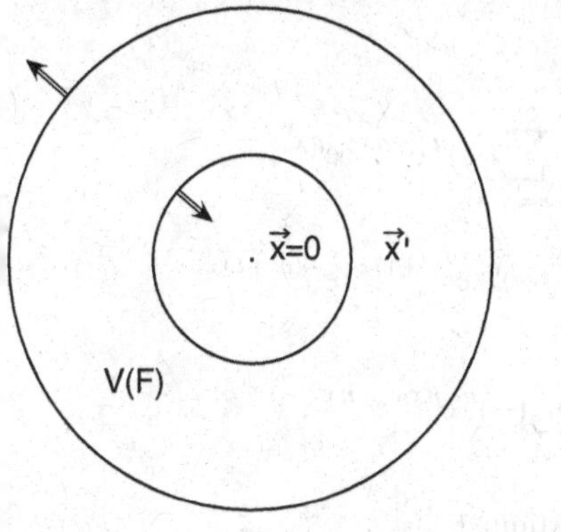

Fig. 1

(b) With $\{\hat{e}_i\}$, $i = 1, 2, 3$ an orthonormal basis in $\mathbb{R}^3$, the first ε-symbol equals the scalar triple product $(\hat{e}_i \times \hat{e}_j) \cdot \hat{e}_k$. Likewise the second ε-symbol equals $\hat{e}_k \cdot (\hat{e}_l \times \hat{e}_m)$. As the sum $\sum_k |\hat{e}_k\rangle\langle\hat{e}_k|$ is equal to 1 the required expression is equal to

$$(\hat{e}_i \times \hat{e}_j) \cdot (\hat{e}_l \times \hat{e}_m) .$$

This is the right-hand side of the equation that was to be proved.

1.6 The space between the two spheres of Fig. 1 defines the volume $V(F)$. Its surface consists of the sphere with radius R_{a} whose normal points outwards, and the sphere with radius R_{i} whose normal points inwards. With $\mathbf{\Delta}(1/r) = 0$ in the intermediate space and with $\mathbf{\Delta}\Phi(\boldsymbol{x}) = -f(\boldsymbol{x})$ Green's second theorem yields

$$\iiint\limits_{V(F)} \mathrm{d}^3x \, \frac{f(\boldsymbol{x})}{r} = \iint\limits_{F} \mathrm{d}\sigma \left\{ -\Phi \frac{1}{r^2} - \frac{1}{r} \frac{\partial \Phi}{\partial r} \right\} ,$$

the centre of the rings being the origin. Both spheres contribute to the right-hand side and we have $\mathrm{d}\sigma = r^2 \mathrm{d}\Omega$. The second term vanishes both in the limit $R_{\mathrm{a}} \to \infty$ and in the limit $R_{\mathrm{i}} \to 0$. While for $R_{\mathrm{a}} \to \infty$ the first term vanishes, too, for $R_{\mathrm{i}} \to 0$ it gives $4\pi\Phi(0)$. This is the answer that was to be proved.

1.7 In order to normalize the given distribution one must calculate the integral

$$I := 4\pi \int_0^\infty r^2 \mathrm{d}r \, \frac{1}{1 + \mathrm{e}^{(r-c)/z}} = 4\pi z^3 \int_0^\infty x^2 \mathrm{d}x \, \frac{1}{1 + \mathrm{e}^{(x-x_0)}} \tag{4}$$

where $x = r/z$ and $x_0 = c/z$. The domain of integration is split into the intervals $[0, x_0)$ and $[x_0, \infty)$, such that in either case the integrand may be written as a geometric series, viz.

$$x < x_0 : \quad \frac{1}{1+\mathrm{e}^{(x-x_0)}} = 1 + \sum_{n=1}^{\infty} (-)^n \, \mathrm{e}^{-nx_0} \, \mathrm{e}^{nx} ,$$

$$x \geqslant x_0 : \quad \frac{1}{1+\mathrm{e}^{(x-x_0)}} = \sum_{n'=0}^{\infty} (-)^{n'} \, \mathrm{e}^{(n'+1)x_0} \, \mathrm{e}^{-(n'+1)x}$$

$$= - \sum_{n=1}^{\infty} (-)^n \, \mathrm{e}^{nx_0} \, \mathrm{e}^{-nx} ,$$

where in the last step $n = n' + 1$ was substituted.

The following two integration formulae are useful for the sequel and are easily derived:

$$I_< := \int_0^a \mathrm{d}x \, x^2 \, \mathrm{e}^x = \mathrm{e}^a (a^2 - 2a + 2) - 2 ,$$

$$I_> := \int_a^\infty \mathrm{d}x \, x^2 \, \mathrm{e}^{-x} = \mathrm{e}^{-a} (a^2 + 2a + 2) .$$

Using these formulae the required integral becomes

$$I = 4\pi z^3 \left\{ \frac{1}{3} x_0^3 + 4x_0 \sum_{n=1}^{\infty} (-)^{n+1} \frac{1}{n^2} - 2 \sum_{n=1}^{\infty} (-)^n \frac{1}{n^3} \mathrm{e}^{-nx_0} \right\} .$$

The infinite sum in the second term can be found, e. g. in [Abramovitz-Stegun; Eq. (23.2.19) and Eq. (23.2.24)]

$$\sum_{n=1}^{\infty} \frac{(-)^{n+1}}{n^2} = -\frac{1}{2} \zeta(2) = \frac{\pi^2}{12} ,$$

where $\zeta(x)$ is Riemann's zeta function. As a result one obtains

$$I = \frac{4\pi c^3}{3} \left\{ 1 + \left(\frac{\pi z}{c} \right)^2 - 6 \left(\frac{z}{c} \right)^3 \sum_{n=1}^{\infty} \frac{(-)^n}{n^3} \mathrm{e}^{-(nc)/z} \right\} \tag{5}$$

and, from this, the formula given in the main text.

In charge distributions of atomic nuclei c, as a rule, is sensibly larger than z (typical values are $c = 6$ fm, $z = 0.2$ fm), i.e. $\exp\{-c/z\} \ll 1$. If one neglects the last term the distribution at $r = c$ takes about half the value it has at $r = 0$. The distance between the radii $r_{0.9}$ and $r_{0.1}$ where it is still 90% and 10% of its value at $r = 0$, respectively, is given by $t = 4 \ln(3) z$. This parameter is customarily quoted as the *surface thickness* of the charge distribution.

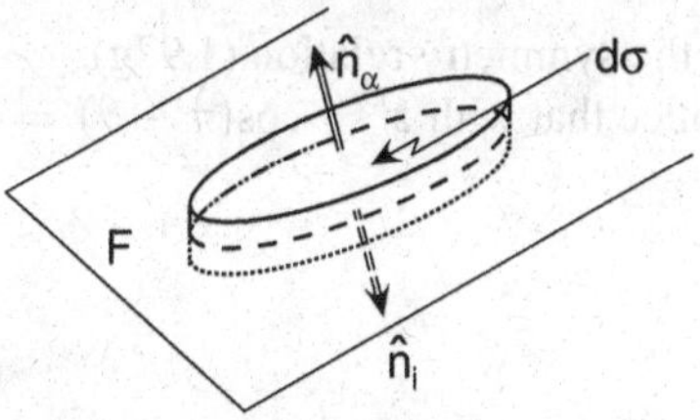

Fig. 2

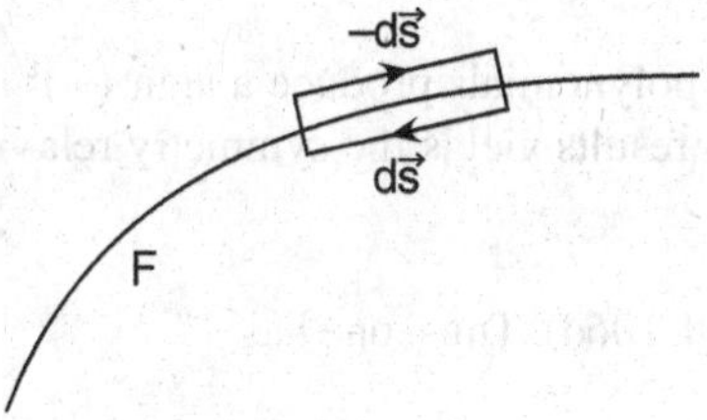

Fig. 3

1.8 The integral over the volume of the "box" in Fig. 2 gives 4π times the surface charge density. The height being assumed to be small of third order, the surface integral receives contributions only from the two end faces of the box which differ by the direction of the normal. Thus, one obtains $(\boldsymbol{E}_{\mathrm{a}} - \boldsymbol{E}_{\mathrm{i}}) \cdot \hat{\boldsymbol{n}}$.

1.9 Choosing the closed path in Fig. 3 such that it cuts through the surface with its short sides, one concludes

$$\oint \boldsymbol{E} \cdot \mathrm{d}\boldsymbol{s} = (\boldsymbol{E}_{\mathrm{a}} - \boldsymbol{E}_{\mathrm{i}}) \cdot \hat{\boldsymbol{t}} = 0 .$$

This shows the continuity of the tangential component.

1.10 At first, the expression (1.97c) for the Legendre functions of the first kind holds only for $m \geqslant 0$. The following alternative representation[6]

$$P_\ell^m(z) = (-)^m \mathrm{e}^{-\mathrm{i}m\pi/2} \frac{(\ell+m)!}{2\pi \ell!} \cdot \int_{-\pi}^{+\pi} \mathrm{d}\psi \left(\cos\theta + \mathrm{i}\sin\theta\cos\psi\right)^\ell \cos(m\psi) , \qquad (z = \cos\theta)$$

holds for all values of m. From this one obtains the symmetry relation

$$P_\ell^{-m}(z) = (-)^m P_\ell^m(z) \frac{(\ell-m)!}{(\ell+m)!} . \tag{6}$$

[6] see e. g. N. Straumann, *Quantenmechanik*, Springer, Heidelberg 2002, Eq. (1.168). English Translation in preparation.

Inserting this into formula (1.97a) one obtains the symmetry relation (1.97g).

For the derivation of the relation (1.97h) notice that with $z' = \cos(\pi - \theta) = -\cos\theta = -z$ one finds

$$e^{im(\phi+\pi)} = (-)^m \, e^{im\phi} \, ,$$

$$P_\ell^m(-z) = (-)^m (1-z^2)^{m/2} (-)^m \frac{d^m}{dz^m} P_\ell(-z) = (-)^{\ell-m} P_\ell^m(z) \, .$$

Here one has used the fact that the Legendre polynomials produce a sign $(-)^\ell$ when $z \mapsto (-z)$. The composition of the two results yields the symmetry relation (1.97h).

1.11 The multipole moments are defined in (1.106d). Thus, one has

$$\begin{aligned} q_{22} &= \iiint d^3x \; r^2 Y_{22}^*(\hat{\boldsymbol{x}}) \varrho(\boldsymbol{x}) \\ &= \frac{\sqrt{15}}{4\sqrt{2\pi}} \iiint d^3x \; r^2 \sin^2\theta \, e^{-2i\phi} \varrho(\boldsymbol{x}) \\ &= \frac{\sqrt{15}}{4\sqrt{2\pi}} \iiint d^3x \; \left(x^1 - ix^2\right)^2 \varrho(\hat{\boldsymbol{x}}) \\ &= \frac{\sqrt{15}}{4\sqrt{2\pi}} \iiint d^3x \; \left\{x^1x^1 - 2ix^1x^2 - x^2x^2\right\} \varrho(\hat{\boldsymbol{x}}) \\ &= \frac{\sqrt{15}}{4\sqrt{2\pi}} \frac{1}{3} \left(Q^{11} - 2iQ^{12} - Q^{22}\right) . \end{aligned}$$

The two other components are calculated in the same manner

$$q_{21} = -\frac{\sqrt{15}}{2\sqrt{2\pi}} \iiint d^3x \; x^3(x^1 - ix^2) \varrho(\boldsymbol{x}) = \frac{\sqrt{5}}{2\sqrt{6\pi}} \left(-Q^{13} + iQ^{23}\right) ,$$

$$q_{20} = \frac{\sqrt{5}}{4\sqrt{\pi}} \iiint d^3x \; \left(3x^3x^3 - r^2\right) \varrho(\hat{\boldsymbol{x}}) = \frac{\sqrt{5}}{4\sqrt{\pi}} Q^{33} \, .$$

In this calculation the symmetry $Q^{ji} = Q^{ij}$ and the definition (1.111c) were utilized.

1.12 One calculates first the integral over the volume V of a ball with radius R and centre at the position of the dipole. This integral is converted to a surface integral over the sphere by means of (1.6):

$$\iiint_V \mathrm{d}^3x\, \boldsymbol{E}_\mathrm{D}(\boldsymbol{x}) = -\iiint_V \mathrm{d}^3x\, \nabla\Phi_\mathrm{D}(\boldsymbol{x}) = -R^2 \iint_{F(V)} \mathrm{d}\sigma\, \Phi_\mathrm{D}(\boldsymbol{x})\,\hat{\boldsymbol{n}}$$

$$= -R^2 \frac{4\pi}{3} \iiint \mathrm{d}^3x'$$

$$\cdot \iint_{F(V)} \mathrm{d}\sigma\, \frac{r_<}{r_>^2} \sum_\mu Y^*_{1\mu}(\hat{x}) Y_{1\mu}(\hat{x}')\,\hat{\boldsymbol{n}}\ .$$

Here the multipole expansion (1.105) was used of which, by the integration over the angular variables, only the term with $\ell = 1$ contributes. Consider the unit vector $\hat{\boldsymbol{n}}$, expanded in terms of spherical harmonics,

$$\hat{\boldsymbol{n}} = \sum_{m=-1}^{+1} a_m Y_{1m}(\hat{x})\ .$$

The surface integral over the sphere is then found to be

$$\left(\iint_{F(V)} \mathrm{d}\sigma \sum_\mu Y^*_{1\mu}(\hat{x})\,\hat{\boldsymbol{n}}\right) Y_{1\mu}(\hat{x}') = \sum_m a_m Y_{1m}(\hat{x}') = \hat{\boldsymbol{n}}'\ .$$

As the dipole is localized, one has $r_< = r'$ and $r_> = R$ so that

$$\iiint_V \mathrm{d}^3x\, \boldsymbol{E}_\mathrm{D}(\boldsymbol{x}) = -R^2 \frac{4\pi}{3}\frac{1}{R^2} \iiint \mathrm{d}^3x'\, \varrho(\boldsymbol{x}') r' \hat{\boldsymbol{n}}' = -\frac{4\pi}{3}\boldsymbol{d}\ .$$

This derivation shows that the formula to be proved holds for every localized dipole distribution and, hence, for the point-like dipole as a special case.

1.13 The capacitor consists of two identical plates in parallel orientation, their surface being F. In the initial state they carry the charges $+q$ and $-q$, respectively. The normal component of the displacement field $\boldsymbol{D} = \varepsilon\boldsymbol{E}$ fulfills the relation $\boldsymbol{D}_n = |\boldsymbol{D}| = 4\pi\eta$, where $\eta = q/F$ is the surface charge density (we are considering the example of the positively charged plate). Short-circuiting the plates causes a current

$$I = \frac{\mathrm{d}q}{\mathrm{d}t} = F\frac{\partial\eta}{\partial t} = \frac{F}{4\pi}\frac{\partial \boldsymbol{D}_n}{\partial t}$$

in the cable joining them. Thus, the current density of the displacement current is

$$\boldsymbol{j}_\mathrm{v} = \frac{1}{4\pi}\frac{\partial \boldsymbol{D}}{\partial t}\ .$$

These relations hold for the vector fields proper because they are perpendicular to the surface of the plates and, hence, are equal to their normal components.

1.14 This example illustrates the method of *image charges*. Choose two point charges $q^{(1)}$ and $q^{(2)}$ on a straight line through the origin and define this to be the 1-axis. The positions of the charges $\boldsymbol{x}^{(1)} = r^{(1)}\hat{\boldsymbol{e}}_1$ and $\boldsymbol{x}^{(2)} = r^{(2)}\hat{\boldsymbol{e}}_1$, respectively, are images of each other with respect to the sphere with radius R and centre the origin which means that $r^{(1)}r^{(2)} = R^2$. Determine the charge $q^{(2)}$ such that the potential vanishes on the sphere. The potential at $|\boldsymbol{x}| = R$ is

$$\begin{aligned}\Phi(\boldsymbol{x})\restriction_R &= \left[\frac{q^{(1)}}{|\boldsymbol{x} - r^{(1)}\hat{\boldsymbol{e}}_1|} + \frac{q^{(2)}}{|\boldsymbol{x} - r^{(2)}\hat{\boldsymbol{e}}_1|}\right]_{|\boldsymbol{x}|=R} \\ &= \frac{q^{(1)}}{R|\hat{\boldsymbol{x}} - (r^{(1)}/R)\hat{\boldsymbol{e}}_1|} + \frac{q^{(2)}}{r^{(2)}|(R/r^{(2)})\hat{\boldsymbol{x}} - \hat{\boldsymbol{e}}_1|}\,.\end{aligned}$$

In this formula the two absolute values in the denominators are equal, viz.

$$\left|\hat{\boldsymbol{x}} - \left(\frac{r^{(1)}}{R}\right)\hat{\boldsymbol{e}}_1\right| = \left|\left(\frac{R}{r^{(2)}}\right)\hat{\boldsymbol{x}} - \hat{\boldsymbol{e}}_1\right| = 1 - 2\frac{r^{(1)}}{R}\hat{\boldsymbol{x}}\cdot\hat{\boldsymbol{e}}_1 + \left(\frac{r^{(1)}}{R}\right)^2 .$$

Therefore, the potential vanishes for $r = R$ if the second charge is chosen to be $q^{(2)} = -q^{(1)}(r^{(2)}/R)$. One can now replace the point charge in the interior of the sphere by the sphere with $\Phi|_R = 0$. This solves the problem.

1.15 Place the dipole in the origin. In the absence of the sphere the potential it creates would be $\Phi_\mathrm{D}(x) = d\ r\cos\theta/r^3$. The presence of the sphere causes an additive term which is such that the potential on the sphere is a constant. Without loss of generality one can choose this constant to be zero. Then the total potential in the interior of the sphere is

$$\begin{aligned}\Phi &= d\cos\theta\frac{1}{r^2} + \sum_{\ell=0}^{\infty} a_\ell r^\ell P_\ell(\cos\theta) \\ &= d\cos\theta\frac{1}{r^2} + a_1 r\cos\theta \\ &= dr\cos\theta\left(\frac{1}{r^3} - \frac{1}{R^3}\right).\end{aligned}$$

Here use was made of the fact that only the term with $\ell = 1$ contributes, the boundary condition $\Phi(R) = 0$ fixes the coefficient a_1 to be $a_1 = -d/R^3$. The radial component and the θ-component of the electric field are, respectively,

$$\begin{aligned}E_r &= -\frac{\partial\Phi}{\partial r} = d\cos\theta\left(\frac{2}{r^3} + \frac{1}{R^3}\right), \\ E_\theta &= -\frac{1}{r}\frac{\partial\Phi}{\partial\theta} = d\sin\theta\left(\frac{1}{r^3} - \frac{1}{R^3}\right).\end{aligned}$$

At $r = R$ the θ-component vanishes. The discontinuity of the radial component follows from (1.87a),

$$(E_r)_a - (E_r)_i = (E_r)_a - d\cos\theta \frac{3}{R^3} = 4\pi\eta\,.$$

By (1.92c) the induced surface charge density on the sphere is

$$\eta = \frac{1}{4\pi}\frac{\partial \Phi}{\partial \hat{n}}\bigg|_{r=R} = -\frac{d\cos\theta}{4\pi}\frac{3}{R^3}\,.$$

(Only the interior contributes to the derivative. The normal of the surface is the negative normal to the sphere, hence the minus sign.) One concludes $(E_r)_a = 0$.

1.16 The potential created by the dipole is calculated as follows:

$$\begin{aligned}\Phi(\boldsymbol{x}) &= \iiint \mathrm{d}^3x \,\frac{\varrho_{\text{eff}}(\boldsymbol{x}')}{|\boldsymbol{x}-\boldsymbol{x}'|} = -\boldsymbol{d}\cdot \iiint \mathrm{d}^3x\,\frac{\nabla_{x'}\delta(\boldsymbol{x}-\boldsymbol{x}^{(0)})}{|\boldsymbol{x}-\boldsymbol{x}'|}\\ &= \boldsymbol{d}\cdot\iiint \mathrm{d}^3x\,\delta(\boldsymbol{x}-\boldsymbol{x}^{(0)})\nabla_{x'}\frac{1}{|\boldsymbol{x}-\boldsymbol{x}'|} = \boldsymbol{d}\cdot\frac{\boldsymbol{x}-\boldsymbol{x}^{(0)}}{|\boldsymbol{x}-\boldsymbol{x}^{(0)}|^3}\,.\end{aligned}$$

This expression coincides with (1.88c). The energy in the external potential is

$$\begin{aligned}W &= \iiint \mathrm{d}^3x\,\varrho_{\text{eff}}(\boldsymbol{x})\Phi_a(\boldsymbol{x})\\ &= -\boldsymbol{d}\cdot\iiint \mathrm{d}^3x\,(\nabla_x\delta(\boldsymbol{x}-\boldsymbol{x}^{(0)}))\Phi_a(\boldsymbol{x})\\ &= \boldsymbol{d}\cdot\iiint \mathrm{d}^3x\,\delta(\boldsymbol{x}-\boldsymbol{x}^{(0)})\nabla_x\Phi_a(\boldsymbol{x}) = -\boldsymbol{d}\cdot\boldsymbol{E}_a(\boldsymbol{x}^{(0)})\,.\end{aligned}$$

This is the known expression for the energy of the electric dipole in an external electric field.

1.17 Without the external field, $\boldsymbol{E}_0 = 0$, the potential would be the potential outside of a spherically symmetric charge distribution, $\Phi(r) = Q/r$; in the absence of the sphere it would be $\Phi(\boldsymbol{x}) = -E_0 z = -E_0 r\cos\theta$. As potentials are additive in the sources one is led to the suggested ansatz, i.e. the sum of a spherically symmetric term and a term with the characteristic $\cos\theta$-dependence of the homogeneous field, viz.

$$\Phi(\boldsymbol{x}) = f_0(r) + f_1(r)\cos\theta\,.$$

Inserting into the Laplace equation $\mathbf{\Delta}\Phi(\boldsymbol{x}) = 0$ one finds

$$\begin{aligned}
\mathbf{\Delta}\Phi(\boldsymbol{x}) &= \frac{1}{r^2}\frac{\mathrm{d}}{\mathrm{d}r}\left(r^2\frac{\mathrm{d}f_0}{\mathrm{d}r}\right) + \frac{1}{r^2}\frac{\mathrm{d}}{\mathrm{d}r}\left(r^2\frac{\mathrm{d}f_1}{\mathrm{d}r}\right)\cos\theta \\
&\quad - \frac{f_1}{r^2\sin\theta}\frac{\mathrm{d}}{\mathrm{d}\theta}(\sin^2\theta) \\
&= \left[\frac{1}{r^2}\frac{\mathrm{d}}{\mathrm{d}r}\left(r^2\frac{\mathrm{d}f_0}{\mathrm{d}r}\right)\right] + \left[\frac{1}{r^2}\frac{\mathrm{d}}{\mathrm{d}r}\left(r^2\frac{\mathrm{d}f_1}{\mathrm{d}r}\right) - \frac{2f_1}{r^2}\right]\cos\theta = 0\,.
\end{aligned}$$

This differential equation holds for all r and all θ. Therefore, the expressions in the big square brackets must both be zero individually:

$$\frac{1}{r^2}\frac{\mathrm{d}}{\mathrm{d}r}\left(r^2\frac{\mathrm{d}f_0}{\mathrm{d}r}\right) = 0\,, \tag{7}$$

$$\frac{1}{r^2}\frac{\mathrm{d}}{\mathrm{d}r}\left(r^2\frac{\mathrm{d}f_1}{\mathrm{d}r}\right) - \frac{2f_1}{r^2} = 0\,. \tag{8}$$

The first of these, equation (7), has the general solution $f_0(r) = A/r + B$. The second differential equation (8) reads $r^2 f_1'' + 2rf_1' - 2f_1 = 0$, and its general solution is $f_1(r) = C/r^2 + Dr$. Thus, the required solution has the form

$$\Phi(\boldsymbol{x}) = \frac{A}{r} + \left(\frac{C}{r^2} + Dr\right)\cos\theta + B\,. \tag{9}$$

The four constants are obtained from the boundary conditions:

a) For $r \to \infty$ the term proportional to $r\cos\theta$ dominates. In this limit only the given external field is felt so that one must have $D = -E_0$.
b) On the sphere the potential must be constant,

$$\Phi(\boldsymbol{x})|_{r=R} = \frac{A}{R} + \left(\frac{C}{R^2} - E_0 R\right)\cos\theta = \text{const.} \quad \forall\ \theta\,;$$

from which follows $C = E_0 R^3$.
c) Gauss' theorem yields the normalization condition

$$\iint\limits_{r=R} \mathrm{d}\sigma\ \boldsymbol{E}\cdot\hat{\boldsymbol{n}} = -\iint\limits_{r=R} \mathrm{d}\sigma\ \frac{\partial\Phi}{\partial r} = 4\pi Q\,.$$

By (9) the same integral is equal to $4\pi A$, giving $A = Q$.
Thus, the solution is

$$\Phi(\boldsymbol{x}) = \frac{Q}{r} + E_0\left(\frac{R^3}{r^2} - r\right)\cos\theta\,. \tag{10}$$

For the case $Q = 0$ the induced surface charge density is calculated to be

$$\eta(\theta) = -\frac{1}{4\pi}\frac{\partial \Phi}{\partial r} = \frac{3}{4\pi} E_0 \cos\theta \,.$$

(Sketch a plane cutting the equipotential surfaces which contains the z-axis, as well as the electric field lines for the example $Q = 0$!)

Remark
The justification for our ansatz is an intuitive one. A more systematic approach makes use of the expansion of the potential in terms of spherical harmonics: By the axial symmetry of the problem the most general ansatz is

$$\Phi(\boldsymbol{x}) = \sum_{\ell=0}^{\infty} f_\ell(r) P_\ell(\cos\theta)$$

with $f_\ell = r^\ell$ or $f_\ell = r^{-\ell-1}$. The potential of the original homogeneous field is axially symmetric and is proportional to $P_1(\cos\theta)$. The added sphere does not modify the angular dependence and therefore causes an additive monopole term only. Thus, the solution must have the form

$$\Phi(\boldsymbol{x}) = \left(\frac{A}{r} + B\right) P_0(\cos\theta) + \left(\frac{C}{r^2} + Dr\right) P_1(\cos\theta)$$
$$\big(P_0(\cos\theta) = 1\,, \quad P_1(\cos\theta) = \cos\theta\big)\,.$$

The constants are determined as before.

1.18 With $\varrho(r) = 3Q/(4\pi R^3)\Theta(R - r)$ the potential and the field strength inside and outside are, respectively,

$$r \leqslant R: \quad \Phi_{\mathrm{i}}(r) = \frac{3Q}{2R^3}\left(R^2 - \frac{1}{3}r^2\right), \quad \boldsymbol{E}_{\mathrm{i}} = \frac{Q}{R^3} r\,\hat{\boldsymbol{e}}_r\,,$$

$$r > R: \quad \Phi_{\mathrm{a}}(r) = \frac{Q}{r}\,, \quad \boldsymbol{E}_{\mathrm{a}} = \frac{Q}{r^2}\hat{\boldsymbol{e}}_r\,.$$

Calculating the energy from the square of the electric field and integrating over the whole space,

$$W_{\mathrm{E}} = \frac{1}{8\pi}\iiint \mathrm{d}^3x\, \boldsymbol{E}^2 = \frac{1}{2}\int_0^\infty r^2\,\mathrm{d}r\, \boldsymbol{E}^2$$
$$= \frac{1}{2}Q^2\left\{\frac{1}{R^6}\int_0^R \mathrm{d}r\, r^4 + \int_R^\infty \mathrm{d}r\, r^{-2}\right\} = \frac{3Q^2}{5R}\,,$$

one obtains the same result as from the given formula, viz.

$$W = \frac{1}{2} \iiint \mathrm{d}^3x\, \varrho(r)\Phi(r) = \frac{9Q^2}{4R^6} \int_0^\infty r^2\, \mathrm{d}r \left(R^2 - \frac{1}{3}r^2\right) = \frac{3Q^2}{5R} .$$

1.19 The energy contained in the field outside the electron is equal to

$$W_\mathrm{a} = \frac{1}{2} e^2 \int_R^\infty \mathrm{d}r\, \frac{1}{r^2} = \frac{e^2}{2R} .$$

Putting this equal to $m_e c^2$ one obtains the given value of R, i.e. the classical electron radius.

1.20 It follows from Maxwell's equations that on a surface carrying the charge density η the tangential component of the electric field is continuous while the normal component of the displacement field changes by the amount $4\pi\eta$. In the problem at stake one has $\eta = 0$, so that the normal component of $\boldsymbol{D}$ is continuous. Thus, the boundary conditions at $r = R$ are

$$\Phi_\mathrm{i} = \Phi_\mathrm{a} ,$$

$$\varepsilon_0 \frac{\partial \Phi_\mathrm{a}}{\partial r} = \varepsilon_1 \frac{\partial \Phi_\mathrm{i}}{\partial r} .$$

The spherical symmetry of the set-up is perturbed only by the external field which is axially symmetric and whose potential is $\Phi(r, \theta) = -E_0 P_1(\cos\theta)$. Thus, both in the inner and the outer regions the problem must have the general solution

$$\Phi_\mathrm{a} = \left(\frac{A}{r^2} + Br\right) P_1(\cos\theta) ,$$

$$\Phi_\mathrm{i} = \left(\frac{C}{r^2} + Dr\right) P_1(\cos\theta) .$$

Letting r go to infinity, $r \to \infty$, one sees that one must have $B = -E_0$; letting $r \to 0$ one concludes that $C = 0$.

The boundary conditions fix the remaining two constants as follows

$$A = \frac{\varepsilon_1 - \varepsilon_0}{\varepsilon_1 + 2\varepsilon_0} E_0 R^3 ,$$

$$D = -\frac{3\varepsilon_0}{\varepsilon_1 + 2\varepsilon_0} E_0 .$$

The two special cases specified in the problem are as follows:

a) $\varepsilon_0 \equiv \varepsilon$, $\varepsilon_1 = 1$: Here the potential is

$$\Phi_{\mathrm{a}} = \left[\frac{1-\varepsilon}{1+2\varepsilon}\frac{R^3}{r^3} - 1\right] r E_0 P_1(\cos\theta)\,,$$

$$\Phi_{\mathrm{i}} = -\frac{3\varepsilon}{1+2\varepsilon} r E_0 P_1(\cos\theta)\,.$$

The field inside the sphere has the modulus

$$E_{\mathrm{i}} = \frac{3\varepsilon}{1+2\varepsilon} E_0\,.$$

As $\varepsilon > 1$ it is *larger* than E_0.
b) $\varepsilon_0 = 1$, $\varepsilon_1 \equiv \varepsilon$: The potential now reads

$$\Phi_{\mathrm{a}} = \left[\frac{\varepsilon-1}{\varepsilon+2}\frac{R^3}{r^3} - 1\right] r E_0 P_1(\cos\theta)\,,$$

$$\Phi_{\mathrm{i}} = -\frac{3}{\varepsilon+2} r E_0 P_1(\cos\theta)\,.$$

The modulus of the field inside is

$$E_{\mathrm{i}} = \frac{3}{\varepsilon+2} E_0$$

and hence is *smaller* than the external field.

Choosing in this case $\varepsilon \gg 1$, one has

$$\Phi_{\mathrm{a}} \simeq \left[\frac{R^3}{r^3} - 1\right] r E_0 P_1(\cos\theta)\,, \quad \Phi_{\mathrm{i}} \simeq 0\,.$$

The field on the inside goes to zero, and one recovers the situation dealt with in Exercise 1.17 (with $Q = 0$).

1.21 The charge density created by the four point charges is

$$\varrho(x) = \frac{e}{2}\{[\delta(x-a) + \delta(x+a)]\,\delta(y)\delta(z) - [\delta(y-b) + \delta(y+b)]\,\delta(x)\delta(z)\}\,.$$

One verifies at once that both the monopole moment

$$q_{00} = \frac{1}{\sqrt{4\pi}} \times \text{ total charge } = 0\,,$$

and all dipole moments

$$d_i = \iiint \mathrm{d}^3x\, x_i \varrho(\boldsymbol{x}) = 0$$

are equal to zero. The entries Q_{ij} of the quadrupole tensor are calculated to be

$$\begin{aligned}
Q_{11} &= \iiint \mathrm{d}^3x\, \left[2x^2 - y^2 - z^2\right] \varrho(\boldsymbol{x}) = e\left(2a^2 + b^2\right) , \\
Q_{22} &= \iiint \mathrm{d}^3x\, \left[2y^2 - z^2 - x^2\right] \varrho(\boldsymbol{x}) = -e\left(2b^2 + a^2\right) , \\
Q_{33} &= \iiint \mathrm{d}^3x\, \left[2z^2 - x^2 - y^2\right] \varrho(\boldsymbol{x}) = e\left(-a^2 + b^2\right) , \\
Q_{12} &= \iiint \mathrm{d}^3x\, 3xy\varrho(\boldsymbol{x}) = 0 , \\
&\text{and, analogously,} \quad Q_{13} = 0 , \ Q_{23} = 0 .
\end{aligned}$$

Thus, $\mathrm{Q} = e\, \mathrm{diag}(2a^2 + b^2, -2b^2 - a^2, -a^2 + b^2)$ and one confirms that Q has trace zero.

The spectroscopic quadrupole moment is

$$\begin{aligned}
Q_0 &= \iiint \mathrm{d}^3x\, r^2(3\cos^2\theta - 1)\varrho(\boldsymbol{x}) \\
&= \iiint \mathrm{d}^3x\, (2z^2 - x^2 - y^2) = Q_{33} = e(b^2 - a^2) .
\end{aligned}$$

In the spherical basis the moments are found to be

$$\begin{aligned}
q_{22} &= \frac{\sqrt{5}}{4\sqrt{6\pi}}(Q_{11} - 2\mathrm{i}Q_{12} - Q_{22}) = \frac{\sqrt{15}}{4\sqrt{2\pi}} e(a^2 + b^2) , \\
q_{21} &= \frac{5}{2\sqrt{6\pi}}(-Q_{13} + \mathrm{i}Q_{23}) = 0 , \\
q_{20} &= \frac{\sqrt{5}}{4\sqrt{\pi}} Q_{33} = \frac{\sqrt{5}}{4\sqrt{\pi}} e(-a^2 + b^2) .
\end{aligned}$$

The moments $q_{2,-1}$ and $q_{2,-2}$ follow by the symmetry relations (1.107).

1.22 The current density is proportional to the surface charge density and to the tangential velocity at the point of reference,

$$\boldsymbol{j}(\boldsymbol{x}) = \eta\omega r \sin\theta\, \delta(r - R)\hat{\boldsymbol{e}}_\phi \equiv j_\phi \hat{\boldsymbol{e}}_\phi .$$

From this one calculates a vector potential by means of the formula (1.116). The unit vector $\hat{\boldsymbol{e}}_\phi$ is decomposed along the 1- and 2-directions, $\hat{\boldsymbol{e}}_\phi = -\sin\phi \hat{\boldsymbol{e}}_1 +$

$\cos\phi\hat{\boldsymbol{e}}_2$. Writing the integral for $\boldsymbol{A}(r,\theta,\phi)$ in terms of spherical coordinates, one has

$$\boldsymbol{A}(r,\theta,\phi) = \eta\omega\frac{1}{c}\int_0^\infty r'^2\,\mathrm{d}r'\int \mathrm{d}\Omega'\, r'\delta(r'-R)\sin\theta'$$
$$\times\left(-\sin\phi'\hat{\boldsymbol{e}}_1 + \cos\phi'\hat{\boldsymbol{e}}_2\right)\sum_{\ell,m}\frac{4\pi}{2\ell+1}\frac{r_<^\ell}{r_>^{\ell+1}}Y^*_{\ell m}(\hat{\boldsymbol{x}}')Y_{\ell m}(\hat{\boldsymbol{x}})\,.$$

The calculation proceeds along the following lines: As the set-up is axially symmetric it is sufficient to calculate $\boldsymbol{A}$ for $\phi = 0$. Conversely, the integral over ϕ' which is proportional to $\hat{\boldsymbol{e}}_1$, is equal to zero. This means that $\boldsymbol{A}(r,\theta,\phi=0)$ is proportional to $\hat{\boldsymbol{e}}_2$ and, hence, is equal to the component A_ϕ. In the integrand make the replacement

$$\sin\theta'\cos\phi' = \sqrt{\frac{2\pi}{3}}\left(-Y_{11}(\hat{\boldsymbol{x}}') + Y_{1-1}(\hat{\boldsymbol{x}}')\right)$$

and calculate the angular integral. The induction field is obtained from the result $\boldsymbol{A} = A_\phi\hat{\boldsymbol{e}}_\phi$.

1.23 As there is neither a current density nor a time-dependent displacement current the field $\boldsymbol{H}$ is irrotational. Thus, it can be represented by a gradient field of a scalar magnetic potential Φ_{M}. In the inner space, inside the smaller of the two spheres, in the intermediate space, and in the outer space one writes down multipole expansions of Φ_{M} as follows,

$$\Phi_{\mathrm{M}}^{(\text{inner})} = \sum_{\ell=0}^\infty a_\ell r^\ell P_\ell(\cos\theta)\,,$$
$$\Phi_{\mathrm{M}}^{(\text{inter})} = \sum_{\ell=0}^\infty\left[c_\ell\frac{1}{r^{\ell+1}}P_\ell(\cos\theta) + d_\ell r^\ell P_\ell(\cos\theta)\right]\,, \qquad (11)$$
$$\Phi_{\mathrm{M}}^{(\text{outer})} = \sum_{\ell=0}^\infty b_\ell\frac{1}{r^{\ell+1}}P_\ell(\cos\theta) + B_0 r P_1(\cos\theta)\,.$$

In this ansatz we have made use of the fact that the potential must be regular in $r = 0$ and that it goes over into the potential of the homogeneous field as the argument goes to infinity, $r \to \infty$.

The boundary conditions are: The potential must be continuous at r and at R; at the two boundary surfaces the tangential component of $\boldsymbol{H}$ is continuous;

furthermore, the normal component of $\boldsymbol{B}$ is continuous, i.e.

$$\Phi_{\mathrm{M}}^{(1)} = \Phi_{\mathrm{M}}^{(2)} , \tag{12a}$$

$$\frac{\partial \Phi_{\mathrm{M}}^{(1)}}{\partial \theta} = \frac{\partial \Phi_{\mathrm{M}}^{(2)}}{\partial \theta} , \tag{12b}$$

$$\mu_1 \frac{\partial \Phi_{\mathrm{M}}^{(1)}}{\partial r} = \mu_2 \frac{\partial \Phi_{\mathrm{M}}^{(2)}}{\partial r} , \tag{12c}$$

where the numbers 1 and 2 stand for any two neighbouring domains and where $\mu_{\text{inner}} = \mu_{\text{outer}} = 1$ in the inner and in the outer spaces, while in the intermediate space $\mu_{\text{inter}} = \mu$. One sees easily that the first two conditions (12a) and (12b) are equivalent. Therefore, it suffices to require continuity of the potentials only. Like in Exercise 1.17 one realizes that only the terms with $\ell = 1$ can contribute. Denoting by r the radius of the smaller, by R the radius of the bigger sphere one obtains the linear system of equations

$$\begin{aligned} a_1 r^3 &= c_1 + d_1 r^3 , \\ c_1 + d_1 R^3 &= b_1 - B_0 R^3 , \\ a_1 r^3 &= \mu \left[-2c_1 + d_1 r^3 \right] , \\ 2b_1 + B_0 R^3 &= \mu \left[2c_1 - d_1 R^3 \right] . \end{aligned}$$

The solution of this system of equations yields the coefficients in the ansatz (11), viz.

$$a_1 = \frac{9\mu R^3}{2(\mu-1)^2 r^3 - (\mu+2)(2\mu+1)R^3} B_0 , \tag{13a}$$

$$c_1 = \frac{3(\mu-1) r^3 R^3}{2(\mu-1)^2 r^3 - (\mu+2)(2\mu+1)R^3} B_0 , \tag{13b}$$

$$d_1 = \frac{3(2\mu+1) R^3}{2(\mu-1)^2 r^3 - (\mu+2)(2\mu+1)R^3} B_0 , \tag{13c}$$

$$b_1 = B_0 R^3 + 3R^3 \frac{(\mu-1) r^3 + (2\mu+1) R^3}{2(\mu-1)^2 r^3 - (\mu+2)(2\mu+1)R^3} B_0 . \tag{13d}$$

As a test of the result consider the case $\mu = 1$, for which the spherical shell is no longer seen. From (13a) to (13d) one obtains $a_1 = -B_0$, $c_1 = 0$, $d_1 = a_1$, $b_1 = 0$; any dependence on either r or R has disappeared.

The magnetic field is obtained by the generic formula

$$\Phi_{\mathrm{M}} = r^\alpha \cos\theta , \quad \boldsymbol{H} = -\boldsymbol{\nabla}\Phi_{\mathrm{M}} = \alpha r^{\alpha-1} \cos\theta \hat{\boldsymbol{e}}_r + r^{\alpha-1} \hat{\boldsymbol{e}}_3 ,$$

the magnetic induction is $\boldsymbol{B} = \mu \boldsymbol{H}$.

In the limit $\mu \to \infty$ the coefficients a_1, c_1, and d_1 go to zero, b_1 goes to $B_0 R^3$.

Chapter 2 (selected solutions)

2.1 The base-k-forms $\mathrm{d}x^{i_1} \wedge \ldots \wedge \mathrm{d}x^{i_k}$ are totally antisymmetric, the indices i_1 to i_k can take all values from 1 up to the dimension n of the space. For fixed k there are

$$\binom{n}{k} = \frac{n!}{k!(n-k)!}$$

such base forms. This is shown as follows. In a first step one counts the number of possibilities to choose k different indices from the set $\{1, 2, \ldots, n\}$. The index i_1 can take any value from 1 to n, and thus there are n possibilities; for i_2 that has to be different from i_1, there are $(n-1)$ possible choices; for i_3 with $i_3 \neq i_1$ and $i_3 \neq i_2$, there remain $(n-2)$ possibilities; up to i_k which can take $(n-k+1)$ values. In total, the number of possible choices is

$$n(n-1)(n-2)\cdots(n-k+1) = \frac{n!}{(n-k)!} .$$

As regards the ordering of k different indices there are $k!$ ways to do this, viz. as many as there are permutations of k elements. Only one of them fulfills the condition $i_1 < i_2 < \cdots < i_k$. Therefore, one must divide the number just obtained by $k!$, thus obtaining the dimension of the space $\Lambda^k(M)$.

2.2 It suffices to consider two fixed values μ and ν different from each other. Then one has

$$S_{\mu\nu} A^{\mu\nu} + S_{\nu\mu} A^{\nu\mu} = S_{\mu\nu} A^{\mu\nu} + S_{\mu\nu}\left(-A^{\mu\nu}\right) = 0 .$$

Equal values of μ and ν do not occur because $A^{\mu\mu} = 0$. The sum over all values of the two indices is the sum over all such pairs.

2.3 The indices α and β must be different from each other. Keeping α and β fixed, the other four indices have values which are not equal to one of these. This implies that one must have either $\sigma = \mu$ and $\tau = \nu$, or $\sigma = \nu$ and $\tau = \mu$. Taking the sum over α and β the term $\varepsilon^{\alpha\beta\sigma\tau}\varepsilon_{\alpha\beta\mu\nu}$ and the term $\varepsilon^{\beta\alpha\sigma\tau}\varepsilon_{\beta\alpha\mu\nu}$ give the same result. Hence the factor 2. With $\varepsilon_{0123} = 1$ one has $\varepsilon^{0123} = -1$, giving a minus sign in the following formula

$$\varepsilon^{\alpha\beta\sigma\tau}\varepsilon_{\alpha\beta\mu\nu} = -2\left(\delta^\sigma_\mu \delta^\tau_\nu - \delta^\sigma_\nu \delta^\tau_\mu\right) .$$

This is the analogue of (1.48a). Take now $\mu = \sigma$ and sum over this index to obtain

$$\varepsilon^{\alpha\beta\mu\tau}\varepsilon_{\alpha\beta\mu\nu} = -2(4-1)\delta^\tau_\nu = -6\delta^\tau_\nu .$$

This is the analogue of (1.48b).

2.4 One calculates the divergence of $\boldsymbol{A}' = \boldsymbol{A} + \nabla\chi$:

$$\nabla \cdot \boldsymbol{A}'(t, \boldsymbol{x}) = \nabla \cdot \boldsymbol{A}(t, \boldsymbol{x}) + \boldsymbol{\Delta}_x \chi(t, \boldsymbol{x}) = \nabla \cdot \boldsymbol{A}(t, \boldsymbol{x}) - \nabla \cdot \boldsymbol{A}(t, \boldsymbol{x}) = 0 \,.$$

In the second step the equation $\boldsymbol{\Delta}(1/\boldsymbol{x} - \boldsymbol{y}) = -4\pi\delta(\boldsymbol{x} - \boldsymbol{y})$ was used and the integral over $\boldsymbol{y}$ was done. Any further transformation with a gauge function ψ that satisfies the homogeneous differential equation $\boldsymbol{\Delta}\psi = 0$, does not modify this result.

A gauge transformation $A''_\mu = A'_\mu - \partial_\mu \psi$ with

$$\psi(x) = \int\limits_0^{x^0} \mathrm{d}t' \, A^0(t', \boldsymbol{x})$$

leads to $A''_0 = 0$, as requested.

2.5 Perhaps the simplest argument is the following: The electron in the initial state has the four-momentum p_i which satisfies $p_\mathrm{i}^2 = m_e^2 c^2$. In the final state it has the momentum p_f, the photon has the momentum k, with $p_\mathrm{f}^2 = m_e^2 c^2$ and $k^2 = 0$. This is in contradiction to energy-momentum conservation which requires $p_\mathrm{i} = p_\mathrm{f} + k$: The condition $p_\mathrm{f} \cdot k = 0$ can only hold true if p_f is lightlike, i.e. if $p_\mathrm{f}^2 = 0$.

Chapter 3 (selected solutions)

3.1 The physical dimensions of the given quantities are

$$[S] = \mathrm{MT}^3 = \frac{\text{energy}}{\text{surface} \times \text{time}} \,,$$

$$[P] = \mathrm{ML}^{-2}\mathrm{T}^{-1} = \frac{\text{momentum}}{\text{volume}} \,,$$

$$[u] = \frac{\text{energy}}{\text{volume}} \,.$$

3.2 Rotations are represented by orthogonal 3×3-matrices, i.e. one has $\mathsf{R}\mathsf{R}^{-1} = \mathbb{1}_3$. Hence

$$\sum_{i,j} R_{mi} R_{nj} \delta_{ij} = \sum_i R_{mi} R^T_{in} = \delta_{mn} \,.$$

The transformation formula for the ε-tensor

$$\sum_{i,j,k} R_{mi} R_{nj} R_{pk} \varepsilon_{ijk} = \varepsilon'_{mnp}$$

is the determinant of R whenever (m, n, k) is an even permutation of $(1, 2, 3)$, and is equal to minus this determinant if (m, n, k) is an odd permutation. The determinant is invariant, its sign can be represented by ε_{mnp}. Hence, $\varepsilon'_{mnp} = \varepsilon_{mnp}$.

Chapter 4 (selected solutions)

4.1 From Maxwell's equations one derives the following boundary conditions: Given a surface separating two different media "1" and "2" which carries the surface charge density η, or in which flows the surface current density $\boldsymbol{j}$; then the following relations hold for normal and tangential components of the fields, respectively,

$$(\boldsymbol{D}_2 - \boldsymbol{D}_1) \cdot \hat{\boldsymbol{n}} = 4\pi\eta\,, \tag{14a}$$

$$(\boldsymbol{B}_2 - \boldsymbol{B}_1) \cdot \hat{\boldsymbol{n}} = 0\,, \tag{14b}$$

$$(\boldsymbol{E}_2 - \boldsymbol{E}_1) \times \hat{\boldsymbol{n}} = 0\,, \tag{14c}$$

$$(\boldsymbol{H}_2 - \boldsymbol{H}_1) \times \hat{\boldsymbol{n}} = -\frac{4\pi}{c}\boldsymbol{j}\,. \tag{14d}$$

Here $\hat{\boldsymbol{n}}$ is the normal unit vector which is oriented such that it points from medium 1 to medium 2. Thus, when there are neither surface charge nor surface current densities the normal components of the fields $\boldsymbol{D}$ and $\boldsymbol{B}$ are continuous, the tangential components of $\boldsymbol{E}$ and the normal component of $\boldsymbol{H}$ are continuous.

4.2 This problem is closely related to Exercise 1.16. The charge density follows from the continuity equation. The vector potential follows from (4.30). The electric field and the magnetic induction field are obtained by means of the standard formulae.

4.3 and **4.4** The charge distribution is

$$\varrho(\boldsymbol{x}) = \frac{q}{2\pi}\left[\frac{1}{a}\delta(r-a) - \frac{1}{b}\delta(r-b)\right]\delta(z)\,,$$

where r denotes the radial coordinate in cylinder coordinates. With $\boldsymbol{v}(\boldsymbol{x}) = \omega|\boldsymbol{x}|\hat{\boldsymbol{e}}_\phi$ the current density reads

$$\boldsymbol{j}(\boldsymbol{x}) = \varrho(\boldsymbol{x})\boldsymbol{v}(\boldsymbol{x}) = \frac{q\omega}{2\pi}[\delta(r-a) - \delta(r-b)]\delta(z)\hat{\boldsymbol{e}}_\phi\,.$$

The magnetic dipole moment follows from formula (1.120b):

$$\begin{aligned}\boldsymbol{\mu} &= \frac{1}{2c}\iiint \mathrm{d}^3x\, \boldsymbol{x} \times \boldsymbol{j}(\boldsymbol{x}) \\ &= \frac{q\omega}{4\pi c}\int_0^\infty r\,\mathrm{d}r \int_{-\infty}^{+\infty} \mathrm{d}z \int_0^{2\pi} \mathrm{d}\phi\, (r\hat{\boldsymbol{e}}_r + z\hat{\boldsymbol{e}}_z) \times \hat{\boldsymbol{e}}_\phi[\delta(r-a) - \delta(r-b)]\delta(z)\end{aligned}$$

$$= \frac{q\omega}{2c}(a^2 - b^2)\hat{e}_r \times \hat{e}_\phi$$
$$= \frac{q\omega}{2c}(a^2 - b^2)\hat{e}_z \,.$$

Chapter 5 (selected solutions)

5.1 Inserting $\phi(\boldsymbol{x})$ into the differential equation (A.4) converts it to an *algebraic* equation

$$(\boldsymbol{k}^2 + \kappa^2)\widetilde{\phi}(\boldsymbol{k}) = -g\frac{1}{(2\pi)^{3/2}} \,,$$

which is easy to solve. The original function which is defined over position space, is obtained by the inverse transformation

$$\phi(\boldsymbol{x}) = -\frac{g}{(2\pi)^3}\iiint \mathrm{d}^3k \, \frac{\mathrm{e}^{\mathrm{i}\boldsymbol{k}\cdot\boldsymbol{x}}}{\boldsymbol{k}^2 + \kappa^2} \,.$$

This integral can be calculated using spherical coordinates. The result is $-g\mathrm{e}^{-\kappa r}/(4\pi r)$.

5.2 We assume $\kappa \equiv \kappa_i$ to be dependent on the generator. By a suitable choice of the basis of the Lie algebra one can choose the $\mathrm{tr}(T_i T_j)$ diagonal, i.e.

$$\mathrm{tr}(T_i T_j) = \kappa_i \delta_{ij} \,.$$

We define then the following totally antisymmetric quantity, with k fixed:

$$\mathcal{E}_{ijk} := \mathrm{tr}\Big(\big[T_i, T_j\big]T_k\Big) = \mathrm{tr}(T_i T_j T_k) - \mathrm{tr}(T_j T_i T_k) \,.$$

This quantity with fixed k can be computed by means of the commutators,

$$\mathcal{E}_{ijk} = \mathrm{i}\sum_n C_{ijn}\,\mathrm{tr}\big(T_n T_k\big) = \mathrm{i}\kappa_k C_{ijk} \,.$$

Exchange the indices j and k to obtain $\mathcal{E}_{ikj} = \mathrm{i}\kappa_j C_{ikj}$, again with fixed j. Both $\mathcal{E}_{ijk}$ and the structure constants C_{mnp} are antisymmetric. Comparison of the last two formulae shows that $\kappa_k = \kappa_j$ as long as the commutator $[T_j, T_k]$ is not equal to zero. Note that in a simple group any two generators are related by nonvanishing commutators. Therefore, all constants κ_i are equal and, hence, are independent of i.

5.3 The adjoint representation of SO(3) is three-dimensional. The gauge fields and the field strengths transform like vectors in $\mathbb{R}^3$. Therefore, the symbolic scalar product in (5.49) is the Euclidean scalar product. A triplet of scalar fields was treated in Example 5.2 so that it is straightforward to write down an SO(3)-gauge invariant Lagrange density (5.51).

5.4 If the structure group is the direct product of two simple Lie groups, every generator of one group commutes with every generator of the other. Gauge potentials, field strengths, and covariant derivatives of the gauge groups SU(p) and SU(q) are independent of each other. Therefore, they can be defined with independent coupling constants q_1 and q_2, respectively. Thus, according to (5.33b) one has for SU(p) and SU(q)

$$A = \mathrm{i}q_1 \sum_{k=1}^{N_p} \mathsf{T}_k \sum_{\mu=0}^{3} A_\mu^{(k)}(x)\,\mathrm{d}x^\mu \qquad (N_p = p^2 - 1)\,,$$

$$B = \mathrm{i}q_2 \sum_{l=1}^{N_q} \mathsf{S}_l \sum_{\mu=0}^{3} A_\mu^{(l)}(x)\,\mathrm{d}x^\mu \qquad (N_q = q^2 - 1)\,.$$

In the Lagrange density (5.49) there are no interaction terms between the gauge bosons of one gauge group and those of the other, as all commutators $[\mathsf{T}_i, \mathsf{S}_k]$ vanish.

5.5 (See the reference quoted in Sect. 5.6 [Constantinescu, Michel, Radicati 1979].)

5.6 In the adjoint representation, using the summation convention for all paired indices, one has

$$\left[\mathsf{U}_{\mathrm{ad}}(\mathsf{T}_i), \mathsf{U}_{\mathrm{ad}}(\mathsf{T}_j)\right]_{ac} = +\mathrm{i}^2\left(C_{ia}^{b} C_{jb}^{c} - C_{ja}^{b} C_{ib}^{c}\right).$$

Using the Jacobi relation (5.21) and the antisymmetry of the structure constants the expression in round brackets on the right-hand side can be rewritten as follows:

$$C_{ia}^{b} C_{jb}^{c} - C_{ja}^{b} C_{ib}^{c} = C_{ia}^{b} C_{jb}^{c} + C_{aj}^{b} C_{ib}^{c} = -C_{ji}^{k} C_{ka}^{c} = +C_{ij}^{k} C_{ka}^{c}\,.$$

On the other hand, writing the above commutator more explicitly, one obtains

$$\begin{aligned} &U_{ab}^{\mathrm{ad}}(\mathsf{T}_i) U_{bc}^{\mathrm{ad}}(\mathsf{T}_j) - U_{ab}^{\mathrm{ad}}(\mathsf{T}_j) U_{bc}^{\mathrm{ad}}(\mathsf{T}_i) \\ &= C_{ij}^{k} C_{ka}^{c} = iC_{ij}^{k}(-\mathrm{i}C_{ka}^{c}) = iC_{ij}^{k} U_{ac}^{\mathrm{ad}}(\mathsf{T}_k)\,. \end{aligned}$$

This yields what had to be shown.

Chapter 6 (selected solutions)

6.1 The construction of an atlas and the proof that the transition maps are diffeomorphisms are analogous to the case $S_R^2 \subset \mathbb{R}^3$. This example is worked out, e. g. in [ME], Sect. 5.2.3.

6.2 On the basis of the argument in example 6.3 in Sect. 6.2.3 one obtains $\Delta\omega/\omega \simeq Hg/c^2$. This yields

$$\frac{\Delta\omega}{\Gamma} = \frac{\Delta\omega}{\omega}\frac{\omega}{\Gamma} = \frac{22.5\,\mathrm{m}\cdot 10\,\mathrm{ms}^{-2}}{(3\cdot 10^8\,\mathrm{ms}^{-1})^2} \simeq 0.7\,\% \,.$$

6.3 The left side of the supposed identity, when written out, reads

$$\begin{aligned} C := {} & XYZ - XZY + YZX - YXZ + ZXY - ZYX \\ & - YZX + ZYX - ZXY + XZY - XYZ + YXZ \,. \end{aligned}$$

The 12 terms cancel pairwise so that one finds $C = 0$, indeed.
For the given vector fields one has

$$\begin{aligned} XY &= y\partial_x(x\partial_y) = y\partial_y + yx\partial_x\partial_y \,, \\ YX &= x\partial_y(y\partial_x) = x\partial_x + xy\partial_y\partial_x \,, \\ XY - YX &= y\partial_y - x\partial_x \,. \end{aligned}$$

In this calculation one used the fact that the base fields commute.

6.4 Evaluating $T^{(1)}$ and $T^{(2)}$ on the two vector fields one finds

$$\begin{aligned} T^{(1)}\left(a^1\partial_1 + a^2\partial_2, b^1\partial_1 + b^2\partial_2\right) &= a^1 b^2 \,, \\ T^{(2)}\left(a^1\partial_1 + a^2\partial_2, b^1\partial_1 + b^2\partial_2\right) &= a^2 b^1 \,. \end{aligned}$$

In general, the answers are indeed different.

6.5 One has $X_{\sigma\,;\mu} = g_{\sigma\rho}X^{\rho}{}_{;\mu}$ where $X^{\rho}{}_{;\mu}$ is taken from the formula (6.57a). The obvious equation

$$\partial_\mu\left(g_{\sigma\rho}g^{\rho\tau}\right) = \partial_\mu\left(\delta_\sigma^\tau\right) = 0 = \partial_\mu\left(g_{\sigma\rho}\right)g^{\rho\tau} + g_{\sigma\rho}\partial_\mu\left(g^{\rho\tau}\right) \,, \tag{15}$$

and the coordinate formula (6.66) are utilized in the following calculation. One computes

$$g_{\sigma\rho}X^{\rho}{}_{;\mu} = g_{\sigma\rho}\left\{\partial_\mu\left(g^{\rho\tau}X_\tau\right) + \Gamma^{\rho}_{\mu\nu}g^{\nu\tau}X_\tau\right\} \,.$$

In differentiating the first term on the right-hand side by means of the product rule, one finds first the expected term $g_{\sigma\rho}g^{\rho\tau}\partial_\mu X_\tau = \partial_\mu X_\sigma$. The other term as well as the remaining terms must yield the Christoffel symbol $(-\Gamma^{\tau}_{\mu\sigma})$. This is verified as follows:

$$g_{\sigma\rho}\left(\partial_\mu g^{\rho\tau}\right) + g_{\sigma\rho}\Gamma^\rho_{\mu\nu} g^{\nu\tau}$$
$$= g_{\sigma\rho}\left(\partial_\mu g^{\rho\tau}\right) + \frac{1}{2} g_{\sigma\rho} g^{\rho\alpha}\left[(\partial_\mu g_{\nu\alpha}) + (\partial_\nu g_{\mu\alpha}) - (\partial_\alpha g_{\mu\nu})\right] g^{\nu\tau}$$
$$= g_{\sigma\rho}\left(\partial_\mu g^{\rho\tau}\right) - \frac{1}{2} g_{\sigma\rho}(\partial_\mu g^{\rho\alpha})\delta^\tau_\alpha - \frac{1}{2} g_{\sigma\rho}(\partial_\nu g^{\rho\alpha}) g_{\mu\alpha} g^{\nu\tau}$$
$$- \frac{1}{2}\delta^\alpha_\sigma(\partial_\alpha g_{\mu\nu}) g^{\nu\tau} .$$

Up to here the relation (15) was used twice in order to shift the derivative to $g^{\rho\alpha}$. If one applies the same trick to the first three terms of the expression obtained in the last step, the first two can be combined so that one finds all in all

$$-\frac{1}{2}(\partial_\mu g_{\sigma\rho}) g^{\rho\tau} + \frac{1}{2}(\partial_\nu g_{\sigma\rho})\delta^\rho_\mu g^{\nu\tau} - \frac{1}{2}(\partial_\sigma g_{\mu\rho}) g^{\rho\tau}$$
$$= -\frac{1}{2} g^{\rho\tau}\left[\partial_\mu g_{\sigma\rho} + \partial_\sigma g_{\mu\rho} - \partial_\rho g_{\sigma\mu}\right] = -\Gamma^\tau_{\mu\sigma} .$$

In the last but one line the summation index ν of the third term was replaced by ρ and, finally, the formula (6.66) was inserted. This proves the formula for the covariant derivative of a tensor of type (0, 1).

One realizes that the proof is sensibly more transparent if one applies the coordinate-free formula (6.40a) and introduces local coordinates there: Let $V = V^\mu \partial_\mu$ be a vector field, and $\omega = X_\lambda \, \mathrm{d}x^\lambda$ a one-form. According to (6.40a) one has

$$(D_V \omega)(W) = D_V(\omega(W)) - \omega(D_V(W)) .$$

Choosing now $V = \partial_\mu$ and $W = \partial_\sigma$ one obtains

$$\left(D_{\partial_\mu}(X_\lambda \, \mathrm{d}x^\lambda)\right)(\partial_\sigma) = \partial_\mu X_\sigma - X_\lambda \, \mathrm{d}x^\lambda \left(\Gamma^\tau_{\mu\sigma}\partial_\tau\right)$$
$$= \partial_\mu X_\sigma - \Gamma^\tau_{\mu\sigma} X_\tau .$$

This is the same formula.

6.6 Consider two overlapping charts (U, φ) and (V, ψ) for the spacetime $(M, \mathbf{g})$ and let $x \in U \cap V$ be a point in their intersection. Using local coordinates the same point is represented by

$$\varphi(x) = \left\{u^0(x), u^1(x), u^2(x), u^3(x)\right\} , \quad \text{and}$$
$$\psi(x) = \left\{v^0(x), v^1(x), v^2(x), v^3(x)\right\}$$

in two copies of $\mathbb{R}^4$. The transition maps $(\psi \circ \varphi^{-1})$ and $(\varphi \circ \psi^{-1})$ are diffeomorphisms so that $v^\mu(x) = (\psi \circ \varphi^{-1}(u))^\mu$. The local representations of the metric in these charts are related by

$$\mathbf{g} = g_{\mu\nu} \, \mathrm{d}u^\mu \otimes \mathrm{d}u^\nu = g'_{\alpha\beta} \, \mathrm{d}v^\alpha \otimes \mathrm{d}v^\beta .$$

The differentials fulfill the linear relation $dv^\alpha = (\partial v^\alpha/\partial u^\mu)du^\mu$, so that

$$g_{\mu\nu} = \frac{\partial v^\alpha}{\partial u^\mu}\frac{\partial v^\beta}{\partial u^\nu}g'_{\alpha\beta} . \tag{16}$$

The comparison of arbitrary coordinates and of normal coordinates $v^\alpha \equiv z^\alpha$ yields the first of the formulae to be shown, viz.

$$g_{\mu\nu} = \frac{\partial z^\alpha}{\partial u^\mu}\frac{\partial z^\beta}{\partial u^\nu}\eta_{\alpha\beta} . \tag{17a}$$

For the inverse of the metric tensor this implies

$$g^{\mu\nu} = \frac{\partial u^\mu}{\partial z^\alpha}\frac{\partial u^\nu}{\partial z^\beta}\eta^{\alpha\beta} . \tag{17b}$$

Now one calculates the Christoffel symbols by means of the formula (6.66) and expresses them in terms of normal coordinates as follows:

$$\begin{aligned}\Gamma^\mu_{\rho\sigma} &= \frac{1}{2}g^{\mu\tau}\left\{\frac{\partial g_{\sigma\tau}}{\partial u^\rho} + \frac{\partial g_{\rho\tau}}{\partial u^\sigma} - \frac{\partial g_{\rho\sigma}}{\partial u^\tau}\right\} \\ &= \frac{1}{2}\left(\frac{\partial u^\mu}{\partial z^{\bar\alpha}}\frac{\partial u^\tau}{\partial z^{\bar\beta}}\eta^{\bar\alpha\bar\beta}\right)\left\{\frac{\partial}{\partial u^\rho}\left(\frac{\partial z^\alpha}{\partial u^\sigma}\frac{\partial z^\beta}{\partial u^\tau}\eta_{\alpha\beta}\right)\right. \\ &\quad \left. + \frac{\partial}{\partial u^\sigma}\left(\frac{\partial z^\alpha}{\partial u^\rho}\frac{\partial z^\beta}{\partial u^\tau}\eta_{\alpha\beta}\right) - \frac{\partial}{\partial u^\tau}\left(\frac{\partial z^\alpha}{\partial u^\rho}\frac{\partial z^\beta}{\partial u^\sigma}\eta_{\alpha\beta}\right)\right\} .\end{aligned}$$

The tensor η is constant and may be taken out of the derivatives. Recalling that all indices except μ, ρ and σ are summation indices one realizes that the three terms in curly brackets can be combined to

$$\Gamma^\mu_{\rho\sigma} = \frac{\partial u^\mu}{\partial z^{\bar\alpha}}\left(\frac{\partial u^\tau}{\partial z^{\bar\beta}}\frac{\partial z^\beta}{\partial u^\tau}\right)\frac{\partial^2 z^\alpha}{\partial u^\rho \partial u^\sigma}\eta^{\bar\alpha\bar\beta}\eta_{\alpha\beta} = \frac{\partial u^\mu}{\partial z^\alpha}\frac{\partial^2 z^\alpha}{\partial u^\rho \partial u^\sigma} , \tag{18}$$

where the factor in parentheses is $\delta^\beta_{\bar\beta}$ so that only $\bar\alpha = \alpha$ contributes.
In (18) the z^α are normal coordinates, u^μ are arbitrary coordinates. This formula applies equally well to the transformation $u \mapsto z(u)$ as to the transformation $v \mapsto z(v)$. From this one derives the transformation formula for Christoffel symbols under the diffeomorphism $u \mapsto v$. One finds

$$\Gamma'^\tau_{\kappa\lambda} = \frac{\partial v^\tau}{\partial u^\mu}\frac{\partial u^\rho}{\partial v^\kappa}\frac{\partial u^\sigma}{\partial v^\lambda}\Gamma^\mu_{\rho\sigma} + \left(\frac{\partial v^\tau}{\partial u^\sigma}\frac{\partial^2 u^\sigma}{\partial v^\kappa \partial v^\lambda}\right) . \tag{19}$$

This affine transformation behaviour shows that the Christoffel symbols cannot be components of a tensor field. Only antisymmetric combinations of them such as they appear in formula (6.80c) for the Ricci tensor or in formula (6.76) for the Riemann tensor, can be tensors. In these cases the inhomogeneous terms cancel.

6.7 In order to show that the Weyl tensor has the same symmetry properties as the Riemann tensor it is sufficient to verify this in the additional terms which contain the Ricci tensor or the curvature scalar. The property $C^{\nu}{}_{\nu\sigma\tau} = 0$ is easily confirmed by a little calculation. For other contractions use the symmetry properties.

Taking into account the symmetry properties of the Weyl tensor one sees that the property

$$C^{\nu}{}_{\nu\sigma\tau} = 0 \tag{20}$$

together with invariance under $\sigma \leftrightarrow \tau$, in dimension n, yields $n(n+1)/2$ constraints. The number of independent components of the Weyl tensor, using (6.78), is

$$\begin{aligned} N_{\mathrm{C}} &= N_{\mathrm{R}} - \frac{1}{2}n(n+1) = \frac{1}{12}n^2(n^2-1) - \frac{1}{2}n(n+1) \\ &= \frac{1}{12}n(n+1)(n+2)(n-3)\,. \end{aligned} \tag{21}$$

In dimension 4 it has ten independent components. In dimension 3 one has $N_{\mathrm{C}} = 0$, the Weyl tensor vanishes.

In the case of a conformally flat metric take $\phi = \mathrm{e}^f$ and calculate the Christoffel symbols by means of (6.66). One finds

$$\Gamma^{\sigma}_{\mu\nu} = (\partial_\mu f)\delta^{\sigma}_{\nu} + (\partial_\nu f)\delta^{\sigma}_{\mu} - (\partial_\rho f)\eta^{\sigma\rho}\eta_{\mu\nu}\,.$$

This is used to calculate R, $\mathsf{R}_{(\mathrm{Ricci})}$, and S, and, eventually, the Weyl tensor. One finds, indeed, that it vanishes.

Index

About the Author

Born 1936 in Berlin, son of Gustav Scheck, flutist. Studied physics at University of Freiburg, Germany. Physics diploma 1962, PhD in theoretical physics in 1964, both at Freiburg University.

Guest scientist at the Weizmann Institute of Science, Rehovoth, Israel, 1964–1966; research assistant at University of Heidelberg, Germany, 1966–1968; research fellow at CERN, Geneva, 1968–1970, as well as corresponding fellow afterwards.

Habilitation at University of Heidelberg 1968. From 1970 until 1976 head of theory group at the Swiss Institute for Nuclear Research (now PSI), Villigen, Switzerland, as well as lecturer and titular professor at ETH Zurich.

Since 1976 professor of theoretical physics at Johannes Gutenberg University in Mainz, Germany; emeritus at this university since 2005. Numerous visits worldwide as guest scientist and visiting professor. Principal field of activity: theoretical elementary particle physics and mathematical physics.